Separations Division 2018

Core Programming Area at the 2018 AIChE Annual Meeting

Pittsburgh, Pennsylvania, USA
28 October - 2 November 2018

ISBN: 978-1-5108-7631-6

Printed from e-media with permission by:

Curran Associates, Inc.
57 Morehouse Lane
Red Hook, NY 12571

CURRAN ASSOCIATES INC. proceedings.com

Some format issues inherent in the e-media version may also appear in this print version.

Printed by Curran Associates, Inc. (2019)

For permission requests, please contact AIChE
at the address below.

AIChE
120 Wall Street, FL 23
New York, NY 10005-4020

Phone: (800) 242-4363
Fax: (203) 775-5177

www.aiche.org

Additional copies of this publication are available from:

Curran Associates, Inc.
57 Morehouse Lane
Red Hook, NY 12571 USA
Phone: 845-758-0400
Fax: 845-758-2633
Email: curran@proceedings.com
Web: www.proceedings.com

TABLE OF CONTENTS

Some abstracts in this book refer to images that were not made available by the publisher

527858 Bioinspired Membranes for CO_2 Capture

Abstract:

Learning from the compositions, structures, formations and functions of natural materials, a myriad of bioinspired membranes have been prepared for efficient CO_2 separation. In this presentation, the following issues and examples will be covered.

Using cell, cell membrane (including aquaporin), marine mussel, carbonic anhydrase and as representative prototypes;

Using biomineralization, bioadhesion, surface segregation as major tools;

Using microcapsules and salts to tune the water environment;

Using GO and functional crosslinker to construct CO_2 transport highways.

538649 Spray-Coated Multilayer Cellulose Nanocrystal â€“ Chitin Nanofiber Films for Barrier Applications

Abstract:

Chitin is an abundant biopolymer whose natural production is second only to cellulose. Similar to cellulose nanocrystals (CNCs) or nanofibers (CNFs), chitin nanofibers (ChNFs) can be isolated and used as sustainable O_2 barrier materials for food, electronics and pharmaceutical packaging. The difficulty in melt processing chitin and cellulose nanomaterials presents an obstacle to their incorporation in packaging. Alternatively, these bio-available nanomaterials are readily dispersed in water, and spray coating is a versatile technique allowing uneven, or delicate surfaces to be coated at high rates. In the present study we demonstrate the successful layer-by-layer spray coating of cationic ChNF and anionic (CNC) suspensions onto poly(lactic acid) (PLA) films. ChNF/CNC multilayers were found to lead to a reduction in the O_2 permeability of the final composite film by as much as 73 % with the largest effects seen in composites with three alternating layers (ChNF-CNC). Multilayer ChNF/CNC coatings were found to have lower O_2 permeability and lower haze than those coated with ChNF or CNCs alone (72 % and 86 % lower haze respectively), pointing to a synergistic effect. The composites had a water vapor transmission rate similar to the PLA substrate.

523491 Fe(III)-Induced Polydopamine Coating withÂ "Deposition-Polymerization" Mechanism forÂ Microfiltration Membrane Hydrophilization

Abstract:

Fe(III)-Induced Polydopamine Coating with "Deposition-Polymerization" Mechanism for Microfiltration Membrane Hydrophilization

Xuehua Ruan, Xuhang Liao, Gaohong He*, Yan Dai, Xiaobin Jiang

State Key Laboratory of Fine Chemicals, School of Petroleum and Chemical Engineering, Dalian University of Technology, Panjin, Liaoning, P.R. China, 124221

Corresponding Author Email: hgaohong@dlut.edu.cn

Polydopamine coating is widely attempted to fabricate high-flux hydrophilic membranes, but subjected to the time-consuming and uneven deposition. In this work, Fe(III)-induced polydopamine coating with "deposition-polymerization" mechanism was proposed for microfiltration membrane hydrophilization, with the intent to accelerate deposition and reduce agglomeration simultaneously. According to the coordination theory, Fe(III) ions could be complexing with threefold dopamine molecules and accelerate deposition effectually. Besides, the multivalent Fe(III)-Fe(II) redox system could activate dissolved oxygen to oxidize dopamine and promote the polymerization. In this instance, the hydrophilic layer was yielded homogeneously and rapidly on membrane surface. In comparison to the regular route, the Fe(III)-induced route can accelerate polydopamine deposition by 5 times, with the rate increased from 11.5 to 62.0 nm/h. The hydrophilic layer with requisite thickness close to 100 nm was constructed rapidly in 2 hours to enhance water permeation flux. Moreover, membrane surface pores could be retained substantially. The PDA-Fe(III)-coated membranes behaved excellent with water flux three times higher than the membranes modified by general routes. Fe(III)-induced polydopamine deposition is a potential way to manufacture high-flux hydrophilic membranes facilely and efficiently.

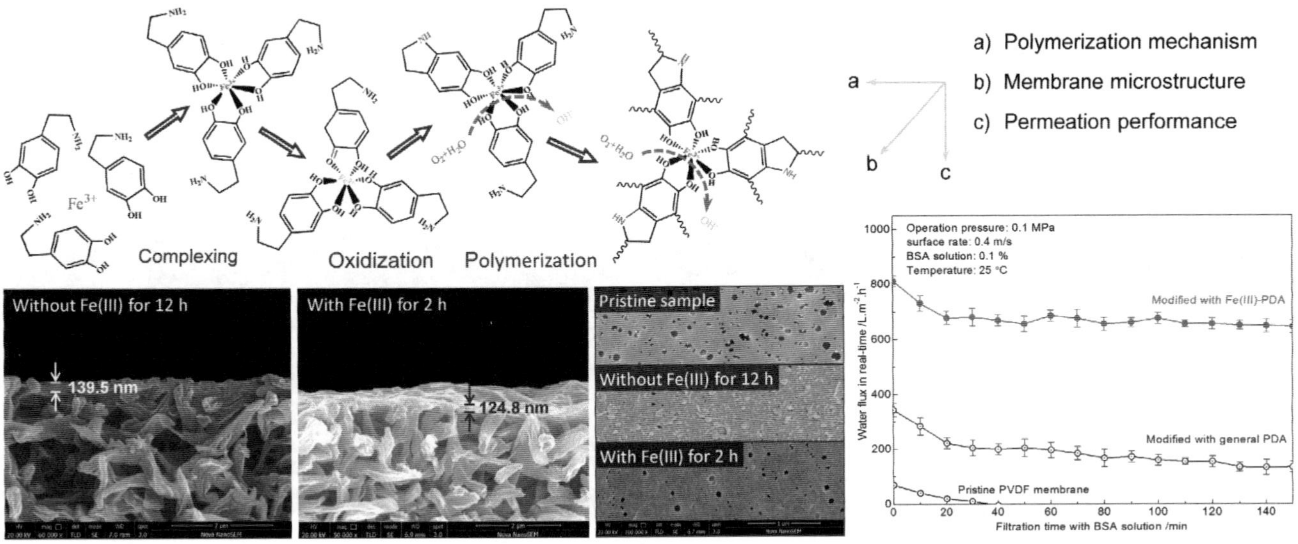

Figure 1 Fe(III)-induced dopamine reaction mechanism, membrane microstructure and performance.

Acknowledgement:

The authors are grateful to the financial support from National Natural Science Foundation of China (No. 21606035, U1663223 and 21676043), the Fundamental Research Funds for the Central Universities (DUT18JC24, DUT16TD19), Liaoning Province S&T Department (Grant no. 201601037) and Education Department (LT2015007), and Changjiang Scholars Program (T2012049).

537162 Creating Biomimetic Membranes with Uniform Subnanometer Pore By Co-Assembly of Well-Oriented Lamellar Block Copolymer with Artificial Channels

Abstract:

Membranes with uniform subnanometer pore sizes are desired for applications such as water purification, small molecule separations and gas separations. Traditional membranes are subject to permeability-selectivity trade-offs due to widely distributed pore size. Excellent transport abilities including both high flux and high selectivity are routinely achieved in natural biological membranes, thus it is a promising strategy to mimic nature for improving membrane performances. Here we present a new fabrication strategy of channel-based membranes through co-assembly of amphiphilic block copolymers and artificial channels. Specifically, cross-linkable triblock copolymers, which is a high molecular weight synthetic analog to lipid, was used to form lamellae-stacked membrane matrix. Through co-assembly, the artificial channels can be incorporated as a subnanometer separation elements. Such a membrane resembles natural lipid membranes in living system with better mechanical properties due to cross-linking of the hydrophobic domains and the bridging hydrophilic domains. The fabrication strategy reported here could promote the construction of more channel-based membranes with desired properties, such as ion separation, responsiveness and sensing.

533525 Bio-Inspired Membrane Based Self-Organizing System

Abstract:

A "Biomembrane" is a highly-organized self-assembly of biomolecules (i.e. lipid, protein etc.) and a key interface for the survival of biological cell. The "Membranome" can be defined as the properties of vesicle (or liposome), which arise from the bilayer molecular assembly of amphiphiles, focusing on "emergent properties" which are not present in the individual components, and is gradually recognized as an important research methodology to investigate the potential functions of vesicles (or liposome) and to apply them for the bioprocess design [1,2]. "Self-Organizing System", such as liposome or vesicle, possesses several benefits in the recognition of (bio)molecules, where it can recognize them with (i) electrostatic, (ii) hydrophobic interaction, and (iii) stabilization effect of hydrogen bonds at its surface. A key of next chemical engineering is the use of "Self-Organizing System", where "enthalpy-driven" nature of chemical process would be converted to "entropy-driven" one. We call this strategy as "Bio-Inspired Chemical Engineering". In this study, the basic and applied aspect of the self-organizing system were reviewed. (1) Phase equibrillium and physicochemical properties of self-organizing systems. Based on the spectroscopic studies (Raman spectroscopy, fluoresce spectroscopy with molecular probes, dielectric dispersion spectroscopy, etc.), the physicochemical properties of self-assembled membranes were characterized focusing on nano/mesoscale [3-6]. (2) Functions of Self-Organizing System. Emergent properties are provided by the designed membranes. For example, chiral recognition of amino acids could be achieved by the ordered membranes [7-10]. Liposomes could be utilized as a platform for enantioselective conversion in aqueous media [11-13]. (3) Its Application to the Development of the Chemical Process Devices. By immobilizing liposomes in hydrogel matrix, optical resolution of amino acids could be achieved [14-15]. In most cases, the microscopic properties of membranes, which were immobilized in matrix, played important roles in providing functions.

References

[1] H Umakoshi and K. Suga, *SERDJ*, **20**, 1-13 (2013). [2] P. Walde *et al.*, *Chem. Commun.*, **50**, 10177-10197 (2014). [3] K. Suga and H. Umakoshi, *Langmuir,* **29**, 4830-4838 (2013). [4] K. Suga *et al.*, *Anal. Chem.*, **87**, 4772-4780 (2015). [5] T. T. Bui *et al.*, *Langmuir,* **32**, 6176–6184 (2016). [6] F. Iwasaki *et al.*, *Langmuir,* **33**, 1984-1994 (2017). [7] T. Ishigami *et al.*, *ACS Appl. Mater. Interf.*, **7**, 21065-21072 (2015). [8] Y. Okamoto *et al.*, *JPCB*, **120**, 2790-2795 (2016). [9] T. Ishigami *et al.*, *Langmuir,* **32**, 6011-6019 (2016). [10] Y. Okamoto *et al.*, *Biomacromol.,* **18**, 1180-1188 (2017). [11] M. Hirose *et al.*, *Langmuir,* **31**, 12968-12974 (2015). [12] F. Iwasaki *et al.*, *ACS Omega*, **2**, 91-97 (2017). [13] F. Iwasaki *et al.*, *ACS Omega*, **2**, 1447-1453 (2017). [14] T. Ishigami *et al.*, *Colloid Surf. B*, **136**, 256-261 (2015). [15] Y. Hiramure *et al.*, *SERDJ*, **25** (1), 37-46 (2018).

519818 Performance Evaluation of Novel Nano-Structured Modified Mesoporous Silica/Polyetherimide Composite Membranes for the Treatment of Oil/Water Emulsion

Abstract:

The threat to the environment due to discharges from industries particularly petroleum industry is of great concern owing to the fact that the threat could be realized at any stage from exploration/drilling to transportation. In this work mesoporous silica (SBA-15) was modified using a bio-inspired coating and was used as filler in the preparation of efficient, robust polyetherimide composite ultrafiltration membranes. The membranes prepared by this facile two-step approach were seen be highly hydrophilic with improved porosity and pore-interconnectivity and a thinner skin layer. The rougher top surfaces of these membranes imparted an oleophobic character under water. The composite membranes exhibited improved emulsion/water flux while maintaining >99% rejection of oil from a synthetic motor oil/water/surfactant emulsion. The composite membranes were investigated for its long-term efficiency in the removal of oil from produced water. The flux decline was only < 15% for up to 9 hours with three intermittent backwashes and the oil/grease content of the permeate was lower than 10 ppm which is within the limits of discharge. The composite membranes exhibited a significant increase in the bacterial anti-adhesion property when evaluated using a gramnegative
bacterium Pseudomonas aeruginosa and gram-positive bacterium, Bacillus subtilis. The results from this study indicate that polydopamine decorated SBA-15 incorporated polyetherimide membranes holds a promising potential to be employed for oil-water emulsion separation.

527669 Bioinspired Biodegradable Superhydrophobic Membrane for Oil-Water Separation Using 3D Printing Technology

Abstract:

In nature, many species show great adaptability by exploiting surface properties derived from unique surface topographical features, such as superwettability. There are countless living things that exhibit wettability; in contrast, the lotus leaf, an ancient but popular plant, exhibits superhydrophobicity, low water adhesion and self-cleaning properties. Inspired by the lotus leaf, materials with superhydrophobicity-superoleophilicity have been proposed as promising candidates for oil-water separation by selectively repelling water from their surface against oil. Many forms of such materials have been fabricated to date, including metal meshes, membranes, aerogels, sponges and textiles, with high selectivity, high flux and low energy consumption. As in nature, superhydrophobicity is typically achieved by either roughening the surface of low surface-energy materials or by modifying the rough surface with low surface-energy compounds. Poly(lactic acid) (PLA) is a biocompatible and biodegradable synthetic polyester and is one of the few synthetic polymers approved by the Food and Drug Administration (FDA) for a variety of human clinical and ecological applications. PLA is a typical hydrophilic polymer comprising ester groups in its skeleton and terminal hydroxyl and carboxyl groups, both of which exhibit water affinity. Hence, it is challenging to construct a superhydrophobic surface out of pure PLA substrate.

Here, we report a superhydrophobic PLA porous membrane with a lotus-inspired surface structure for gravity-driven oil-water separation through 3D printing and facile two-step post-treatments. The membrane surface show a hierarchical roughness. Specifically, the micro-scale feature of the superhydrophobic PLA membrane was fabricated via 3D printing and chemical etching, leading to an ultra-high water adhesion (380 Î¼N). The nanoscale features were achieved by further decoration of PS nanospheres. The hierarchical structure provided a large surface roughness and led to low water adhesion (21.8 Î¼N). Furthermore, 3D printing technology provides an efficient, precise and customized route to fabricate membranes with specific shape and pore sizes. By tailoring the membrane pore sizes and wettability, the PLA membrane exhibited an excellent oil-water separation efficiency of over 99% while retaining a high flux of 60 kL m^{-2} h^{-1}. The relatively low cost, excellent chemical tolerance and mechanical durability demonstrate its promising application in sophisticated and practical situations, such as the recovery of oil spills in waste water.

534713 Design, Synthesis, and Characterization of Artificial Water Channel Based Polymeric Membranes

Abstract:

Naturally occurring biological protein channels such as aquaporins have recently gained attention as promising materials for high efficiency membranes. This interest is driven by their, high water permeability and selectivity. However, lower than expected improvement in baseline performance, high costs, difficulties in fabrication and achievement of high packing channel density, and low stability associated with biological materials are perhaps holding back the large-scale applications of protein water channel-based membranes. To overcome these challenges, artificial water channels have emerged as an alternative to biological channels which provide stability while maintaining their distinctive molecular transport property. In this talk, an overall strategy of design, synthesis, and characterization of artificial channel based polymeric membrane will be presented. Experimental and simulation data will be presented on peptide-appended pillar[5]arene (PAP) artificial water channels incorporated into poly(butadiene)-poly(ethylene oxide) (PB-PEO) block copolymers as complete synthetic models of biological membranes. These PAP channel-based membranes preserved single channel molecular transport property even at the macro-scale. Also, they presented remarkably enhanced membrane performance compared to other polymeric membranes, in terms of the expected permeability/selectivity trade-off.

532671 Design and Development of Advanced Automotive Fuel Cell Membranes

Abstract:

Proton exchange membrane fuel cells (PEMFC) have been regarded as one of the most promising power sources for automotive applications. Automotive fuel cell systems have requirements that differ from other fuel cell applications. The challenge is to maintain its high efficiency during dynamic operation over the wide range of operating conditions experienced by the vehicle during its target life. This presentation will review automotive PEMFC system demands, and the design and development strategies for advanced membrane materials.

As one of the most important components of PEMFC, the proton exchange membrane, or PEM, plays a critical role in proton transport, electrical isolation and reactant gas separation. Membrane proton transport resistance is an important factor influencing ohmic loss in fuel cells. Gas crossover leads to fuel efficiency loss and fuel cell material degradation, which in turn increases the cost of vehicle ownership. This presentation will also discuss the performance considerations for PEMs, as well as evaluation methodologies on proton transport resistance, gas crossover as well as fuel cell polarization.

For fuel cell membrane durability consideration, this presentation will focus on development of mechanical support options and the methodology for trading off between mechanical stability and transport (proton, gas, water) properties. It will also discuss fuel cell membrane chemical degradation phenomena induced by impurity contamination, operating conditions, and membrane chemical structure. In addition, this presentation will discuss systematic characterization methods for evaluating membrane mechanical and chemical durability.

Beside the performance and durability discussion on PEMs in fuel cell stack system, this presentation will also discuss membrane cost analysis from point view of fuel cell system production and integration. The presentation will include analysis of fuel cell vehicle cost of ownership as a function of membrane properties such as resistance and gas crossover. Criteria for selection of membrane materials based on proton conductivity, water uptake, gas permeance, chemical stability, mechanical properties and cost will be discussed.

524937 Design of 2-D Novel Materials for Anhydrous Proton Transport

Abstract:

Water management and hydration in proton exchange membrane (PEM) fuel cells are well-known limiting factors for for proton transport. We have used first-principles theory to develop novel 2-D materials that can transport protons anhydrously. These materials are graphamine (amine functionalized graphane) and hydroxyl functionalized boron nitride. Our molecular dynamics simulations show that the proton transport self-diffusivity increases with temperature. Moreover, our thermodynamic stability calculations indicate that these materials are appropriate for intermediate-temperature fuel cells. Proton transport barriers were found to be remarkably low and the diffusion profile was observed to be in the Fickian regime. These observations from dynamics and energetics indicate that facile anhydrous proton conduction can be realized with appropriate functionalization of these 2-D materials that can mimic the behavior of hydrogen bonding in water.

535711 Quantitative Structure Analysis of Polymerized Ionic Liquids with Atomistic Molecular Simulations

Abstract:

The design of solid-state electrolytes for electrochemical applications that utilize polymerized ionic liquids (polyiLs) would require the elucidation of molecular-level understanding of structure-property relationships. Atomistic molecular dynamics simulations are used to investigate the structural properties of a homologous series of poly(nalkyl-vinylimidzolium bistrifluoromethylsulfonylimide) poly(CnVim Tf2N) and present the direct S(q) comparison with experiment. Excellent agreement is found in terms of peak position and shape. The quantitative cluster analysis along with color-coded snapshots vividly demonstrates that discrete nonpolar islands first form within the continuo polar network, then nonpolar domains grow beyond the percolation threshold, finally interconnect with the polar network into a bicontinuous 'sponge-like' nanostructure. Moreover, we exploit the selective labeling technique of neutron scattering using hydrogen/deuterium substitution to afford further insight into the morphology of poly(CnVi Tf2N). The neutron scattering profiles markedly depend on the isotopic substitution pattern. The total neutron structure factors of the backbone deuterated samples reveal the most noticeable low-q peak and are much more intense than those observed in single isotope neutron scattering. We also investigate a model ammonium based polyiLs and anion effect. We hope these insights will lead to a fundamental understanding into the structure and morphology of polylls and pave a path forward towards the rational design of future polyiLs for electrochemical devices.

527892 Saturated *N*-Heterocyclic Cationic Polymers As Anion Exchange Membranes in Alkaline Fuel Cells

Abstract:

Alkaline fuel cells (AFCs) hold the promise of a low-cost (no platinum) energy source for electric vehicles with no point-of-use carbon emissions. For this promise to be realized, AFCs require solid-state anion exchange membranes (AEMs) or separators that possess high chemically stability in alkaline conditions and also possess additional multiple desired properties of high hydroxide ion conductivities, robust mechanical properties, and are low in cost, easily processable, and can easily be integrated into AFCs. Polymers containing unsaturated *N*-heterocyclic cations (*i.e.*, imidazolium) have been synthesized and investigated for various applications, including alkaline fuel cells. Recently, results have shown higher alkaline chemical stability for polymers bearing a saturated *N*-heterocyclic cation (pyrrolidinium) versus an unsaturated *N*-heterocyclic cation (imidazolium). Additionally, several studies on small molecules have reported increasing stability with increasing saturated *N*-heterocyclic cation size. However, to date, few studies have systematically investigated polymers containing saturated *N*-heterocyclic cations, specifically *N*-heterocyclic cations with larger ring sizes (*i.e.*, azepanium, azocanium, azonanium). In this study, styrene-based saturated*N*-heterocyclic cationic (SNHC) homopolymers with various covalent attached cations (methylpyrrolidinium, methylpiperidinium, methylazepanium, methylazocanium) were successfully synthesized to investigate the influence of ring size/strain on the properties of SNHC polymers. Alkaline chemical stability, temperature and humidity dependent water sorption, and ion conductivity of these SNHC polymers were measured and evaluated in regards to the relationship with ring size/strain. Additionally, multiblock polymers containing styrene-based saturated *N*-heterocyclic cationic chemistry were synthesized and investigated as functional polymer membranes. This work provides a fundamental understanding of the impact of the ring size/strain in SNHC polymers and furtherintroduces a promising chemistry for producing highly stable, ion conductive solid-state AEMs for AFCs.

529870 Highly Durable Direct Formate Solid Alkaline Fuel Cells Using New Aromatic Anion Exchange Polymer and Carbon Free Electro-Catalysts

Abstract:

Solid alkaline fuel cells (SAFCs) have attracted attention as next-generation energy conversion devices. Despite numerous advantages including the usages of non-noble metal catalysts and liquid fuels with high energy density, the practical applications of these devices are limited by drastically decreasing cell performance during the operating at high temperature and alkaline condition due to the severe degradation of membrane-electrode assemblies (MEAs). Therefore, the development of durable MEAs is critical to improve cell performances and stabilities in this system.

In this study, we developed a new spirobifluorene-based aromatic ionomer with no ether-linkage. The anion-conductive ionomer exhibited high alkaline and oxidative durability because the backbone did not contain ether linkages, heteroatoms, and benzylic C-H bonds. It was also demonstrated that the thin-layer coating (< 1 μm) of spirobifluorene-based ionomer on the anion-exchange membrane improved alkaline and oxidative stability and suppressed the permeability of potassium formate (HCOOK) as fuel.

A carbon-free catalyst consists of a nanosized beaded network formed by the connection of Pt-Fe nanoparticles.[1–3] The beaded metal network is electrically conductive, enabling the removal of carbon supports from catalyst layers. Carbon-free catalyst layers can eliminate carbon corrosion problems, leading to high durability against start-stop operations. We demonstrated the high durability of this carbon-free catalyst in alkaline electrolyte solution. In addition, the specific activity of a connected Pt-Fe catalyst for oxygen-reduction-reaction (ORR) is about 5 times higher than that of a commercial Pt-nanoparticle catalyst supported on carbon black.

We prepared the MEA using these durable materials of the spirobifluorene-based ionomer, and anion conductive pore-filling membrane with the durable thin layers and the carbon-free connected Pt-Fe catalyst. The MEA operated at high temperature of 80 C, and 4M HCOOK and 2 M KOH aqueous solution was used as liquid fuel. As shown in Fig.1, the MEA exhibited high power density of 220 mW cm^{-2} and high open circuit voltage of about 1.0 V. Importantly, the high performance retained even after the operation at 80 °C and 0.2 A cm^{-2} for 150 h.

This study demonstrated, for the first time, long-lasting MEA at high temperature and alkaline condition in direct formate SAFCs.

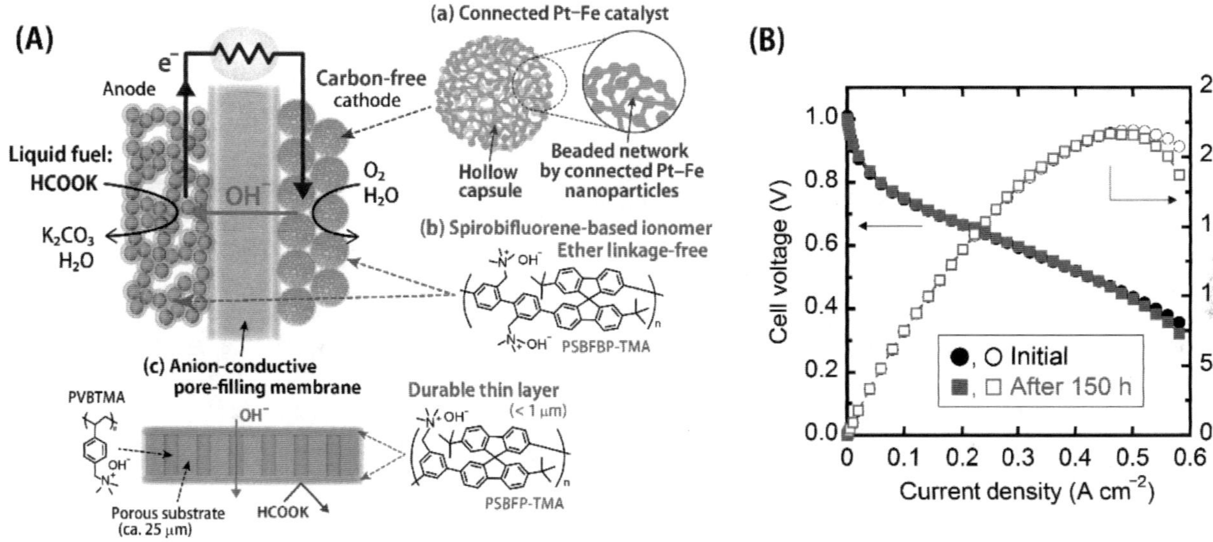

Fig. 1. Schematic of (A) highly-durable MEA using (a) carbon-free connected Pt–Fe catalyst, (b) ether linkage-free spirobifluorene-based ionomer, and (c) pore-filling membrane with durable thin layers. (B) IV curves of the MEA before and after 150 h operation at 80 °C and 0.2 A cm^{-2}.

References

[1] T. Tamaki, H. Kuroki, S. Ogura, T. Fuchigami, Y. Kitamoto, and T. Yamaguchi, *Energy Environ. Sci.*, 8, 3545–3549 (2015).

[2] H. Kuroki, T. Tamaki, and T. Yamaguchi, *J. Electrochem. Soc.*, 163(8), F927–F932 (2016).

[3] H. Kuroki, T. Tamaki, M. Matsumoto, M. Arao, Y. Takahashi, H. Imai, Y. Kitamoto, and T. Yamaguchi, *ACS Appl. Energy Mater.*, 1(2), 324–330 (2018).

532145 Oxidatively Stable Borate-Containing Membranes for H_2 Purification for Fuel Cells

Abstract:

Borate-containing membrane was developed for H_2 purification for fuel cell application. The oxidatively stable membrane contained quaternaryammonium hydroxide as a mobile carrier for the facilitated transport of CO_2, quaternaryammonium fluoride-containing polymer as a fixed-site catalyst for the reaction between CO_2 and water molecules, a borate-based additive as another catalyst for the same reaction in a membrane matrix made of crosslinked polyvinylalcohol-polysiloxane. The optimized membrane with a borate-based additive of tetrafluoroboric acid reached a CO_2 permeance of 100 GPU and CO_2/H_2 selectivity >100. The membrane demonstrated oxidative stability at 120°C for at least 144 hours with humid air as the sweep gas. As the membrane thickness was reduced from 15 μm to 10 μm, a significant drop in CO_2/H_2 selectivity was observed due to the increase of the H_2 permeance; while the CO_2 permeance was not significantly improved. Addition of the number of the substrate layers underneath the membrane resulted in a reduction of water permeation through the membrane. As the number of the substrate was increased up to 3 pieces, the water permeance was reduced almost proportionally. A continuous roll-to-roll membrane fabrication machine was used to scale up the fabrication of a 14-inch wide flat-sheet membranes with > 1400 feet in total length. Compared to the lab-scale membranes, a similar performance was obtained from the scale-up membrane. The membrane developed in this work can be used in membrane processes with air as the sweep gas, such as the H_2 purification for fuel cell applications.

Free volume enhanced ion transport in anion exchange membranes for fuel cell application

Kuibo Zhang, Baoqiang Zhang, Shoutao Gong and Fengxiang Zhang*

School of Petroleum and Chemical Engineering, Dalian University of Technology

2 Dagong Road, Liaodongwan New District, Panjin, China, 124221

Alkaline electrolyte membrane fuel cell (AEMFC) development has attracted tremendous attention in the past decades, but its development toward practical application is still hindered by the insufficient conductivity of alkaline electrolyte membrane (AEM). To address this issue, we here report a strategy that is different from the conventional protocols of high ion exchange capacity (IEC) and microphase separation. We first tuned backbone structure of the membrane materials to facilitate hydroxide ion transport, and found that the polysulfone AEM containing isopropylene groups at the backbone showed a higher conductivity than that without such a group although the two membranes have the same cation (imidazolium) and similar IEC values. Such a difference is due to the free volume imparted by the isopropylene groups, which are non-coplanar relative to their neighboring benzene rings and thus result in less compact chain packing.

Based on the above finding, we further designed and fabricated AEMs of bent-twisted non-coplanar block copolymer. The bent-twisted polymer backbone creates high free volume size so that hydroxide ion transport is facilitated. The free-volume effect can also facilitate microphase separation due to enhanced chain segment motion. With the free-volume holes working synergistically with microphase-separation enabled ion channels, the membrane showed a high conductivity (35 mS cm^{-1} at room temperature) at a relatively low ion exchange capacity (IEC, 1.25 meq g^{-1}); it also exhibited a good dimension stability, the swelling ratio being 4.7 % at room temperature and kept virtually unchanged with temperature because of relatively low IEC. A H_2/O_2 fuel cell employing the fabricated AEM yielded a high power density (262 mW cm^{-2} at 50 °C). Our work opens up a new route to fabrication of highly conductive AEM with relatively low IEC.

537965 A Systematic Approach for Membrane-Based Hybrid Separation Network Synthesis

Abstract:

Membrane separations play an indispensable role in many areas of chemical industry via facilitating (i) molecular separations (e.g., reverse osmosis), (ii) chemical transformations (e.g., membrane reactors), and (iii) enhanced mass and energy transfer between different phases through integration of hybrid separation techniques in a single unit (e.g., membrane contactors) [1-2]. Recent trends in novel process design via process intensification also highlight the importance of utilizing membrane-based separations rather than costly standalone distillation-based operations [3]. However, limited studies exist that consider the design and optimization of hybrid networks involving membrane and other separation techniques [4-6].

Recently, we have introduced a novel superstructure representation for process intensification based on building blocks [7], which has been demonstrated for automatic identification of novel process alternatives including membrane reactors, reactive distillation, etc. The block-based representation is also applied for process synthesis [8] and integration [9-10]. The proposed superstructure is a collection of blocks positioned on a two-dimensional grid which can be used to represent several physical and chemical phenomena. The search for optimal processing routes is automated via a mixed integer nonlinear programming (MINLP) formulation. In this work, we use the building block representation for design and synthesis of hybrid separation networks. We represent each membrane module as a collection of neighboring blocks separated by a membrane boundary. Similarly, vapor-liquid contact is represented as two neighboring blocks sharing a common boundary for phase contact and transition which constitutes a tray for a distillation operation. The material flow in the superstructure is achieved through inter-block material streams that connect different blocks, e.g., membrane and vapor-liquid contact modules, with each other. Systematic arrangement of the building blocks in a grid formation paves the way for the representation of different membrane flow patterns, i.e. cross-flow, counter-current, co-current, and different recycle alternatives between separation unit alternatives.

The proposed method is generic in the sense that it has the potential for designing separation systems involving membranes, distillation and/or membrane distillation. We will present one such case study on H_2/CO separation that resulted in more than 40% decrease in the total annual cost compared to the base-case design reported in the literature [11] when the membrane-based design is optimized.

References:

[1] Drioli, E., Stankiewicz, A.I. and Macedonio, F., 2011. Membrane engineering in process intensificationâ€"An overview. *Journal of Membrane Science*, *380*(1-2), pp.1-8.

[2] Gorak, A. and Stankiewicz, A., 2011. Intensified reaction and separation systems. *Annual review of chemical and biomolecular engineering*, *2*, pp.431-451.

[3] Tian, Y., Demirel, S. E., Hasan, M. M. F., Pistikopoulos, E. N. An Overview of Process Systems Engineering Approaches for Process Intensification: State of the Art. *Chemical Engineering and Processing: Process Intensification*, Submitted.

[4] Marquardt, W., Kossack, S. and Kraemer, K., 2008. A framework for the systematic design of hybrid separation processes. *Chinese Journal of Chemical Engineering*, *16*(3), pp.333-342..

[5] Caballero, J.A., Grossmann, I.E., Keyvani, M. and Lenz, E.S., 2009. Design of hybrid distillationâˆ' vapor membrane separation systems. *Industrial & engineering chemistry research*, *48*(20), pp.9151-9162.

[6] Tula, A.K., Befort, B., Garg, N., Camarda, K.V. and Gani, R., 2017. Sustainable process design & analysis of hybrid separations. *Computers & Chemical Engineering*, *105*, pp.96-104.

[7] Demirel, S. E., Li, J., and Hasan, M. M. F., (2017). Systematic Process Intensification using Building Blocks, *Computers and Chemical Engineering*, 105, 2-38.

[8] Li, J.; Demirel, S.E.; Hasan, M.M.F. Process Synthesis using Block Superstructure with Automated Flowsheet Generation and Optimization. *AIChE Journal*, 2018, under review.

[9] Li J., Demirel S.E., Hasan M.M.F. Process Integration using Block Superstructure. *Industrial & Engineering Chemistry Research*, 2018, 57: 4377â€"4398.

[10] Li J., Demirel S.E., Hasan M.M.F. Fuel Gas Network Synthesis Using Block Superstructure. *Processes*. 2018, 6(3): 23.

[11] Uppaluri, R.V., Linke, P. and Kokossis, A.C., 2004. Synthesis and optimization of gas permeation membrane networks. *Industrial & engineering chemistry research*, *43*(15), pp.4305-4322.

533412 Immobilised Solvent Systems: Evaluating the Potential of High Surface Area Membrane/Solvent Hybrid Sorbent Materials

Abstract:

Microencapsulation of liquid solvents (MECS) [1] and Solvent Impregnated Polymers (SIPs) [2] are two recently developed hybrid materials that aim to exploit the selectivity, versatility and thermodynamic properties of liquid solvents for gas separation, while simultaneously increasing the surface area available for mass transfer. In each technology, liquid solvents are immobilised inside high surface area solids. In MECS, the solvent is trapped inside a capsule a few hundred microns in diameter. The shell is composed of a thin, permeable polymer membrane, and the liquid solvent is held in the capsule core. In SIPs, the solvent is trapped inside a polymer matrix, forming a gel that may be shaped into thin sheets or particles. Immobilised solvents have 1-2 orders of magnitude greater surface area than the liquid in a traditional absorption column, and they may in principle enable thermodynamically favourable solvents with slow absorption kinetics or high viscosity, volatility or corrosivity to be used in a practical way.

In practice, however, the technology faces several obstacles. The most obvious is the need to upscale the manufacture of these materials. MECS particles are currently manufactured inside microfluidic devices at a rate on the order of 1 g/hr, and the production of tonnes of particles for industrial-scale gas separation will be a major challenge. SIPs, which were specifically developed to overcome this barrier, are manufactured via a scalable one-pot method which bypasses the need for any microfluidic chips. Instead, a fine solvent-in-polymer emulsion is first created, and after it is formed into the desired shape the polymer is crosslinked to create a solid gel. Particles may be manufactured by dispersing the single emulsion as a double emulsion prior to crosslinking, while thin sheets may be manufactured by spreading the emulsion on a supporting surface. Though SIPs may prove simpler to manufacture at scale than MECS particles, selection of appropriate polymers and surfactants remains a major challenge. Initial efforts focused on immobilising concentrated K_2CO_3 solutions inside PDMS for carbon capture applications. Several PDMS-based surfactants were identified which were able to stabilise 50/50wt% K_2CO_3(aq)-in-PDMS emulsions, and SIP particles with internal droplet size on the order of 20-50 microns were created. However, these emulsions were quite viscous, and when dispersed as a double emulsion the mean particle size was approximately 2mm, substantially larger than MECS particles. Furthermore, the relatively large internal droplet size means very fine particles or layers cannot be produced without significant solvent leakage. One possible means of overcoming these challenges is to use ethanol to reduce the internal surface energy. By adding 1-5wt% ethanol to the aqueous phase, a low-viscosity emulsion may be created with internal droplet size on the order of 1-5 microns. These emulsions are very stable, however when cross-linked the solid gels were sticky and not as rigid as other SIP materials. While SIPs may be a feasible pathway to large-scale production of immobilised solvent systems, further work is required to identify ideal polymers, emulsifers and manufacturing conditions.

A second question relates to the need to rigorously quantify the enhancement in absorption rate that solvent immobilisation can provide, as compared to traditional absorption operations. To date, estimates for this kinetic enhancement have been based upon the increase in surface area alone, and have not accounted for differences in the hydrodynamics and transport mechanisms, shell resistance, or absorber voidages. Moore et al [3] used surface renewal theories and well-established mass transfer correlations to rigorously compare absorption rates inside MECS with absorption rates inside traditional packed columns. General expressions for the increase in overall absorption rate were derived, and both chemical and physical solvents were considered. Two model systems - chemical absorption of CO_2 into 30wt% K_2CO_3, and physical absorption of CO_2 into Selexol - were investigated in depth.

It was found that, for most practical chemical solvents, the suppression of liquid mixing inside immobilised solvents will not significantly reduce the gas flux, relative to a traditional absorption column. For physical solvents, suppression of liquid mixing may reduce the flux, as mass transfer is more reliant upon convective mixing to the bulk, however this effect is counterbalanced with an increase in gas flux associated with the increase in concentration gradients that occurs as spatial scales are reduced. Different effects will dominate for different systems, however for CO_2 absorption into Selexol, the gas flux into 100 micron MECS particles was roughly 4 times larger than in a traditional packed column, leading to more than a 200-fold increase in the overall gas absorption rate.

The increase in surface area that microencapsulation can provide was also calculated, using reliable correlations for surface area in absorption columns. When the differences in absorber voidage and the very high surface area of structured packings such as Mellapak 500Y were accounted for, the increase in surface area was 3-10 times smaller than previous publications had suggested, though it was still more than an order of magnitude for the finest particles. Overall, MECS may be expected to increase the gas absorption rate per unit volume of absorber by up to an order of magnitude for chemical solvent systems, and over two orders of magnitude for physical solvent systems.

This analysis must be modified for SIP particles, as the mean gas flux is more dependent upon the specific size of the particles. As the solvent is not free to move inside the material, a dead-zone forms near the surface, similar to that inside a shrinking-core model. In order to better understand this behaviour, a volume-averaged model for gas absorption inside SIPs materials was developed and experimentally verified. If SIPs are too large, the diffusion resistance through the outer core dominates, and mass transfer is poor. On the other hand, for very concentrated solvents (such as 40wt% K_2CO_3 for CO_2 capture) the permeability of the gas through the polymer matrix may be 1-2 orders of magnitude larger than the permeability inside the liquid, and for appropriately sized particles this can lead to a 2-4 fold *increase* in the mean gas flux during absorption.

Immobilised solvent systems such as MECS and SIPs are a promising means of increasing gas uptake rates, and they may find application in a wide range of gas separations. Plausible pathways for upscaling their manufacture are available, and the mass transfer enhancement such materials may provide are substantial. However the technology is still in its infancy, and further work is required to identify the processes and solvents to which it is best suited, to refine manufacturing methods and material selection, and to quantify the thermodynamic enhancement that the use of otherwise impractical solvents may provide.

References

[1] Vericella, J. J. et al. (2015) Encapsulated liquid sorbents for carbon dioxide capture. Nat. Commun. 6:6124 doi: 10.1038/ncomms7124

[2] Moore, T. et al. (2017) Solvent Impregnated Polymers for Carbon Capture. Poster Session presented at Australia CCS Conference, Melbourne, Australia

[3] Moore, T. et al. (2018) Enhancement in Specific Absorption Rate by Solvent Microencapsulation. Submitted for publication.

537993 A Synthesis Framework for Hybrid Separation Sequences Based on Reduced Directed Graph Superstructure

Abstract:

Chemical engineering process usually involves separation of multi-component mixtures, the energy consumption and capital cost of separation process takes a great part in the whole investment, thus separation process is of great importance in process system engineering. Distillation, which is still the major separation technique in chemical engineering, with the advantages of the low capital investment, the operational flexibility, and the low operational risk. However, when distillation boundaries or low relative volatilities complicate distillation, it is desirable to substitute distillation with unit operations like extraction, crystallization, adsorption, or membrane separation, if possible. The key problem of this hybrid separation synthesis is to choose appropriate separation techniques and determine optimal separation sequence.

Previous work mainly focused on synthesis between two separation techniques[1-4], when extraction, crystallization, adsorption, and membrane separation are introduced into the synthesis framework simultaneously, as there are many candidates which can be found in separation techniques above, problem of combinatorial explosion becomes apparent. This problem was usually solved using heuristic sequential search strategy[5], however, this strategy usually leads to a suboptimal solution. Besides, rules of thumb in the past were not applicable in this problem because of their weakness in handling with too many choices.

In this work, a synthesis framework is proposed to reduce the space of separation directed graph by introducing matrix analysis. First, maturity of separation techniques, requirement for feed composition and product purity was considered in this methodology, aiming to preliminary screen feasible techniques. Then for the candidates of techniques which can complete the same separation task, shortcut models were introduced to determine the cost of different candidates. In this way, disadvantaged candidates would be removed from set of technique, and the space of set decreased.

Considering the separation characters of technique, separation techniqueâ€"component incidence matrix analysis was proposed in this methodology: directed graph of distillation sequence was generated first to build basic set of distillation task. Then the separation tasks of non-distillation techniques were derived from top to bottom, from more components to fewer components, using the incidence matrix mentioned above. It was worth noticing that new generated tasks were added into the basic set of task continually. After this, the entire separation directed graph was completed, all illegal separation tasks were removed, and solution space decreased significantly.

To improve the efficiency of searching solution space, based on ordered branch search strategy[6], the separation sequence was searched in the order: the non-distillation separation which separate multiple components at a time would be ranked first; for components which were separated from mixture, if the relative volatility between them was large, this task would be ranked front; the non-distillation separation prior to distillation separation. After this, the advantaged set of separation sequence were obtained. Then rigorous model was introduced to verify the cost of these sequences.

Two illustrative examples were presented to demonstrate the validity and advantages of the proposed approach. Mixtures which contained twelve components from C_6-C_8 and mixtures which contained fifteen components from C_8-C_{10} were going to be separated, where solvent extraction, pervaporation, adsorption and crystallization techniques were introduced. The number of sequence in the new directed graph was less than 1% of the original one.

Reference:

[1] Errico M, Sanchez-Ramirez E, Quiroz-Ramìrez J J, et al. Synthesis and design of new hybrid configurations for biobutanol purification[J]. Computers & Chemical Engineering, 2016, 84: 482-492.

[2] Beierling T, Micovic J, Lutze P, et al. Using complex layer melt crystallization models for the optimization of hybrid distillation/melt crystallization processes[J]. Chemical Engineering and Processing: Process Intensification, 2014, 85: 10-23.

[3] Ritter J A, Wu F, Ebner A D. New approach for modeling hybrid pressure swing adsorptionâ€"distillation processes[J]. Industrial & Engineering Chemistry Research, 2012, 51(27): 9343-9355.

[4] Harvianto G R, Ahmad F, Lee M. Vapor permeationâ€"distillation hybrid processes for cost-effective isopropanol dehydration: modeling, simulation and optimization[J]. Journal of Membrane Science, 2016, 497: 108-119.

[5] Floquet P, Pibouleau L, Aly S, et al. Fuzzy heuristic approach for sharp separation sequence synthesis[J]. Computers & chemical engineering, 1994, 18(9): 785-795.

[6] Rodrigo B, Fernando R, Seader J D. Synthesis of separation sequences by ordered branch search[J]. AIChE Journal, 1975, 21(5): 885-894.

515689 Kinetic Study of Degradation of Pharmaceutical Drugs By Ozone Microbubbles

Abstract:

KINETIC STUDY OF DEGRADATION OF PHARMACEUTICAL DRUGS BY OZONE MICROBUBBLES

Surabhi Patel[1], Rohit Agarwal[1], S. K. Majumder[1*], P. Das[2] and Pallab Ghosh[1]

[1]Department of Chemical Engineering, Indian Institute of Technology Guwahati, Guwahati-781039, India, Email: p.surabhi@iitg.ernet.in; agarwal.rohit@iitg.ernet.in; skmaju@iitg.ernet.in; pallabg@iitg.ernet.in;

[2]Department of Chemical Engineering, Jadavpur University, Kolkata-700032, India, Email: papitasaha@gmail.com

*Corresponding and presenting author's email: skmaju@iitg.ernet.in

The frequent occurrence of pharmaceuticals in the aquatic environment as well as in drinking water has raised a concern about their potential impact on environmental and public health. However, their elimination using conventional treatment methods has not been an easy task. The study on the kinetics of ozonation by ozone microbubbles for pharmaceutical drugs at room temperature is reported in this present work. The results showed that at pH 5, the ozonation of carbamazepine in aqueous solution reduced the concentration to about half of its initial value which was 6.35×10^{-5} M. As per model, the rate constant for carbamazepine is found to be 5.638×10^{-1} $M^{-1}s^{-1}$. The concentration of ibuprofen (IBP) in aqueous solution reduced from 3.6×10^{-6} M to 0.26×10^{-6} M in 41 munites which reflects 92.2% removal. The estimated value of second order rate constant for IBP is 11.58 $M^{-1}s^{-1}$ at pH 5. In case of methylisothiazolone (MIT) it is observed that its concentration reduced from 10×10^{-6} M to 0.32×10^{-6} M in 50 s. In this case the calculated value of second order rate constant for MIT is 138.12 $M^{-1}s^{-1}$at pH 5. In case of ranitidine the value of pseudo first order rate constant 'k_{obs}' is increasing from 0.01505 s^{-1} to 0.03961 s^{-1} with increasing inlet ozone concentration at pH 5 while the value of second order rate constant 'k' is decreasing from 144.43 $M^{-1}s^{-1}$ to 95.03 $M^{-1}s^{-1}$with increasing inlet ozone concentration. The present study may be helpful for further understanding the advanced oxidation of phermaceutical drugs to impliment ahead the wasteawter treatment processes.

KEY WORDS: Ozonation, microbubble, pharmaceutical drugs.

REFERENCES:

[1] Hübner, U.; Seiwert, B.; Reemtsma, T.; Jekel, M. Ozonation products of carbamazepine and their removal from secondary effluents by soil aquifer treatment–indications from column experiments. Water Research 2014, 49, 34-43.

[2] Rosal, R.; Rodríguez, A.; Gonzalo, M.; García-Calvo, E. Catalytic ozonation of naproxen and carbamazepine on titanium dioxide. Applied Catalysis B: Environmental 2008, 84 (1-2), 48-57.

[3] Sotelo, J. L.; Beltran, F. J.; Benitez, F. J.; Beltran-Heredia, J. Ozone decomposition in water: kinetic study. Industrial & Engineering Chemistry Research 1987, 26 (1), 39-43.

[4] Quero-Pastor, M.; Garrido-Perez, M.; Acevedo, A.; Quiroga, J. Ozonation of ibuprofen: a degradation and toxicity study. Science of the Total Environment 2014, 466, 957-964.

527621 Porous Frozen Material Approach to Freeze-Drying of Instant Coffee

Abstract:

Instant coffee is commercially prepared by either spray drying or freeze-drying. Although spray drying provides most instant coffee production due to its high capacities, it often brings possibilities of losing some heat-sensitive ingredients as a typical thermal drying. Freeze-drying is a unique form of drying that removes moisture mainly by sublimation. It can lead to the best quality of dried products among almost all drying techniques. Freeze-drying yields a better quality product of instant coffee than spray-drying because it can preserve aroma and antioxidant ingredients as much as possible. However, freeze-drying is an energy intensive and expensive process due to relatively long drying time and poor overall energy efficiency. Reducing freeze-drying time so as for lowering energy consumption and raising productivity therefore becomes a worldwide challenge during the past decades. Conventional freeze-drying of aqueous solutions is to freeze a liquid material into a solid frozen material without initial porosity. Mass transfer within this frozen material is limited largely during drying. If an aqueous solution to be dried is first frozen into a porous material with a certain initial porosity like ice-cream and then freeze-dried, mass transfer in such a material is expected to promote significantly. This innovative technique has the advantages of better product quality, faster drying rate and easy operation.

To reduce the drying time and increase the energy utilization efficiency, an experimental investigation is presented aimed at verifying the proposed idea. Initially porous frozen materials with prefabricated porosity were prepared to explore its influence on freeze-drying experimentally. Coffee powder was selected as the solute in aqueous solution. Liquid nitrogen ice-cream making method was employed to prepare frozen materials with different initial porosities. Results showed that freeze-drying can be remarkably enhanced with the initially porous frozen material, and more than 30% of drying time can be saved compared with the traditionally solid frozen material. SEM images of dried products revealed that the initially porous frozen material has the spherical and uniform porous structure, which is favorable to migration of sublimated/desorbed vapor and improvement of the drying rate. Chamber pressure has little effect on the freeze-drying process and increasing ambient temperature can be beneficial to the process. Combining radiation with conduction heating can reduce the drying time further. When the temperature was above 45°C, the solid frozen material was found to collapse leading to the failure of freeze-drying. In contrast, the porous one can bear higher operation temperature resulting in a relatively high drying efficiency. The conventionally solid material experienced the raising rate stage, constant rate stage and falling rate stage during freeze-drying, while the porous material was always in the falling rate stage throughout the process.

Acknowledgement

The authors are grateful to the financial supports from National Natural Science Foundation of China (21676042) and Natural Science Foundation of Liaoning Province (201602167).

References

[1] Burmester K, Fehr H, Eggers R. A comprehensive study on thermophysical material properties for an innovative coffee drying process. Dry Technol. 2011, 29, 1562–1570.

[2] Wang W, Chen G H, Mujumdar A S. Physical interpretation of solids drying: an overview on mathematical modeling research. Drying Technol. 2007, 25, 659–668.

[3] Ishwarya S P, Anandharamakrishnan, C. Spray freeze-drying approach for soluble coffee processing and its effect on quality characteristics. J. Food Eng. 2015, 149, 171–180.

[4] Wang W, Chen M, Chen G H. Issues in freeze drying of aqueous solutions. Chin. J. Chem. Eng. 2012, 20, 551–559.

[5] Wang W, Hu D P, Pan Y Q, Zhao, Y Q, Chen G H. Freeze-drying of aqueous solution frozen with prebuilt pores. AIChE J. 2015, 61, 2048–2057.

[6] Ratti C. Hot air and freeze-drying of high value foods: a review. J. Food Eng. 2001, 49, 311–319.

[7] Wang W, Hu D P, Pan Y Q, Niu L J, Chen G H. Multiphase transport modeling for freeze-drying of aqueous material frozen with prebuilt porosity. Int. J. Heat Mass Transfer 2018, 122, 1353–1365.

[8] Wang W, Yang J, Hu D P, Pan Y Q, Wang S H, Chen G H. Experimental and numerical investigations on freeze-drying of porous media with prebuilt porosity. Chem. Phys. Lett. 2018, 700, 80–87.

539686 Unique Properties of Graphene Oxide Membranes for Membrane Distillation Desalination

Abstract:

In this study the properties of graphene oxide (GO) membranes for membrane distillation (MD) are characterized based on parameters such as pH, GO thickness, and drying method. The performance of the MD process for desalination of brackish water is evaluated by taking into consideration these membrane parameters. The impact of the membrane properties on the degree of scaling in the MD process is also evaluated. GO is an interesting material that recently has been proposed as a component for membrane formation and resilience. This study will contribute to understand the contribution of this unique material.

531048 Improved Antifouling Performance of Polysulfone Ultrafiltration Membrane Via Peptoid Molecules

Abstract:

The development of antifouling membranes to minimize nonspecific protein adsorption has relevance in various biomedical applications. In the present study, electrically neutral NMEG peptoids containing 2-methoxyethyl side chains were attached to polysulfone (PSU) membranes via polydopamine. NMEG peptoids with varying length (NMEG5, NMEG10, NMEG15, and NMEG20) were synthesized and attached to PSU membranes and antifouling performance was assessed. The effect of surface composition, surface roughness, hydrophilicity, hydration capacity, and electrical neutrality on antifouling behavior of membranes were determined. NMEG peptoids presented a high hydrophilicity and hydration capacity as compared to unmodified PSU membranes. The antifouling performance of the membranes was evaluated using a bovine serum albumin filtration test and platelet adhesion. According to the cross-flow filtration results, NMEG modified membranes showed a significant improvement in antifouling ability. Furthermore, flux recovery ratios obtained from NMEG modified membranes were much higher than unmodified membranes. These studies provide a convenient strategy to improve antifouling, hydrophilicity and hemocompatibility of PSU membranes for use in biomedical and environmental applications.

529973 Fouling Free Hydrophilic Polysulfone Ultrafiltration Membranes

Abstract:

Abstract

This work describes the preparation of hydrophilic polysulfone (PSF) ultrafiltration membranes having reduced fouling tendency and increased separation performance. The various additives used in this study are hydrophilic polymers, namely poly(2-acrylamido-2-methyl-1-propanesulfonic acid) (AMPS), poly(ethylene glycol) (PEG) and polymer nanocomposites viz. titanium dioxide- poly(2-acrylamido-2-methyl-1-propanesulfonic acid) (TiO_2-AMPS). Direct blending technique is used for these additives and membranes are prepared by phase inversion method. The membranes are then characterized for their morphological and permeation characteristics by using different characterization techniques. Field emission electron microscopy (FESEM) is used for the structural analysis of the membranes and analysis of flux profiles for the permeability assessments. Fourier transform infrared spectroscopy (FTIR) is used for the confirmation of the presence of the additives in the prepared membranes. Lastly, antifouling and separation efficiencies are tested with humic acid (HA) solutions. The membranes modified with hydrophilic polymers (entitled M1) have shown narrow pore size (2.31-7.91 nm) as compared to the membranes (entitled M2) modified using polymer nanocomposites (16.27-21.30 nm). These differences can also be seen from the FESEM micrographs of the membranes. The FTIR analysis of the membranes confirms the successful incorporation of the additives in the membranes. Also, the hydrophilicity of the membranes increases for both M1 and M2 membranes on addition of hydrophilic additives. This is confirmed by the decreasing water contact angle of the membranes, for M1 membranes it is decreased from 72.12° (M1_0) to 40.32° (M1_4) and 71.71° (R1) to 36.56° (R4) for M2 membranes. Further, the M1 shows better separation characteristics, when analyzed with HA (98.93% removal) solutions as compared to M2 membranes (92% removal). This is due to the better surface presence of the polymeric additives and narrow pores size range of the membranes. On the other hand, M2 membranes shows better antifouling nature (F_{RR} from 47.23 (R1) - 67.25 (R4)) compared to M1 membranes (F_{RR} from 48.63 (M1_0) - 52.11 (M1_4)). However, the overall comparison shows that the membranes modified with polymer nanocomposites are best in performance due to the synergistic effect of both the polymer as well as the nanoparticles present in a polymer nanocomposite. The membranes prepared in this study are suitable for the treatment of contaminated drinking water and industrial wastewater as well.

Keywords: *Polysulfone, ultrafiltration, phase inversion, hydrophilic and antifouling, polymer nanocomposites*

529839 Facile Fabrication of Sulfonated Polyphenylenesulfone (sPPSU) Membranes with High Separation Performance for Organic Solvent Nanofiltration

Abstract:

Two types of organic solvent nanofiltration (OSN) membranes with high solvent permeances and good solute rejections have been fabricated through membrane modifications of (1) electrostatic interactions between polyethyleneimine (PEI) and sulfonated polyphenylenesulfone (sPPSU) as well as (2) subsequent glutaraldehyde (GA) cross-linking. The PEI-coated sPPSU membrane fabricated under the optimized dip-coating condition has a molecular weight cut-off (MWCO) of about 500 Da in ethanol, while the GA-cross-linked membrane has a MWCO of 300 Da. The former has a rejection of 99.8% to Rose Bengal and a permeance of > 40 $Lm^{-2}h^{-1}bar^{-1}$ in the methanol system because methanol has a small kinetic diameter and a low viscosity. The permeance drops to about 10 $Lm^{-2}h^{-1}bar^{-1}$ and the rejection dips slightly to 99.1% in the ethanol system because ethanol has a large kinetic diameter and a high viscosity. The latter possesses an ethanol permeance of 1.4 $Lm^{-2}h^{-1}bar^{-1}$ and shows stable OSN performance in 14-day tests to Methylene Blue in methanol, ethanol and isopropanol (IPA) systems. The GA-cross-linked membrane and the commercially available DuraMem 300 membranes have a similar MWCO, but the GA-cross-linked membrane has a 3.5-time higher ethanol permeance than DuraMem 300. This study may provide useful insights to design sPPSU membranes with high separation performance for OSN applications.

538862 Fabricating Polyurethane/Zeolite Mixed Matrix Membranes for Pervaporation of Dilute Aqueous Organic Solvents

Abstract:

Acetone, butanol, and ethanol (ABE) are excellent alternative renewable liquid fuels that have the potential to partially supplement fossil fuels. Currently bioconversion of biomass is the leading technology for the production of renewable ABE. After a series of mechanical and chemical pretreatments, biomass liquor is enzymatically hydrolyzed to release monomeric sugars. The biomass hydrolysate is then forwarded to the fermentation unit for the conversion of monomeric sugars to ABE. The the final product is more than 95 wt.% water and less than 5% ABE. Water content should be reduced to less than 2 wt.%. Currently, the leading technology for concentration of organics in the fermentation broth is distillation.

The successful deployment of economically feasible renewable ABE is highly dependent on a cascade of thoroughly engineered unit operations. Phase-changing separations are known to be energy-intensive, and replacing such separation could lead to energy savings of up to one order of magnitude. Here we present our recent work on fabricating mixed matrix membranes with commercial Thermoplastic Polyurethane (TPU, Texin DP7-1199) and the high-silica ZSM-5 zeolites. To date, polyurethane has not been well evaluated for pervaporation applications. A significant advantage of polyurethane over polydimethylsiloxane (PDMS) membranes is that it does not require crosslinking. Our results show that mixed matrix of polyurethane-zeolite membranes are more permeable than PDMS mixed matrix membrane. Easier processing and higher fluxes are promising for future manufacturing.

524660 Investigation of the Impacts of Dope Solution Viscosity on Membrane Morphological and Operational Characteristics

Abstract:

In non-solvent induced phase separation (NIPS), organic solvents are generally used to dissolve polymers. Organic solvents derived from petroleum sources, such as N-methyl-2-pyrrolidone (NMP) and dimethylacetamide (DMAc), have been traditionally used to fabricate polymeric membranes. These solvents are often volatile and/or hazardous, and regulatory pressure has been rising in Europe and China to stop the use of these. Therefore, using renewable solvents derived from biomass is of great interest in order to make membrane fabrication more sustainable. In this study, the viscosities of different dope solutions made by dissolving polysulfone (PSf) in DMAc and in NMP were measured as a function of polymer concentration and dope temperature. Morphological characteristics of membranes prepared using the tested dope solutions monitored included contact angle, zeta potential, surface roughness, surface pore size distribution and pure water permeability. Correlations between dope solution viscosity and membrane properties were analyzed to develop a mathematical model that could be extrapolated to provide information regarding casting dope viscosity to achieve desirable membrane properties when using new solvents.

523793 High Performance Thin-Film Composite Forward Osmosis Membrane with Novel PVDF/PFSA Substrate

Abstract:

In this work, a small amount of hydrophilic perfluorosulfonic acid (PFSA) is incorporated into the polyvinylidene fluoride (PVDF) substrate to develop the high-performance thin-film composite (TFC) membrane for forward osmosis applications. Because of the hydrophilicity and excellent compatibility with PVDF, PFSA not only significantly improves the wettability of the modified membrane substrate, but also well optimizes the pore size by changing the membrane morphology of PVDF/PFSA substrate. These two factors effectively mitigate the water transport resistance and favor the better formation of the polyamide (PA) layer during the interfacial polymerization, thus improving both the water flux and membrane selectivity of the resultant TFC membranes simultaneously. Effects of the PFSA concentration on the overall properties of the modified substrates and corresponding PVDF/PFSA TFC membranes are systematically investigated via various characterizations. Consequently, the PVDF/PFSA TFC membrane achieves a best water flux (J_V) of 54.4 LMH and a reverse salt flux (J_S) of 10.9 gMH in AL-DS mode, and a J_V of 27.0 LMH with J_S of 8.4 gMH in AL-FS mode using deionized (DI) water and 1M NaCl aqueous solution as the feed and draw solutions respectively. This work may provide a new orientation in developing TFC membranes on the hydrophobic substrate with its intrinsic advantages maximized.

527486 Enhanced CO_2 Electroreduction in Selectivity Tuned By Anion Modification of the Ionic Liquids

Abstract:

Enhanced CO_2 electroreduction in selectivity tuned by anion modification of the ionic liquids

Jianpeng Feng [a,b], Shaojuan Zeng[a], Suojiang Zhang [a], Xiangping Zhang [a,*]

[a] *Institute of Process Engineering, Chinese Academy of Sciences, 100190, Beijing, China*

[b] *University of Chinese Academy of Science, Beijing 100049, China*

**Corresponding author:* xpzhang@ipe.ac.cn

In the past hundreds years, the unrestrained combustion of non-renewable fossil fuels leads to an aggravated energy crisis and a continuously rising level of atmospheric carbon dioxide (CO_2). which has created serious environmental, social and global ecological issues[1, 2]. However, CO_2 is also a cheap, abundant and renewable C1-feedstock that can be converted into useful chemicals. Electrochemical reduction was extensively studied owing to their mild reaction conditions and easy implementation for practical applications. Moreover, they can directly use the seasonal, geographical, and intermittent energy (tied, wind, and solar et al.). However, the low solubility of CO_2 and complicated reaction process in aqueous solution render CO_2 reduction inefficient. Ionic liquid (IL) has been proven to be a promising reaction medium for CO_2 electrochemical reaction because of its unique physicochemical properties (e.g., high CO_2 solubility, high intrinsic ionic conductivity, wide electrochemical potential windows)[3, 4]

In this work, Herein, we present an efficient CO_2 electroreduction system based on a diverse class of imidazolium ILs, including a novel superbase IL 1-Butyl-3-Methylimidazolium 1,2,4-Triazolide ([Bmim][124Triz]). The superbase IL has a superior capacity of 0.726 mol CO_2 per mol IL, and the high concentration of dissolved CO_2 in the electrolyte is favorable for the fast transport of CO_2 to the cathode surface. On Pb electrode, the applied potential for CO_2 reduction to formic acid (HCOOH) was lowered from circa -1.97 V to -1.78 V (vs. Ag/Ag^+) compared to [Bmim][PF_6] IL. For now it was widely reported that the reduced form from imidazolium cation, namely, imidazolium neutral radical serves as the reaction site for CO_2 reduction, nevertheless, we found that in the HCOOH formation on Pb electrode, the chemical binding of the $[124Triz]^-$ anion and CO_2 molecule can decrease the activation energy, significantly.

Furthermore, within the same strategy, the highest Faradaic efficiency (98.4%) at partial current density of 11.7 mA cm^{-2} for CO was obtained on Ag in ILs system to date. Our finding about the role of the anion on CO_2 electroreduction will stimulate to develop more novel anions to enhance this reaction process and open a potential way for the efficient utilization of CO_2.

1. S. Zeng, X. Zhang, L. Bai, X. Zhang, H. Wang, J. Wang, D. Bao, M. Li, X. Liu and S. Zhang, *Chem. Rev.*, 2017, **117**, 9625-9673.
2. M. MitroviÄ‡ and A. Malone, *Carbon capture and storage (CCS) demonstration projects in Canada*, Energy Procedia, Canada, 2011.
3. B. A. Rosen, A. Salehi-Khojin, M. R. Thorson, W. Zhu, D. T. Whipple, P. J. A. Kenis and R. I. Masel, *Science*, 2011, **334**, 643-644.
4. P. Markewitz, W. Kuckshinrichs, W. Leitner, J. Linssen, P. Zapp, R. Bongartz, A. Schreiber and T. E. Muller, *Energy & Environmental Science*, 2012, **5**, 7281-7305.

529535 Understanding the Role of Ionic Liquids in the Enzyme Catalyzed Breakdown of Cellulose Using Molecular Dynamics Simulations

Abstract:

Because of their potential for single-pot cellulose dissolution and enzymatic degradation, Ionic liquids (ILs) afford an opportunity to optimize the biocatalytic conversion of non-food plant biomass into biofuels. Established methods for the pretreatment of biomass are costly, inefficient, and often use harsh processing conditions and chemical solvents resolving in unwanted by-products.[1] ILs possess unique properties (e.g. negligible flammability, and vapor pressure), demonstrate wide chemical and thermal stability, and can be easily recycled.[2] However, they have also been shown to negatively affect the activity of many cellulases. With the goal of rapid conversion of biomass to glucose, molecular level insight can assist in the rational design of enzymes and ILs.

Progress has been made to further understand IL-induced unfolding and deactivation[3,4], but still little is known about how ILs influence the catalytic mechanisms of these cellulases. Using molecular dynamics, these mechanisms can be probed at an atomic level. The key challenge in modeling these systems is to find the balance needed between computational cost and accuracy. The hybrid quantum mechanics/molecular mechanics (QM/MM) approach is a method that can be used to probe these enzymatic reactions with the strength of QM accuracy, and the speed of MM forcefields. Still, the timescales needed to observe these phenomena occur on much larger time scales than are computationally affordable. Enhanced sampling techniques can be used to circumvent these timescale issues. For example in metadynamics[5], a powerful enhanced sampling algorithm, a time dependent bias is applied to a few coarse slow degrees of freedom to enhance phase space exploration by discouraging the system from visiting previous states.

This talk presents our research using QM/MM molecular dynamics to model the biocatalytic breakdown of cellulose. Using the enhanced sampling method well-tempered metadynamics[6] we study the effect of ILs on the thermodynamics of enzyme catalyzed reactions and determine the role ILs play on the catalytic mechanism. These simulations feature an exemplary Cel5A retaining endoglucanase, in complex with a cellotetraose ligand modeled in an aqueous IL 1-butyl-3-methylimidazolium chloride [BMIM][Cl] solution, Figure 1. The results indicate the presence of an ILs in the active site in both the holo and apo forms of the enzyme. The free-energy landscape for glycosylation suggests IL inhibition of the intermediates formed in this two-step reaction. Broadly, this provides us with a basis for understanding IL inhibition of Cel5A activity.

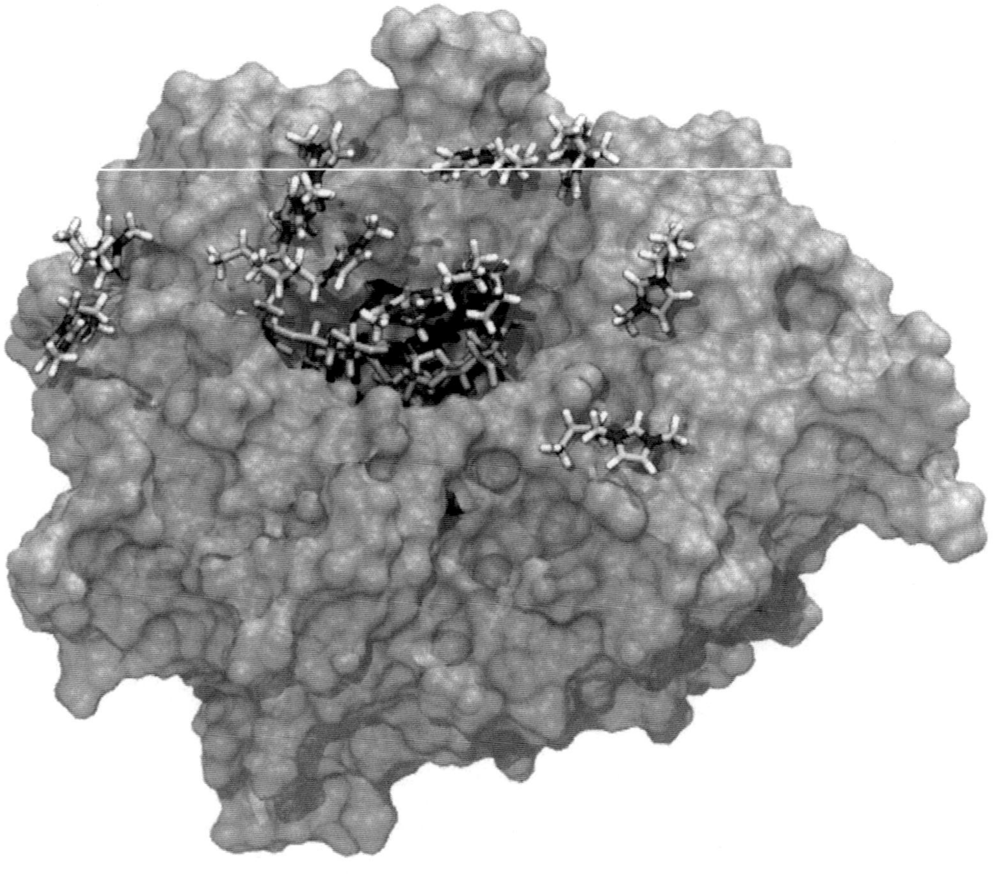

Figure 1. Showing the presence of [BMIM][Cl] (colored by atom name in licorice), in the active site (orange,surf) of Cel5A in complex with a cellulose ligand (green, licorice).

References

[1] P. Engel, R. Mladenov, H. Wulfhorst, G. Jager, & A. C. Spiess. "Point by point analysis: how ionic liquid affects the enzymatic hydrolysis of native and modified cellulose", *Green Chem.*, **2013**, 12, 1959-1966.

[2] J. Serra Moreno, S. Jeremias, A. Moretti, S. Panero, S. Passerini, B. Scrosati, G.B. Appetecchi. "Ionic liquid mixtures with tunable physiochemical properties", *Electrochimica Acta.*, **2015**, 599-608.

[3] S.R. Summers, K. G. Sprenger, J. Pfaendtner, J. Marchant, M. F. Summers, J. L. Kaar, "Mechanism of Competitive Inhibition and Destabilization of Acidothermus cellulolyticus Endoglucanase 1 by Ionic Liquids", *J. Phys. Chem. B.*, **2017**, 121, 10793–10803.

[4] V. Jaeger, P. Burney, J. Pfaendtner, "Comparison of Three Ionic Liquid-Tolerant Cellulases by Molecular Dynamics", *Biophys. J.*, **2015**, 108(4) 880–892.

[5] A. Laio and M. Parrinello. "Escaping free energy minima", *Proc. Natl. Acad. Sci. USA*, **2002**, 99:12562–12566.

[6] A. Barducci, G. Bussi, M. Parinell, "Well-Tempered Metadynamics: A Smoothly Converging and Tunable Free-Energy Method", *Phys. Rev. Lett*, **2008**, 100(2), 020603.

533405 Novel Ionic Liquids Preparation and Application in Gas Separation Process

Abstract:

As a new solvent, the unique structures and properties of ionic liquids make them show significant advantages in the separation fields. That offers new opportunities for new separation technology development. Previously, we have done some interesting work for understanding the fundamental nature of ILs and developing new gas separation process. New task-specific ILs were designed and synthesized by modifying the structure of the anion and cation of the ILs. These task-specific ILs were used for CO_2 capture from flue gas, amonia separation from tail gas of ammonium molybdate process and so on, which showed obviously higher solubility and selectivity. The large-scale preparation process of task-specific ILs was established, and the qualified absorbent was prepared by using industrial raw materials.

The continuous experiments were all completed to evaluate the absorbent performance by using gas separation unit ($10Nm^3/h$). The optimum operating parameters of the process were obtained. Process simulation was also developed to optimize the gas seperation performance using pure ILs (Used for NH_3 separation) and ILs mixture (Used for CO_2 capture). The results showed that ILs process saved heat energy consumption and separation cost compared to the traditional process. At present, we have completed the engineering design. The demonstration plant is under construction by cooperating with company.

Acknowledgement:

This work was supported by the National Key Projects for Fundamental Research and Development of China (2016YFB0600903) and the National Natural Science Foundation of China (No. 21776277).

533653 Isobutane Alkylation with C4 Olefin Catalyzed By Combination of so₃h-Functionalized Ionic Liquids and Sulfuric Acid

Abstract:

Isobutane Alkylation with C4 Olefin Catalyzed by Combination of SO_3H-Functionalized Ionic Liquids and Sulfuric Acid

Weizhong Zheng[1], Piao Cao[1], Weizhen Sun[1], and Ling Zhao[1]

[1] *State Key Laboratory of Chemical Engineering, East China University of Science and Technology, 200237, Shanghai, China, wzzheng728@foxmail.com, 1621756797@qq.com, sunwz@ecust.edu.cn, zhaoling@ecust.edu.cn*

Ionic liquids (ILs), also called room temperature ionic liquids (RTILs), are a class of organic salts that are liquid at near room temperature. Recently, ILs have attracted significant attention because of their unique properties, such as negligible vapor pressure, high chemical, thermal, and photocatalytic stability, nonflammability, and thus become good alternatives to traditional liquid solvents and catalysts.[1] In particular, the ILs have been widely investigated in the C4 alkylation, because the alkylate, produced by the C4 alkylation, a desirable blending component for high quality gasoline.[2] Currently, the alkylation processes for commercial importance still rely heavily on either sulfuric or hydrofluoric acid as catalyst. However, the equipment corrosion, high acid consumption, safety, and environmental issues related to the disposal of spent acid are still challenges to overcome.[3] To cope with these problems, SO_3H-functionalized ionic liquids (SFILs) are regarded as a promising candidate, playing a dual solvent-catalyst role in extensive applications for considerable chemical processes due to their excellent properties, including tunable acidity, high catalytic activity, and better dissolving capacity for a wide range of organic and inorganic compounds.[4] Therefore, it is necessary to deeply investigate the catalytic performances of SFILs as co-catalyst for the H_2SO_4 alkylation, especially the structure-property relationship.

In this work, the SFILs with different alkyl chain length were investigated as co-catalyst mixed with the concentrated H_2SO_4 for the isobutane alkylation. The SFILs with longer alkyl chain show a better catalytic performance with the C8 selectivity up to 75.73 and RON up to 95.66, respectively. The better catalytic performance can be attributed to the better dispersion of isobutane in the SFIL/H_2SO_4 system led by the SFILs with longer alkyl chain, which is further correlated with the nanostructured-aggregation feature of the longer alkyl chain confirmed by MD simulation, as shown in Figure 1. In addition, the reusability of SFIL/H_2SO_4 mixture can reach up to 22 runs, outclassing the pure H_2SO_4, which can be attributed to cationic clusters formed by the strong hydrogen bonds between the sulfonic acid group and the H_2SO_4. The fundemental understanding in this work can bring us important insights into the physicochemical properties of the alylation process and provide a novel approach to screen and design SFILs for isobutane alkylation.

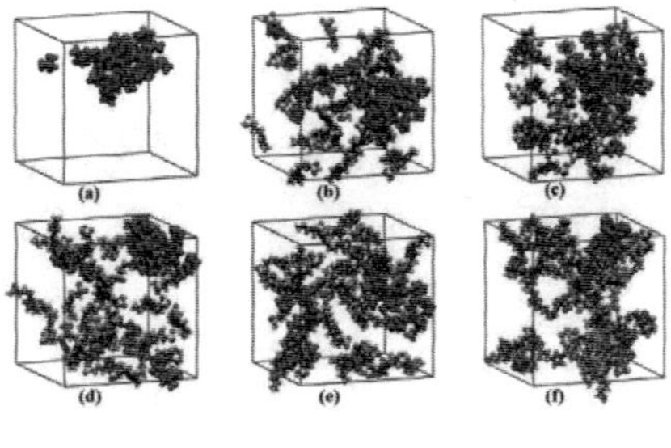

Figure 1. Snapshots of the isobutane dispersion behaviors in different simulation systems.

(a) Pure H_2SO_4 (b) [MPSIm][HSO_4] (c) [EPSIm][HSO_4] (d) [BPSIm][HSO_4] (e) [HPSIm][HSO_4] (f) [OPSIm][HSO_4]. Molecules in red color represent isobutane, and molecules with the blue (N), grey (C), yellow (S), and red (O) atoms stand for the cations of the SFILs. The concentrated H_2SO_4 and anions are removed.

The financial support by the National Natural Science Foundation of China (91434108) is gratefully acknowledged.

[1] Welton T. Room-temperature ionic liquids. Solvents for synthesis and catalysis[J]. Chemical reviews, 1999, 99(8): 2071-2084.

[2] Busca G. Acid catalysts in industrial hydrocarbon chemistry[J]. Chemical reviews, 2007, 107(11): 5366-5410.

[3] Corma A, Mart"ªnez A. Chemistry, catalysts, and processes for isoparaffin-olefin alkylation: actual situation and future trends[J]. Catalysis Reviews-Science and Engineering, 1993, 35(4): 483-570.

[4] Amarasekara A S. Acidic ionic liquids[J]. Chemical reviews, 2016, 116(10): 6133-6183.

533765 Superoxide-Derived CO_2 Reduction at Low over-Potentials and Ultra-Fast: A General Approach in Ionic Liquids

Abstract:

Accomplishing highly selective and efficient electrochemical CO_2 reduction is essential for a sustainable utilization of fossil energy resources. Here, we report ultra-fast CO_2 reduction in ionic liquids (ILs) performed in aerobic environments. CO_2 reduction to CO is observed at a very low over-potentials, namely, >90% faradic efficiency and near 100% carbon selectivity. In the present article, we show that this general approach can be conducted three conventional electrodes (Pt, Au and carbon). The reduction process is initiated by an electrochemical generation of superoxide radicals ($O_2^{\cdot-}$) and followed by an efficient CO_2 activation through N-heterocyclic carbene (NHC) CO_2 adducts in IL. The observed experimental results, supported by a first principle computational modeling suggest an ECE mechanism, where $O_2^{\cdot-}$ superoxide efficiently transforms NHC structures, and activates CO_2 into NHC-CO_2. This work offers a green approach for CO_2 reduction under simple conditions, with potential applications in a broad range of platforms, catalytic systems, and biological processes involving O_2

534675 Biocatalysis in Anhydrous Ionic Liquids

Abstract:

Enzymes can perform many industrially relevant reactions, such as esterification, hydrolysis, oxidation, reduction, and C-C bond formation, with high specificity and under significantly milder conditions than their chemical counterparts. They can perform these reactions on a wide range of substrates, making them highly attractive for many applications. As a result, research into the use of enzymes as industrial biocatalysts has been gaining ground, particularly in conjunction with emerging solvent systems such as ionic liquids. However, enzymes often have very low solubilities in nonaqueous environments and are frequently unstable, limiting the window of usability. Consequently, there is a need to develop new biotechnologies that improve solubility and stability of biocatalysts in nonaqueous media.

Surface modification of proteins, to yield protein-polymer surfactant nanoconjugates, has been demonstrated as a robust method for synthesizing protein-rich biofluids, devoid of any solvent. This new class of material has been shown to be a promising new technology where enzymes have been stabilized in non-aqueous environments. Using a variety of spectroscopic and scattering techniques, these novel biomaterials have been shown to allow for extreme enzyme thermal stability[1], stability against aggregation[2], and retained dynamics[3] and function[4]. Importantly, biofluids of the industrially relevant enzyme lipase showed that enzyme activity was not only retained in the absence of water, but actually enhanced at temperatures up to and including 150 °C[5].

Recently, we showed that protein-polymer surfactant nanoconjugates are soluble in both hydrophilic and hydrophobic ionic liquids, and we demonstrated that biomolecule architecture can be preserved in the non-aqueous environment. Furthermore, the solubilized protein displayed improved thermal stability as compared to aqueous solutions[6]. Here, we show recent results involving nanoconjugates of the enzyme β-glucosidase in anhydrous ionic liquids. Particularly, we demonstrate that the enzyme has significantly improved activity at 120 °C, and importantly, activity towards water insoluble cellulose. As a result, this nascent technology could provide a platform for biocatalysis in industrially relevant nonaqueous solvent systems.

(1) Brogan, A. P. S. *et al. Chem. Sci.* **2012**, *3*, 1839–1846.

(2) Brogan, A. P. S. *et al. J. Phys. Chem. B.* **2013**, *117*, 8400–8407.

(3) Gallat, F.-X.; Brogan, A. P. S. *et al. J. Am. Chem. Soc.* **2012**, *132*, 13168–13171.

(4) Perriman, A. W.; Brogan, A. P. S. *et al. Nat. Chem.* **2010**, *2*, 622–626.

(5) Brogan, A. P. S. *et al. Nat. Commun.* **2014**, *5*, 5058.

(6) Brogan, A. P. S., and Hallett, J. P. *J. Am. Chem. Soc.* **2016**. *138*, 4494-4501.

526252 Porous Crystalline Molecular Sieve Membranes for Kr/Xe Separation

Abstract:

Chemical separations account for approximately half of US industrial energy use, and up to 15% of the nationâ€™s total energy consumption. Non-thermally driven membrane based separations can make industrial relevant separations more energy efficient. In particular, the separation of Krypton (Kr) from Xenon (Xe) is an industrially challenging separation. Kr and Xe are widely used in fluorescent light bulbs. High-purity Xe, has been used in commercial lighting, medical imaging, anaesthesia and neuroprotection. Effectively separating Kr from Xe in nuclear reprocessing plants, would lead to a considerable reduction in storage costs, and in potential revenue generated from the sale of pure Xe. The conventional method to separate these two gases is fractional distillation at cryogenic temperatures, which is an energy intensive process. In this respect, membrane technology could play a key role in making this separation less energy intensive and therefore economically feasible. Membrane separation processes have several advantages over conventional fractional distillation; for instance, it is a viable energy-saving method, since it does not involve any phase transformation, furthermore, the required membrane process equipment is simple, easy to operate, control and scale-up. In particular, if prepared in membrane form, microporous crystalline molecular sieves (such as zeolites, metal organic frameworks, and aluminophosphates) combine highly desirable properties, such as uniform micropores, high surface areas, and exceptional thermal and chemical stability, making them ideal candidates for challenging molecular gas separations, such as Kr/Xe separation. In this presentation, I discuss three recent successful examples of crystalline microporous molecular sieve membranes (SAPO-34, ZIF-8, and AlPO-18) that effectively separate Kr/Xe gas mixtures. To our best knowledge these membranes are the first examples of any microporous crystalline membrane displaying effective separation for Kr/Xe gas mixtures at industrially relevant feed compositions. Fundamentally, I discuss the separation mechanisms involved: molecular sieving, competitive adsorption, and difference in diffusivities.

525436 3D Printing Thin Film Composite Membranes

Abstract:

Interfacial polymerization, while effective at making thin film composite desalination membranes of high permeability and selectivity, is a relatively uncontrolled process that prevents tuning of polyamide film thickness or roughness. A new approach is presented which uses electrospraying to deposit monomer solutions in nanoscale droplets directly onto a substrate where they subsequently react to form the crosslinked polyamide film essentially on any surface. This printing approach enables film thickness control to 4 nm resolution (~20 nm total thickness) and achieves films with a root mean square roughness as low as 2 nm. The higher performing membranes in this study exceeded both the permeance (as much as a 10-fold improvement) and selectivity (notably higher salt rejections) of commercially available reverse osmosis membranes in addition to their substantially lower roughness.

523484 Bio-Inspired Low Biofouling Nanocomposite Membranes: From Batch-Scale to Continuous-Scale Membrane Fabrication

Abstract:

According to the International Water Management Institute (IWMI), one in three people globally endures some form of water scarcity, one-quarter of the world's population lives in areas where water is physically scarce, and over a billion people live where water is economically scarce. Existing water supplies may be limited in quantity or quality for meeting the increasing demands from population growth and industry expansion. In many arid and semi-arid areas, providing the large volume of water required for industrial, agricultural, recreational and potable applications is especially difficult. The search for new technologies for water treatment is an increasingly important research topic. Depending on the application, water is desired in varying degrees of purity. For some applications, such as desalination, a very high degree of purification is desired.

Membrane processes are now a proven and reliable method of providing high-quality, cost-effective water. Membrane technologies have immediate applications to treatment of fresh, brackish and seawaters, as well as wastewater reclamation. With innovative module design and engineering, micro- and ultra-filtrations have become effective and economical for drinking water production, particularly for removal of microorganisms. Desalination is becoming more and more a commodity to satisfy the growing demand for water. This study shows the results of low-biofouling nanocomposite membranes, when using batch-scale fabrication and testing techniques, and when using continuous-scale fabrication and testing techniques. This holistic study begins with nanoparticle manufacturing and selection, then focuses on nanocomposite membrane synthesis and fabrication, and ends with testing and characterization. Nanocomposite membranes loaded with casein-coated silver nanoparticles (Casein-AgNPs) were cast using two approaches, doctor-blade extrusion (batch-scale) and slot-die casting (continuous-scale), to determine their biofouling control properties. Overall, filtration experiments and membrane testing following biofouling tests showed that laboratory-scale composite membranes operated for 24 hr were effective in mitigating biofouling formation. Conversely, continuous-scale nanocomposite membranes operated for 26 days did not show clear improvement in biofouling control, however there was visible damage to cells accumulated on the membrane.

537709 Enabling Widespread Use of Microporous Membranes for Challenging Organic Solvent Separations

Abstract:

The industrial production of chemicals and fuels relies heavily on established separation technologies such as distillation, crystallization, and absorption. Currently, these thermally-driven separations consume anywhere up to 50% of the total energy required for the production processes associated with fuels, chemicals, pharmaceuticals, and water. By avoiding these energy-intensive phase changes during separation, new low energy, low carbon emission separations could be realized. Membranes are one such technology that have been developed to meet these needs. Membrane separations can utilize as little as 10% of the energy required for thermal-based desalination technologies, as seen in membrane-based water desalination. The earliest filtration membranes were able to differentiate between molecules of 10−100× difference in size; today, membranes can separate molecules with just 1.05−1.5× difference in size. Indeed, membrane technology is now the state of the art for seawater desalination facilities and is also competitive in certain gas separations such as air separation, hydrogen recovery, and natural gas processing and is emerging as a leading technology for olefin/paraffin separations. Membrane-based organic solvent separations could revolutionize the field of hydrocarbon separations by enabling new low energy, low carbon emission technologies. Microporous materials are potential game changers for these challenging separations due to their ability to provide superb size and shape discrimination; however, large-scale application of these materials has yet to be adopted. This talk will cover the current state of the art in the separation of similarly sized organic molecules via microporous membranes, challenges and strategies for scaling out these materials into real devices, and critical needs in this emerging area of membrane science.

537736 Batch Distillation Simulation, Optimization, and Control: Past, Present, and Future

Abstract:

The recent increase in the production of high-value-added, low-volume specialty chemicals and biochemicals has generated a renewed interest in batch processing technologies. Batch distillation is a critical unit operation in the batch processing industry and is widely used. The flexibility of batch distillation combined with the inherent unsteady nature of the process poses challenging design and operation problems. This talk presents the overview of simulation, optimization, and control starting from the first theoretical development in 1902 with Rayleigh equation. This talk covers:

1. The various operating modes of batch distillation
2. Examines challenges involved in rigorous modeling
3. Need for the hierarchy of models
4. Optimization and optimal control problems encountered
5. Recent developments related to complex columns, complex systems, and uncertainties, and the role of computer-aided molecular design in batch distillation

541778 Food-Energy-Water Nexus Systems Engineering

Abstract:

Due to population growth, economic development, urbanization, climate change, and natural resource degradation, the global demand for food, energy and clean/freshwater is rapidly increasing [1]. As demand grows the stresses and interdependences between these three resource systems, which are commonly referred to as Food-Energy-Water Nexus (FEW-N), are becoming more apparent [2,3]. Decision making can be very challenging due to multiple dimensions of the biophysical water, energy and food systems, and the players and stakeholders connected with them [4]. A holistic systems engineering approach is thus clearly needed to navigate the multi-faceted FEW-N space, identify opportunities for synergistic benefits and systematically explore interactions and trade-offs [5].

In this work, we present the foundations of a systems engineering framework and quantitative decision-making tools for the analysis and trade-off optimization of stressed interconnected resource FEW-N networks. The framework combines data analytics and mixed-integer modeling and optimization methods establishing the interdependencies and potentially competing interests amongst the food, energy and water elements in the system, along with policy, sustainability and feedback from the various stakeholders. A multi-objective optimization strategy is followed for the analysis of the trade-offs empowered by the introduction of composite FEW-N metrics as means to facilitate decision making and compare alternative process and technological options. The versatility, potential and applicability of the proposed framework is demonstrated by three nexus case studies, (i) a complex power generating system in the South-Central Texas region [6], (ii) a dairy production and processing plant [7], and (iii) a land use allocation problem on a regional farming system in China [8].

References

[1] D'Odorico, P., Davis, K. F., Rosa, L., Carr, J. A., Chiarelli, D., Dell'Angelo, J., ... & Rulli, M. C. (2018). The Global Foodâ€Energyâ€Water Nexus. Reviews of Geophysics.

[2] McCarl, B. A., Yang, Y., Schwabe, K., Engel, B. A., Mondal, A. H., Ringler, C., & Pistikopoulos, E. N. (2017). Model Use in WEF Nexus Analysis: a Review of Issues. *Current Sustainable/Renewable Energy Reports*, *4*(3), 144-152.

[3] McCarl, B. A., Yang, Y., Srinivasan, R., Pistikopoulos, E. N., & Mohtar, R. H. (2017). Data for WEF Nexus Analysis: a Review of Issues. *Current Sustainable/Renewable Energy Reports*, *4*(3), 137-143.

[4] Daher, B., Mohtar, R.H., Pistikopoulos, E.N., Portney, K.E., Kaiser, R. and Saad, W. (2018) Developing Socio-Techno-Economic-Political (STEP) Solutions for Addressing Resource Nexus Hotspots, Sustainability 2018, 10, 512; doi:10.3390/su10020512

[5] Garcia, D. J., & You, F. (2016). The water-energy-food nexus and process systems engineering: a new focus. *Computers & Chemical Engineering*, 91, 49-67.

[6] Diangelakis, N. A., & Pistikopoulos, E. N. (2017). A multi-scale energy systems engineering approach to residential combined heat and power systems. *Computers & Chemical Engineering*, 102, 128-138.

[7] Avraamidou, S., Milhorn, A., Sarwar, O., & Pistikopoulos, E. N. (2018). Towards a quantitative Food-Energy-Water Nexus metric to facilitate decision making in process systems: A case study on a dairy production plant. *28th European Symposium on Computer-Aided Process Engineering*: Accepted.

[8] Nie, Y., Avraamidou, S., Li, J., Xiao, X., & Pistikopoulos, E. N. (2018). Land use modeling and optimization based on food-energy-water nexus: a case study on crop-livestock systems. *13th International Symposium on Process Systems Engineering*: Accepted.

Renewable Carbons from Food Waste for Separation and Catalysis Technologies

This study explores the production of renewable and sustainable carbons from waste biomass resources, for adsorption and catalytic applications. Specifically, the research will explore the production of carbons derived from **food waste** and their utilization as sorbents of aromatic contaminants in **water** and catalyst supports for the production of clean **energy and fuels**.

The idea of using renewable and sustainable resources to replace the carbon derived from fossil fuels is not new. Researchers have been working on the utilization of various biomass resources, such as rice husk, coconut shells and others for years.[1–5] However, the utilization of food waste as a carbon resource has not been extensively explored. Food waste could be the ultimate resource for carbon production, since it is produced in excessive volumes and it represents one of the most significant environmental problems, which is urgently looking for a solution. The critical question is whether the activated carbons derived by food waste have the potential to be used for purification and/or catalysis purposes. Food waste is heterogeneous in nature and contains various types of minerals, which may provide advantages or disadvantages, depending on the application.

In our lab, we receive the food waste from the Dining Halls of the University of Connecticut and after pretreatment, that includes washing, grinding and sieving, we perform pyrolysis and activation to turn the waste into carbons. The produced carbons are then characterized and they are tested as sorbents and catalyst supports. Our goal is to understand how the conditions of pyrolysis and activation affect the properties of the carbons and, consequently, their effectiveness as sorbents and catalyst supports. During the pyrolysis reaction, temperature and residence time are the most important parameters that affect the yield and the properties of the produced biochar. Activation of the biochar is performed either using steam, physical activation, or using chemicals (e.g. H_3PO_4), chemical activation. During the former method of activation, temperature and residence time are, again, among the most important parameters that affect the surface area, the carbon content and the porosity of the produced carbons. During the latter method, the impregnation ratio between the chemical and the biochar is the most critical.

During this talk we will discuss the properties of the activated carbons produced by food waste and how these properties are affected by the reaction conditions and the method of activation. Surface area, porosity, carbon content, surface functional groups, morphologies are the properties of interest which will be discussed. Then, we will discuss the potential of using the activated carbons produced by food waste, as sorbents of aromatic hydrocarbons in water and catalyst supports for the hydrodeoxygenation of bio-oil.

References

(1) Tripathi, M.; Sahu, J. N.; Ganesan, P. Effect of Process Parameters on Production of Biochar from Biomass Waste through Pyrolysis: A Review. *Renew. Sustain. Energy Rev.* **2016**, *55*, 467–481.

(2) Kwapinski, W.; Byrne, C. M. P.; Kryachko, E.; Wolfram, P.; Adley, C.; Leahy, J. J.; Novotny, E. H.; Hayes, M. H. B. Biochar from Biomass and Waste. *Waste and Biomass Valorization* **2010**, *1* (2), 177–189.

(3) Ahmad, M.; Rajapaksha, A. U.; Lim, J. E.; Zhang, M.; Bolan, N.; Mohan, D.; Vithanage, M.; Lee, S. S.; Ok, Y. S. Biochar as a Sorbent for Contaminant Management in Soil and Water: A Review. *Chemosphere* **2014**, *99* (November 2013), 19–23.

(4) Tan, Z.; Lin, C. S. K.; Ji, X.; Rainey, T. J. Returning Biochar to Fields: A Review. *Appl. Soil Ecol.* **2017**, *116* (September 2016), 1–11.

(5) Azargohar, R.; Dalai, A. K. Biochar as a Precursor of Activated Carbon. *Appl. Biochem. Biotechnol.* **2006**, *131* (1–3), 762–773.

541779 A Computational Framework for Sustainable Waste Management and Simultaneous Recovery of Nutrients and Energy

Abstract:

To protect human health and the environment, communities should have sustainable infrastructures to manage waste. Critically, livestock waste may cause some air quality degradation from ammonia and methane emissions, soil quality detriment due to in-excess nutrients and acidification, and water pollution issues resulting from nutrient and pathogens runoff to water bodies, which leads to eutrophication, hypoxia, and harmful algal blooms (HABs).

In addition, the recovery of carbon from organic waste offers an alternative for replacing fossil fuels with renewable sources for energy production. Despite the significant environmental benefits by performing pollution management of these organic materials, the deployment of technologies for obtaining value-added products and renewable energy from livestock waste is not a current practice due to the high investment costs and low product market values.

This contribution describes a multi-scale approach for the development of a supply chain framework to identify and evaluate techniques and logistics for the management of solid waste generated from primary sources of nutrient pollution. This effort will inform stakeholdersâ€™ waste management options while assessing potential economic and environmental benefits. Other potential benefits include the reduction of nutrient pollution and HABs, recovery of valuable products, and adoption of nutrient-reducing technologies in different locations (e.g., community, state, regional, etc.). In addition, the proposed framework captures techno-economic and logistical issues and can accommodate diverse types of material management incentives obtained at federal and state levels.

540773 2018 Outlook for Energy: A View to 2040

Abstract:

The Outlook for Energy is ExxonMobil's global view of energy demand and supply through 2040. It also highlights the dual challenge of providing affordable energy to support prosperity while reducing environmental impacts – including the risks of climate change. We use the Outlook to help inform our long-term business strategies and investment plans.
The Outlook anticipates significant changes through 2040 across the world to boost living standards, reshape the use of energy, broaden access to abundant energy supplies, and accelerate decarbonization of the world's energy system to address the risks of climate change.

- Global energy needs will rise about 25 percent over the period to 2040
- Electricity demand will nearly double in non-OECD countries
- Worldwide electricity from solar and wind will increase about 400 percent
- Natural gas will expand its role, led by growth for electricity generation
- Growth in oil demand will be driven by commercial transportation and chemical needs
- Efficiency gains and growing use of less-carbon intensive energy sources will contribute to a nearly 45 percent decline in the carbon intensity of global GDP

Energy is fundamental to modern life, and as the world's population approaches 9 billion people in 2040, we are challenged to improve living standards everywhere. Understanding the factors that drive the world's energy needs – and likely choices to meet those needs – is the mission of the Outlook. By sharing the Outlook with the public, we hope to broaden that understanding. Energy matters to everyone, and we all play a role in shaping its future.

541543 Energy Decarbonisation Scenarios

Abstract:

With the energy sector being associated with 2/3 of the annual greenhouse gas emissions, decarbonizing the energy is a key step to reduce their level. The National Determined Contributions (NDC's) announced so far by the participating countries do not meet the objectives of the Paris-climate agreement, as far as reaching a peak in emissions in the near future or decreasing to a level that is compatible with the stated long-term targets.

Using different scenarios for energy demand and supply, options for energy sector decarbonisation are evaluated. Meeting the 2-degree target requires a major transformation in the power sector, efficiency improvements – specially in the buildings and in the industry, and a revolution in transport towards fuel and modal shifts. With a significant increase in renewable energy, system integration and optimization will also be key in reducing the cost of the energy transition. Meeting more ambitious climate goals ("well below 2 degrees") will push further the technological innovation and implementation requirements throughout the entire energy value chain. A scenario aiming at zero net-emissions by 2060 is presented, along with a comparison to the 2-degree case.

542625 Fundamental Research Needs to Advance Energy Technologies

Abstract:

The U. S. Department of Energy (DOE), Office of Science, Office of Basic Energy Sciences (BES) supports fundamental research in chemical and materials sciences to provide the foundations for new energy technologies and to support DOE missions in energy, environment, and national security. This presentation will discuss opportunities for fundamental research to impact DOE's energy mission "to catalyze the timely, material, and efficient transformation of the nation's energy system and secure U.S. leadership in energy technologies" with a focus on the way we generate, store and use energy nationally. I will provide an overview of BES strategic planning over the past decade that identified priority research directions for advancing energy applications, highlight key scientific advances in these areas, and discuss some future opportunities for modern science, particularly chemical sciences, to accelerate the transformation of the U. S. energy portfolio.

531297 On the Flexibility of Porous Materials

Abstract:

In this plenary session we celebrate and commemorate contributions of Prof. Peter Monson to the field of adsorption in porous materials. They are numerous, ranging from phase equilibria under confinement to water adsorption in activated carbons to connectivity effects and adsorption hysteresis. What unites all these themes is the elegance and often simplicity of the employed models to explain complex behaviour and intricate thermodynamics of fluids in porous structures.

Inspired by this approach, my group has been working on several problems in the field of adsorption in porous materials. In this talk, I will focus specifically on developing simple mechanical models to explain flexibility and structural changes in several classes of crystalline porous materials. In the most recent developments, we employed classical density functional theory as an efficient tool to explore complex free energy landscapes and emerging structural transitions in porous materials as a result of adsorption processes.

534648 MOF and Zeolite Membranes and Catalysts

Abstract:

An overview of our recent efforts to prepare MOF-based membranes and catalysts and comparisons of their performance with zeolites will be presented. Examples include (i) the use of zeolite and MOF-nanosheets for the preparation of selective membranes and (ii) sugar isomerization performance of modified MOF catalysts and Lewis acidic zeolites.

539948 Size-Independent Separation of Functionalized Nanoparticles on Porous Substrates

Abstract:

Emerging technologies of production and processing of functionalized nanoparticles (NP) requite advanced methods of nanoparticle characterization and separation. While various methods are available for NP separation by size, there are no efficient methods for NP separation by surface chemistry. It is highly desirable to develop a technique for size-independent separation of functionalized NPs that would be similar to the method of liquid chromatography at critical conditions (LCCC) used for molecular weight independent separation of polymers by their composition and morphology.
In this work, we explore the conditions of chromatographic separation of NPs on porous substrates grafted with polymer brushes (PB) in a binary mixture of "good" and "poor" solvents. This process involves NP flow through pore channels with walls covered by PBs. NP separation is controlled by the competition between enthalpic attraction due to adsorption of polymer chains to NP surface and entropic repulsion caused by geometrical confinement. Using dissipative particle dynamics simulations in conjunction with the ghost tweezers free energy calculation technique, we calculate the free energy landscapes of NP-PB adsorption interaction and determine the Henry constant that characterizes adsorption equilibrium and NP partition between the stationary phase of PB and the mobile phase of flowing solvent. We find that with the increase of the poor solvent fraction and respective PB contraction, NP separation exhibits a transition from the hydrodynamic (size exclusion) to adsorption regimes of separation with the reversal of the sequence of elution. At good solvent conditions. the entropic repulsion dominates, causing the size exclusion separation with larger particles having shorter retention time. As the solvent quality worsens, the enthalpic attraction becomes dominant and the adsorption regime of separation is observed with the reverse order of elution with smaller particles moving faster along the column. This finding suggests a possibility of the existence of an analogous to LCCC regime in NP chromatography, at which NPs with like surface properties elute together disregarding of their size. The latter has important practical implications: NPs can be separated by surface chemistry rather than by their size employing the gradient mode of elution with controlled variation of the solvent composition.

548632 The Rigid Adsorbent Lattice Fluid Model for Stepped Isotherms

Abstract:

Adsorption systems exhibiting stepped isotherms are of practical interest for a number of reasons. A potential advantage is that the steps allow for a large working capacity over small temperature and pressure ranges, which can be exploited in pressure or temperature swing adsorption processes. The steps can also be used to aid selectivity during separation of a multicomponent system. Although the causes underlying the emergence of non Type I isotherms are understood qualitatively, such isotherms are usually modelled by empirical functions for the purpose of process simulation. Such functions rely heavily on experimental data, may require many fitting parameters and are not predictive. Molecular simulations on the other hand, have become a powerful quantitative tool in describing such systems, but are computationally demanding and therefore not suitable in process modelling.

Here we present the prediction of stepped isotherms using a lattice fluid based model, the Rigid Adsorbent Lattice Fluid (RALF) model. The RALF model is a relatively simple model, with few parameters, which all have a physical meaning, such as interaction energies and densities. These are fixed for any given molecule and the number of adjustable parameters is therefore kept to a minimum.

For non-flexible adsorbents, the RALF model is able to predict type V inflections in isotherms based on the pore volume of the solid and the strength of the intermolecular interaction of the guest molecules. As observed experimentally, the model predicts that highly coordinating molecules such as water or ethanol show this behaviour more readily than weakly interacting aliphatic organics in frameworks with the same porosity. It is remarkable that a relatively simple model is capable of reproducing both type I and type V shapes.

Stepped isotherms may also originate from flexibility in the adsorbent, for instance resulting from host-guest interactions, thermal or mechanical stimuli. A typical example is the breathing behaviour of the MIL-53 family of metal organic frameworks (MOFs), whereby the material reversibly swaps between two phases with open and closed porosity, respectively. This structural transition leads to a step akin to a double type I isotherm. We show that the RALF model is also able to capture this effect, simply by treating the two solid phases as identical but with different densities. Some variation of the density with respect to amount adsorbed is allowed within the phases to reflect the materialâ€™s large flexibility. Phase stability is determined by evaluating the total Gibbs energy of the two phases during adsorption. Structural transitions occur under those conditions at which the Gibbs energy profiles cross-over due to differences in the amount adsorbed for the two phases. Due to the difference in densities of the two solid phases, the RALF model reproduces a step in the isotherm upon the structural transition. Predictions for breathing of MIL-53 in CO_2, CH_4 and xenon are in good agreement with experimental results, but currently do not include hysteresis effects.

525182 Multi-Scale Modelling of the Synthesis of Nanoporous Silica Materials

Abstract:

Computational material design offers the possibility of tailoring the properties of materials to suit particular applications. To achieve this, however, one must surmount a major challenge - obtaining a fundamental, molecular-level understanding of the material synthesis process. Such knowledge would allow us to predict how each set of synthesis conditions relates to the structural and chemical properties of the final material. For the particular case of nanoporous silica materials, like zeolites and periodic mesoporous silica (PMS), the synthesis process relies on a delicate interplay of chemical reactions, self-assembly and phase equilibrium, taking place over a wide range of time and length scales. This makes it quite difficult, if not impossible, to achieve a unique view of the mechanism by experimental means alone. This talk will report recent advances by our group in the development of a new multi-scale modelling approach to describe the synthesis of nanoporous silica materials, that bridges time and length scales from the quantum to the coarse-grained levels.

Our simulations of the synthesis of MCM-41 [1], the archetypal example of a mesoporous silica material, have revealed unprecedented details of the molecular-level interactions that control the process. The addition of silica to an aqueous alkylammonium surfactant solution initially gives rise to micelle size increase and micelle fusion [2], promoting a shape transition from spherical to rod-like aggregates [3]. The strong adsorption of silicates at the micelle surface facilitates silica polymerisation. Once silica oligomers are present in the system, they act as multi-dentate binders inducing the formation of a hexagonally-ordered mesophase [4]. It is this mesophase that ultimately determines the pore network structure of MCM-41, upon further condensation and template removal. Before this takes place, however, it is possible to change the mesophase structure by altering the synthesis conditions (e.g. pH, temperature or composition). The results of our multi-scale model correctly describe experimental observations on this system, and make useful predictions for future design of these materials.

We have also extended our model to describe the synthesis of bioinspired silica. Inspired by the natural process of biosilicification, this strategy makes use of polyamine templates to produce porous silica materials under much more environmentally friendly conditions (ambient temperature and pressure, near-neutral pH) than currently employed in industry [5]. Compared to traditional templates used in industrial porous silica manufacture, polyamines bring the added complexity of being pH-responsive, given that their degree of ionisation changes with pH. We have applied our multi-scale model to the first example of bioinspired silica, which was heralded at the time as the first example of a â €œneutral templating route to porous silicaâ€ [6]. Crucially, our results show that such a hypothesised mechanism, based on hydrogen-bond interactions between silicates and templates, is not viable under the experimental synthesis conditions. Instead, the mechanism hinges on charge-matching interactions between precursor molecules [7]. Preliminary experimental data supports our interpretation of the synthesis mechanism. Our work suggests routes for increased control over the properties of this class of materials, paving the way for computational material design of bioinspired silica.

[1] Beck, J. S.; Vartuli, J. C.; Roth, W. J.; Leonowicz, M. E.; Kresge, C. T.; Schmitt, K. D.; Chu, C. T. W.; Olson, D. H.; Sheppard, E. W.; McCullen, S. B.; Higgins, J. B.; Schlenker, J. L. *J. Am. Chem. Soc.* 114 (**1992**) 10834.

[2] Jorge, M.; Gomes, J. R. B.; Cordeiro, M. N. D. S; Seaton, N. A. *J. Am. Chem. Soc.* 129 (**2007**) 15414.

[3] Pérez-Sánchez, G.; Gomes, J. R. B.; Jorge, M. *Langmuir* 29 (**2013**) 2387.

[4] Pérez-Sánchez, G.; Chien, S.-C.; Gomes, J. R. B.; Cordeiro, M. N. D. S.; Auerbach, S. M.; Monson, P. A.; Jorge, M. *Chem. Mater.* 28 (**2016**) 2715.

[5] Manning, J. R. H.; Yip, T.; Centi, A.; Jorge, M.; Patwardhan, S.V. *ChemSusChem* 10 (**2017**) 1683.

[6] Tanev, P. T. and Pinnavaia, T. J. *Science* 267 (**1995**) 865.

[7] Centi, A.; Jorge, M. *Langmuir* 32 (**2016**) 7228.

535398 Advanced Characterization of Nanoporous Materials: *Effect of Pore Size and Temperature on the Adsorption and Phase Behavior of Wetting and Nonwetting Fluids*

Abstract:

During the last decades major advances in understanding the adsorption and phase behavior of fluids confined in ordered nanoporous materials have been made, which led to significant progress in the physisorption characterization methodology (summarized in the 2015 IUPAC report on physisorption characterization [1]). Here we discuss progress and challenges for the physisorption characterization of nanoporous solids exhibiting various levels of porosity from micro- to macropores [2]. While physisorption allows one to assess micro- and mesopores, a widely employed method for textural analysis of macroporous materials is based on the intrusions/extrusion of mercury, which does not wet the majority of surfaces at room temperature. Furthermore, in order to assess aspects of surface chemistry, adsorption methods have been applied which combine wetting fluids (e.g. argon and nitrogen at their boiling temperature) with adsorbates where the wetting behavior is appreciable controlled by surface chemistry (e.g. water adsorption at room temperature. see ref [2). We demonstrate that an improved understanding of confined geometry effects on the adsorption and phase behavior of wetting and non-wetting fluids allows one to arrive at a more reliable structural characterization of complex, nanoporous pore networks.

.

References

[1] M. Thommes, K. Kaneko, A.V. Neimark, J.P. Olivier, F. Rodriguez Reinoso, J. Rouquerol and K.S.W Sing, "Physisorption of gases, with special reference to the evaluation of surface area and pore size distribution " (IUPAC Technical Report), Pure Appl. Chem. 87, 1051 (2015)

[2] K. A. Cychosz, R. Guillet-Nicolas, J. Garcia-Martinez, J., M. Thommes, "Recent Advances in the textural characterization of hierarchically structured nanoporous materials" Chem. Soc. Rev. 46, 389, (2017)

532051 Combined Molecular- and Process-Level Modeling to Evaluate Metal-Organic Frameworks for Post-Combustion CO_2 Capture

Abstract:

In order to reduce global greenhouse gas emissions in the near-term, it is imperative to develop economically promising carbon capture and sequestration (CCS) technologies. Post-combustion CCS, where CO_2 is separated from N_2 and other gases in the flue gas, has attracted much interest recently since the costs of retrofitting it to current power plants is cheaper than other methods. Of the technologies currently available for post-combustion CCS, pressure swing adsorption (PSA) is one of the more promising options due to low energy requirements and short cycle times compared to other adsorption based technologies. One attractive group of adsorbents for CCS applications are metal-organic frameworks (MOFs) due to their tuneable pore structure and chemical functionality. Because of these features, thousands of MOFs have been synthesized in recent years, with various proposed applications, including post-combustion CCS. Many authors measure CO_2 isotherms for MOFs and suggest that the materials are potentially promising for CCS, but metrics for rigorously comparing materials are limited. The objective of this work is to use a combination of molecular modeling, database screening, and process-level modeling to investigate promising MOFs that have been reported in the literature and calculate their process-level and economic performance.

We assessed the process performance of selected MOFs using a 5-step Skarstrom cycle with a heavy reflux step. We also determined the productivity and the energy requirements of the MOFs to determine their economic feasibility. Our PSA model consists of a system of partial differential algebraic equations incorporating mass and energy balances, pressure drop across the column, competitive multi-site Langmuir isotherms and the linear driving force model.[2] The MOFs investigated include Ni- and Mg-MOF-74, UTSA-16, and SIFSIX-2-Cu-i. In addition, we also evaluated two MOFs from the Computational-Ready Experimental (CoRE) MOF database[3] identified as top candidates during our previous investigations.

1. *The Future of Coal*; 2007. http://web.mit.edu/coal/.
2. Leperi KT, Snurr RQ, You F. Optimization of Two-Stage Pressure/Vacuum Swing Adsorption with Variable Dehydration Level for Postcombustion Carbon Capture. *Eng. Chem. Res.* 2016;55:3338-3350.
3. Chung YG, Camp J, Haranczyk M, Sikora BJ, Bury W, Krungleviciute V, Yildirim T, Farha OK, Sholl DS, Snurr RQ. Computation-Ready, Experimental Metalâ€“Organic Frameworks: A Tool To Enable High-Throughput Screening of Nanoporous Crystals. *Mater.* 2014;26:6185â€“6192.

540999 Insulin Exits Skeletal Muscle Capillaries By Fluid-Phase Transport

Abstract:

Before insulin can stimulate myocytes to take up glucose, it must first move from the circulation to the interstitial space. The continuous endothelium of skeletal muscle (SkM) capillaries restricts insulin's access to myocytes. The mechanism by which insulin crosses this continuous endothelium is critical to understand insulin action and insulin resistance; however, methodological obstacles have limited understanding of endothelial insulin transport in vivo. Therefore, we developed a fully bioactive, fluorescent insulin probe and intravital microscopy platform to enable direct visualization of insulin efflux from SkM capillaries in vivo. Mathematical modeling of insulin efflux kinetics was applied to identify fluid-phase transport as the major mode of transendothelial insulin efflux in mice. Model-independent experiments demonstrating that insulin movement is neither saturable nor affected by insulin receptor antagonism confirmed the modeling results. Our finding that insulin enters the SkM interstitium by fluid-phase transport may have implications in the pathophysiology of SkM insulin resistance as well as in the treatment of diabetes with various insulin analogs.

541001 Pharmacometric Model Guided Control for Improved Therapeutic Exposure

Abstract:

An adaptive model-predictive controller was simulated which, directed by a semi-PBPK model, reduces the cardiotoxic effect of doxorubicinol (DOXol), the primary metabolite of doxorubicin (DOX), while maintaining the effectiveness of DOX.

A PBPK model of DOX [1] was expanded to include multiple species and a heart compartment that captures the known myocardial toxicity of DOXol. Model parameters were fit using allometric scaling of subclinical DOX/DOXol tissue data combined with human clinical tissue data. For the adaptive model-predictive control, the DOX and DOXol concentrations in the blood are measured in real-time using a photoplysmograph [2]; the PBPK model predicts the DOXol heart concentration and DOX AUC; and the controller gathers heart concentration predictions and adjusts the infusion based on the continuous measurements model predictions to stay within a target AUC range without exceeding a maximum heart concentration. A simulated study was conducted with a control group and an intervention group that implemented the model-guided controller.

The intervention group showed a 46.9% improvement in maintaining therapeutic AUC levels (control 29.0%, intervention 75.9%). The intervention group also never exceeded the preset maximum heart concentration, indicating the simulation did not allow for the concentration to reach a toxic point. The control group exceeded maximum concentration in 19.8% of simulation iterations.

A PBPK model coupled with the controller code has shown promise in reducing the occurrence of over-prescribing DOX. Simultaneously controlling these values through infusion manipulation will help achieve optimal therapeutic dose with minimal side effects. Tunable user-inputs and physical parameters make this system potentially versatile for broader application.

References:

[1] Dubbelboer, et al. (2017). *Mol. Pharmaceutics*, 14(3):686-698.

[2] Michalak, et al. (2010). *Optics Express*, 18(25):26535-26549.

541002 Spatiotemporal Dynamics of the Human Gut Microbiome

Abstract:

The human gut is host to a diverse and complex ecosystem of over a thousand microorganisms that are collectively known as the "gut microbiome". The gut microbiome has already been correlated with over a hundred diseases and has created a revolution in medical science. These microbes play a major role in providing nutrition for our body and maintaining a strong immune system. Modeling of the spatiotemporal dynamics of the microbes is useful to investigate how the gut microbiome may influence "disease progression". The models could also help in designing strategies to dynamically manipulate the system trajectory away from a "disease" state to a "healthy" state. The importance of microbial dynamics will be illustrated in the context of our research on Autism Spectrum Disorders (ASD). System level modeling approaches using connectivity maps, rules, and mathematical equations have provided insights into the dynamic changes in the gut microbiome in ASD. Such system-level approaches can potentially be used for personalized and predictive diagnoses and for preventive nutritional recommendations

541000 Sequence Specific Modeling of *E. coli* Cell-Free Protein Synthesis

Abstract:

Cell-free protein synthesis (CFPS) is a widely used research tool in systems and synthetic biology. However, if CFPS is to become a mainstream technology for applications such as point of care manufacturing, we must understand the performance limits and costs of these systems. Toward this question, we used sequence specific constraint based modeling to evaluate the performance of E. coli cell-free protein synthesis. A core E. coli metabolic network, describing glycolysis, the pentose phosphate pathway, energy metabolism, amino acid biosynthesis, and degradation was augmented with sequence specific descriptions of transcription and translation and effective models of promoter function. Model parameters were largely taken from literature; thus the constraint based approach coupled the transcription and translation of the protein product, and the regulation of gene expression, with the availability of metabolic resources using only a limited number of adjustable model parameters. We tested this approach by simulating the expression of two model proteins: chloramphenicol acetyltransferase and dual emission green fluorescent protein, for which we have data sets; we then expanded the simulations to a range of additional proteins. Protein expression simulations were consistent with measurements for a variety of cases. The constraint based simulations confirmed that oxidative phosphorylation was active in the CAT cell-free extract, as without it there was no feasible solution within the experimental constraints of the system. We then compared the metabolism of theoretically optimal and experimentally constrained CFPS reactions, and developed parameter free correlations which could be used to estimate productivity as a function of carbon number and promoter type. Lastly, global sensitivity analysis identified the key metabolic processes that controlled CFPS productivity and energy efficiency. In summary, sequence specific constraint based modeling of CFPS offered a novel means to a priori estimate the performance of a cell-free system, using only a limited number of adjustable parameters. While we modeled the production of a single protein in this study, the approach could easily be extended to multiprotein synthetic circuits, RNA circuits, or the cell-free production of small molecule products.

541004 Viewing Crystallization from the World of Population Balances

Abstract:

Crystallization is one of the early applications of population balances in chemical engineering. Nucleation of new crystals followed by their growth along with breakage and agglomeration in an agitated suspension fit naturally into the population balance framework towards making predictions of the size distribution, a prime determinant of the product quality. This talk is a commentary on the historical evolution of how various features of this modeling tool engaged us in the pursuit of its application to solving crystallization problems.

537530 Silver Recovery from Catalysts Using a Leaching and Emulsion-Liquid-Membrane Hybrid Process

Abstract:

Silver is a precious metal that is widely used in many industries including the petrochemical, electrical, electronics, jewelry, and photography[1]. Recovery of silver from waste materials such as photographic and X-ray films, jewelry components, and deactivated catalysts is of economic and environmental interests[2]. To extract silver from solid wastes, many methods have been used[3-6]. Among them, leaching processes are more popular[7-8]. However, a leaching process transfers all types of metal ions present in the solid phase into a liquid phase. So, this process must be combined with a complementary process to separate silver ions from other types of metal ions present in the leach solution. In recent years, the membrane technology, in particular the emulsion liquid membrane (ELM) technology, has received more attention as a promising technology that is more environmentally friendly, is more efficient, and has lower operating and capital costs[9-11].

This paper introduces a hybrid process for an efficient silver recovery from the deactivated industrial catalysts. The hybrid process combines leaching and ELM processes. Leaching first transfers silver from a catalyst to an aqueous solution. In a single separation step, ELM then extracts silver ions from the leach solution. 2-ethylhexyl phosphoric acid ester (MEHPA, a mixture of 55% mono- and 45% di-esters) is used as the carrier in membrane phase (facilitated transport) to extract silver ions from the aqueous solution. A paraffinic and naphthenic solvent is used as the diluent. As a major problem with ELM is emulsion instability resulting from poor surfactants like Span 80, a new surfactant, polypropylene glycol (PPG)-PEG-PPG triblock copolymer, is synthesized. Two stability methods are used to evaluate the ELM stability and the compatibility of the ELM process components (diluent, surfactant, and carrier). Finally, effects of process operating conditions such as carrier, surfactant and internal phase (receiving phase) concentrations, pH of the external phase (feed phase), striping speed, volume ratio of the internal phase to the membrane phase, and volume ratio of the emulsion phase to the external phase (emulsion phase holdup) on the silver recovery are investigated to arrive at the optimal operating conditions that maximize the silver recovery.

References

1. Park, Y. J.; Fray, D. J., Recovery of high purity precious metals from printed circuit boards. *Journal of Hazardous Materials* **2009,** *164* (2-3), 1152-1158.
2. Ajiwe, V.; Anyadiegwu, I., Recovery of silver from industrial wastes, cassava solution effects. *Separation and Purification Technology* **2000,** *18* (2), 89-92.
3. Nasser, I. I.; Amor, F. I. E. H.; Donato, L.; Algieri, C.; Garofalo, A.; Drioli, E.; Ahmed, C., Removal and recovery of Ag (CN) 2-from synthetic electroplating baths by polymer inclusion membrane containing Aliquat 336 as a carrier. *Chemical Engineering Journal* **2016,** *295*, 207-217.
4. Desai, K. R.; Murthy, Z., Removal of silver from aqueous solutions by complexationâ€"ultrafiltration using anionic polyacrylamide. *Chemical Engineering Journal* **2012,** *185*, 187-192.
5. Ahmad, A.; Kusumastuti, A.; Derek, C.; Ooi, B., Emulsion liquid membrane for heavy metal removal: An overview on emulsion stabilization and destabilization. *Chemical engineering journal* **2011,** *171* (3), 870-882.
6. Wang, X.; Zhao, Y.; Li, X.; Ren, Y., Performance evaluation of a microfiltration-osmotic membrane bioreactor (MF-OMBR) during removing silver nanoparticles from simulated wastewater. *Chemical Engineering Journal* **2017,** *313*, 171-178.
7. Alvarado-MacÃas, G.; Fuentes-Aceituno, J.; Nava-Alonso, F.; Lee, J.-c., Silver leaching with the nitriteâ €"copper novel system: A kinetic study. *Hydrometallurgy* **2016,** *160*, 98-105.
8. Hu, P.; Zhang, Y.; Huang, J.; Liu, T.; Yuan, Y.; Xue, N., Eco-friendly leaching and separation of vanadium over iron impurity from vanadium-bearing shale using oxalic acid as a leachant. *ACS Sustainable Chemistry & Engineering* **2017**.
9. Chakrabarty, K.; Saha, P.; Ghoshal, A. K., Separation of lignosulfonate from its aqueous solution using emulsion liquid membrane. *Journal of Membrane Science* **2010,** *360* (1-2), 34-39.
10. Zeng, L.; Yang, L.; Liu, Q.; Li, W.; Yang, Y., Influences of axial mixing of continuous phase and polydispersity of emulsion drops on mass transfer performance in a modified rotating disc contactor for an emulsion liquid membrane system. *Industrial & Engineering Chemistry Research* **2015,** *54* (40), 9832-9843.
11. Park, H.-J.; Chung, T.-S., Removal of phenol from aqueous solution by liquid emulsion membrane. *Korean Journal of Chemical Engineering* **2003,** *20* (4), 731-735.

520974 Upgrading of Bio-Oil As Transportation Fuel Using Water/Oil Based Extraction

Abstract:

Conversion of biomass into biofuels is a promising approach for the production of renewable fuels to replace liquid fuels in aviation and other modes of transport. Although the share of biomass in the total final energy consumption by end-use sector is 14%, majority of it is in heating and cooling (12.6%) and only 0.8% is in transportation. The challenge is the incompatible properties of bio-oil including low heating value, thermal instability, and high corrosivity that necessitates upgrading. Water-oil extraction is investigated as a simple and inexpensive upgrading method that can improve the properties of bio-oil by extracting corrosive and thermally unstable acidic compounds by using their greater polarity and preferential solubility in water. In this paper, pine wood is used as bio-oil and water-oil extraction is performed by mixing water and bio-oil which forms an aqueous phase (water soluble fraction) and an organic water-insoluble fraction (or "pyrolytic lignin"). Removing the aqueous phase which is composed of high-polarity compounds such as formic acid, acetic acid, and ketone results in "upgraded" bio-oil. The water-insoluble fraction is expected to have notably lower acidity, increase its pH, reduced oxygen and its water content. Parameters in water extraction are investigated for optimization of the water extraction process for the bio oil which includes water temperature, stirring time, and water-bio oil ratio. The properties of water-insoluble fraction are characterized and compared with the initial bio oil without water extraction.

Keyword:

Biomass renewable energy, fast pyrolysis, bio oil upgrading, water extraction, pyrolytic lignin

536790 Integration of Enzymatic Reactions into Continuous Countercurrent Extraction Processes with Mixed Surfactant Solutions

Abstract:

Recently, the interest in aqueous surfactant two-phase systems for extraction purposes has grown. The fields of application lie thereby in the extraction and purification of metal ions, proteins, biomaterials, organic compounds and toxic aromatics from wastewater. Aqueous surfactant systems provide mild conditions for biomolecules sensitive to organic solvents or high temperatures. Thus, surfactant-basedsystems with an in-situ product removal to increase conversion are especially interesting for biocatalytic and biotransformation processes with regard to environmental friendliness. Reactive micellar processes are only rarely described in literature so far. Thereby, the utilization of aqueous surfactant two-phase systems for biocatalytic and biotransformation processes requires an operating temperature below the deactivation temperature of the biocatalyst and thus can be beneficial.

In this work, the feasibility of the continuous extractive biocatalysis with cloud-point extraction using mixed surfactant solutions is demonstrated for the first time. Thereby an enzymatic reaction (one step out of penicillin production) was integrated into a continuous extraction process. Since the reaction product was continuously removed, higher yields in comparison to batch experiments were reached. The flexibility of the mixed surfactant systems allows to tune the extraction systems to get the highest separation efficiency. Therefore basic studies regarding the phase behavior, phase separation kinetics and temperature-dependency of the mixed surfactant systems in regards of the continuous (reactive) extraction processes were performed. Further, the interaction between the surfactant and the enzyme in the reactive separation unit were studied.

The general applicability of the heterogeneous enzymatic catalysts in extraction columns, thus preventing the enzyme loss (a disadvantage of the homogeneous enzymatic catalysis) was demonstrated in this work. Thereby different concepts of integration of a heterogeneous biocatalysis in the extraction processes with surfactant-based systems and in continuous extraction columns in general were suggested. The concept of continuous heterogeneous extractive bioactalysis was extended to different types of columns and internals, whereas the usage of structured packings coated with the enzyme-containing gels seemed to be the most promising. Here the key influences of physical and structural properties in mixed surfactant systems were investigated. Additionally, the integration of micellar extraction with the whole cell biocatalysis was examined, using microalgae as a model system.

Summarizing, in this work the potential of the integration of biocatalytic reactions with enzymatic catalysis as well as whole cell biotransformations into continuous extraction with mixed surfactant solutions is revealed.

531771 Enhanced Wet Extraction of DHA Using Room-Temperature Ionic Liquids

Abstract:

Polyunsaturated Fatty Acids (PUFAs), especially omega-3 fatty acids such as docosahexaenoic acid (DHA), are significant in modulation and prevention of various diseases. Therefore increasing attentions to PUFAs extraction are paid for biotechnological industries. Marine fish are still the major but not sustainable source of PUFAs due to the growing demand. Thraustochytrids, as heterotrophic microalgae, have been targeted for alternative source because of the robust growth rates and high docosahexaenoic acid (DHA). Recently, ionic liquids (ILs) have been extensively investigated to improve the efficiency of solvent extraction of lipids from microalgae and yeast. Moreover, ILs allow lipid extraction from wet biomass at ambient temperatures in less time than traditional extraction methods. This work aimed to study PUFAs extraction from *Thraustochytrium sp.* (T18) under pretreatment of the phosphonium-based IL (tetrabutylphosphonium propanoate [P_{4444}][Prop]). The promising ionic liquid was first demonstrated to be applied in this microalgae via gravimetric analysis of total extractable oils. The resultant oil extracted after IL pretreatment was ulteriorly characterized with respect to fatty acid methyl ester (FAME) composition. The effect of process parameters were investigated for the ionic liquid, for instance the ratio of ionic liquid to co-solvent, the mass ratio of algae to the mixture of ionic liquid and co-solvent. The results indicate that the ionic liquid possesses the ability to disrupt the cells of *Thraustochytrium sp.* by itself and allow a facile recovery of oils over a large degree of wet microalgae (77.2 wt % water) in a short time at room temperature, providing the exploitation of a water compatible, energy saveable, PUFAs production process.

522141 Deep Desulfurization of Fuels Using Imidazolium Anion-Based Ionic Liquids

Abstract:

An extractive desulfurization method was developed using the imidazolium anion (Im)-based ionic liquids (ILs), 1,8-diazabicyclo[5.4.0]undec-7-ene Im ([DBU][Im]) and 1,1,3,3-tetramethylguanidine Im ([TMG][Im]), as extractants. The DBU and TMG cations with strong alkaline and Im anion which is a weak proton donor show high affinities for sulfur compounds. Effects of extraction time, initial sulfur content in model oil, the mass ratio of IL to model oil and extraction temperature on the removal efficiencies of sulfur compounds were investigated. The removal efficiencies of DBT, 4,6-DMDBT and BT with [DBU][Im]/[TMG][Im] as extractants under the selected conditions (IL:model oil (w/w) = 1:1, room temperature) were 79.2%/69.4%, 68.4%/57.7% and 61.2%/54.4%, respectively. The extra heterocyclic aromatic ring in [DBU] cation and the imidazolium ring in [Im] anion could enhance the Ï€-Ï€ interaction between [DBU][Im] and sulfur compounds, resulting in obtaining higher removal efficiency of sulfur compounds with [DBU][Im] as extractant than that with [TMG][Im]. Both the extraction processes with these two extractants were fast (extraction equilibrium was reached within 10 min). After five multi-stage cross-current extraction processes under the selected conditions ([DBU][Im]:model oil (w/w) = 1:3, room temperature), the sulfur content (DBT) in the model oil decreased from 500 ppm to 5 ppm. The reusability of [DBU][Im] and [TMG][Im] and reactive extraction model for desulfurization were studied too. Moreover, the characterization methods of ^{1}H NMR and FT-IR spectra and the calculation methods of optimized geometry and interaction energy between sulfur compounds and extractants by Gaussian 09 software were used for the study of extraction mechanism. Furthermore, the oxidation-extraction two-step deep desulfurization method was applied in model oil system and real diesel fuel. The results showed that the removal efficiencies of DBT and 4,6-DMDBT in model oil reached up to 98.2% and 99.9%, respectively. Meanwhile, the sulfur removal efficiency of real diesel fuel reached up to 97.0% by one-stage oxidation and three-stage cross-current extraction processes with the mass ratio of IL to real diesel fuel as 1:2.

532463 Isosteric Heats of Adsorption of Water on Zeolites

Abstract:

In adsorption based technologies, it is vital to consider the heat released by the adsorption process, because it will influence the overall process in terms of the cost of energy that is required to control column temperature, a change in adsorption equilibrium to lower adsorption capacity and the energy cost of desorption during adsorbent regeneration.

The heat evolved by adsorption can be described as a differential heat of adsorption which is defined in thermodynamics models as the isosteric heat q_{st}, which is the heat released by the transfer of gas at fixed adsorbate loading at a certain temperature and pressure [1,2]. The isosteric heat of adsorption can be obtained from adsorption isotherms measured in a narrow temperature rage via Clausius-Clapeyron equation. In present study we want to highlight Gravimetric Dynamic Vacuum Vapor Sorption Analyzer (DVS Vacuum) that is employed to measure water adsorption/desorption isotherms in a broad temperature range in order to calculate the isosteric heat of asdorption.

532958 Hierarchical Bayesian Estimation for Adsorption Isotherm Parameter Determination and Applications to CO_2Â capture

Abstract:

To slow down the current rise in the temperature of the earth and the abnormal changes of the climate and ease the impact of the greenhouse effect, researchers around the world are developing many separation techniques to recover greenhouse gases. Among the greenhouse gases, the impact of carbon dioxide (CO_2) is particularly significant, which has become one of the most active areas in separation research.

Among many separation techniques, adsorption has many advantages such as high energy utilization efficiency and reusability of the adsorbent. Because of these advantages, adsorption processes are promising and attract researchers to develop many novel adsorbents. Adsorption isotherms are measured as a way to evaluate conventional and novel adsorbents in adsorption processes. These large-scale experimental results are being collected in the database at the National Institute of Standards and Technology (NIST). The isotherm data on this database could be a crucial component in developing process models for separation, which can be utilized to develop and analyze the economics of the adsorption process.

However, a critical problem has been recently reported by Park et al. [1], where substantial mismatches of the adsorbed amount of have been reported by different researchers even for the same adsorbent, specifically for adsorbents called metal-organic frameworks (MOFs), at the similar ranges of temperature and pressure. These discrepancies could be caused by different measurement techniques and procedures, but more importantly by variations in the underlying materialâ€™s properties such as inconsistent surface areas and reaction yield due to different adsorbent synthesis procedures. In their study, a statistical technique was applied to find upper and lower bounds where the adsorbed amount is believed to be reliable. Nevertheless, the problem of determining an isotherm model and incorporating parametric uncertainty upon the consensus bounds have not been addressed yet.

In this study, we resolve the issue stated above by applying the hierarchical Bayesian estimation. In this method, statistical inference by Bayes' theorem is used to update the probability for hypothesis as more data or information becomes available. By hierarchical Bayesian estimation we can quantify the differences in the adsorption amount reported by different researchers, while obtaining a single set of the probability density distributions of isotherm parameters simultaneously. The application of hierarchical Bayesian estimation allows researchers to utilize the isotherm data on the NIST database to develop a reliable isotherm model, which can save the measurement cost and time substantially, which leads to efficient development and economic evaluation of adsorption processes. This method can be applied to a wide variety of applications of adsorption not only for capture.

Reference:

[1] J. Park, J. D. Howe, D. S. Sholl, How Reproducible Are Isotherm Measurements in Metalâ€"Organic Frameworks? , Chem. Mater. 29 (2017) 10487-10495.

536104 The Influence of Compositions and Defects on Vibrational and Optical Properties of Sodalite through Density Functional Theory

Abstract:

Zeolites are well-known for their applications in processes such as catalysis, adsorption, and ion-exchange. Recently, zeolites have been explored for applications in optoelectronics due to their uniform micropores (< 2nm) and regular arrays of defined channels. For example, different types of extra-framework charge- compensating cations have been shown to have different external quantum efficinecies. In addition, the water content can also have dynamical emission color change in some silver-exchanged zeolites. These intriguing optical properties, which can be tailored by changing the composition and synthesis conditions, are typically poorly interpreted. Understanding and characterizing the relationships between chemical composition, synthesis conditions, and the resulting structural, vibrational, and optical properties are important to utilizing these materials more efficiently in optoelectronics. Among the characterization tools available for zeolitic materials, vibrational spectroscopy (infrared and Raman) and photoluminescence spectroscopy have been used extensively. Sodalite (framework code **SOD**) is a relatively simple framework that can be used as a model system to explore optical and ifrared behavior. We explore the effects of compositional variation on the vibrational properties of sodalite through density functional theory (DFT) calculations with periodic boundary condintions (PBC). DFT and experiments are in good agreement in terms of predicting the characteristic vibrational modes of sodalite. However, in our experimental IR, we observed substantial peak broadening in the asymmetric stretching region and additional peaks between the bending mode and symmetric/asymmetric stretching regions. We hypothesize that this comes from the existence of hydroxyl groups and/or water content. This suggests that we should work toward determining the effects of other types of defects (e.g., hydroxyl groups, water, charge-compenstating cations). The response of zeolites to adsorbates may expand their utility to humidity sensors and other applications.

536747 Pore Structure Characterization of MOF-Based Materials By Gas Adsorption

Abstract:

Adsorption of test adsorbates, such as N_2, Ar and CO_2, is the most practical method for pore structure characterization of MOF-based materials, which in many cases display composite hierarchical structures. The pore size distribution analysis is commonly based on conventional NLDFT or QSDFT methods developed for carbon and silica microporous solids. The kernels of references isotherms employed in these methods do not take into account the specifics of MOF pore morphologies, which differ significantly from one MOF family to another. In some cases, application of these methods provides reasonable agreement with the crystallographic pore dimensions, however, in many cases the discrepancies are striking. To remedy this situation, we develop dedicated kernels of reference â€œfingerprintâ€ isotherms capable of distinguishing between different MOF structures and access the defects, inter-crystalline pores, and secondary micro-mesopore structures of binding materials. The fingerprint isotherms are built from MC simulated isotherms verified by experiments on ideal non-defective MOF crystals. The efficiency of the proposed approach is demonstrated by drawing on the example of Cu-BTC composites.

536366 Assessment of Options for Determining the Total Adsorption Uptake from Liquid Solution: Alkane-Î±,Ï‰-Diols/(Water or Ethanol) Onto Silicalite-1

Abstract:

Experiments and molecular simulations are employed to probe the total uptake of binary mixtures containing an alkane-Î±,Ï‰-diol with four or five carbon atoms solvated in water or ethanol onto all-silica MFI-type zeolite. One of several assumptions (ranging from no solvent adsorption to complete pore filling) must be made to determine the total uptake of solution-phase adsorption in batch experiments, as only the bulk solution-phase is open to direct measurement. The diol loadings from aqueous solution obtained using different assumptions differ significantly (more than 10% relative deviation) at both low and high concentrations (exceeding one molecule per unit cell at high concentrations), with consensus among approximations only occurring at intermediate concentrations. Different assumptions yield water loadings at low diol concentrations that vary from 0 up to 60 molecules per unit cell and can even result in negative values at intermediate and high concentrations, making a determination of the adsorption selectivity impossible. The diol loadings obtained from Gibbs ensemble simulations with transferable force fields agree very well with those determined experimentally. The water loadings from these molecular simulations follow chemical intuition for adsorption onto a hydrophobic, all-silica zeolite with zero defects. This includes almost no water loading at low diol loading (or concentration), significant water co-adsorption at intermediate diol loading, and displacement of water as saturation loading of the diol is approached. It is demonstrated that a co-adsorption model based on the simulated equilibria can be used to map the experimental adsorption of water without any significant change in diol loading over the entire concentration range. Ideal adsorbed solution theory is found to vastly overestimate the adsorption selectivity for the aqueous diol solutions. The experimental determination of the total uptake for ethanolic solutions is less fraught with errors (with exception of the no-solvent-adsorption and volume-change-by-solute-adsorption assumptions) where the solvent fills the pores until being displaced by the solute at higher concentrations. This work is expected to inform future experimental approaches and thermodynamic models for determining the total uptake from liquid solution onto nanoporous materials.

530007 Measurement of Water Adsorption Kinetics Using the Zero Length Column

Abstract:

In this contribution a zero length column (ZLC) system is adapted to run experiments with water at high relative humidity (>75%) and the results are interpreted to determine the kinetics of adsorption for a mesoporous material.

Results for an SBA-15 sample show that to in order achieve equilibrium conditions very low flowrates are needed in the hysteresis region. As the flowrate is increased progressively kinetic limitations become dominant, but the mass transfer mechanism is complex. In order to interpret the experimental results it is first necessary to correlate the equilibrium curve with an appropriate analytical isotherm. This information allows to determine from the experimental ZLC curves the apparent linear driving force (LDF) constant at different adsorbed phase concentrations. The experimental results, show that this apparent LDF constant is dependent on flowrate and therefore cannot be used to represent the system accurately. A diffusion model is introduced with a concentration dependent diffusivity and this is shown to provide a good match to the observed kinetic responses, without the need to introduce additional parameters.

The kinetic experiments have been carried out with different sample masses, and the talk will also cover how to interpret correctly the response of smaller sample masses, where the correction for the blank becomes increasingly important. A signal deconvolution approach is shown to correctly transform the signal from the detector to the water concentration in the vapor phase.

534505 Impact of the Addition of Polyethylene Glycol (PEG) on Amine Efficiency and Heat of Sorption in Polyethylenimine (PEI)/Silica Sorbents

Abstract:

In CO_2 capture, solid amine sorbents have lower heat capacity and require less energy for regeneration than aqueous amine solutions. However, due to significant diffusional limitations that increase up during CO_2 sorption due viscosity excursions, the buried amine sites are inaccessible and therefore not utilized, resulting in amine efficiencies significantly lower than theoretical limits. To increases amine efficiency, efforts have been made to decrease diffusional barriers and alter theoretical CO_2: amine from 0.5 to 1. The latter requires a change in mechanism of adsorption, while the former requires a reduction in intermolecular interactions between amines.

We examined the impact of polyethylene glycol (PEG) addition on the performance of PEI sorbents for CO_2 capture, with an emphasis on determining the impact diffusion on measured capacity and heat of sorption values. A combined breakthrough reactor (BTR) with mass spectrometer (MS) and differential scanning calorimeter (DSC) enabled simultaneous measurement of both sorption capacity and heat flow in a packed bed column configuration, simulating a cycle in the pressure/temperature swing adsorption (PTSA) process. We synthesized a series of PEI-PEG/SiO_2 sorbents where the total polymer content is 40 wt.%, and performed breakthrough experiment at five different temperatures. Breakthrough results illustrated the influence of PEG addition on CO_2 diffusion, amine efficiency and the heat of CO_2 binding to confined amines. The CO_2 breakthrough curve becomes less steep with increasing PEG content or at elevated temperatures, indicating a reduced diffusional barrier and CO_2 capability to access buried amine sites. Amine efficiency increases with PEG content, and this increasing effect is much more significant at low temperatures. The sample with the highest PEG content exhibits amine efficiency two times higher than 40 wt% PEI/SiO_2 at 25â„ƒ. The heat of sorption data also reveals energetically less favorable interactions at lower PEI content and low temperatures, indicating the possible formation of the less stable carbamic acid and the shift in the CO_2: amine stoichiometric ratio. At high temperatures, the unstable species no longer exists and the heat of sorption is similar amongst samples. This work studied the role of PEG in promoting CO_2 diffusion and possible formation of less stable carbamic acid, which greatly increases the amine efficiency of PEI-impregnated silica. The breakthrough instrument with the simultaneous capability to measure heat flow approximates the industrial PTSA separation process, and the combination of capacity and heat flow data enables more robust spatiotemporal modeling of CO_2 sorption in a packed bed.

539721 Gas Separation Properties of Novel Poly(benzimidazole)s

Abstract:

Gas and water vapor separation properties of a systematic series of sulfonyl-containing poly(benzimidazole)s are reported as a function of pressure and temperatrue. Gas separation properties are reported over temeratures ranging from ambient up to 190C. Additionally, separation properties of PBIs blended with polyimides are also reported, with some blends exhibiting transport properties above the 2008 upper bound.

528573 Graphene Oxide Membranes for Gas Separation â€“ Gas Transport Mechanism

Abstract:

Graphene oxide (GO) membranes have shown promising gas separation characteristics specially for hydrogen that make them of interest for industrial applications. However, the gas transport mechanism for these membranes is unclear due to inconsistent permeation and separation results reported in literature. It appears that synthesis method and characteristics of GO sheets (size or defects) affect the permeability and separation characteristics of these membranes. This talk will give an overview of our recent progresses on GO membranes for gas separation with a focus on clarifying the transport mechanism of these membranes. An inter-sheet and intra-sheet two-pathway model is proposed to explain the permeation and separation results of GO membranes obtained in our study. We found that gas permeation and separation through GO membranes is controlled by transport through inter-sheet space of extremely high tortuosity and intra-sheet defects of extremely low porosity. The GO membrane structure that defines both transport pathways is difficult to control leading to variation of permeation and separation properties reported in the literature. We also investigated structure changes and gas permeation properties of GO membranes made by different synthesis methods, sheet sizes, and GO sheet prepared by different oxidation methods (Hummers vs Brodie). We also, demonstrated a scalable fabrication of GO membranes on polyester track etch polymer substrates by easily scalable spray coating technique. Our recent study on membrane fabrication with narrow interlayer spacing height show improved (H_2/CO_2) separation performance.

533557 Ionic Liquid-Based Gel Membranes with Tough Double-Network for CO_2 Separation

Abstract:

Concerns on CO_2 emissions from coal-fired power plants are rapidly increasing because of their major contribution to global warming and climate change. Hence, the development of an economically viable CO_2 capture technology is inevitable. Recently, ionic liquids (ILs) have been proposed as attractive media for CO_2 separation membranes owing to their distinct properties such as high CO_2 absorption capacity, negligible vapor pressure, huge chemical diversity, and high thermal stability. Therefore, the use of ILs as a CO_2 separation medium in a membrane has been extensively investigated. As a type of IL-based CO_2 separation membrane, IL-based gel membranes have both good CO_2 solubility selectivity and high CO_2 permeability, and they have attracted considerable attention. Among several kinds of IL-based gel membranes, the IL-based gels with polymer networks and polymer/ILs composites have been investigated as a material for CO_2 separation membranes. However, because conventional IL-based gel membranes had high polymer network composition to give enough mechanical strength, the CO_2 permeability as well as the CO_2 diffusivity in the gel was limited. In other words, the CO_2 permeability could be improved by decreasing the network content of the gel layer; the CO_2 permeability of the IL-based gel membrane could be increased by increasing the IL content in the gel. Thus, it is important to overcome the trade-off between the mechanical strength and solute diffusivity in a gel without increasing the network composition. One promising way to overcome this trade-off is to utilize tough gels.

Recently, we successfully developed a novel IL-based gel membrane composed of 80 wt% of room temperature ILs and a specific inorganic/organic composite network, termed inorganic/organic double-network ion gel membrane.[1)] Although normal IL-based gel membranes show a trade-off relationship in the mechanical strength and gas permeability, our developed double-network ion gel membrane having low composition of loosely cross-linked organic polymer network indicated both high mechanical strength and high gas permeability. The prepared DN ion gel with ca. 80 wt% of IL showed an extraordinarily high mechanical strength with a compressive fracture stress of more than 25 MPa. Owing to the high mechanical strength, the DN ion gel membrane exhibited excellent pressure resistance. The CO_2 permeability and N_2 barrier property of the DN ion gel membrane were also good, and were comparable to those of the supported IL membrane containing the same ILs. Our prepared DN ion gel membranes have the potential to overcome the trade-off limitation with respect to the mechanical strength of the ion gel and the solute diffusivity in the gel.

Here, we present a novel ion gel membrane with an extraordinarily high mechanical strength and high CO_2 permeability. The ion gel membranes consisted of an inorganic/organic composite network and a large amount of RTILs up to 80 wt%. The developed ion gel membranes can be prepared via a single-pot preparation process. For the synthesis of the inorganic/organic composite ion gel membrane, the inorganic component was formed by thermally induced polycondensation of tetraethylorthosilicate (TEOS), and the polymer network was formed by UV-initiated free-radical polymerization of N,N-dimethylacrylamide (DMAAm). The TEOS polycondensation resulted in the formation of silica nanoparticles, which are inter-connected to form a three-dimensional (3D) silica particle network in the IL. The silica particle network could be formed by weak physical interactions such as hydrogen bonding and van der Waals attraction between the silica nanoparticles in the IL. Since the fragile silica particle networks acted as the sacrificial bond, it could dissipate the loaded energy by destruction during the application of a force. On the other hand, polydimethylacrylamine (PDMAAm) was selected as the polymer network because of the good compatibility with several RTILs. After the formation of silica nanoparticles in the IL, a ductile PDMAAm network was formed to make an inorganic/organic composite network. The inorganic/organic composite network can be synthesized in the same IL pot, which allows the preparation of freely shapeable ion gel, including a film shape. In addition, the mechanical strength of the inorganic/organic ion gel can be varied by controlling not only the cross-linking density of the organic network but also the inorganic/organic network composition. As mentioned above, the mechanical strength of conventional ion gels could be controlled by the network content and the cross-linking density of the network, although both of them have the trade-off limitation between the mechanical strength and solute diffusivity in the gel. Here, we propose an inorganic/organic composite ion gel membrane, which can overcome the trade-off limitation. Without changing the IL content in the inorganic/organic composite ion gel membranes, the mechanical strength could be controlled by changing the cross-linking degree of the polymer network and the composition of inorganic/organic networks.

The mechanical strength of the composite ion gel membrane could be increased by increasing the inorganic network composition as well as decreasing the cross-linking degree of the organic polymer network. Increasing of the inorganic network composition, in other words the decrease of the organic network composition, directly enhanced the energy dissipated by the internal rupture of the inorganic networks. Decreasing the cross-linking degree of the polymer network allowed the ion gel membrane to elongate higher degree, which caused additional internal rupture of the inorganic network during further elongation and increased the energy dissipation. As the results, the ion gel membrane having low composition of loosely cross-linked polymer network showed high mechanical strength. On the other hand, because the decrease of the content and cross-linking degree of the polymer network enhanced the diffusivity of the gases absorbed in the ion gel membrane, the ion gel membrane having low composition of loosely cross-linked polymer network showed high CO_2 permeability. The composite ion gel membrane with optimized network composition and 80 wt% ionic liquid sustained about 1200 barrer of CO_2 permeability and 25 of CO_2/N_2 selectivity for more than 300 h at 50 ^oC under humid condition.

In this presentation, we demonstrate the good pressure resistance, IL holding property, and excellent CO_2 permeability of the fabricated composite ion gel membrane with the optimized inorganic/organic composite network.

Reference:

1) E. Kamio T. Yasui Y. Iida J. P. Gong H. Matsuyama, Inorganic/Organic Doubleâ€Network Gels Containing Ionic Liquids, Advanced Materials, 29, 1704118, 2017

527055 Designing Sorption-Enhanced Mixed Matrix Membranes for H_2/CO_2 Separation Using an Integrated Experimental and Modeling Approach

Abstract:

Membrane technology is an energy-efficient and low-cost approach for pre-combustion CO_2 capture and H_2 purification in the integrated gasification combined cycle (IGCC) processes. Conventional membranes are based on rigid polymers with strong size sieving ability, such as poly[2,2'-(m-phenylene)-5,5'-bisbenzimidazole] (PBI) that provides high H_2/CO_2 diffusion selectivity. In this study, we demonstrate enhanced H_2 sorption and diffusion in PBI films with embedded palladium (Pd) nanoparticles, which have strong affinity towards H_2. Pd nanoparticles with uniform diameters of 6 - 8 nm are prepared via a hot-injection colloidal synthesis. The loading of Pd nanoparticles in PBI increases H_2 sorption by almost 1,000 times, and at high Pd loadings, the Pd nanoparticles may form fast channels allowing the H_2 molecules to jump from one particle to another and thus increasing the effective H_2 diffusivity. For example, adding 58 wt.% Pd in PBI increases H_2 permeability from 25 to 60 Barrers, and H_2/CO_2 selectivity from 13 to 34 at 175 °C. Such performance is above the Robeson's upper bound for H_2/CO_2 separation, demonstrating the potential of these new materials for industrial H_2/CO_2 separation. The gas transport in these PBI-Pd nanocomposites is being modeled using computational fluid dynamics (CFD) to elucidate the mechanisms for the facilitated H_2 transport. This presentation will also discuss the visualized microstructure of the nanocomposites and provide a unified view of the H_2 and CO_2 transport in the nanocomposites using an integrated experimental and simulation approach.

535835 Optimizing Carbon Capture with Membrane Processes

Abstract:

Carbon capture and sequestration (CCS) is envisioned as one approach to mitigate the potential environmental impact of carbon dioxide emissions from fossil fuel power plants. The US Department of Energy (DOE) has set aggressive CCS targets of 90% CO_2 capture at 95% purity followed by compression to 140 bar for transport and storage without increasing the levelized cost of electricity (LCOE) more than 35%.

A promising approach utilizing a hybrid membrane-cryogenic process has been proposed by Membrane Technology and Research, Inc. (MTR). The process utilizes a two-stage membrane pre-concentration step followed by a cryogenic flash. The boiler feed air is used as a sweep in the second membrane stage to achieve capture targets. The improvement in capture comes at the expense of reduced oxygen concentration in the boiler feed air.

The design space for this process is systematically explored to identify an optimal design. Of special interest is the trade-off between capital and operating expenses associated with considering membrane materials that lie along the upper bound for transport.

Results are presented for designs that achieve carbon capture targets with a boiler feed air oxygen concentration of 18% - the lowest level being considered in proposed applications. A broad range of membrane materials, membrane operating pressures, and cryogenic flash conditions yield comparable LCOE near the minimum. This observation suggests that efforts to reduce membrane manufacturing costs may have the greatest potential to improve process economics.

540290 Haifeng Power Plant CO_2 Capture Demo Unit

Abstract:

Membrane Technology has been identified as the second generation technology for CO2 capture from coal fired power plant flue gas. Compared with conventional solvent absorption technology, CO2 capture with membrane process showed a number of advantages. Recently MTRâ€™s membrane CO2 capture technology has been selected for demonstration at Haifeng Coal Power Plant test platform. A three-stage membrane unit was used to produce more than 95% purity CO2 which will be further treated at a CPU next the membrane unit. The CO2 product will be used in several different ways. This demonstration unit will be the biggest one in Asia. In this paper, an overview about this demonstration project will be presented.

529512 PVDF Hollow Fibers with Novel Sandwich Structure and Superior Wetting Resistance for Vacuum Membrane Distillation

Abstract:

PVDF hollow fiber membranes with a unique sandwich structure and superior wetting resistance have been designed for vacuum membrane distillation (VMD) of seawater desalination in this study. The sandwich structure consists of two sponge-like inner and outer porous layers and a thin middle layer full of small-size macrovoids. The sponge-like structure in the two porous layers are specially designed because it enhances the mechanical strength of hollow fibers with high wetting resistance. This structure is derived from the spinodal liquid-liquid decomposition during phase inversion. Meanwhile, the middle layer of macrovoids is beneficial for VMD flux because they perform as highways for water transport. Comparing to the large size macrovoids, the small-size macrovoids in PVDF hollow fibers are preferred for VMD applications because they would not significantly reduce the membrane mechanical strength. As a result, the newly developed membranes possess impressively high liquid entry pressures (LEPs) of greater than 3 bar. This study may provide valuable guidelines in designing next generation PVDF hollow fiber membranes for practical VMD applications.

521851 A Celebration of My 15+ Year Collaboration with Peter Monson on Simulating the Self Assembly of Nanoporous Materials: Are We Sipping from the Holy Grail?

Abstract:

Zeolites are the most used catalysts by weight on earth and offer the potential for new applications in carbon dioxide capture, biofuel production, and nano-electronics. The applications of zeolites arise from their nanoporous crystalline structures and stabilities. Despite the great importance of zeolites, zeolite chemists still rely heavily on trial-and-error in their search for new materials, because the mechanisms controlling zeolite formation remain poorly known. Understanding such mechanisms will be critical to one of the "holy grails" of materials science -- rational design of tailor-made nanoporous materials. In this lecture, I celebrate my 15+ year collaboration with Peter Monson by sharing the story of the development of a multi-scale molecular modeling program in search of this holy grail.

I begin by reviewing present-day understanding of zeolite synthesis with a focus on the role of "structure directing agents" (SDAs). Peter and I have studied this problem through a multi-scale application of Density Functional Theory (DFT) and Monte Carlo (MC) methods. Our DFT calculations provide key structural and energetic parameters, and explain the curiously wide range of bulk moduli observed for silica polymorphs. Our MC simulations of silica polymerization have elucidated the formation of amorphous silica, silica-SDA nanoparticles, zeolites, and mesoporous silica. Here I emphasize the uniqueness and power of Peterâ€™s research vision in developing both insightful models and efficient simulations. Our MC simulations reproduce NMR signatures of silica polymerization; predict that such polymerization is not diffusion controlled; and reveal sought-after structures of silica-SDA nanoparticles. Replica exchange MC is found to be essential for modeling zeolite crystallization, allowing predictions of SDA sizes that optimize zeolite yield and crystallization rate. I conclude with remarks about the potential for molecular modeling to peer even more deeply into the atomic dance of nanopore formation, a potential largely created by the brilliance of Peter Monson.

537132 Confinement-Induced Compression and High Pressure Phases in Nanopores

Abstract:

There is an abundance of anecdotal evidence that nanophases adsorbed within nanoporous materials can exhibit high pressures as a result of the confinement[1,2]. The pressure in the pore is a second order tensor, and for simple pore geometries has both a normal pressure component (normal to the walls) and one or more tangential components (parallel to the walls).

For simple fluids in pores that are up to a few nanometers in width, molecular simulations show that both the normal and tangential pressures can be locally very high (thousands or tens of thousands of bars) in the pore, even though the bulk phase in equilibrium with the pore is at a pressure of one bar or less. The cause of these high in-pore pressures will be discussed.

When the molecules in the confined nanophase react with each other chemically it may be possible to achieve even higher tangential pressures, in the megabar range. Thus, Kaneko et al.[3] have shown that when sulfur atoms are confined within a narrow carbon nanotube they covalently bond to form a one-dimensional phase that is metallic. In the bulk phase sulfur forms a metallic phase only at pressures above 95 GPa. In our recent molecular dynamics simulations of this system we find that the sulfur atoms are covalently bonded in the pore and that they experience tangential pressures in excess of 100 GPa as a result of the strong confinement[4]. Recently, Medeiros et al.[5] have reported experiments in which they observe a similar covalently bonded one-dimensional phase, and an insulator-metal transition, for tellurium in single-walled carbon nanotubes.

1. Yun Long, Jeremy C. Palmer, Benoit Coasne, Małgorzata Śliwinska-Bartkowiak and Keith E. Gubbins, MACROBUTTON MTEditEquationSection2 Equation Chapter 1 Section 1 SEQ MTEqn \r \h * MERGEFORMAT SEQ MTSec \r 1 \h * MERGEFORMAT SEQ MTChap \r 1 \h * MERGEFORMAT "Pressure enhancement in carbon nanopores: A major confinement effect", *Physical Chemistry Chemical Physics*, **13**, 17163-17170 (2011).
2. Yun Long, Jeremy C. Palmer, Benoit Coasne, Małgorzata Śliwinska-Bartkowiak, George Jackson, Erich A. Müller and Keith E. Gubbins, "On the Molecular Origin of High Pressure Effects in Nanoconfinement: Effects of Surface Chemistry and Roughness", *Journal of Chemical Physics*, **139**, 144701 (2013)
3. Y. Fujimori, A. Morelos-Gómez, Z. Zhu, et al., "Conducting Linear Chains of Sulphur Inside Carbon Nanotubes", *Nature Comm.*, **4**, 2162 (2013).
4. K.E. Gubbins, C.A Addington and J.M. Mansell, "Computer Simulation of Conductive Linear Sulfur Chains Confined in Carbon Nanotubes", *Molecular Simulation*, **43**, 519-525 (2017).
5. P. V. C. Medeiros *et al.*, "Single-Atom Scale Structural Selectivity in Te Nanowires Encapsulated Inside Ultranarrow, Single-Walled Carbon Nanotubes," *ACS Nano*, **11**, 6178–6185 (2017).

531657 My Research Adventures with Pete Monson

Abstract:

In the mid-1980s, Pete Monson and I collaborated for several years on some interesting questions surrounding the numerical solution of the Ornstein-Zernike equation for simple fluids in the two-phase region and near the critical point. This was at a time when there was some significant interest in, and quite a bit of controversy, about the critical behavior of simple fluids predicted from liquid-state theories. Pete and I approached this from complementary viewpoints â€“ his from the point of view of numerical solution, mine from insights gained from solving integral equations analytically. Our collaboration led to three papers [1]-[3] that I thought were very insightful. In this talk, I will try to convince you why!

1. Cummings, P. T. Monson, P. A., "Solution of the Ornstein-Zernike Equation in the Vicinity of the Critical-Point of a Simple Fluid," *Journal of Chemical Physics*, **82**, 4303-4311 (1985). 10.1063/1.448820
2. Monson, P. A. Cummings, P. T., "Solution of the Percus-Yevick Equation in the Coexistence Region of a Simple Fluid,"*International Journal of Thermophysics*, **6**, 573-584 (1985). 10.1007/bf00500330
3. Cummings, P. T. Monson, P. A., "A New Method for the Numerical-Solution of Integral-Equation Approximations,"*International Journal of Thermophysics*, **11**, 97-107 (1990). 10.1007/bf00503862

531803 Monte Carlo Simulation of Percolation Properties, Including Cluster Numbers and Elastic Backbones

Abstract:

We describe various simulation procedures to simulate properties in percolation, including the Leath Cluster Growth method, the Hoshen-Kopelman method, and the Newman-Ziff method. We first give an introduction and overview into percolation including its relation to the Potts model and thermodynamic systems. Then we discuss specific problems studied using these Monte Carlo methods, including crossing probability and cluster density, for systems at and near the critical point. We also talk about the importance of the geometric embedding and show how in two dimensions the isoradial construction can be used to find the correct embedding to have isotropic growth. These embeddings are important in practical problems to get a correct understanding of the anisotropy of the flow, especially for problems where there is anisotropic bond probabilities.

532503 Cyclic and Polymeric Rotaxanes

Abstract:

Among the most important advances in chemistry in the last 60 years is the synthesis of molecules featuring a mechanical bond or the mechanical interlocking of components. A molecule featuring a mechanical bond has additional mechanical degrees of freedom, namely the relative position and orientation of the interlocked components, that can have dramatic and controllable effects upon the shape and size of the molecule. An example of such an interlocked molecule is a linear rotaxane, comprised of a ring threaded with an axle that is stoppered at each end to prevent the ring from falling off. As the ring and axle are not covalently linked, the ring is free to translate the length of the axle and provides a continuum of accessible mechanical conformers, which can be unlocked to do work, as in a drug delivery system, or to dissipate energy, as in a molecular shock absorber.

Here we describe the mechanical conformers of (1) cyclic rotaxanes, and (2) branched polymers. We demonstrate that with only a few mechanical bonds, we can achieve significant fluctuations in molecular size and shape for mechanical bonds that are unbiased (as in freely mobile rings or an entropy-dominated molecule), as well as those biased with attractive stations (as in a molecular switch with added enthalpy control). Our statistical mechanics models suggest that the cyclic rotaxane can be used as a switchable host in an inclusion complex and that a mechanically-branched polymer provides switchable colloidal stabilisation via an adjustable depletion force.

527607 The Treatment of Pair Correlations in an Augmented Mean-Field Density Functional Theory of a Simple Model Liquid Crystal

Abstract:

We present a density functional theory (DFT) for fluids composed of anisometric molecules that may form ordered (liquid-crystalline) phases. Our DFT is based on an augmented modified mean field treatment of pair correlations assuming that the structure of ordered and disordered phases in our model is characterized principally by the repulsive interactions between the spherically symmetric cores of the liquid-crystal molecules. By comparison with corresponding Monte Carlo simulations both in the grand canonical and the isothermal-isobaric ensembles we demonstrate the improvement achieved with our novel approach to describe the pair correlations. The validity of our assumption concerning the structure of the liquid crystal is demonstrated by assessing various orientation dependent pair correlation function by expanding it in terms of rotational invariants.

531457 Molecular Modeling of Nanoparticle Permeation in Lipid Membranes for Drug Delivery Applications

Abstract:

Permeation of small molecules through phospho-lipid bilayer membranes is a fundamental biological process that is important to understand in developing drug delivery applications. Nanoparticles have recently become important carriers for drug delivery in such therapeutic applications, in which drugs are encapsulated within nanoparticles. We describe coarse-grained simulation methods applicable to such large systems and use examples where the permeants are bare gold nanocrystals, gold-core nanoparticles with hydrophobic ligands (alkanethiol ligands of various lengths), and gold-core nanoparticles with hydrophilic ligands (methyl-terminated polyethylene glycol ligands of various lengths). In addition to spherical gold nanoparticles, we also examine the characteristic permeation mechanisms of the differently shaped gold nanorods with polyethylene glycol ligands, where the aspect ratio different from 1 makes the permeation event dependent on the angle of the nanorod axis relative to the membrane surface. Our results indicate that several phenomena accompany nanoparticle permeation including lipid flip-flop, water and ion penetration as well as lipid loss.

References

1. Oroskar, C.J. Jameson, and S. Murad," Molecular Dynamics Simulations Reveal How Characteristics of Surface and Permeant Affect Permeation Events at The Surface of Soft Matter", Molecular Simulation, 43,439 (2017).
2. Oroskar, C.J. Jameson, and S. Murad," Rotational behavior of PEGylated Gold Nanorods in a Lipid Bilayer System", Molecular Physics, 115, 1122 (2017)
3. Oroskar, C.J. Jameson, and S. Murad, "Simulated permeation and characterization of PEGylated gold nanoparticles in a lipid bilayer system", Langmuir, 32, 7541 (2016)
4. Oroskar, C.J. Jameson, and S. Murad, "Surface-functionalized Nanoparticle Permeation Triggers Lipid Displacement and Water and Ion Leakage", Langmuir, 31,1074 (2015)

533305 Phase Diagrams By Computer Simulations: A Good School for Force Fields

Abstract:

After the initial steps of computer simulations when simulations were performed mostly for just a certain thermodynamic state, the determination of phase equilibria from computer simulations has been an area of growing interest. Although the fluid-solid equilibria of hard spheres was the first example, most of the work was devoted to the determination of vapor-liquid equilibria and often force fields parameters were determined by fitting vapor-liquid equilibria. However the idea that fluid-solid equilibria could be useful to learn the basic physics about molecular systems was gaining interest after the work of Peter Monson and coworkers using simple potential models.
The work with him moved me to consider the possibility that fluid-solid equilibria could also be used to determine the parameters of the intermolecular potential.
In the last three decades it has been shown that fluid-solid equilibria can be determined by computer simulations using several methods (direct-coexistence, free energy calculations).
That allowed to propose reasonable potential models for molecules as important as water. In the last years it has been realized that the determination of the solubility of a solid in a solvent by computer simulation is feasible and we will shown how this can also be used to propose improved pair potentials for salts in water. Obviously the determination of the equilibrium conditions is the first step for addressing nucleation studies.
Thus, this is a short summary of a trip that I started in Amherst with Peter , twenty five years ago.

521904 Experimental Investigation and Simulation of the CO_2 Removal with Indirectly Heated and Cooled Adsorbers

Abstract:

Temperature Swing Adsorption (TSA) processes are applied for the removal of impurities from various gas streams. The limiting factor of the TSA process is the regeneration step, which is characterized by long heating and cooling times. To reduce the regeneration time and the energy consumption, indirectly heated and cooled adsorbers were developed [1]. This new TSA process is able to remove higher quantities of impurities, e.g. CO_2 >2 vol-%. Therefore, new fields of application arise, such as CO_2 removal from flue gas or as prepurification unit in a combined membrane - PSA plant for Helium recovery from natural gas. A pilot plant has been built up to investigate the adsorptive CO_2 removal with indirectly heated and cooled adsorbers: Adsorbent inside the tubes and a heat transfer fluid on the shell-side.

A three adsorber process to remove CO_2 from N_2 was realized. The process consists of an adsorption and a regeneration step. The regeneration step was divided into heating, purging and cooling. Using three adsorbers, a continuous process could be established. In this process at least one adsorber was in adsorption, while the others run through the regeneration steps. The temperature profile of the three adsorbers and especially inside one adsorber tube will be presented and discussed. Additionally, the energy consumption and the CO_2 trend during the regeneration steps, heating and purging, will be illustrated.

In a previous work [2], a mathematical model was developed and implemented in COMSOL Multiphysics to describe adsorption experiments in one adsorber tube. This model was modified and expanded to describe the laboratory experiments in a tube bundle adsorber. The results of breakthrough measurements and heating and cooling experiments were taken to validate the model and to adjust model parameters. A comparison between the simulated temperature profile and the CO_2 concentration during regeneration and the experimental values will be presented. Furthermore, the results of a parameter study on the influence of the CO_2 feed concentration and purge gas flow on the process will be discussed.

References:

[1] Mérel J., Clausse M. and Meunier F.: *Carbon Dioxide Capture by Indirect Thermal Swing Adsorption Using 13X Zeolite*; Environmental Progress, 2006, Vol. 25, No. 4, p. 327-333

[2] Salazar Duarte G.: *Carbon dioxide removal from industrial gases using an indirectly heated and cooled temperature swing adsorption process*; Ph.D. Thesis, University Duisburg-Essen, 2017

537399 Structured Adsorbent Pressure Temperature Swing Adsorption Cycles for Metabolic CO_2 Removal from Spacecraft Cabins

Abstract:

NASA has a great deal of interest in the use of structured adsorbents in environmental systems for space applications for a variety of reasons. The most important reason perhaps is to minimize or even eliminate the formation of sorbent dust. Sorbent dusting most recently caused the failure of a TSA unit on board the International Space Station (ISS). The use of a Catacel structured 13X adsorbent may alleviate the dusting problem associated with the TSA unit in the carbon dioxide removal assembly (CRDA) and possibly provide even better performance characteristics.

The overall goal of this project is to design a temperature swing adsorption (TSA) process that utilizes this structured 13X adsorbent. This TSA cycle design would be perhaps a drop-in replacement for the TSA unit on board the ISS or it may be a grass root s design for future applications. This design is based on simulations being carried out with the USC dynamic adsorption process simulator (DAPS) and validated experimentally using the USC single bed apparatus in a TSA mode.

A corrugated structured sorbent consisting of a steel foil support coated with a layer of 13X is being experimentally evaluated for potential use in CDRA in lieu of traditional pelletized zeolite systems. Interest in using this type of column stems from faster kinetics that allow better bed utilization, lower pressure drop and stronger resilience against mechanic stress over pelletized systems. An 18 in long by 1.5 in diameter column containing the structured sorbent, with an effective zeolite 13X bulk density of around 150 g/L is being cycled between $T_{ambient}$ (20 °C) and $P_{ambient}$ (14.7 psia) during feed and 170 °C and 0.5 psia during regeneration without purge. This study is evaluating the performance in terms of CO_2 removal as a function of cycle time and other design variables. The latest results of this project will be presented.

532288 Comparison of Different Intensified Alternatives for the Downstream Separation in the OCM Process

Abstract:

The oxidative coupling of methane (OCM) has been investigated for over 40 years, owing to this process is an alternative method for ethylene production using natural gas and biogas as feedstocks. Despite its economic potential, and the recent implementation of an OCM demonstration plant by Siluria technologies and Braskem, there are many technical challenges to overcome for a sustainable implementation. Major limitations for the OCM industrialization are the low conversion of methane into ethylene, and the high energy consumption in the downstream processing.

In addition to ethylene and the unconverted methane, the reactor effluents contain a variety of byproducts such as water, ethane, CO_2, N_2, and other minor impurities, which must be removed. The current separation operations involve CO_2 removal through amine absorption, followed by high pressure cryogenic distillation to split the light hydrocarbons and the remaining impurities. However, according to some reports, only the demethanizer unit accounts for nearly 60% of the total energy consumption of the process, and the distillation trains consume 97% of the total energy use in the plant [1]. In this regard, the downstream purification of the OCM products requires the development of enhanced and intensified separations processes to achieve a cost-effective implementation.

In this direction, this work developed a comparison between the current separation technology with two different intensified processes: Dividing wall column (DWC) under cryogenic conditions, and a pressure swing adsorption (PSA) system using zeolite molecular sieves. The process was modeled based upon corroborated thermodynamic models, and validated absorption data. The models were implemented, solved and optimized in Aspen Plus and Aspen Adsorption. According to results, CO_2 removal is a key step in the separation process, allowing to reduce distillation heat duties, and permitting a high selectivity during the PSA separation. After CO_2 removal, the dividing wall column configuration can reduce energy consumption up to 30% with respect to the conventional process, allowing to obtain polymer grade ethylene. Comparatively, and despite it involves high level of complexity for the dynamic operation, the PSA system reduces energy consumption up to 70% with respect to the traditional process. High purity methane, ethane and ethylene can be obtained as major products of the PSA system. The obtained and validated models can be used for further process up-scaling and preliminary economic evaluations.

References

[1] X. S. Nghiem, â€œEthylene Production by Oxidative Coupling of Methane: New Process Flow Diagram Based on Adsorptive Separation,â€ Technische UniversitÃ¤t Berlin, 2014.

537604 Improved Kinetic Pressure Swing Adsorption Process

Abstract:

With the advent of new zeolitic kinetic adsorbents having attractive rate and kinetic selectivity, separations using such zeolitic kinetic adsorbents for removing oxygen from argon, become more efficient. Standard kinetic PSA processes, such as those disclosed in US Patent 7,501,009, benefit from replacing carbon molecular sieve with these adsorbents. Improved kinetic PSA processes, such as those disclosed in US 7,645,324, where the adsorbents have been converted into laminate structure to reduce deleterious effect of mass transfer resistance and flow friction pressure drop, are able to take full advantage of the fast kinetics of such adsorbents.

Even with the improved adsorbents, traditional kinetic PSA processes still suffer from low recovery when high product purity is required. While the laminate configuration can achieve high productivity, it introduces significant amounts of void volume into the bed that can further reduce overall recovery.

In this work we will discuss the limitations of both types of processes with these adsorbents and how to meet the needs of high recovery and productivity to match the adsorbentsâ€™ potential.

532473 Enrichment of Low Concentration CH_4/N_2 Mixture By Adsorption Process with Displacement Chromatography Technology

Abstract:

Enrichment of Low concentration CH4/N2 mixture by adsorption process with displacement chromatography technology

Ying Yang[1], Kai Lu[1,2], Donglei Qu[1,2], Ping Li[1,2], Jianguo Yu[1,2]**

1. State Key laboratory of Chemical Engineering, National Engineering Research Center for Integrated Utilization of Salt Lake Resources, East China University of Science and Technology, Meilong Road 130, Shanghai, China, 200273.

2. State Key Laboratory of Chemical Engineering, College of Chemical Engineering, East China University of Science and Technology, Shanghai, 200237, China

yyang@ecust.edu.cn, liping_2007@ecust.edu.cn, jgyu@ecust.edu.cn

Abstract:

The recovery and utilization of coal mine methane (CMM) provide a number of significant energy, economic, safety and environmental benefits. Adsorption-based processes are recognized to be a promising method for the enrichment of CMM because of the advantages of low capital investment, reusable nature of the adsorbents and straightforward automatic operation. However, the typical pressure swing adsorption (PSA) processes have thermodynamic constraints (low adsorbed amount and low CH_4/N_2 selectivity) on the purity of the heavy product, thus resulting in the low performance when PSA is adopted in the enrichment of CMM.

In this work, the principles of displacement chromatography are employed, and different PSA processes implemented with displacement concept have been investigated experimentally and theoretically for the enrichment of CMM.

The adsorption equilibrium isotherms of CH_4, N_2 and CO_2 on activated carbon(AC) are gravimetrically measured at 273 K-338 K up to 1000 kPa in a magnetic suspension balance. The Sips model, Toth model and Multisite Langmuir model can fit the isotherms quit well. The diluted breakthrough curve experiment and corresponding isothermal bidisperse pore model is used to study the adsorption kinetics of $CH_4/N_2/CO_2$ on AC. It is found that the micropore resistances of CH_4, N_2 and CO_2 are the dominate resistance within AC particle. There are little differences between the micropore diffusivities of CH_4 and N_2.

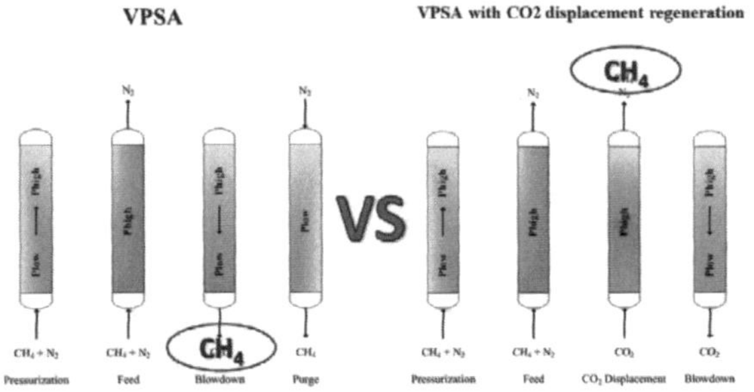

Figture 1. Schematic diagram of four-step VPSA and CO2 displacement-VPSA cycle

The four-step Skarstrom-type VPSA process and the four-step CO_2 displacement VPSA process on AC packed bed are compared for the separation of low concentration CH_4/N_2 system. It is found that the performance of CO_2 displacement VPSA process is better than that of typical VPSA process. Furthermore, a five-step CO_2 displacement VPSA process is proposed to solve the problem that some CO_2 would be vented to the atmosphere in the four-step cycle. The simulations of five-step CO_2 displacement VPSA process are carried out for CH_4/N_2 separation. With the five-step cycle, 79.3% CH_4 is captured with a purity of 86.9% from 10% CH_4 feed gas, the productivity and energy consumption are 0.21 molCH$_4$/(kgads.¡¤h) and 139.2 kJ/molCH$_4$, respectively; even the 1% CH_4 feed gas can be increased to 45.8% with 77.5% CH_4 recovery, 0.019 molCH$_4$/(kgads.¡¤h) productivity and 1537.3 kJ/molCH$_4$ energy consumption. The results demonstrate that CO_2 displacement is a promising technology for the enrichment of CMM.

Keywords: Activated carbon; IAST-Sips model; displacement chromatography technology; Vacuum pressure swing adsorption

529987 Simulation Study of Concentrating High Purity CO_2 from Syngas after Oxy-Fuel Combustion By Pressure Swing Adsorption Process

Abstract:

Abstract:

Exhaust emissions from industrial activities contain large amounts of carbon dioxide, which causes serious greenhouse effect, resulting in climate change which will endanger our living environment. According to the investigation of Intergovernmental Panel on Climate Change (IPCC), CO_2 is the most important greenhouse gas from human activities. Therefore, to reduce the emission of CO_2 is a major task nowadays.

Pressure swing adsorption (PSA) plays an important role in the separation of gas mixtures and concentration. Comparing to other separation method such as absorption, membrane separation, and cryogenic separation, PSA has the advantages of low cost and being easy to operate. As more and more novel adsorbents are synthesized, PSA is a potential method to slow down the global warming.

The system of capturing CO_2 can be divided into three categories by combustion types: pre-combustion capture, post-combustion capture and oxy-fuel combustion.

This simulation research studies concentrating carbon dioxide from the syngas of a gasifier after oxy-fuel combustion and dehydration with composition of 95% CO_2 and 5% N_2 by pressure swing adsorption process, so that the concentrated carbon dioxide can be captured and utilized/stored to reduce greenhouse gas emission. High purity carbon dioxide is mainly applied in the areas of laser, semiconductor wafer cleaning, and critical extraction of medicine, etc. The zeolite 13X is used as adsorbent. First, the experimental adsorption isotherm data were regressed to obtain the parameters of Langmuir-Freundlich isotherm equation. The mass transfer coefficient of linear driving force (LDF) model was calculated by theory and verified by breakthrough curve and desorption curve experiments, shown in Figure 1 and Figure 2. Then we verified the simulation program by comparison with the data of a single-bed four-step process experiment. The verification shows the accuracy of the simulation program.

Finally, the four-bed twenty-four-step and four-bed twenty-step PSA processes for syngas after oxy-fuel combustion and dehydration are studied to find the optimal operating conditions. The PSA processes and operating variables such as feed pressure, feed temperature, step time, bed length, and vacuum pressure were discussed.

The results show that the four-bed twenty-step PSA process without tank and with recycle shown in Figure 3 is the best process for getting high purity of carbon dioxide. When feed pressure increases, total amount of feed gas increases and so does the amount of gas be adsorbed, which makes CO_2 product purity increase, but makes recovery decrease because of more gas exhaust from top stream. While temperature increases, the selectivity of CO_2 to N_2 increasing with temperature, which makes CO_2 purity increase. As bed length increases, more gas is adsorbed and less gas exits at cocurrent depressurization step, which makes CO_2 purity decrease but recovery increase. When vacuum pressure increases, CO_2 purity increases because more N_2 exits at cocurrent depressurization step, but recovery decreases because CO_2 also exits at cocurrent depressurization step.

After discussing the operation variables, the optimal operating conditions are feed pressure 5.00 atm, vacuum pressure 0.238 atm, adsorption time 231 s, cocurrent depressurizaton time 129 s, vacuum time 486 s, pressurization equilibrium time 204 s, and temperature 338.14 K. The simulation results of optimal operating conditions are 99.99994% purity and 9.81% recovery of carbon dioxide at bottom product of four-bed twenty-step PSA process without tank and with recycle shown in Figure 4.

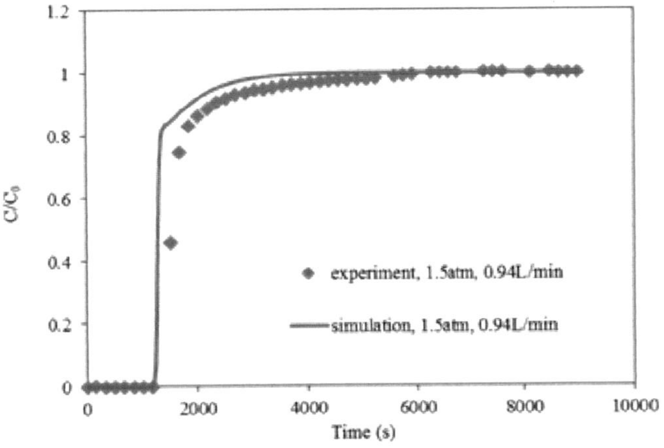

Figure 1. Breakthrough curve of CO_2 for 50% CO_2 and 50% N_2 gas mixture.

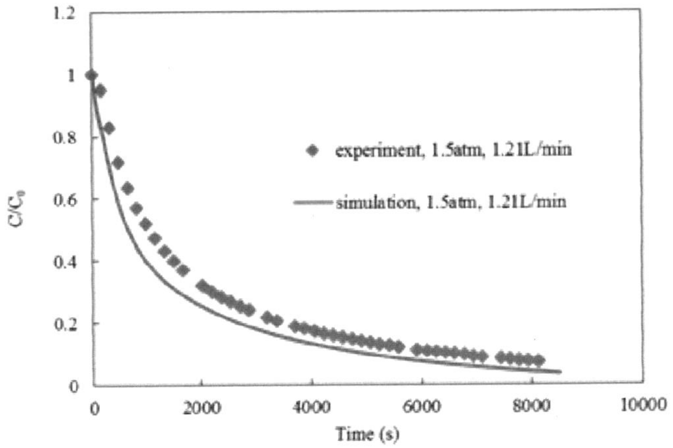

Figure 2. Desorption curve of CO_2 by pure helium.

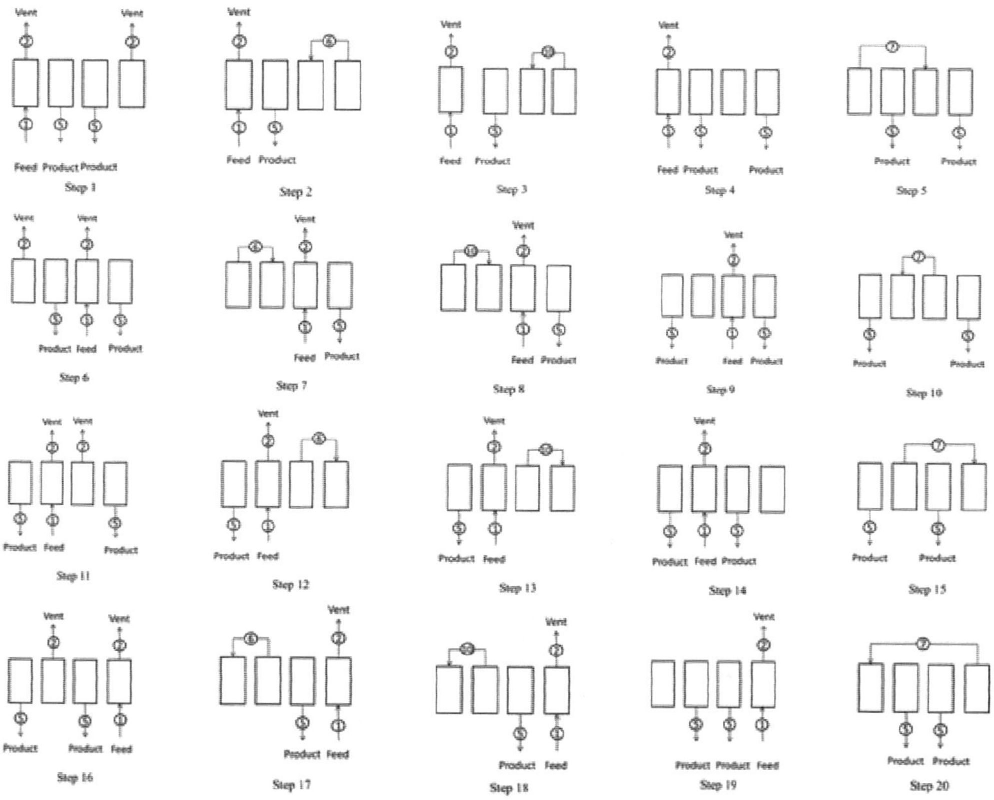

Figure 3. Schematic diagram of four-bed twenty-step PSA process.

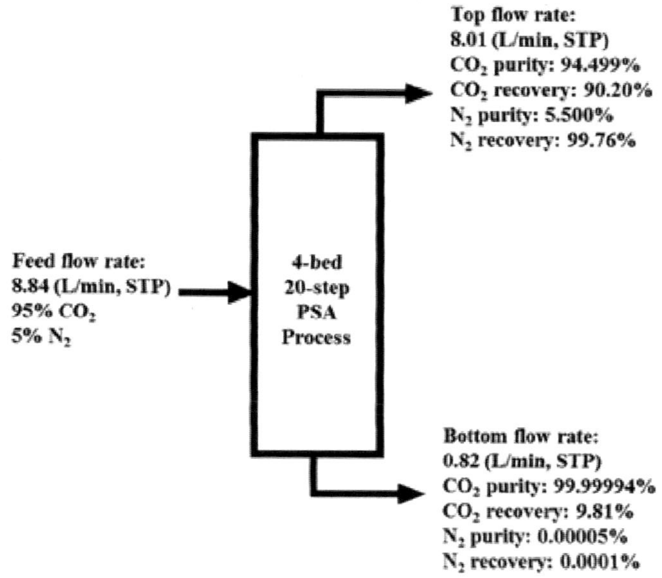

Figure 4. Schematic diagram of optimal results for CO_2 capture from syngas feed.

525863 Patterning Various Commercial Nanofiltration and Reverse Osmosis Membranes

Abstract:

There have been differing views on the ability to pattern polyamide thin-film composite (TFC) membranes directly by nano-imprint lithography. The goal of this study was to understand what factors control patternability, working towards a set of heuristics for use by the membrane community to pattern any polyamide TFC membrane. Despite completing a comprehensive set of experiments to investigate the roles played by membrane chemistry, surface properties, mechanical properties, and performance properties on pattern peak heights for thirteen commercial nanofiltration and reverse osmosis membranes, we found no correlation between the variables studied and patternability of *individual* membranes. We did discover significant differences in patternability between membranes grouped by polyamide class, with those prepared by interfacial polymerization of *m*-phenylenediame and trimesoyl chloride having the largest pattern peak heights. We further discovered that the humectant (pore filler) used for membrane preservation plays a role on the patternability of the membranes. Upon replacement of the original humectants used by the membrane manufacturers with a 15 wt% glycerol solution, the pattern peak heights approached a similar value for each membrane class. Tests performed to elucidate the role of the glycerol on patternability were inconclusive. Thus, while the humectant clearly contributes to membrane patternability, the reason why remains unknown.

533582 Membrane Surface Modification Using Acrylate- and Thiol-Containing Zwitterionic Materials Via Bio-Adhesive Polydopamine

Abstract:

Fouling is one of the major challenges for the use of membrane technology for water purification. One effective way to mitigate the fouling is to chemically modify membrane surface to reduce its favorable interactions with foulants, such as grafting with superhydrophilic zwitterions. However, zwitterionic materials are water-soluble, and it is challenging to graft or coat zwitterions for long-term underwater operation. In this work, we demonstrate facile covalent grafting of zwitterionic materials (functionalized by acrylate and thiol) on the surface of ultrafiltration (UF) membranes using dopamine. Specifically, dopamine is deposited with sulfobetaine methacrylate (SBMA) or a thiol-containing zwitterionic polymer (p(MPC_{160}-*co*-$DTMAEL_{42}$, or PMD). In the presence of oxygen, dopamine forms polydopamine (PDA) adhering onto the membrane surface and covalently grafts SBMA or PMD via Michael addition to form a robust thin superhydrophilic layer, as confirmed by contact angle measurement and X-ray photoelectron spectroscopy (XPS). Interestingly, the coating does not completely cover the pores (as evidenced by the SEM) and thus, the modified membranes still exhibit reasonably high water permeance. The antifouling properties of the membranes were determined using a crossflow system and bovine serum albumin (BSA) as a model foulant. The modified UF membrane with SBMA and dopamine demonstrates up to 30% higher water flux than the uncoated one when tested with 1 g/L BSA. Moreover, the membrane modified with PMD and dopamine exhibits less flux decline (38% reduction) than the unmodified membrane (53% reduction). The facile approach of membrane modification employing SBMA/dopamine is also adapted for postâ€'modification of a commercial nanofiltration (NF) membrane module, which demonstrates enhanced antifouling properties when tested with real wastewater at a wastewater treatment plant.

538832 High-Flux, High Capacity Adsorptive Membranes Based on Polysulfone and Block Polymer Composites

Abstract:

In the coming decades, the increasing global population, improved standards of living, and the expansion of irrigated agriculture will make meeting the demand for fresh water a challenge. One critical aspect of this challenge, is ensuring the adequate removal of heavy metal contamination that enters fresh water resources through a variety of mechanisms. The design and implementation of adsorptive membranes that target the capture of heavy metal ions is one potential strategy for helping to ensure a continuous supply of potable drinking water. Herein, we report a high-flux, high capacity adsorptive membrane fabricated from commercially-available polysulfone and block polymer materials. Through the appropriate selection of casting parameters, membranes with a bi-continuous network of pores ~1 μm in diameter were fabricated using a surface-segregation and vapor induced phase separation methodology. Compared with traditional ion-exchange resins, which are ~300 μm in diameter, the membranes exhibit reduced mass transfer limitations due to their uniform pore size and shorter diffusion length. After casting, hydraulic permeability measurements demonstrate that the as-cast parent membranes are pH-responsive with the hydraulic permeability ranging from 3.3×10^4 L m^{-2} h^{-1} bar^{-1} in a solution at pH 1 to 1.9×10^4 L m^{-2} h^{-1} bar^{-1} in a solution at pH 13. This behavior provides evidence that the poly(acrylic acid) (PAA) moieties of the block polymer segregate to the pore wall during membrane fabrication. Moreover, these PAA brushes allow the pore wall chemistry to be tailored for heavy metal removal via straightforward solid state coupling reactions. Specifically, the covalent attachment of metal ion chelating groups enables the highly efficient purification of simulated ground water or seawater solutions by capturing 99^{+}% of the cations dissolved in them. In this effort, to develop an adsorptive membrane capable of efficiently removing metal ions under conditions where the contamination is present at trace concentrations (*e.g.*, < 50 ppm), a tailor-made terpyridine group was incorporated along the pore wall of the membranes. The high metal binding affinity of the terpyridine moiety results in an adsorptive membrane that reaches its saturation capacity (1.2 mmol g^{-1}) at bulk ion concentrations less than 1 mM. The heavy metal binding performance of these membranes are further examined in a broad spectrum of ions and simulated background electrolyte conditions (*e.g.*, deionized water, ground water or sea water) at different ion concentrations. Using the terpyridine-functionalized membranes, metal ion removals greater than 99% were observed in single stage equilibrium experiments. Additionally, the cation binding performance is coupled with a modest change in the hydraulic permeability for the composite membranes (2.8×10^4 L m^{-2} h^{-1} bar^{-1}). With their readily tunable surface chemistry, these microfiltration membranes are a potential platform for the design and fabrication of membrane devices that can be implemented in advanced separation and purification applications.

536004 High Performance Electrospun Nanofiber Membranes for Protein Purifications

Abstract:

Membrane based cation-, anion-exchange and hydrophobic interaction chromatography (HIC) for protein purifications are often limited by the capacity in the flow through mode to remove impurities and aggregates, or capacity, selectivity and recovery in the bind-and-elute mode for protein fractionation. Electrospun nanofiber membranes are characterized by high surface to volume ratio and high permeability. Cationic and anionic ligands are grafted on the electrospun regenerated cellulose membrane substrates using UV-initiated and atom transfer radical polymerization. Static and dynamic binding capacities for model proteins such as BSA, IgG and lysozyme were determined under appropriate bind and elute buffer conditions. Static binding capacities in the order of several hundred mg/mL and dynamic binding capacities in the order of one hundred mg/mL can be obtained. In addition, thermo- and ionic strength responsive poly(vinylcaprolactam) (PVCL) ligands were grafted on the membrane surface for HIC applications. Compared to commercial HIC capacity in the range of 10-20 mg/mL, a much enhanced protein binding capacity was obtained due to the high surface-to-volume ratio of the electrospun membranes.

534693 Pnipam Functionalized Temperature Responsive Membranes and Pollutant Adsorption

Abstract:

This study is aimed at evaluating the effect of temperature on the degradation of chlorinated organics in water streams. PVDF microfiltration membranes are functionalized with poly-N-isopropylacrylamide (PNIPAm), and its temperature responsive behavior is studied as it relates to water flux and partitioning of toxic pollutants. Effective pore size can be predicted based on the swelling extent of the temperature-responsive PNIPAm polymer around its LCST. PNIPAm is known to show a sharp transition to a hydrophobic state at its lower critical solution temperature (LCST) at around 32°C, making PNIPAM attractive for industrial applications. The entrapment of reactive Fe/Pd nanoparticles in a poly-acrylic acid (PAA) polymer domain has been reported for the effective degradation of chlorinated organic compounds. The NPs are also entrapped in the temperature responsive PNIPAm-co-PAA polymer network for dechlorination and contaminant degradation of PCBs and TCEs. This study aims to evaluate PCB and TCE degradation by reactive immobilized nanoparticles in a PNIPAm functionalized PVDF MF membrane. The ability of PNIPAm particles to selectively adsorb various contaminants, specifically PCBs, TCEs, and PFOA, at various temperatures is also studied, as well as adsorption kinetics and ability to desorb through temperature changes. Increasing the surrounding temperature affects inter-particle spacing, solute adsorption and diffusion because of the changing PNIPAm conformation, as well as the intrinsic rate constant for the reaction. These effects are predicted, and studied by experimental degradation results. This research is supported by the NIEHS-SRP grant P42ES007380, and by the NSF KY EPSCOR program. Full-scale PVDF membranes were developed through collaborative work with Nanostone-Sepro (Oceanside, CA, USA).

536123 PVDF Membrane Pore Functionalization Approaches with Applications to Pollutant Remediation

Abstract:

Responsive membranes provide a highly tunable platform through ease of functionalization, accessibility with charged groups (-COOH, -OH) and tuning of pore size. Specifically, polyvinylidene fluoride (PVDF) microfiltration membranes have distinct advantage for open structure in terms of high internal surface area and ease of access in the pore domain. This give an edge to functionalize PVDF membrane with suitable responsive (pH, temperature) polymer and further incorporating Pd-Fe bimetallic nanoparticles for degradation of chlorinated organic compounds (COCs). However, making responsive membranes with specific responsive factor and precise amount of nanoparticle loading is a function of type of membrane and polymerization conditions. Specially, role of polymerization condition during making membrane is very critical to control membrane mass gain, water permeability, nanoparticle size and loading and finally,
the rate of degradation of COCs. Unfortunately, in literature this aspect is somehow overlooked. To address this aspect of making pH responsive membrane this study has demonstrated the effects of changing monomer and cross-linker concentrations matrix on the mass gain, water permeability, Pd-Fe nanoparticle (NP) loading, and the rate of degradation of one of COCs. Here, 3,3',4,4',5-pentachlorobiphenyl (PCB 126) was used as a model compound. The results revealed precise information on mass gain, water permeability, NPs size and loading on pH responsive functionalized membrane. The NPs functionalized membranes were then tested for use as a platform for the degradation of PCB 126. The observed batch reaction rate (K_{obs}) for PCB 126 degradation for per mg of catalyst loading was found 0.08~0.1 hr^{-1}. Degradation study in convective flow mode shows 98.6% PCB 126 is degraded at a residence time of 46.2 seconds. The corresponding surface area normalized reaction rate (K_{sa}) is found about two times higher than K_{sa} of batch degradation; suggesting overcoming the effect of diffusion resistance for degradation of PCB 126 in convective flow mode operation. Lastly, statistical analysis based on experimental results allows us to portray responsive model behavior of functionalized membrane.

We like to thank NIH-NIEHS-SRC and Chevron for funding this research project.

537462 On the Use of the Dual Process Langmuir Model for Gas Mixture Components That Exhibit Single Process or Linear Isotherms

Abstract:

There are two interesting cases that arise when one of the components in a gas mixture is described by either the single process Langmuir (SPL) model or the linear isotherm (LI) model and the other component is described by the dual process Langmuir (DPL) model. These two cases are very subtle and have never been addressed in the literature.

For the first case, a methodology will be presented based on three approaches to predict the mixed gas isotherms of a binary system, wherein one of the components is described by the DPL model (usually the heavier component) and the other one is described by the SPL model (usually the lighter component). These approaches differ on whether the individual processes of the isotherms correlate in a perfect positive, perfect negative or indifferent fashion. For the second case, a methodology will be presented based on two approaches to predict the mixed gas isotherms of a binary system, wherein one of the components is described by the DPL model (the heavier component) and the other one is described by the LI model (the lighter component). These approaches differ on whether the individual processes of the isotherms correlate in a perfect positive or perfect negative fashion. For both cases, a consistent set of single and binary isotherms from the literature that behave these ways are used as the basis for comparison to demonstrate these concepts.

530020 The Rigid Adsorbent Lattice Fluid Model: A New Thermodynamic Framework for Multicomponent Adsorption

Abstract:

In this contribution a new Rigid Adsorbent Lattice Fluid (RALF) model is presented and simple expressions for adsorption of mixtures are derived from a corrected version of the Lattice Fluid (LF) model of Sanchez and Lacombe [1-3]. The LF model, which formulates the properties of the mixture in its close-packed state, is specialised to the case in which the total volume of the lattice corresponds to the volume occupied by a microporous adsorbent, ie. the solid volume including the micropores [4]. The RALF model bears some similarities to the Non Equilibrium Lattice Fluid (NELF) of Doghieri and Sarti [5], which has proven to be a very successful thermodynamic framework to model pure and mixed sorption of light gases and solvents in glassy polymers, but the modifications to the LF model lead to different expressions for the fugacity coefficient of the adsorbed phase.

In presenting the RALF framework, the special case of the *frozen solid* is discussed in detail and shown to correspond to a type I isotherm. Explicit expressions for the Henry law constant, the zero loading adsorption energy and the saturation capacity are derived. These can be used to determine from adsorption data the 3 model parameters for an adsorbent. As a case study adsorption on silicalite is considered, given the availability of data in the literature. The parameters for silicalite are obtained from Henry law constants, zero loading adsorption energies and saturation capacities for 6 molecules [6] and data on n-paraffins are then predicted from the remaining two binary interaction parameters present in the RALF model. This should be compared to a standard Langmuir model that requires 3 parameters (Henry law constant, saturation capacity and the adsorption energy).

One of the advantages of the RALF framework is that it is derived as a multicomponent model, therefore prediction of adsorbed mixtures can be carried out either estimating the interaction parameter from fluid phase data or assuming that the interaction parameter between adsorbed molecules is zero. Literature data for CH_4-C_2H_6 on silicalite [7] will be used to demonstrate the application of the RALF framework to binary system.

The talk will conclude with a discussion on how to extend the RALF framework to heterogeneous adsorbents.

References

[1] Sanchez, IC and Lacombe, RH Elementary Molecular Theory of Classical Fluids - Pure Fluids. *J. Phys. Chem.* 80, 2352-2362 (1976).

[2] Lacombe, RH and Sanchez, IC Statistical Thermodynamics of Fluid Mixtures. *J. Phys. Chem.* 80, 2568-2580 (1976).

[3] Sanchez, IC and Lacombe, RH Statistical Thermodynamics of Polymer Solutions. *Macromolecules* 11, 1145-1156 (1978).

[4] Brandani, S, Mangano, E, and Sarkisov, L Net, Excess and Absolute Adsorption of Helium. *Adsorption* 22,261-276 (2016).

[5] Doghieri, F and Sarti, GC Non-Equilibrium Lattice Fluids: A Predictive Model for the Solubility in Glassy Polymers. *Macromolecules* 29, 7885-7896 (1996).

[6] Golden, TC and Sircar, S Gas Adsorption on Silicalite. *J. Colloid Interface Sci.* 162, 182-188 (1994).

[7] Dunne, JA, Rao, M, Sircar, S, Gorte, RJ, and Myers, AL Calorimetric Heats of Adsorption and Adsorption Isotherms. 3. Mixtures of CH_4 and C_2H_6 in Silicalite and Mixtures of CO_2 and C_2H_6 in NaX. *Langmuir* 13, 4333-4341 (1997).

533724 Constant-Pattern Design Method for Separating Ternary Mixtures of Rare Earth Elements Using Ligand-Assisted Displacement Chromatography

Abstract:

The rare earth elements (REEs) consist of the 15 elements of the lanthanide series, yttrium, and scandium. REEs are important components of many high-tech products, and as a result, the demand for REEs is growing. Traditionally, REEs are mined from ores that contain a mixture of several different elements. For most applications, high purity of a single REE is required, so the ores must be separated and purified. However, since REEs have similar physicochemical properties, the separation and purification is a challenging and expensive process. Ligand-assisted displacement chromatography (LAD) was developed to separate REEs in the 1950s and 1960s using commercially available ion exchange sorbents and ligands. While previous studies have demonstrated the feasibility of LAD for separating REEs with high purity (>99%) and high ligand efficiency, the productivity for these experiments were very low. Without a systematic process to design and optimize LAD systems with mass transfer effects, LAD systems were designed using a trial and error approach.

In this study, a constant-pattern design method is developed to separate REEs with high purity and high productivity for a given target yield. When the mixed band regions in an LAD system reach a constant pattern state, the yields of high purity products will be maximized. A general map was developed to show the relationship between the minimum column length to reach a constant pattern state and key dimensionless groups, which include selectivity, loading fraction, cut, and overall mass transfer coefficients. Given a target yield, intrinsic parameters, and material properties, operating conditions were calculated using the general map without trial and error. The design method was verified experimentally for different target components, yields, ligand concentrations, and feed compositions. The yields of the target components were within 3% of the expected values, demonstrating the robustness of the design method. High purity (>99%) of REEs were obtained from experiments with two orders of magnitude (x800) higher productivity than literature results. In cases with a minimum required yield, the component with the lowest feed concentration can be the limiting component, even in cases where it does not have the lowest selectivity.

530271 The Effect of Aluminum Ordering on Carbon Dioxide Adsorption in Zeolites

Abstract:

The distribution of Al in cationic zeolites is important in controlling the performance of these materials for catalysis and separations. However, it is difficult to measure this distribution experimentally. In this work, we examine the effect of Al distribution on CO_2 adsorption in seven industrially-relevant zeolite topologies using molecular simulations based on DFT/CC-derived force fields. A Reverse Monte Carlo method is used to generate Al distributions with varying degrees of order or randomness. Grand Canonical Monte Carlo simulations for Na^+-exchanged zeolites indicate that CO_2 adsorption isotherms can be used to distinguish between Al distributions. These simulation results agree with experimental results in the limited number of cases where information is independently available about Al ordering. Differences in CO_2 adsorption isotherms can be understood in terms of the existence of dual and single cation adsorption sites. Our results indicate that CO_2 adsorption measurements combined with GCMC simulations can be used to determine the Al distribution in zeolites without relying on more challenging experimental techniques.

537172 Ethane Diffusion in a Mixed Linker ZIF-7-8 By Pulsed Field Gradient NMR in Combination with a Single Crystal IR Microscopy

Abstract:

Metal organic frameworks (MOFs) and zeolitic imidazolate frameworks (ZIFs), a class of MOFs, present an exciting opportunity to work with a robust library of materials whose metal ions and organic linkers can be tailored. As these materials become increasingly more complex, there is a growing need to understand their transport properties.

Here we report an experimental study where two complimentary techniques, pulsed field gradient (PFG) NMR and infrared microscopy (IRM), are applied to investigate ethane diffusion inside crystals of mixed linker ZIF-7-8. This material is formed by a combination of ZIF-8 and ZIF-7 imidazolate linkers within the same framework, and is expected to have a smaller aperture size in comparison with that in ZIF-8. Ethane diffusivities in ZIF-7-8 are compared with those in ZIF-8 and ZIF-7.

An application of ^{13}C and ^{1}H PFG NMR in combination with IRM for studies of ethane diffusion revealed that, as expected, the ethane self- and transport diffusivities were significantly lower in ZIF-7-8 than in ZIF-8. The data obtained by both techniques indicate that ethane diffusivities are different in different ZIF-7-8 crystals, and the diffusivity ratio for different crystals can be as large as 8. Based on the IRM data, this observation was assigned to a difference in the fraction of the ZIF-7 and ZIF-8 linkers in different ZIF-7-8 crystals. A satisfactory agreement was observed between the corrected ethane diffusivities from IRM studies and the corresponding self-diffusivities measured by PFG NMR under similar conditions.

527806 Evaluation of Carbon Monoliths for CO_2 Separation

Abstract:

In order to efficiently capture CO_2 from gas streams, rapid processes have to be developed. In this work, we have studied carbon monoliths for the separation of CO_2 from flue gas. The effect of wall thickness on separation efficiency was studied. Breakthrough experiments were performed at gas velocities between 0.1 and 10 m/s (resulting in contact times as low as 0.01 s) in order to assess mass transfer limitations. Thermocouples inside the monolith channels allowed to follow the thermal wave in the channels during adsorption and desorption. Significant thermal effects were observed during both adsorption and desorption, with a strong impact on the separation efficiency and cycle time. Experiments were performed at different levels of relative humidity, demonstrating that the carbon monoliths are suitable for gas streams containing relative large amounts of water vapour as a result of the hydrophobic nature of the material.

Various regeneration methods were tested in order to optimise the balance between cycle time and dynamic adsorption capacity.

529382 A Critical Review on the Merits and Fallacies of Adsorption Kinetics Models

Abstract:

Although literature on adsorption kinetics modeling is rich the majority of models employed are empirical, based on chemical reaction kinetics or oversimplified versions of diffusion models. In this lecture the most used batch reactor adsorption kinetics models are presented and compared to a two-phase homogeneous diffusion model equipped with a double-selectivity isotherm. The merits and pitfalls of all models are discussed with focus on repeated mistakes in the literature, which can help to prevent their propagation in scientific knowledge.

537702 Development of an Automated Multi-Stage Continuous Reactive Crystallization System with in-Line Pats for High Viscosity Process

Abstract:

An automated multi-stage continuous reactive crystallization system with in-line PATs for a high viscosity process was developed in the present study. Data Acquisition (DAQ) hardware and Labview software were used as the local control system. A â€œforward-backwardâ€ burst pumping strategy was developed to smoothly transfer the highly viscous hot slurry from one vessel to the next. In addition, a comparative analysis between a PFR and CSTRs in series revealed that to achieve the same conversion, the latter would require more volume than the former, but less than a single CSTR. For a second-order reaction, the value of the DamkÓ§hler number necessary to achieve conversion of 90.0% in a single CSTR is 90. Thus, it is reasonable to approximate a PFR using CSTRs in series to obtain a high yield with a smaller reaction volume (compared with a single CSTR). In-line FBRM and React IR were used to monitor the crystal size and reactant concentration in the vessel containing cold slurry, as the probes could not be positioned in the hot slurry due to fouling issues. E-factors of batch and continuous processes were also compared, and the continuous reaction could obtain a lower E-factor because less waste was generated.

532114 Disturbance Studies of Continuous Cooling Crystallization of Carbamazepine

Abstract:

Continuous manufacturing (CM) is an emerging technology in the pharmaceutical manufacturing sector. Continuous crystallization is important in an integrated CM process and may significantly affect the drug substance critical quality attributes (CQAs), such as particle size distribution and polymorphism. However, few publications have been focused on potential manufacturing disturbances, the impact on drug substance quality and how the risks can be mitigated. Here, we designed and built an automated two-stage Mixed Suspension Mixed Product Removal (MSMPR) crystallization platform for a model compound (Carbamazepine, CBZ) that exhibits multiple polymorphs. The crystallization process includes integration of Process Analytical Technology (PAT) tools (on-line Raman microscopy and Focused Beam Reflectance Microscopy, FBRM) for real-time monitoring. We performed continuous runs and evaluated the effects of typical manufacturing disturbances, such as temperature fluctuations, agitation rate change, and feed concentration hikes, on process dynamics. This talk will discuss our results and findings.

532752 Interface Nucleation in Segmented Continuous Crystallization Process

Abstract:

The continuous crystallization behavior of segmented tubular crystallization process was investigated using glycine as a model compound. In our previous work, air-liquid slug flow crystallization was studied systematically through experiments coupled with a Realizable kâ€“Îµ model in a computational fluid dynamics platform and obtained optimal results. In order to alleviate the phenomenon of wall crystallization occurred in the air-liquid segmented tubular crystallization system, liquid-liquid slug flow crystallization was investigated using dodecane as the oil phase. Process parameters such as surface active agent, ultrasonic were studied experimentally for their effect on the growth rate, size distribution, and morphology of the produced crystals as well. Static nucleation experiment under microscope of air-liquid interface and liquid-liquid interface was specially conducted and compared, and the results indicated that the oil restrained the nucleation compared to gas. The binding energy between glycine and different tubing materials were simulated using the material studio software and found that the binding energy between glycine and silicon tube was the strongest.

531977 Model Based Process Design on Continuous Cooling Crystallization

Abstract:

The use of first-principle models in the pharmaceutical industry presents great promise, as the models may significantly help on process design and control. Models can enhance scientific understanding of the process dynamics, and predict the behavior of a system at various conditions. As part of the effort to understand emerging technologies such as continuous manufacturing,[1-2] the development and verification of a population balance model (PBM) for the continuous cooling crystallization of Carbamazepine (CBZ) is studied. A systematic experimental design will be followed to estimate the kinetic parameters for the stable and metastable form (CBZ form III and II, respectively); a series of batch cooling crystallization experiments were performed to determine the parameters for nucleation, growth and dissolution kinetics.

In addition, we will perform a design of experiment (DOE) to systematically determine the design space to produce products of a certain size and polymorphic composition range. The population balance model (PBM) will be used to simulate our continuous crystallization processes. Five factors considered in the determination of the design space are: feed concentration, initial concentration, seed loading, seed crystal size distribution, and residence time. Various points within the design space and at the boundaries were evaluated experimentally.

1. Lee, S. L.; Oâ€™Connor, T. F.; Yang, X.; Cruz, C. N.; Chatterjee, S.; Madurawe, R. D.; Moore, C. M.; Lawrence, X. Y.; Woodcock, J., Modernizing pharmaceutical manufacturing: from batch to continuous production. *J Pharm Innov* **2015,** *10* (3), 191-199.
2. Yang, X.; Acevedo, D.; Mohammad, A.; Pavurala, N.; Wu, H.; Brayton, A. L.; Shaw, R. A.; Goldman, M. J.; He, F.; Li, S.; Fisher, R. J.; Oâ€™Connor, T. F.; Cruz, C. N., Risk Considerations on Developing a Continuous Crystallization System for Carbamazepine. *Organic Process Research & Development* **2017**.

534345 High Throughput Continuous Crystallization of Lysozyme in an Oscillatory Flow Baffled Crystallizer with Real Time Process Monitoring

Abstract:

High throughput continuous crystallization of lysozyme in an oscillatory flow baffled crystallizer with real time process monitoring

Biopharmaceuticals are increasingly used for medicinal therapies and now make up roughly 20% of the pharmaceutical market share. While most (small molecule) pharmaceutical products are purified with at least one crystallization step, biopharmaceuticals (specifically proteins) are very difficult to crystallize due to extraordinarily slow growth kinetics[1]. Moreover, large protein chains require subsequent folding operations in order to reach a desired activity level. Until now, oscillatory flow strategies have primarily been used in batch mode to improve the process kinetics of biopharmaceuticals through enhanced micromixing and heat transfer[2]. Known to have high surface area to volume ratios, oscillatory flow baffled reactors (OFBR) generate turbulence by superimposing piston driven oscillatory flow onto the net flow generated by a pump. This combination approach allows for adequate solid suspension with greatly improved micromixing compared to the traditional stirred tank approach[3]. Herein, lysozyme was chosen as a model compound for both crystallization and folding applications, as its kinetic parameters are relatively well studied in literature[4].

Using the commercially available DN6 (Alconbury West Ltd) with the new Blaze probe (Blaze 900, Blaze Metrics), the oscillatory system was able to significantly reduce the time required to crystallize lysozyme to several sizes (varying between 50-200μm). The Blaze Metrics probe allowed for real time monitoring of the system, both in terms of steady/transient state analysis as well as monitoring the critical quality attributes of the crystals (size and shape). These dynamic measurements allow for real time feedback control strategies to be implemented on this new commercially available platform.

In a separate application, the DN6 was also used to continuously fold the newly purified lysozyme. Proper folding of proteins is essential for their activity, and, for enzymes, activity can be measured using a colorimetric assay utilizing fluorescently tagged substrates. To ensure the activity of lysozyme was not affected by the crystallization process, renatured lysozyme was examined for its ability to degrade fluorescein-labeled *Micrococcus lysodeikticus* cell walls and compared to pre-crystallization lysozyme. It was found that the recrystallization step had little to no impact on the lysozyme's activity. Continuous oscillatory flow systems with proper monitoring offer advanced approaches to generating quality crystalline proteins for high throughput applications.

References

1. Gavira, J. A. (2016). Current trends in protein crystallization. *Archives of Biochemistry and Biophysics*, *602*, 3–11. http://doi.org/10.1016/j.abb.2015.12.010

2. Abbott, M. S. R., Harvey, A. P., Perez, G. V., & Theodorou, M. K. (2013). Biological processing in oscillatory baffled reactors: operation, advantages and potential. *Interface Focus*, *3*(1), 20120036.

3. McDonough, J. R., Phan, A. N., & Harvey, A. P. (2015). Rapid process development using oscillatory baffled mesoreactors - A state-of-the-art review. *Chemical Engineering Journal*, *265*(1), 110–121. http://doi.org/10.1016/j.cej.2014.10.113

4. Liu, Y., Wang, X., & Ching, C. B. (2010). Toward further understanding of lysozyme crystallization: Phase diagram, protein-protein interaction, nucleation kinetics, and growth kinetics. *Crystal Growth and Design*, *10*(2), 548–558.

525444 Dynamic Modeling of a Continuous Reactive Crystallization Process

Abstract:

Pharmaceutical and fine chemical companies often employ reactive crystallization to make intermediates and final products. The particle formation rate, particle size and shape, and growth rate are influenced by two temperature and concentration dependent phenomena: A) the reaction, via temperature and concentration dependent kinetic and reaction rate and activation energy; B) crystallization, via temperature and concentration dependent supersaturation ratio and nucleation and crystal growth rate. Therefore, the critical process parameters in productivity (reaction yield) and control of particle size distribution could be manipulated to achieve desirable quality and economics.

Determining the impact of operating temperature on the particle size distribution could be challenging when higher temperature would advance the reaction rate (more solute generation/concentration), but may impact supersaturation ratio and crystal growth rate (more solubility at higher temperature). Beside temperature, the reagents concentration (flow-rate) would impact the reaction rate and generating solute molecule that would lower supersaturation. Effect of micromixing and local supersaturation is not studied in this project.

Most of the reactive crystallization processes have been in batch mode. The reactive crystallization can be used in continuous fashion for API synthesis. The continuous reactive crystallization in a MSMPR (CSTR) method has been less studied and published. In this work, a dynamic process simulation was developed for a continuous reactive crystallization, where the effect of temperature and concentrations (reagentsâ€™ flow-rates) were studied on the system response and particle size distribution. The dynamic mode case studies cover a variation of CPPs disturbances and demonstrate system dynamic in terms of the particle size distribution and response rate in comparison to the RTD of the continuous reactive crystallizer.

528957 Separation of Volatile Organic Acids from Fermentation Using Non-Ionic Surfactants

Abstract:

Volatile organic acids (VOAs), such as acetic, propionic, butyric, and lactic acids, are high value compounds that can be used as precursor for a wide range of applications. These applications include food additive and preservative, solvent, plasticizer, perfume, polymer synthesis, synthetic fabrics and fibers, pharmaceutical and biofuels. VOAs can be produced through anaerobic fermentation, and is typically recovered by precipitation followed by re-acidification using sulfuric acid. This process, however, produces large amount of solid slurry and calcium sulfate as waste. Alternatively, liquid-liquid extraction can be applied to recover VOAs from fermentation broth. Among the solvents studied for VOAs extraction include long-chain aliphatic amines and fatty alcohols (or their mixture) due to their high distribution coefficients. Previous study by the authors indicated that non-ionic surfactant, particularly a mixture of alcohol ethoxylates, can also be used for extraction of butyric acid from aqueous medium. This study extends the previous work by establishing the phase equilibria of lactic acid with various non-ionic surfactants. This will be done to evaluate the applicability of these surfactants for liquid extraction of lactic acid, as indicated by distribution coefficients and separation factors. The non-toxic and biodegradable characteristics of non-ionic surfactants could be useful as well for in situ extraction or extractive fermentation of VOAs.

535583 Integrated Solvent Design for the Separation of Aromatics from Aliphatics with Ionic Liquids

Abstract:

Current processes for the separation of aromatics from aliphatics are plagued with challenges including: feed composition limitations, low aromatics/aliphatics selectivity, low aromatics capacity, and high energy consumption. Collectively, these limitations call for the development of new technologies to improve separation. Solvent design of ionic liquids (ILs) for the separation process is one of the most promising research directions in this field, because ILs offer several advantages including nonvolatility, thermal stability, and tunability [1,2,3]. In this study, we integrate conceptual process design and predictive thermodynamic models into the molecular design of an IL for this separation process, using toluene/heptane as a model system for aromatic/aliphatic.

Process design and predictive thermodynamic models are integrated into the IL design using an iterative process. First, we apply computational chemistry methods to characterize the candidate IL molecule. This involves using Gaussian to do molecular geometry optimization and to do Conductor-like Polarizable Continuum Model (CPCM) calculations to characterize solvation effects. These results are then used to do Conductor-like Screening Model (COSMO) calculations to determine a screening charge density profile for the IL, which is used by the COSMOtherm software to obtain molecular parameters and physical properties. Second, we evaluate the candidate IL using a simple process simulation (Aspen Plus) model of the IL-based separation process with a simple flash drum as the solvent recovery unit. We use COSMO - Segment Activity Coefficient (COSMO-SAC) as the activity coefficient model, with the molecular parameters determined in the first step. From the process simulation, we can obtain separation process results such as toluene purity, recovery, energy duty, capital cost, etc. After the evaluation of the second step, the IL molecule can be modified, and the two-step procedure repeated. For comparison of the IL process modeling results with the current commercial sulfolane process we have developed an Aspen Plus model of this process using the non-random two liquid (NRTL) approach as the activity coefficient model. The goal of this strategy is to design the best IL that not only meets the separation product criteria but also maintains a lower energy and capital cost compared to the current commercial process.

At this time, we have tested multiple ILs based on the design strategy described above. Some are previously tested ILs in the aromatic/aliphatic separation with modified functional groups and alkyl chain length, such as bis(trifluoromethylsulfonyl)imide ($[Tf_2N]^-$) anion paired with "modified" pyridinium, imidazolium, thiolanium cations. Others are newly designed ILs that can potentially interact with aromatics via electromagnetic interactions such as hydrogen bonding, σ-π interactions, π-π interactions, etc. The process modeling results of tested ILs are compared with other published IL processes using the same solvent recovery method [4,5], as well as to the commercial sulfolane process, to show their applicability.

[1] Lin, W.-C., Tsai, T.-H., Lin, T.-Y., Yang, C.-H. J. Chem. Eng. Data2008; 53: 760-764.

[2] Cháfer, A., de la Torre, J., Font, A., Lladosa, E. J. Chem. Eng. Data2015; 60: 2426-2433.

[3] Brennecke. J. F., Maginn. E. J. AIChE J. 2001; 47: 2384-2389.

[4] Meindersma G. and Haan A. de. Chemical Engineering Research and Design 2008; 86(7): 745–752.

[5] Ferro, V. R., de Riva, J., Sanchez, D., Ruiz, E., & Palomar, J. Chemical Engineering Research and Design 2015; 94(1):632–647.

529920 A Study on Thermodynamic Properties of Binary Mixtures of Sesame OIL with Aliquat 336 and Tributyl Phosphate

Abstract:

Binary mixtures of sesame oil with Aliquat 336 and Tributyl Phosphate (TBP) have immense application as extractant in liquid membrane based separation processes. This research work is intended to investigate the molecular interactions and transport properties of miscible binary mixtures of sesame oil + Aliquat 336 (Mixture A) and sesame oil + TBP (Mixture B) by determining the density and viscosity over a temperature range of 25°C-70°C under atmospheric pressure. Basic parameters such as density and viscosity of mixtures lead to an understanding of solute-solvent interactions through the calculations of excess molar volume and viscosity deviation. The data of excess molar volume and viscosity deviation further throw light on molecular interactions as they have applications in solution theory and molecular dynamics. Yousra Litaiem and Mahmoud Dhahbi studied the physicochemical and transport properties of binary mixtures of Aliquat 336 and formamide [1]. Srirachat et al., carried out similar experimental work to study the effect of polarity and temperature on binary mixtures of Di-(2-ethylhexyl) phosphoric acid (D2EHPA) with organic solvents (kerosene, n-heptane, chlorobenzene and 1-octanol) [2].

The formation of ideal mixture hardly causes any change in volume however; real mixture undergoes either an increase or decrease in volume due to molecular interactions between the components of mixtures. Excess molar volume and apparent molar volume indicate the changes or differences in the binary mixtures and specific interactions between them. Apparent molar volume is usually less than the molar volume. A positive deviation in excess molar volume indicates volume expansion on mixing thus repulsive interaction of mixing solvents or weaker interactions than the interactions of molecules in pure solvents. While a negative trend in the data of excess molar volume shows strong interactions of molecules in the mixture than individual molecules before mixing. Similarly, a negative deviation in apparent molar volume indicates a good molecular interaction between the solute and solvent. As the molecules of solute are strongly attracted to the solvent molecules they occupy less space owing to the decrease in apparent molar volume and vice versa. As temperature increases the apparent molar volume increases due to volume expansion caused by weak bonding leading to lesser molecular interactions.

From the measurement of density, the excess molar volume and apparent molar volume have been calculated for mixture A and mixture B. The viscosity data has been used to determine the viscosity deviation of the mixtures (A and B) to understand the attractive-repulsive forces between the interacting molecules of the solvent mixtures. An empirical relation given by Parthasarathy and Bakshi [3] and Vogel-Fulcher-Tammann equation [2] have validated the density and viscosity data, respectively. The excess molar volume and viscosity deviation data have been fitted into the Redlich-Kister polynomial equation [2]. Further, thermal expansion coefficient, excess Gibbs' free energy (ΔG^E), change in enthalpy (ΔH) and change in entropy (ΔS) of activation of the liquids have also been calculated. The volumetric thermal expansion coefficient shows the change in the volume of the liquid with a change in temperature at constant pressure. With increase in bond energy, the value of the coefficient decreases. While it increases with increase in temperature due to weak bonding. A decreasing trend in the values of the volumetric thermal expansion coefficient with increase in mole fraction further indicates that the intermolecular forces in the mixtures are stronger than in pure solvents.

The density (ρ) of sesame oil lies in between Aliquat 336 and TBP (ρ_{TBP} {0.973 g/mL} > $\rho_{Sesame\ oil}$ {0.914 g/mL} > $\rho_{Aliquat}$ {0.886 g/mL} at 25°C). In mixture A, density decreases with increase in mole fraction of Aliquat and temperature. Whereas in mixture B, density increases with increase in mole fraction of TBP and declines with rise in temperature. The expansion in volume due to increase in kinetic energy and weak bonding with rise in temperature leads to decrease in density. The excess molar volume (V_M^E) data for aliquat mixture is negative. This negative trend indicates strong interactions of molecules in the mixture than individual molecules before mixing. While in mixture B, the excess molar volume data changes from negative to positive with increase in mole fraction showing that the strong interactions of molecules in the mixture shift to repulsive interaction or weaker interactions as the concentration of TBP increases. The apparent molar volume trend is same for both the mixtures. It increases with temperature and mole fraction. This is because rise in temperature leads to volume expansion due to weak bonding leading to lesser molecular interactions. The viscosity (η) of Aliquat is more than that of sesame oil and TBP ($\eta_{Aliquat}$ {1.048 Pa.s}> $\eta_{Sesame\ oil}$ {0.039 Pa.s}> η_{TBP} {0.0025 Pa.s} at 25°C). While viscosity of mixture A decreases with temperature, it increases with increasing mole fraction. In case of mixture B, viscosity decreases for both, temperature and increasing mole fraction of TBP. The viscosity deviation ($\Delta\eta$) data is negative and decreases with increase in temperature for both the mixtures. The volumetric thermal expansion coefficient (β) is almost constant with temperature and mole fraction for both the binary mixtures. For pure aliquat, β = 0.0007°C^{-1} and β value for pure TBP is 0.0009°C^{-1}. The trend of ΔG^E data is same for both aliquat and TBP mixtures. ΔG^E increases with temperature to give a maximum value at 0.5 mole fraction of Aliquat for mixture A while in mixture B, ΔG^E value is maximum at 0.58 mole fraction of TBP. ΔH increases while ΔS decreases with mole fraction for mixture A whereas in mixture B, ΔH decreases while ΔS increases with mole fraction. Further, V_M^E and $\Delta\eta$ is minimum at 0.47 - 0.76 mole fraction of Aliquat showing that the molecular interaction is comparatively stronger in binary mixture than in pure solvents. From this, it can be concluded that the binary mixture of Aliquat and sesame oil (mixture A) may be applied in future for separation or extraction processes.

Keywords: Sesame oil, Excess molar volume, Viscosity deviation.

References

[1] Y. Litaiem and M. Dhahbi, Physicochemical Properties of an hydrophobic ionic liquid (Aliquat 336) in a polar protic solvent (formamide) at different temperatures. J. Disper. Sci. Technol 36 (5) (2015) 641-651.

[2] W. Srirachat, T. Wannachod, U. Pancharoen S. Kheawhom, Effect of polarity and temperature on the binary interaction between D2EHPA extractant and organic solvents (kerosene, n-heptane, chlorobenzene and 1-octanol): Experimental and thermodynamics. Fluid Ph. Equilibria 434 (2017) 117-129.

[3] C. M. Rodenbush, F. H. Hsieh, D. S. Viswanath, Density and viscosity of vegetable oils. J. Am. Oil Chem.' Soc 76 (12) (1999) 1415-1419.

529819 Enhanced Oil-Solid Separation By Multifunctional Switchable Solvents

Abstract:

A kind of process aids, switchable-hydrophilicity tertiary amines (SHTA), have been synthesized and applied to enhancing oil-solid separation with or without assistance of hydrocarbon solvent. This solvent can switch between its hydrophilic and hydrophobic states with the addition or removal of CO_2, respectively. When working together with the hydrocarbon solvent, it is observed that the addition of SHTA solution (e.g., triethylamine, N, N-dimethylcyclohexylamine, etc.) facilitates the recovery of heavy hydrocarbons and improves the quality of bitumen product by surface modification. When working individually or with water, the SHTA solution (e.g., N, N, N′, N′-tetraethyl-1, 3-propanediamine , TEPDA) is found to perform as two different roles: as an organic solvent and can be ionized in water to form cations. With these properties, TEPDA was proven to perform well with water in separating heavy oil from a carbonate solid surface, resulting in over 10% additional oil recovery. A mechanistic study shows that in this aqueous/non-aqueous hybrid process, TEPDA acts mainly as a solvent in softening and dissolving heavy hydrocarbons (dissolution effect). Meanwhile, a small amount of ionized TEPDA in the aqueous phase acts as a surface-active material accumulating at the oil-water and water-solid interfaces (interfacial modification). Both of these effects work to enhance the detachment of oil from solid surfaces. Similar to those in above hydrocarbon system, the quality of the recovered oil and the cleanliness of the residual solids are both improved. Further tests show that this water-assisted TEPDA process also performs well in detaching oil from other solid surfaces, such as metal and silica. The above results show that the switchable solvent would be a promising ‘green chemicals’ in oil production or soil remediation.

KEYWORDS: Switchable solvent; Solvent extraction; Unconventional oil; Oil-solid separation; Diamine

References:

[1] Xingang Li, Ziqi Yang, Hong Sui, Ashish Jain and Lin He, A hybrid process for oil-solid separation by a novel multifunctional switchable solvent, *Fuel*, **2018**, 221: 303-310.

[2] Hong Sui, Lin Xu, Xingang Li and Lin He, Understanding the roles of switchable-hydrophilicity tertiary amines in recovering heavy hydrocarbons from oil sands, *Chem Eng J*, **2016**, 290: 312–318.

533962 Liquid-Liquid Extraction and Theoretical Binding Studies of Crown Ether Derivatives with Mixed Heteroatoms Towards Platinum Group Metals

Abstract:

Platinum-group metals (PGM) especially palladium (Pd) and platinum (Pt) are extremely scarce in comparison to other precious metals due to their low natural abundance and the complex processes required for their extraction and refining. Currently, the separation and purification of PGMs are carried out using commercial extractant having low selectivity, which require multiple stages of extraction process. Therefore, development of effective extractants with high selectivity will improve its sustainability especially for PGMs recovered from secondary sources such as spent automotive catalysts, industrial catalyst and electronic waste.

Herein, the design of novel selective extractant was based on fine tuning the coordination specificity by changing the heteroatoms, structure of crown ether (CE) and cavity size ($Ø_{CE}$) was realized with the aid of experimental extraction and theoretical Density Functional Theory (DFT) studies. Efficient green synthesis of 15-14 membered crown ethers with mixed heteroatoms was achieved via intramolecular cyclization of bulky or rigid bis-epoxides with aromatic N or S di-nucleophiles using trimethylamine (Et_3N) and lithium chloride template in water to give crown ether derivatives with mixed heteroatoms (O, N, S). Initial experimental liquid-liquid extraction and DFT studies were implemented to evaluate the performance of each synthesized CE towards Pt/Pd extraction. Among the CEs synthesized, mixed thiacrown ether having dibenzo moieties, $2HDBS_2 14C4$, with $Ø_{CE}$=1.58 Å shows highest selectivity towards Pd ($Ø_{Pd}$=1.52 Å) even in the presence of other platinum (Pt, Ru and Rh) or precious group metals (Sn, Sb, Hf, Ir and Au). In order to determine the effect of the structure and the heteroatoms towards Pd or Pt coordination, optimized DFT structures of the free CE using the *ab initio* Jaguar program B3LYP with LACVP** basis set was employed. From the DFT results, the cavity size match relationship, where the CE cavity size is similar to the metal diameter, as well as changing the heteroatoms plays major role in synthesizing Pd selective CEs.

The novel thiacrown ethers synthesized were applied in liquid-liquid extraction process using simulated PGM solutions. The thiacrown ethers were first methylated to increase its solubility in the extraction solvent toluene prior to liquid-liquid extraction studies. The effects of HCl and thiacrown ethers concentration to the distribution of Pd and Pt were performed subsequently to determine the optimum HCl concentrations, CE-Pd stoichiometry, and extraction equilibrium coefficients (K_{ex}). Re-extraction of Pd in the solvent phase using thiourea-HCl and its recyclability through repeated extraction and re-extraction were done to determine its stability and applicability in real extraction system. Finally, the thiacrown ethers were used as extractants to recover Pd and Pt from simulated spent automotive catalyst leach solutions.

This work was supported by NRF funded by the Korea government funded by the Ministry of Science and ICT (2017R1D1A1B03028102 and 2017R1A2B2002109) and Ministry of Education (2009-0093816 and 22A20130012051(BK21Plus)).

.

522899 Minimum Reflux Behavior of Multicomponent Mixture Separation Using Complex Distillation Columns

Abstract:

To separate a multicomponent mixture, a series of distillation columns are usually required. Each sequence or arrangement of distillation columns generates a distillation configuration. In many distillation configurations, one or more columns involve more than one feed stream and/or one or more sidedraw streams. These multi-feed, multi-product columns, which are often referred to as complex columns, are very common in many industrial applications, such as multi-effect distillation systems. On the other hand, the minimum reflux ratio of a column is related to its energy consumption and capital cost, and thus is a key parameter in distillation design. Therefore, to design and build energy efficient and cost effective distillation configurations, it is practically very important to an efficient and accurate method to determine the minimum reflux ratio for complex columns.

Despite its practical significance, this problem is intellectually difficult to solve. Most existing approaches involve either rigorous tray-by-tray calculations or iterative guessing of reflux ratio which are computationally expansive. Here, we present a simple and easy-to-use algorithmic method to determine the minimum reflux ratio for a general complex distillation column for ideal or close-to ideal multicomponent mixture separations. The classic Underwoodâ €™s shortcut method is applicable for simple columns with exactly one feed and a top and a bottom product. Our method uses Underwoodâ€™s ideas as a starting point and then formulates new concepts and incorporates them to create an algorithm that can solve for minimum heat duty of a distillation column with multiple feeds and multiple products including those from intermediate locations of the column. Compared with other existing shortcut based approaches, our method does not require guessing of pinch location or controlling split, which may lead to incorrect solutions.

Through several case studies, we demonstrate the accuracy and effectiveness of our new approach. Furthermore, we show that this method can be easily incorporated into a global optimization framework to facilitate the calculation of the minimum heat duty requirement for a multicomponent distillation configuration under minimum reflux. Finally, we present some useful heuristics to help industrial practitioners design energy efficient and cost effective distillation systems.

533198 Control of Non-Key Compositional Bulges in Industrial Distillation Unit Operations

Abstract:

In modern chemical plants distillation units are increasingly put under pressure with the requirement to achieve greater throughput while ensuring higher product purities. Most of these distillation units also contain with multi-component feed streams and often has demanding high purity requirements both at the distillate and bottoms streams. These tight product and bottoms purity specifications together with the multi-component nature of the feed streams can result in the accumulation of non-key components in the distillation columns. This is particularly an issue when the feed stream to the column consists of both polar and non-polar components, which can create polar and non-polar regions with the column. These regions can potentially â€œtrapâ€ the non-key component in the middle of the distillation column between the polar and non-polar regions of the column. From a design point of view, this accumulation of non-key components is managed by introducing a side draw where the accumulated non-key middle boiling components can be extracted to maintain steady state operations.

However, from a process control point of view managing this accumulation and maintaining a steady state operation can be challenging, and requires a sound understanding of process fundamentals, process dynamics and control. This work will employ three industrial case studies on high purity methanol distillation, Propylene Oxide/ Ethyl Benzene separation and C3/C4 as the basis to identify the practical difficulties faced in industrial operations when managing the on specification operation of these type of distillation units. To this end, the shortfalls in employing classical measurement techniques such as temperature measurements and the inadequacies of classical distillation control configuration in operating these columns would be discussed. This will be followed by the historical reasons behind the use of these measurement techniques and control structures would be discussed followed by a discussion on the relevance of these reasons with the recent advances made in both process monitoring and control domains. This work will finally present alternative process control schemes to the standard classical distillation control configurations that also takes advantage of the recent advances in measurements techniques.

538072 Synthesis of Distillation-Based Separation Networks Using Block Superstructure

Abstract:

Distillation is the most widely-used technology for mixture separation among other alternatives, accounting for about 40% of the total energy consumption in chemical industries [1]. Process intensification enables the advancement of the process performance through improvements in controllability, sustainability and energy efficiency [2-3]. Using intensification principles, distillation operations are evolving into more energy-efficient processes through distillation sequence screening [4], heat integration [5], thermal coupling [6] and etc. Furthermore, novel distillation column configurations, such as divided-wall columns, provide more degrees of freedom for sustainable operation of distillation columns [7]. Although many advances have been made in modeling distillation column at task level and phenomena level [8-9], systematic methods for identifying rigorous distillation networks involving various alternatives, e.g., thermally coupled columns, divided-wall columns, are in high demand.

Previously, building block superstructure [2] has been applied to address systematic process intensification, e.g., membrane reactors, reactive distillation systems, etc., by utilizing fundamental phenomena building blocks instead of equipment-based representation. Later on, this approach is applied to address general process synthesis and integration problems [9-11]. In this work, we apply the block-based approach for systematic synthesis of distillation networks. Block superstructure is comprised of blocks positioned on a two-dimensional structure. Interaction of blocks with material and energy flow is facilitated by the direct connecting streams between adjacent blocks and 'jump connecting streams' among all blocks. The boundary between each block can be used to represent vapor-liquid equilibrium phenomena, or designated as completely restricted boundary indicating restriction of material flow (i.e. wall). Also this boundary can be unrestricted for achieving splitting composition requirements. With this, distillation column trays and dividing walls separating different regions of the distillation columns can be represented with multiple adjacent blocks. To achieve heat integration among all separation alternatives, each stream connecting adjacent blocks is assigned with heaters/coolers. The ensemble of these blocks leads to various type of distillation network representations. We formulate this block superstructure for distillation network synthesis as a mixed integer nonlinear programming optimization (MINLP) problem to minimize the total annual cost. The proposed model is applied to various distillation network synthesis problems, involving simple distillation sequences and heat-integrated distillation networks. Among these case studies, various type of distillation alternatives can be identified such as dividing wall columns and thermally coupled columns.

References:

[1] Humphrey, Jimmy L. Separation processes: playing a critical role. *Chemical Engineering Progress*, 1995, 91:10.

[2] Demirel, S. E., Li, J., Hasan, M. M. F. Systematic process intensification using building blocks. *Computers & Chemical Engineering*, 2017, 105: 2–38.

[3] Tian, Y., Demirel, S. E., Hasan, M. M. F., Pistikopoulos, E. N. An Overview of Process Systems Engineering Approaches for Process Intensification: State of the Art. *Chemical Engineering and Processing: Process Intensification*, Submitted.

[4] Torres-Ortega, C. E., & Segovia-Hernandez, J. G. The importance of the sequential synthesis methodology in the optimal distillation sequences design. *Computers & Chemical Engineering*, 2014, 62: 1-9.

[5] Leeson, D., Fennell, P., Mac Dowell, N., & Shah, N. Simultaneous design of separation sequences and whole process energy integration. *Chemical Engineering Research and Design*, 2017, 125: 166-180.

[6] Caballero, J. A., & Grossmann, I. E. Optimal synthesis of thermally coupled distillation sequences using a novel MILP approach. *Computers & Chemical Engineering*, 2014, 61: 118-135.

[7] Madenoor Ramapriya, G., Tawarmalani, M., & Agrawal, R. A systematic method to synthesize all dividing wall columns for n component separation—Part I. *AIChE Journal*, 2018, 64(2): 649-659.

[8] Lutze, P., Gani, R., Woodley, J. M. (2010). Process intensification: a perspective on process synthesis. *Chemical Engineering Processing: Process Intensification*, 49(6): 547–558.

[9] Li, J.; Demirel, S.E.; Hasan, M.M.F. Process Synthesis using Block Superstructure with Automated Flowsheet Generation and Optimization. *AIChE Journal*, 2018, under review.

[10] Li J., Demirel S.E., Hasan M.M.F. Process Integration using Block Superstructure. *Industrial & Engineering Chemistry Research*, 2018, 57: 4377–4398.

[11] Li J., Demirel S.E., Hasan M.M.F. Fuel Gas Network Synthesis Using Block Superstructure. *Processes*. 2018, 6(3): 23.

521482 Optimization of Heat-Integrated Multicomponent Distillation Sequences

Abstract:

Separation by distillation accounts for roughly 3% of the total energy consumption in the US. Almost all the above ambient distillation configurations are operated by heat. In multicomponent distillation, a common strategy to reduce the overall heat duty is to supply the rejected heat from a higher temperature condenser to a lower temperature reboiler [1]. Without heat integration of this sort, the Fully Thermally Coupled (**FTC**) configuration is known to be among the best in terms of minimum vapor duty, which serves as a proxy for heat duty. But with this mode of heat integration, even a simple sharp split configuration could perform better than the FTC configuration [2].

To check the feasibility of heat transfer, one would normally require relations to calculate the temperatures of condensing vapor and boiling liquid. Building upon the work of Agrawal and Herron (1997), we derive a condition that allows the comparison of the temperature levels of two streams, without calculating temperatures explicitly. This condition requires only information about the composition, component relative volatilities and pressure ratios of the streams.

The derived condition is extremely useful for optimization models for the identification of energy efficient distillation configurations. For sloppy-split configurations, the optimal vapor flows and component distributions are usually difficult to determine without solving an optimization problem [4]. Consequently, heat integration options are difficult to identify prior to optimization.

Our group has designed an optimization model incorporating Underwoodâ€™s shortcut method to minimize the vapor duty of a configuration [5]. This program can be run to quickly evaluate and rank list thousands of regular column configurations. We extend the framework by including the developed condition to optimize the vapor duty of heat-integrated configurations. In addition to the standard McCabe - Thiele assumptions, we assume that the changes in relative volatility with pressure are negligible, for small changes in pressure. The proposed framework predicts a 45% reduction in vapor duty for a four-component mixture of light alkanes, when the basic direct split configuration is heat integrated. Aspen Plus simulations show a good agreement with the results. Furthermore, our optimization model predicts that the vapor duty of this heat-integrated configuration is lower than that of the FTC configuration.

1. Agrawal, R. (2000). Multieffect distillation for thermally coupled configurations. AIChE Journal, 46(11), 2211-2224.
2. Rév, E., Emtir, M., Szitkai, Z., Mizsey, P., & Fonyó, Z. (2001). Energy savings of integrated and coupled distillation systems. Computers & Chemical Engineering, 25(1), 119-140.
3. Agrawal, R., & Herron, D. M. (1997). Optimal thermodynamic feed conditions for distillation of ideal binary mixtures. AIChE journal, 43(11), 2984-2996.
4. Nallasivam, U., Shah, V. H., Shenvi, A. A., Tawarmalani, M., & Agrawal, R. (2013). Global optimization of multicomponent distillation configurations: 1. Need for a reliable global optimization algorithm. AIChE Journal, 59(3), 971-981.
5. Nallasivam, U., Shah, V. H., Shenvi, A. A., Huff, J., Tawarmalani, M., & Agrawal, R. (2016). Global optimization of multicomponent distillation configurations: 2. Enumeration based global minimization algorithm. AIChE Journal, 62(6), 2071-2086.

528233 Rigorous Modeling of Chlorine Drying Using Sulfuric Acid

Abstract:

Sulfuric acid absorption is used to dry chlorine manufactured by chloralkali processes to very low levels, and packed columns are customarily the column internal of choice. Pumparounds are usually used for temperature control and because the feed rate of sulfuric acid is relatively low. Design of this process requires both an accurate vapor-liquid equilibrium (VLE) correlation and a rate-based model to evaluate the effects of column height, packing type, pumparound rates, etc. Development of the VLE correlation is challenging since the sulfuric acid concentration varies from about 98 wt% at the top of the column to about 50 wt% in the bottom of the column, and the system is a complex electrolytic mixture. We have developed an accurate VLE correlation using correct speciation and the Electrolyte NRTL model. The process model is often simplified to one using ideal stages and HETP (height equivalent of a theoretical plate), but this simplification renders the model inadequate to evaluate the effect of column height, pumparound rates and packing types. We demonstrate that the rate-based model in Aspen Plus enables reliable estimates of these variations, and therefore of capital and operating costs.

533208 Molecular Tracking: An Alternative Computer-Aided Concept for Multi-Component Distillation Column Design

Abstract:

In this work a new design methodology for distillation columns is proposed based on the concept of molecular tracking. This is a novel computer-aided design tool for distillation processes. Molecular Tracking methodology works by tracking one single molecule at a given time from the feed location to outlet streams of column (distillate, bottom and probable side streams). The method is based on a probability function, which is highly correlated with system thermodynamics and operations such that it produces a random pathway for each molecule that is then visualized. Combining these â€œrandom pathwaysâ€ allows generating a detailed visualization of the internal movement of components within the column. In comparison to the traditional approach design methodologies and corresponding simulations, molecular tracking approach provides a number of benefits such as intuitive visualization and the ease of reconfiguration with respect to additions such as side-draws as well as divided wall, which are becoming increasingly common.

This study first analyzes in detail the influence of thermodynamic properties and operational conditions on molecular tracking by demonstrating the concept on a simple binary distillation unit. This is followed by an industrial case study of a high purity methanol distillation unit consisting multi-component methanol and water feed with trace levels of ethanol. In this case study, molecular tracking is employed to find the optimal location of the column side draw. The analysis is based on thermodynamic and operational information gathered column with classical column configuration (only a distillate and bottoms). The generated results are then compared with results obtained from a validated process simulation and industrial observations. Finally, this work will discuss in detail the pros and cons of the concept of molecular tracking, its coordinates in relation to existing computer-aided process design tools and the future developments required to exploit in full its potential.

532794 Intelligent RO Systems: Advances and Challenges

Abstract:

Considerable progress has been made since the early days of Reverse Osmosis (RO) water treatment/desalination in reducing energy consumption through the development of more efficient membranes, effective process configurations and energy recovery devices. There have also been effective developments of low fouling membranes, high performance water treatment chemicals (e.g., antiscalants and coagulants) and feed pretreatment processes to mitigate membrane mineral scaling and fouling. Given the growing interest in inland desalination, there is a growing need for effective management of geographically distributed desalination plants. Such plants must effectively handle temporal variability of water quality and production demand. Reducing membrane fouling and mineral scaling, reducing energy consumption, decreasing plant footprint, and reducing operational costs are all critical elements to future development of inland membrane-based water treatment and desalination for increasing local water portfolios. In order to address the above, a multi-pronged approach was undertaken to assess the prospects for effective intelligent (autonomous) self-adaptive operation of reverse-osmosis desalination that includes a series of feed water pretreatment units. Translation of theoretical and laboratory studies to pilot field demonstration included: (a) optimization of process operating conditions as aided by a new class of membrane monitoring approaches for early detection of fouling/mineral scaling and its integration with plant control, (b) self-adaptive operation of two-stage RO that optimizes system operation so as to minimize the energy consumption, and (c) deployment of machine-learning tools for plant fault detection and isolation. Field results from both seawater and brackish water RO plants, designed for self-adaptive operation, will be presented to demonstrate improved operational efficiency through advanced monitoring and control strategies toward a path to high recovery (toward zero liquid discharge) and handling the variability of feed water quality.

533013 Membrane Materials for Liquid-Based Separations in Water and Organic Solvents

Abstract:

Membranes have a key role in water purification. Their relevance for chemical separations is growing. While reverse osmosis is well-established for decades, the development of multilayer membranes has been essential to promote the implementation of processes such as nanofiltration, forward and pressure retarded osmosis in emerging fields of application.[1, 2] The collaboration between our group at KAUST and Chungâ€™s group at NUS has been focused on new materials for hollow fibers and flat-sheet membranes with selective layers added by dip-coating, interfacial polymerization and surface functionalization by grafting.[3-6] Tailoring selectivity and permeance, improving fouling resistance, tuning shape and porosity of hollow fibers and the exploring new materials for interfacial polymerization, such as dendrimers, porphyrin and zwitterionic segments, are among the more recent developments in our group.[7-11]

References

1. N. L. Le, S. Nunes, Sustainable Materials and Technologies 2016, 7, 1.
2. Y. K. Ong, G. Shi, N. L. Le, Y. Tang, J Zuo, S. Nunes, T. S. Chung, Progress in Polymer Science 2016, 57, 1.
3. P. Duong, S. Chisca, P. Hong, H. Cheng, S. Nunes, T. S. Chung, ACS Appl. Mater. Int. 2015, 7 (7), 3960.
4. P. Duong, S. Nunes, T. S. Chung, J. Membr. Sci. 2016, 520, 840.
5. N. L. Le, N. Bettahalli, S. Nunes, T. S. Chung, J. Membr. Sci. 2016, 505, 157-166.
6. N. L. Le, M. Quilitzsch, H. Cheng, P. Hong, M. Ulbricht, S. Nunes, T. S. Chung, J. Membr. Sci. 2017, 522, 1.
7. P. Duong, J. Zuo, S. Nunes, Ind. & Eng. Chem. Res. 2017, 56, 14337-14349.
8. N. L. Le, D. Kim, S. Nunes, Scientific Reports 2017, 7, 9171.
9. P. Duong, P. Hong, V. Musteata, K. Peinemann, S. Nunes, Chem. Select 2017, 2, 6612-6616.
10. N. L. Le, S. Nunes, J. Membr. Sci. 2017, 533, 171-178.
11. N. L. Le, M. Ulbricht, S. Nunes, Ind. & Eng. Chem. Res. 2017, 56 (23), 6785-6795.

532719 A New Insight into the Formation and Impact of Roughness Features in Polyamide Reverse Osmosis Membranes

Abstract:

Water scarcity is one of the major global challenges of our time. Reverse osmosis technology plays a crucial role in providing alternative sources of water through seawater desalination and water reclamation. Thin film composite polyamide membranes are the gold standard for reverse osmosis technology, but tuning their permeability and selectivity remains a major challenge because of the inherent permeability-selectivity trade-off. The polyamide rejection layer of a typical reverse osmosis membrane contains a large number of discrete nano-sized voids occupying as much as 30% of its total volume. Despite the common belief that these nanovoids regulate the membrane water permeability and salt rejection, little is known about their formation mechanism. In this presentation, we show that these nanovoids are formed due to the degassing of nanobubbles from the monomer solution during the interfacial polymerization process to fabricate the polyamide selective film. Simple nano-foaming methods, such as addition of sodium bicarbonate and ultrasound application, can result in remarkable improvement in both membrane water permeability and salt rejection, thus overcoming the long-standing permeability-selectivity trade-off of desalination membranes. Our study further establishes a framework for the fundamental understanding of the structure-property-performance relationship of polyamide desalination membranes.

525514 High-Performance Ceramic Supported Thin Film Composite Membrane for Organic Solvent Nanofiltration

Abstract:

Aromatic polyamide is a common material used as selective layers for thin film composite (TFC) reverse osmosis (RO) membranes. This material has also demonstrated excellent performance in organic solvent nanofiltration (OSN). RO membranes, however, are built on solvent intolerant polymers that can dissolve or swell in the presence of many solvents. This necessitates the use of solvent tolerant supports for TFC membranes designed for OSN. Ceramic membranes offer solvent resistance, but forming polyamides on ceramic substrates has been challenging because ceramic membrane pore sizes have typically been too large to support the thin and fragile polyamide layer. With the recent advances in ultrafiltration ceramic membranes by Fraunhofer IKTS, the pore sizes have been reduced to a size that is more in line with conventional polymeric membranes used to support polyamide selective layers in RO,. In this work, we demonstrate the formation of a polyamide selective barrier layer *in-situ* onto a ceramic UF membrane and evaluate the TFC for use in OSN. We have selected ceramic tubular UF membranes provided by Fraunhofer IKTS as substrates. The resulting membranes have demonstrated a pure methanol permeance of 26.3 $Lm^{-2}h^{-1}bar^{-1}$, and dye rejection for methyl orange (327.3g/mol), acid fuchsin (585.5g/mol), chlorophyllin (724.2g/mol) and brilliant blue (825.6g/mol) of 30 %, 90%, 95%, and 91%, respectively. This work represents the first time that polyamide selective layers have been directly formed onto a tubular ceramic membrane and tested for OSN.

524488 Novel Polyimide Membranes and Their Modification for Pervaporation Applications

Abstract:

Membrane is one of the critical factors for the separation efficiency of the pervaporation technology. A desirable pervaporation membrane should possess high flux, superior separation factor, and long-term chemical and mechanical stability. Common hydrophilic polymeric membrane materials, such as polyvinyl alcohol, chitosan and sodium alginate etc., have been widely used for pervaporation dehydration because of their high selectivity towards water, but they generally suffer severe swelling in the aqueous solution, leading to the poor operation stability. Alternatively, polyimide materials have drawn more and more attentions as the pervaporation membrane material in recent years, with their superior mechanical strength and chemical stability. Here we mainly introduce our research works on the polyimide membranes for the pervaporation separations in recent years, including the molecular design of membrane materials and the structural optimization of membrane morphologies. We start with the commercial polyimide materials with suitable chemical structures and physical properties. Further chemical modifications (crosslinking, surface grafting, etc.) and morphology optimization are also employed to achieve the improved pervaporation performance and the long-term stability of the membranes. Additionally, based on the practical separation system, we also design and synthesize several new types of polyimide membranes with crosslinking and thermally-cyclodehydrated modifications. Compared to the commercial polyimide materials, the synthetic polyimides have the optimized molecular structures and superior separation efficiency.

524024 Alignment and Immobilization of Functionalized Aquaporins on Polybenzimidazole Nanofiltration Membranes

Abstract:

The overarching objective of this project is to form a biomimetic membrane incorporated with functional aquaporins dispersed in a polymeric membrane selective layer and capable of operation under high hydraulic pressure. Polybenzimidazole (PBI) membranes were surface modified with mutated aquaporins in order to improve the rejection of protons, ions and other impurities, aquaporin channels were aligned with the direction of water flow. Cysteine functional groups were installed on N-terminus of AqpZ (now called Aqp-SH) for covalent attachment to the polymer matrix so that the proteins could be immobilized to the membranes and aligned in the direction of the flow. Modified membranes showed lower MWCO as compared to unmodified membranes. Depth profiling using XPS analysis showed the presence of functional groups corresponding to mutated aquaporins in modified membranes. FTIR analysis showed confirmation of an amphiphilic polymer (PVA-alkyl) which was used to seal the gaps in between protein molecules immobilized onto the membrane surface.

In agreement with pore size distributions, charge interactions, and added resistance to flow due to modification, as the pore size of the PVA-alkyl membranes decreased as compared to unmodified PBI membranes, the flux values were lower while salt rejection was higher. On the other hand, Aqp-SH modified membranes had a similar pore size to unmodified PBI membranes, but with the added resistance to flow, displayed lower flux values as compared to unmodified PBI but higher as compared to PVA-alkyl modified membranes. Furthermore, contrary to expectations due to pore size distributions, Aqp-SH modified membranes displayed the highest salt rejection values among all membranes analyzed in the study. Aqp-SH modified membranes displayed a nearly constant salt rejection irrespective of the salt concentration of the feed solution, while unmodified PBI and PVA-alkyl modified PBI membranes showed a decrease in rejections as feed salt concentration increased. However, due to the hindrance of detergent or PVA-alkyl in aquaporin solutions, the surface of the membrane was not completely covered with immobilized and aligned aquaporins, which in turn led to rejection values lower than 100%. If the concentration of functional aligned aquaporins on the membrane surface could be optimized, the performance of membranes could be further be improved.

532974 Surface Functionalization and Peeling-Off Manipulation Generating Diversified Janus Membranes Toward Multifunctional Applications

Abstract:

Janus membranes (JMs) exhibit cross-sectional unidirectional liquid transportation feature with the aid of the asymmetric surface wetting property, viz. liquid â€œdiodeâ€. It demonstrates tremendous potential for applying in liquid manipulation, lab-on-a-chip, etc. Nevertheless, many barriers such as complicated fabrication process and high cost have to be removed to endow as-prepared JMs with more practical applications. Herein, surface functionalization and peeling-off manipulation was used to generate diversified JMs toward multifunctional applications. As a proof of concept, based on the structure and hydrophobicity of the commercial-available polytetrafluoroethylene (PTFE) composite microfiltration membranes, hydrophobic type of membranes were used to fabricate JMs with unidirectional water or oil transport behaviors in this study. Totally hydrophilic treatments on pristine membranes were conducted to prepare water unidirectional transport JMs via directly peeling off top layer or subsequent peeling after multifunctional surface treatment. On the other hand, floating substrates on dopamine solution resulted in another type of oil unidirectional transport JMs with thin hydrophilic layer and unidirectional oil transport behavior. Afterwards, the corresponding asymmetric chemistry and morphology, and extended applications were studied. We exploited JMs with unidirectional oil transport behaviors for separation of the oil-in-water emulsion, and exploited the feasibility on collection and lock of liquids using JMs, explored mechanism based on the contrast hydrophilic or hydrophobic membranes. As proof of concept, the combination of the surface morphology, chemical composition, wettability, responsiveness and the Janus performance towards Janus membranes were explored. It exhibited that the tape peeling-off could tear the PTFE skin layer to generate new hydrophobic surface derived from the nature of the exposed PTFE. Unilateral deposition of polydopamine could endow hydrophobic membrane with a thin hydrophilic layer. Pre-wetting of the thin layers of JMs could reduce the channel sizes, the Laplace pressure would be enhanced. It resulted in much higher hydrostatic pressures of liquids to penetrate the Janus film, which can achieve the locking behaviors. Magnetic JMs can be tuned by magnetic field to collect and purify the dye waste water, pH response JMs can be controlled via adjusting the pH of droplet or aqueous environment on the unidirectional water transport behavior, which demonstrates the benign controllability of multi-functionalization technology and Janus performance.

542620 The Impact of Shale Gas and Oil on the Chemical Industry

Abstract:

Lower alkanes and aromatics are very useful as both fuels and chemical raw materials. Occasionally supply disruptions constrained availability but also encouraged investigation of alternative feedstocks and chemistries. Recent new technologies for the production of unconventional gas and oil sources from tight shale formations significantly increased availability of both methane and condensate, radically altered process economics, resulted in the construction of much new intermediates capacity, and multiplied the ways traditional and alternative chemistries might be exploited. This presentation will outline historically how natural gas components became useful as chemical feedstocks, how government regulations occasionally led to perturbations in availability and price, how supply crises led to the development of alternative chemistries, catalysts, and ultimately the exploitation of shale resources. More specifically, will be discussed how excess coproducts, especially condensate, from increasing shale gas and oil production for energy are having significant unexpected impacts on the chemical process industries and opportunities for new process chemistry research.

541186 Sustainable Energy and Chemicals: Past, Present, and Future

Abstract:

The last 40 years has seen the U.S. energy situation vacillate from gasoline rationing and concerns over Peak Oil, to projected energy and feedstock prices that were noncompetitive in global markets for chemicals production, to the economic revolution of frack gas and light tight oil. The importance of low cost energy and feedstocks in driving economic opportunity will be discussed relative to this history, along with the importance of sustainability in decision making to gain stakeholder buy-in and acceptance. Strategies for managing career paths and competencies despite changing targets and high uncertainty will be addressed.

542618 Disruptions: What the Future May Hold

Abstract:

Disruptive forces are all around us. They change the way we communicate and work. They change the way we develop goods and services. They also change the types of products that are manufactured. Examples such as the way in which we produce and consume energy or the feedstocks we use to produce product have seen many changes over the course of the last few decades, and these disruptions are expected to continue. The chemical engineering discipline and the expectations for a chemical engineerâ€™s skills has also changed and adapted as a result of these disruptive forces. A discussion of the broad themes associated with these forces and the implications that they may have to chemical engineers in the future will be discussed.

542621 Geopolitical Factors Influencing the Evolution of the Chemical Industry

Abstract:

The basic chemicals and plastics industry is built on a simple principle: find a cheap feedstock and convert it into something more valuable. However, the right feedstock depends on place and time. Certain policies and actions of various governments have influenced the evolution and profitability of the chemical industry during the past 50 years, sometimes in surprising ways. For example, high oil prices and US government funding encouraged the development and commercialization of coal gasification technology during the late 1970s and early 1980s. This enabled China to develop a competitive basic chemical industry using coal as a feedstock during the last 25 years. Similarly, an abrupt increase in oil prices during the first decade of the 21st century encouraged the present shale gas revolution, causing a major shift in the economy of producing basic chemicals and polymers in the US. If history is our guide, then what governments choose to fund and regulate may be as important as market forces in determining the shape of our industry and profession in the future.

542622 Agility & Resilience: How to Maintain Career Competitiveness in the Changing Chemical Industry

Abstract:

The world around is changing. The shift to protectionism, global instability and price volatility of raw materials and feedstocks has the potential to wreak havoc on the chemical industry. With the revolution of shale gas in North America and the sudden drop in feedstock costs, the chemical sector in the US has reasserted its historic dominance. Learn how the role of a chemical engineer can contribute to industry growth and resilience, while adapting to allow for personal agility throughout the career life cycle.

532159 Kinetic Study for Comprehensive Understanding of Solid State Phase Transitions of Nicotinamide/Pimelic Acid Co-Crystals

Abstract:

For many material, understanding of solid-state phase transformation (SSPT) is beneficial not only for the fundamental reasons but in the case of pharmaceuticals, for the avoidance of product issues due to the change in physicochemical properties. Despite the fact that this topic has been regarded as an interesting research subject, the mechanism of SSPT is still not fully understood. It is especially surprising that most co-crystal studies have been biased to identify novel polymorphs of co-crystals and to reveal their crystal structures with very little effort on understanding the kinetics. Herein we presented the comprehensive kinetic study of the polymorphic system of 1:1 Nicotinamide (NA)-pimelic acid (PA) co-crystals with the combination of different experimental techniques. The bulk kinetic modeling with powder x-ray diffraction (PXRD) indicated that polymorphic transition in the solid state of NA/PA co-crystals occurs by nucleation and growth mechanism, and no phase separation of components in the co-crystals was observed during the transitions. In-situ atomic force microscopy (AFM) observation was performed to monitor the surface morphology change during transitions at the nanoscale to supplement the bulk kinetic analysis.

Differential scanning calorimetry (DSC) thermal analysis was also conducted to obtain insight into the dependence of the crystal sizes and the defects on the kinetics of transitions. The deviation in the activation energies of SSPT for both powder and polycrystalline form were achieved respectively. In addition, the change in the tendency of transition point with increase in the number of transition recycling was observed (initially decrease then after increases), which might be attributed to the change in crystal sizes and defect inclusions during the transition process.

533858 Case Study: Modifying API Crystal Habit to Improve Flowability

Abstract:

Case Study: Modifying API Crystal Habit to Improve Flowability

Daniel A. Green, Jennifer L. Tansey, & Yan T. Sun â€“ GlaxoSmithKline, Upper Merion, PA, USA.

Abstract

An Active Pharmaceutical Ingredient (API) precipitates as acicular crystals from the original crystallization solvent mixture: water and isopropyl alcohol (IPA). Extremely poor flowability of the solid API was noted after shipping a sample from the US to the UK. Our hypothesis was that the crystal habit exacerbates the observed poor flowability upon shipping. To test this hypothesis, we investigated altering the crystal habit by crystallizing from different solutions. We varied the ratio of water and IPA in the original water/IPA solvent system, and investigated the acetone/water system. We compared bulk density and powder flow properties for several cases. We find both crystal habit change, density, and flowability improvement. Specifically, the average aspect ratio of the crystals is significantly less in acetone/H_2O compared to IPA/H_2O. The aspect ratio depends on the acetone/H_2O ratio as well: increasing as the fraction of acetone increases over the range studied (50 - 90% acetone), while the dependence on the IPA/H_2O ratio is much weaker.

527117 Solution Coating of Pharmaceutical Nanothin Films and Multilayer Nanocomposites with Controlled Morphology and Polymorphism

Abstract:

With the advent of combinatorial chemistry and high throughput screening, active pharmaceutical compounds (APIs) with low aqueous solubility have become increasingly pervasive in drug development pipeline. The formulation challenges of these drug candidates, that exhibit poor bioavailability, often encumber their route to pharmacy shelf from lab-bench. This issue is one of the major contributing factors for the skyrocketing cost of drug development today, which has reached over $1.3 billion per drug on average.[1]

In recent years, nanosizing is rapidly emerging as an effective and versatile method for enhancing solubility and thus the bioavailability of poorly aqueous soluble APIs. Previous studies have indicated that confining crystalline APIs to the nanoscale dimension exhibit increased solubility and enhanced dissolution, owing to large surface area and high surface curvature.[2] However, current industrially compatible nanonization processes mostly utilize top-down approaches where bulk crystals are broken down via mechanical forces. These methods of nanosizing are energy inefficient and require long processing time while other methods such as formation of polymeric micelles and nanoemulsions have poor stability and their processing mandates expensive instrumentation and high concentration of emulsifiers, respectively. In this scenario, industrially compatible techniques for large-scale manufacturing of API crystals with consistent crystal size distributions smaller than few hundred nanometers are urgently needed.

In this talk, we present meniscus-guided solution coating as a new technique of producing pharmaceutical films and multilayered composites at the nanoscale. This facile technique has previously been employed for depositing semi-conductor nanothin films for printed electronic devices and enables easy modulation of film morphology and polymorphism.[3-5] While API crystal morphology determines its dissolution rate, the packing modes of API molecules play crucial role in controlling the bioavailability, processability and stability of the APIs.[6] By changing simple process parameters such as API solution concentration and coating speed in order to modulate the driving forces involved in the film deposition, we demonstrate control of a) film thickness over more than two orders of magnitude (30 nm to 1.5 μm), b) thin film morphology and c) crystal polymorphism. In fact, solution coating technique also allowed us to systematically trap metastable polymorph in the film of a poorly soluble API. The controlled crystallization of metastable polymorph is advantageous in the case of poorly aqueous soluble APIs because metastable polymorphs usually exhibit higher solubility rate owing to their higher free energy.

Furthermore, we performed layer-by-layer solution coating of alternating nanometer-thick films of an API and a biocompatible polymer in a multilayered fashion, in a variety of substrates. This demonstrates the potential of solution coating technique in formulation of API doses consisting of multiple drugs which enables controlled and sequential release of each of the components as per patientÕs need, catering to the enormous enthusiasm of modern medical research towards personalized medicine.

References

1. DiMasi, J. A.; Grabowski, H. G.; Hansen, R. W., Innovation in the pharmaceutical industry: New estimates of R&D costs. *J Health Econ* **2016,** *47*, 20-33.

2. Kaptay, G., On the size and shape dependence of the solubility of nano-particles in solutions. *International Journal of Pharmaceutics* **2012,** *430* (1-2), 253-257.

3. Diao, Y.; Shaw, L.; Bao, Z. A.; Mannsfeld, S. C. B., Morphology control strategies for solution-processed organic semiconductor thin films. *Energ Environ Sci* **2014,** *7* (7), 2145-2159.

4. Diao, Y.; Tee, B. C. K.; Giri, G.; Xu, J.; Kim, D. H.; Becerril, H. A.; Stoltenberg, R. M.; Lee, T. H.; Xue, G.; Mannsfeld, S. C. B.; Bao, Z. N., Solution coating of large-area organic semiconductor thin films with aligned single-crystalline domains. *Nature Materials* **2013,** *12* (7), 665-671.

5. Diao, Y.; Lenn, K. M.; Lee, W. Y.; Blood-Forsythe, M. A.; Xu, J.; Mao, Y. S.; Kim, Y.; Reinspach, J. A.; Park, S.; Aspuru-Guzik, A.; Xue, G.; Clancy, P.; Bao, Z. N.; Mannsfeld, S. C. B., Understanding Polymorphism in Organic Semiconductor Thin Films through Nanoconfinement. *J Am Chem Soc* **2014,** *136* (49), 17046-17057.

6. Singhal, D.; Curatolo, W., Drug polymorphism and dosage form design: a practical perspective. *Advanced Drug Delivery Reviews* **2004,** *56* (3), 335-347.

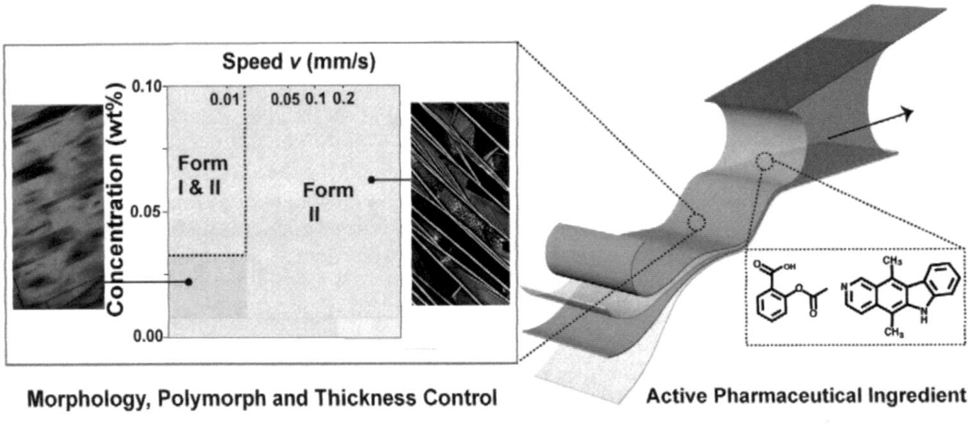

538726 Optimization of Cooling Crystallization of an Active Pharmaceutical Ingredient Undergoing Degradation

Abstract:

Crystallization is one of the most important unit operations in pharmaceutical manufacturing. Crystallization is the â €œmissing linkâ€ between Drug substance and Drug product design and thereby is very important to the successful operation of downstream unit operations. The most important performance indicators of any crystallization process are the process yield and the end-of-batch mean particle size. Frequently Active Pharmaceutical Ingredients(APIs) degrade at higher temperatures due to various reasons viz. solvolysis, etc. This leads to loss of process yield since the crystallization process needs to be carried out from a lower starting temperature. Sometimes degradation of APIs can be so problematic that cooling crystallization has to be discarded as a crystallization strategy and alternate crystallization strategies viz. Anti-solvent crystallization, etc need to be considered.

Unseeded Cooling Crystallization processes typically follow a parabolic cooling profile where initially the system is nucleated very fast and then allowed to grow until the temperature is decreased sharply towards the end of the batch. In this work, the crystallization process has been modeled mathematically using population balance equations for an API undergoing degradation at high temperatures. The optimal cooling profile for the crystallization process has been obtained for different optimization objectives viz. Minimization of concentration of degraded product and Maximization of Mean Crystal Size at end of batch. It has been seen that the typical parabolic cooling profile is not always the optimal cooling profile when the API degrades and a completely counterintuitive optimal cooling profile results when the objective is to minimize the degraded product concentration. An online Ultra High Performance Liquid Chromatography was also used to monitor the concentrations of the API and the degraded impurities. It was seen that adsorption of the impurity on the crystal faces occurs for a particular class of cooling profiles. The process model was modified to account for the simultaneous degradation of API and adsorption of impurity and the parameters of the rate processes were estimated using the on-line measured concentrations.

530374 Antisolvent and Cooling Crystallization of Pharmaceuticals Using a Continuous-Flow Microfluidic Platform

Abstract:

Solid materials with different crystalline structure may be problematic when it comes to bioavailability, chemical stability or physical property. Controlling the crystal structures requires systematically study on the crystallization route and influential factors. Antisolvent and cooling crystallizations are two well-known routes for crystallization and are widely employed by pharmaceutical industries. Understanding the effect of experimental parameter such as supersaturation ratio, solvent type/composition, antisolvent and temperature on polymorphic solids is vital for synthesis of the desired polymorph. Moreover, identifying the growth rate as well as size distribution of crystals is an ultimate goal for any screening process. Here we are proposing a microfluidic crystallizer to study the polymorphism of crystals under Hydrodynamically controlled condition. The micro scale of this crystallizer is ideal to use significantly small amount of drugs particularly at early stage of development while screening numerous conditions. Our proposed microfluidic platform is fabricated using cheap and affordable material, eliminating the need of expensive lithography procedure. Using this platform antisolvent and cooling crystallizations have been studied for Acetaminophen and Glutamic acid have been selected for this study. In the antisolvent crystallization effects of different organic solvents on polymorphs of glutamic acid have been studied however for acetaminophen water is the only antisolvent. High throughput screening has also been performed within single device by increasing the number of stagnant zone and applying various ratios for entering flows. Finally, the synthesized crystals are compared with the reference materials to clarify the type of the polymorphs in different experimental condition.

522653 A Case of a Twisty Route to First Crystals

Abstract:

Nowadays, polymorph and salt screening are usually conducted in a high throughput (HT) fashion were hundreds of conditions are explored simultaneously. The HT studies are typically conducted by varying solvent systems, temperature programs, ways of creating supersaturation and counter ions. Quite often, HT polymorph and salt screening lead to several crystalline forms and the selection of the crystalline form to progress in development is based mostly on stability and purity attributes. However, in some cases HT polymorph and salt screening results in mostly oiling out and / or solids with poor crystallinity. In that situation, the search of crystalline solids is approached with a focused scientific rationale and conducted under narrower but better targeted conditions. In this presentation we present a case study where a focused scientific approach was utilized to identify and isolate a crystalline salt of a substance for which high throughput salt screening of hundreds of conditions only led to oiling out and/or amorphous solids in the. The presentation describes how a path to 1st crystals can be identified from learnings about the system behavior in terms of solubility and liquid-liquid phase separation (LLPS). The causes leading to oiling out and to the reluctance to crystal nucleation are discussed. In addition, an approach to perform a clarifying filtration when the solublities of both parent and salt do not permit a full dissolution is presented.

537696 Experimental Study and Results of Liquid Distribution ProfilesÂ in Large Scale Distillation Column with Structured Packing

Abstract:

This paper presents results of the study carried out in experimental large scale distillation column with diameter of 1.9m. Experiments were performed on the structured packing Mellapak 750.Y which is series of 25 elements with total height of 5.2m. The working fluid is HCFC225cb (1,3-dichloro-1,1,2,2,3-pentafluoropropane) which has low surface tension and viscosity. And the gas phase was nitrogen containing saturated HCFC225cb. The column has 33 compartments at the bottom which enable measurement of local liquid distribution profiles.

Measuring liquid distribution profile on bottom of cross-sectional area in column with wide range of gas and liquid condition, we observed the results that deviation of liquid distribution remained constant without maldistribution in the region below loading point. While in the region above loading point, steep increase of deviation of liquid distribution was observed.

548226 Understanding Distillation Curves: Crude Oils

Abstract:

Refineries process streams with thousands of components. Crude oil is one of these streams. Crude oil makes conventional analytic methods for defining precise chemical compositions of feeds extremely complex and time consuming. The refining industry has developed standardized methods for using batch distillation data from the laboratory for designing continuous distillation units. The normal procedure for these methods is to convert laboratory data to a true boiling point (TBP) distribution for the crude oil or other refinery stream.

Over time the conversion methods have evolved with better laboratory equipment and more data. Modern laboratory technique includes both physical distillations and simulated distillations using chromatography.

This paper includes:

- Review of laboratory physical distillations
- Review of laboratory simulated distillations
- Conversion of different laboratory techniques to true boiling point curves
  - Different operating pressures
  - Different number of stages and reflux
- Introduction to non-standard methods for physical distillations

531343 Optimal Design and Operation of Four-Product Dividing-Wall (Kaibel) Distillation Column

Abstract:

Optimal Design and Operation of Four-Product Dividing-Wall (Kaibel) Distillation Column

Abdallah Alshammari* & Farrukh Ilyas Abid

alshammari@kfupm.edu.sa

Due to the increased market competition, many chemical industries try to optimize their operations by maximizing productivity and minimizing operational cost, mainly energy consumption. Distillation operation alone accounts for a significant amount of the world's energy consumption since it is widely used in refineries and petrochemical industries. It has been estimated that approximately 3% of the world's energy consumption is utilized by this distillation operation alone [1]. This energy consumption can be greatly reduced by using an efficient complex distillation column configuration, such as a fully thermally coupled distillation column or divided-wall distillation column, over conventional arrangements. The divided-wall column (DWC) offers potential energy and capital investment savings. For multi-product separation, conventional distillation column configurations (direct, indirect and distributed sequences, etc.) require a greater number of heat exchangers, distillation columns, etc., compared with complex distillation column arrangements (side stripper, side rectifier, fully thermally coupled distillation column and divided-wall distillation column). Also, compared with conventional column configurations, complex distillation configurations lead to lower capital investment and energy consumption as well. However, design and operation remain a challenge due to the large number of design and operational degrees of freedom. In this study, it was shown that energy savings of up to 23% can be achieved by using a four-product divided-wall distillation column (Kaibel) compared with conventional column arrangements, Table 1. This work demonstrates the steady-state and dynamic behavior of a Kaibel distillation column using commercial simulation software i.e., Aspen HYSYS. Three different mixtures (Alcohols, Hydrocarbons, and Aromatics) are used to conduct the study and the comparison.

The optimal operation and control of dividing-wall distillation columns are more difficult than those of a simple distillation column. Thus, various industries have been reluctant to adopt this energy-efficient distillation column, as a greater number of operational degrees of freedom must be specified than for a conventional column. The distillate (D) and bottom (B) are assumed to be used as level controls for the condenser and reboiler, respectively. Thus, only six operational degrees of freedom remain for a Kaibel distillation column i.e., liquid split ratio (R_L), vapor split ratio (R_V), reflux flow rate (L), vapor flow rate (V) and the flow rates of the two side draw locations (S_1, S_2). The liquid split ratio (R_L) can be defined as the ratio of the amount of liquid introduced at the top of the prefractionator section to the overall amount of liquid sent to the main column as reflux, and the vapor split ratio (R_V) can be defined as the ratio of the amount of vapor sent to the bottom of the prefractionator section to the overall amount of vapor obtained at the bottom of the main column from the reboiler.

Therefore, it is important to operate the dividing-wall column at optimal values of the liquid and vapor split ratios, as energy efficiency will be lost if the distillation column is not operated at the optimum vapor split ratio (Figures 1-3). The main objective of this study is to examine the steady-state and dynamic behavior of a four-product Kaibel distillation column with the help of process simulation software; and it discusses the optimal operation parameters of DWC. Operation of the DWC was studied under two different modes:

i. To achieve the defined product purities by providing minimum energy.

ii. To achieve the maximum product purities with a fixed available energy value.

The first mode requires specifying all the required product purities, which should be obtained with minimum available energy. In the second mode, the available energy should be specified, and maximum product purities should be obtained with this available energy. The objective functions of the two modes can be defined as follows, respectively;

$J = V$ where product purities ≥ 0.95 **(Eq. 01)**

$J = D(1 - xD) + S1(1 - xS1) + S2(1 - xS2) + B(1 - xB)$ **(Eq. 02)**

The optimal operation is studied using both of the above-described modes. The dividing-wall distillation column is used to reduce energy consumption and to obtain nearly pure products by determining liquid and vapor split ratio.

Table 1: Energy comparison between conventional & complex configurations for alcohol case

Column Arrangements	Methanol	Ethanol	1-Propanol	1-Butanol	Heat Energy (kW)	% Energy Saving
Feed Properties	0.25	0.25	0.25	0.25	--	--
Direct Sequence	0.95	0.82	0.95	0.95	3055.2	22.1 %
Indirect Sequence	0.95	0.86	0.95	0.96	3095.5	23.1 %
Distributed Sequence	0.95	0.92	0.92	0.95	3041.5	21.7 %
Divided Wall Column	0.96	0.92	0.92	0.95	2381	---

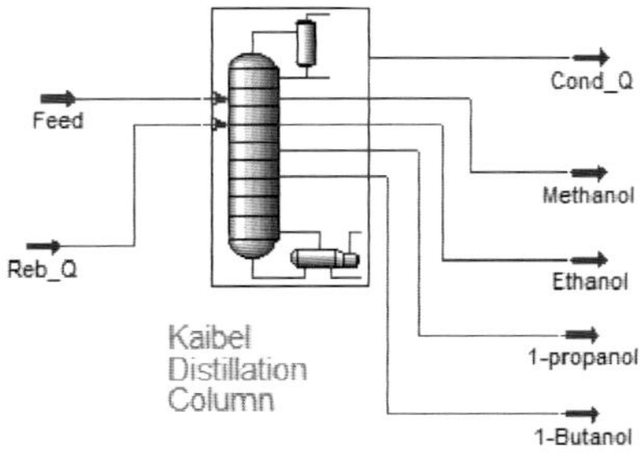

Figure 1: Four-product divided-wall (Kaibel) distillation column.

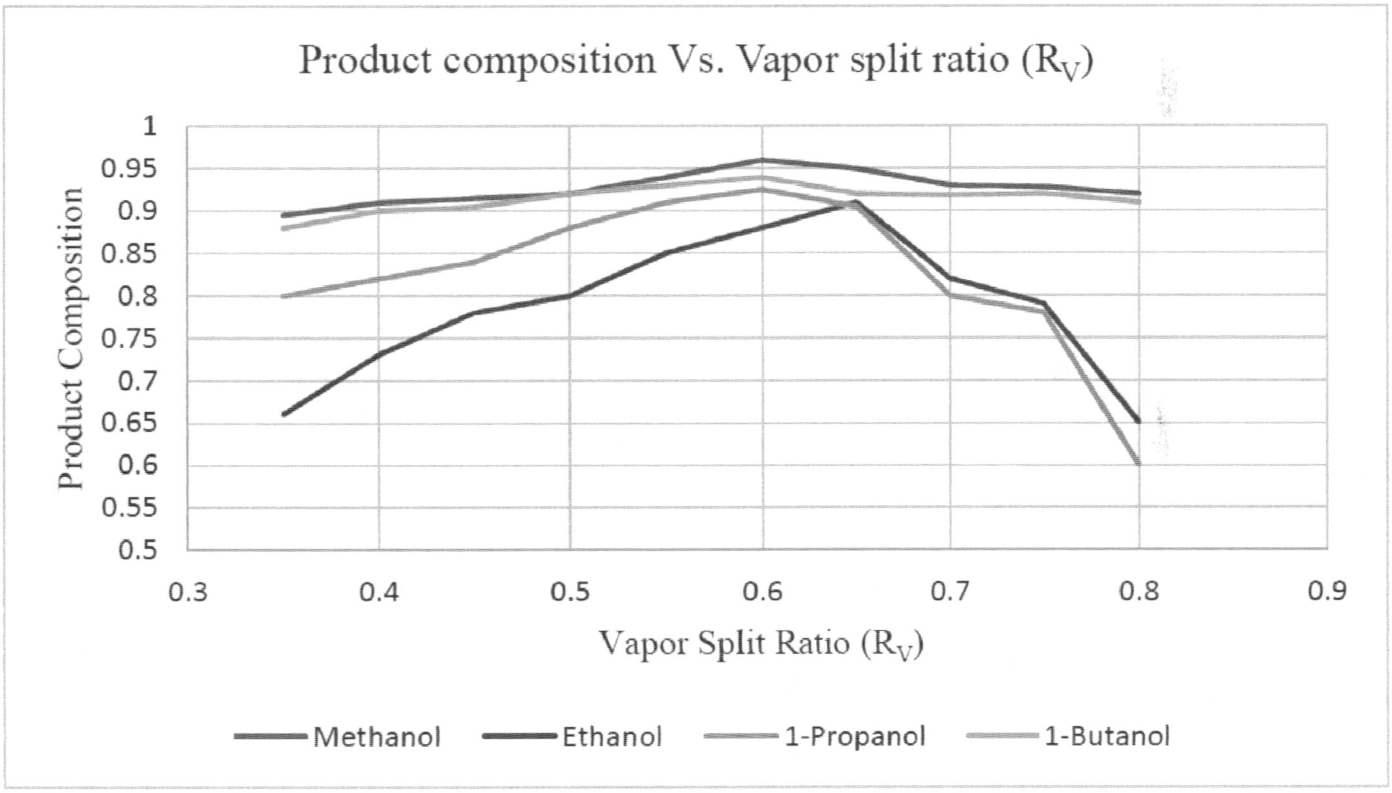

Figure 2: Effect of vapor split ratio on product composition

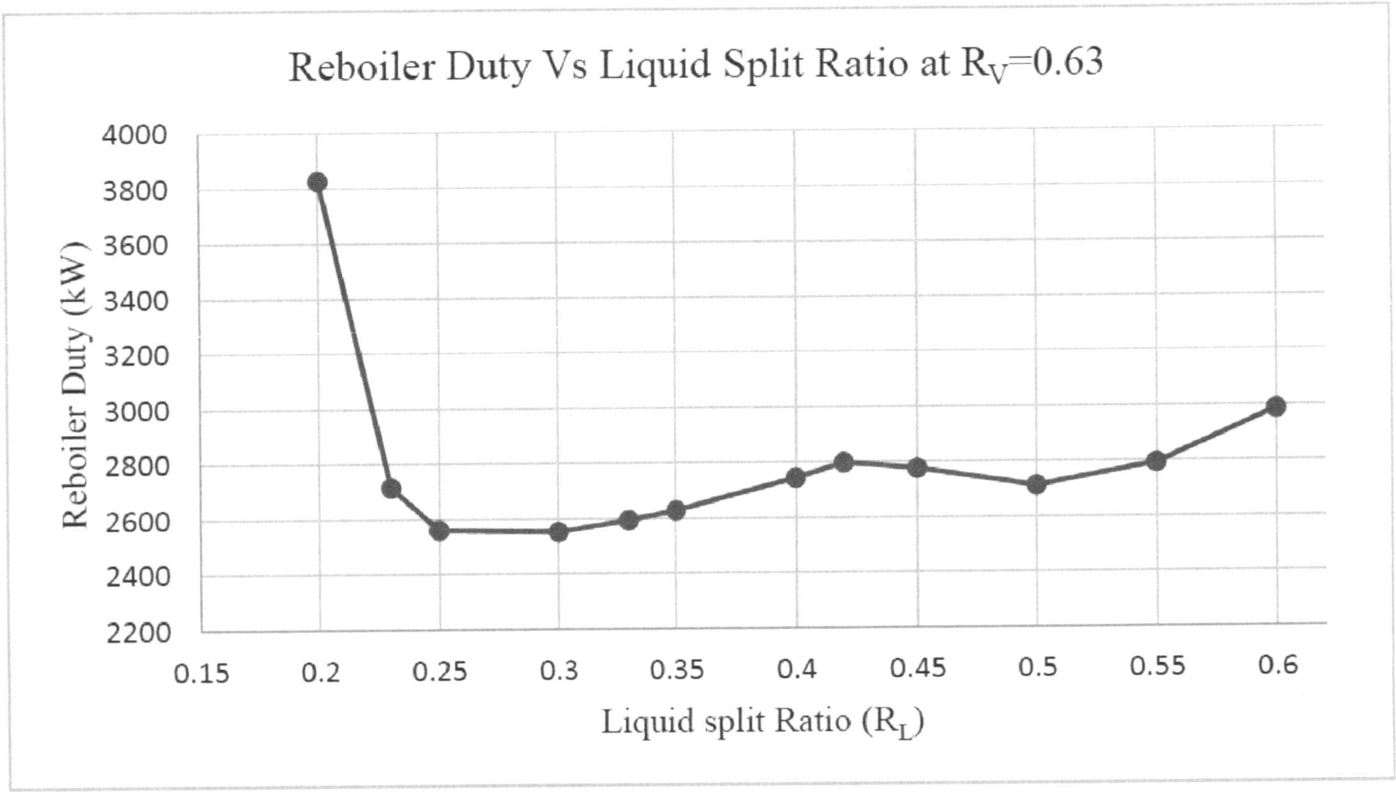

Figure 3: Effect of liquid split ratio on reboiler duty

[1] J. Humphrey and A. Siebert, "Separation technologies; An opportunity for energy savings," *Chem. Eng. Progress;(United States)*, 1992.

530315 Dual-Dividing Wall Column R&D and Deployment

Abstract:

Dual-Dividing Wall Column R&D and Deployment

Robert Piszczek, Michael L. Hergenrother, Sundar Narayanan, Rustom Billimoria

ExxonMobil Research and Engineering (EMRE) Company

EMRE discovered simplified design configurations for Multi-Partition Dividing Wall Columns (MP-DWC) that contain two dividing walls and are referred to as Dual-Dividing Wall Columns (DDWC). DDWC is the next generation of Dividing Wall Columns (DWC) technology that utilizes Process Intensification (PI) to produce up to six high purity products (>99 wt%) in one vessel at reduced CAPEX, OPEX, and plot space compared to multiple conventional columns and/or DWCs in series.

Multiple process configurations with DDWCs were developed and screened for a potential first commercial deployment for a new chemicals project. Numerous technical challenges were encountered, mitigated, and no showstoppers are anticipated.

Cost estimates indicate total erected cost (TEC) savings of 30% for a 6-Product DDWC compared to the next best alternative consisting of two DWCs and one conventional column in series. A risk mitigation strategy was developed to reduce the potential risk exposure for the first deployment of DDWC technology. As a result, a lower consequence 5-Product DDWC configuration with prorated CAPEX savings of 20% was selected as the lead option for further development, evaluation, and detailed design in parallel to the Stage 3 base process configuration for the project. Also, energy savings of up to 25% were projected but depend on each specific application and DDWC design philosophy.

Robert Piszczek, P.E. graduated from Rutgers University with B.S. and M.S. in chemical engineering. He has over 17 years of experience in process engineering in the areas of distillation (9 yrs), process control (6 yrs), gas treating and wet treating (2 yrs). He is currently on assignment as the Distillation Specialist for ExxonMobil Asia Pacific PTE LTD for Singapore.

533138 *Modeling of Liquid Distribution in Random Packed Columns*

Abstract:

The phenomenon of liquid maldistribution in random packed columns has long been known and has been the subject of numerous scientific investigations. In many cases, maldistribution affects the separation efficiency of packed columns and must be taken into account in the design. So far, more accurate predictions about the maldistribution can only be made based on experimental studies. A new approach in this field is the WelChem cell model [1]. The idea behind this approach is to accurately predict the maldistribution based on virtual irrigation experiments that require only the CAD data of a single packing element.

The validation and further development of this model is the aim of the project "Cell Model for Design of Packed Columns and Liquid Distributors" [2-4] funded by the Bavarian Research Foundation (AZ-1033-12). Within the project maldistribution is experimentally investigated by varying the experimental parameters column diameter, packing type, packing element nominal size, packing height, initial liquid distribution, liquid load and gas load. Most of the experiments are performed in a column operated with air/water with a diameter of 1.2 m and a packing height up to 3.0 m where the liquid distribution is measured in a liquid collector at the bottom of the column [5]. The experimental raw data gained in the project are also publicly available for research purposes (www.apt.mw.tum.de/maldistribution).

The original version of the WelChem cell model [1] predicts liquid distribution from cell to cell using directional dispersion coefficients independent of the liquid load. These are obtained from a virtual 3D irrigation experiment using the CAD model of a single packing element.

In order to map the influence of the liquid load in the WelChem cell model, which has not been considered previously, the original approach is extended by an additional distribution effect, which occurs in packing areas with a high local hold-up. The influence of the gas load is implemented locally via an amplification of the distribution effects coupled to column fluid dynamics. Furthermore, the dimensions of the unit cell for each packing type are defined uniformly as a function of the bulk density. The influence of the column wall is included by void cells randomly distributed in the outermost ring of the cell layer. These void cells have no distribution function and pass all liquid to the cell directly below.

Comparative calculations with experimental data show that the liquid distribution for the considered packing types can be predicted in the technically relevant operating range with very good agreement. The modified model is implemented in the column design software *TrayHeart* and can be applied to a wide range of random packing types.

[1] A. Wild and V. Engel: *Ein neuartiges Zellenmodell zur Maldistributions- und Verteilgütenbestimmung.* FVT-Meeting, Karlsruhe, Germany, **2007**.

[2] F. Hanuš, R. Hoffmann, S. Rehfeldt und H. Klein: *Experimentelle Untersuchung der Flüssigkeitsmaldistribution in Füllkörperkolonnen mit unterschiedlichen Durchmessern.* FVT-Meeting, Bremen, Germany **2015**.

[3] F. Hanuš, S. Rehfeldt and H. Klein, *Chem. Ing. Tech.* **2015**, 87, No. 8, 1063-1064. https://doi.org/10.1002/cite.201550041

[4] F. Hanuš, J. Solá, K. Hoffmann, S. Rehfeldt und H. Klein: *Experimentelle Untersuchung der Flüssigkeitsverteilung in Füllkörperpackungen mit verschiedenen Nenngrößen und Kolonnendurchmessern.* FVT-Meeting, Garmisch-Partenkirchen, Germany, **2016**.

[5] F. Hanusch, S. Rehfeldt and H. Klein, *Liquid Maldistribution in Random Packed Columns: Experimental Investigation of Influencing Factors. Chem. Ing. Tech.* **2017**, 89, No. 11, 1550-1560. https://doi.org/10.1002/cite.201700015

529719 Challenges and Opportunities of Nano-Aerosol Filtration Using Nanofiber Filter

Abstract:

Wallace Woon-Fong LEUNG

Chair Professor, Mechanical Engineering

The Hong Kong Polytechnic University, Hung Hom, Hong Kong

Fellow (AICHE, ASME, AFS, HKIE, HKAES)

Challenges and opportunities of nano-aerosol filtration using nanofiber filter

Abstract

Nano-aerosols/ultrafine particles (<100nm) in high concentration (50-500+ million/m^3) are found in polluted air both outdoors and indoors. They are from combustion of fossil fuel from vehicles, industries and power plants. Another source of these nano-aerosols is the photochemical reaction under sunlight between the emitted NOx and VOC from vehicles, resulting in smog and haze, which becomes problematic for most congested cities. Air-borne seasonal flu and epidemic viruses are also in the same size range. By their small sizes, these invisible killers can infiltrate readily into our bodies from breathing leading to health problems and diseases, from cardiovascular diseases to mental/behavioural disorders.

Filters made of nanofibers, with diameter 100-300nm, are very effective to capture these invisible killers. Despite their large specific surfaces are advantageous providing effective diffusion and interception capture, they can also induce high pressure drop. This presentation discusses various innovative nanofiber technologies being researched, developed, and some of which are being commercialized on filtration of nano-aerosols to achieve high capture efficiency, low pressure drop, and high storage capacity.

For light aerosol loading as used in respirator and facemasks, to achieve high capture efficiency for a given aerosol size, a certain specific amount of nanofibers per square meter of filter is required. When the same specific amount of nanofibers is divided into multiple layers, each separated by a permeable scrim material, the pressure drop of the entire filter is much reduced as compared to the case with all nanofibers in a single layer. This is because the effective flow pore is much reduced when many layers of randomly oriented nanofibers are lay over each other. When permeable material is introduced between thin layers of nanofibers, this introduces macro-pores, which provides a relief to flow in an otherwise strict micro-pore structure, thus relieving the pressure drop. Another approach is to charge electrostatically the nanofibers, which is also quite challenging.

For extended aerosol loading as used in cabin filter, vehicles and HVAC, other than high efficiency and low pressure drop, the aerosol storage capacity becomes important. In fact, a cake forms on the nanofiber filter surface relatively quickly. The transition from depth to surface/cake filtration has been studied experimentally, theoretically and with CFD model. It has been found that more aerosols are captured and deposited at the upstream end of the filter forming a skin layer, followed by plugging of pores in the skin layer by aerosol deposit, building of dendritic bridges above the plugged pores of the capillaries in the skin layer, and finally cake forming above the filter. The cake formed above a nanofiber filter is quite dense, resulting in higher pressure drop across the entire filter. It can be made less dense and more permeable when a microfiber filter with specific fiber diameter and packing is installed upstream of the nanofiber filter. The cake formed from such composite filter is rather complicated. This interesting technology has been proven both experimentally and by CFD modelling.

Depth filtration on submicron aerosols and nano-aerosols using nanofiber filter as function of face velocity, filter thickness, and fiber packing density has been well established. This includes low Peclet number (Pe<10) condition for which diffusion dominates over convection/interception capture. Most experiments were almost exclusively ran based on sodium chloride aerosols or using organic liquid droplets. An interesting challenge is to demonstrate on the diesel aerosols from emission with even lower Peclet number, Pe<1, corresponding to highly diffusion dominant.

Another interesting challenge is to clean the filter after it is being loaded with nano-aerosols, we have developed backpulse-and-backblow to clean the filter without breaking the nanofibers. The filter can be reuse for loading again until it needs to be cleaned. We have also attempted to understand the complicated loading behaviour of the reuse filter after it has being cleaned.

All-in-all, there are many interesting challenges in filtering nano-aerosols using nanofiber filter, but they also present as possible opportunities for development of break-through technologies.

<u>Biography of Wallace Woon-Fong LEUNG</u>

Dr. Wallace W-F Leung received his MS and ScD both from MIT. For the past 40 years, he has worked on filtration and separation from ultrafiltration of protein solution, centrifugation of process stream, vacuum filtration of minerals, to air filtration. For 18 years, he was with Bird/Baker Hughes as Senior Research Scientist and Director of Process Technologies directing centrifugal separation/filtration. Earlier, he has worked for Gulf Oil and Schlumberger on flow in petroleum reservoirs. For the past 13 years, he is Chair Professor of Innovative Products and Technologies at The Hong Kong Polytechnic University. Besides air filtration, he is also interested in applications of nanofiber technologies in renewable energy (photovoltaics, hydrogen production using water spliting), clean air and water (photocatalysis), and health technologies (wound dressing).

Dr. Leung is a fellow of American Institute of Chemical Engineers, American Society of Mechanical Engineers, American Filtration and Separations Society, Hong Kong Institute of Engineers, and Hong Kong Academy of Engineering Sciences. He has 49 US patents and two books to his credit. He is the Chairman for the World Filtration Congress in 2004 (New Orleans) and 2020 (San Diego). He is also Chairman for the International Delegation on Filtration 2016-2020 with 12 member countries, overseeing the world Filtration Congress and promoting filtration and separation activities worldwide.

531585 The Advancement and Challenges of Solid/Liquid Separation in the Process Industry

Abstract:

Solids/liquid separation is used in all parts of a chemical manufacturing process, from raw material purification, product separation, and all the way to waste handling. Although filtration and separation is nothing new to this industry, the growing emphasis on process efficiency, reliability and waste minimization has brought new challenges to this seems to be mature technology.

Solids/liquid separation covers a wide spectrum of technologies. In addition to the core filtration and separation, technologies in particle/liquid property characterization, agglomeration, flocculation, filter aid, slurry handling, and auxiliary equipment are all crucial parts of a solid/liquid separation process. Examples of challenges, new developments and future directions will be discussed to illustrate this vibrant and important field of chemical engineering.

535090 Membrane Technology in Bioprocessing -- Current ApplicationsÂ andÂ Future Opportunities

Abstract:

Membrane systems are used extensively for the purification of a wide range of important biological products. This presentation will review current applications of membrane technology in bioprocessing, highlighting recent developments that have enabled membrane processes to address critical challenges facing the biopharmaceutical industry in the production of a range of biological products. This includes the development of robust virus filtration membranes that are now a standard component of the overall virus clearance strategy. In addition, increasing demand for biopharmaceutical products and competition from biosimilars has generated interest in transitioning to continuous biopharmaceutical processes. Countercurrent staged membrane processes for continuous diafiltration have recently been explored that could provide new opportunities for the development of membrane technology for bioprocessing applications

538817 Replacing Phase Changing Separations: Revising Ammonia Manufacturing

Abstract:

With more than 150 million metric tons annual production, ammonia is the second most produced commodity chemical, and its production is the largest energy-consumer in the chemical industry. Ammonia production is complicated, and it is carried out at high temperature (>400 °C) and high pressure (150-300 bar.) High temperature is required to accelerate splitting the nitrogen triple bonds, while high pressure is required to ensure reasonable single-pass conversions. Ammonia separation relies on phase-changing separations, which requires high pressure processing for a cost-effective separation. High pressure processing imposes substantial operating (e.g., compression) and capital (advanced costly alloys, thick reactor casing, etc.) expenses. The gas mixture that leaves the reactor at ~450 °C should be cooled down to -20 °C in order to separate ammonia, necessitating significant heat exchanging. These are major issues that complicates ammonia manufacturing, which directly roots in the nature of phase-changing separation.

In this presentation I will discuss an innovative process for ammonia manufacturing: high temperature reaction-absorption process. In this process, we replaced the condenser with absorber column, packed with cheap abundant metal halides. These materials have extraordinary capacity for ammonia (e.g. 100 g of $MgCl_2$ can absorb up to six moles of ammonia), even at temperatures as high as 300 °C. This separation unit allows lower pressure operation. Unlike condensation which is a well-mixed unit operation, the proposed absorber column is considered a staged unit. It allows more complete separation of ammonia at lower pressure, unless breakthrough occurs. I will present our recent findings about the mechanism of reaction-absorption process and novel absorbent materials that we have developed for ammonia separation.

523850 Hydraulics of Taylor-Couette Disc Contactors

Abstract:

Hydraulics of Taylor-Couette Disc Contactors

Annika Grafschafter, Matthäus Siebenhofer

A global and sustainable implementation of renewable bioresources has become a high priority target nowadays. Downstream processing in the biobased industry often faces highly dilute mixtures making product isolation a challenge. In separation and isolation of value added products from highly dilute aqueous feed, liquid-liquid extraction has established as a basic unit operation. The Taylor-Couette Disc Contactor (TCDC) [1, 2] is a stirred liquid-liquid extraction column with a simple design of internals and thus suitable for applications in bioseparations. The internals of the TCDC are similar to the Rotating Disc Contactor (RDC), but without stator rings and with increased shaft and disc diameter. Without stator rings, crud accumulation and fouling along the active mixing zone of the column is avoided, making the TCDC particularly suitable for harsh operation conditions. The increase of the shaft diameter up to an optimum ratio of $D/D_{Sh} = 0.5$ prevents formations of hydrodynamic dead zones along the shaft and does induce banded two-phase flow as observed in Taylor-Couette reactors. Compared to Taylor-Couette reactors, the TCDC provides a much higher hydraulic capacity beyond 30 m^3 m^{-2} h^{-1}.

For successful design and operation performance of extraction columns, the comprehensive knowledge of hydraulics is needed. Therefore, data of the dispersed phase holdup (Ï†), drop size distribution (DSD) and residence time distribution (RTD) were determined in a 0.1 m and 0.3 m diameter pilot plant scale TCDC. For hydrodynamic characterization, the effect of the hydraulic load, rate of rotation and phase ratio was investigated. The minimum rate of rotation necessary for formation of fully developed toroidal vortexes in a single compartment can be predicted for specified hydraulic load. The influence of the hydraulic load on DSD is nearly negligible. The effect of the phase ratio on Ï† at low hydraulic load is insignificant but at high hydraulic load, the phase ratio has a strong impact on Ï†. Analysis of the RTD of the continuous phase suggests application of the tank-in-series model. The number of corresponding vessels in series was deduced from the maximum of the exit age distribution [3]. For given rate of rotation, the number of corresponding vessels in series increases with increasing hydraulic load. A reversed trend was observed for fixed hydraulic load and varying rate of rotation. Increasing rate of rotation contributes to backmixing. Summing up the operation experience in pilot scale Taylor-Couette Disc Contactors provide high load performance and flexibility in operation. Due to the simple design principle, it can withstand harsh operation conditions, even significant solids load in the feed.

[1] E. Aksamija, C. WeinlÃ¤nder, R. Sarzio, M. Siebenhofer, *Sep. Sci. Technol.* **2015**, 50 (18), 2844â€"2852. DOI: 10.1080/01496395.2015.1085406

[2] A. Grafschafter, M. Siebenhofer, *Chemie-Ingenieur-Technik* **2017**, 89 (4), DOI: 10.1002/cite.201600142

[3] O. Levenspiel, *Chemical Reaction Engineering*, Wiley, **1999**.

535124 Non-Invasive Analysis of Hydrodynamics and Mass-Transfer in Liquid-Liquid Extraction By Means of Computed Tomography

Abstract:

Flow phenomena inside packed liquid-liquid extraction columns are poorly understood due to numerous effects that influence droplet motion and restricted experimental access in such opaque systems. While single droplets were studied extensively with optical measurement techniques, these are not applicable in opaque packed columns or at high disperse hold-up. Advanced measurement techniques as X-ray computed tomography allow for non-invasive observations of fluid dynamics in those systems at high spatial resolution.

In this work fluid dynamic parameters such as drop size distribution and disperse holdâ€'up are investigated inside a packed liquid-liquid extraction column. Three different non-invasive measurement techniques are applied: X-ray radiography is used to measure disperse holdâ€'up at high temporal resolution (500 fps). The local hold-up is validated by integral hold-up values measured by the differential pressure method. Data on drop sizes, drop distribution and deformation is obtained from tomographic images with a high spatial resolution (80 µm) and a temporal resolution of 1 fps.

The experiments are conducted in a glass column with internal diameter of 50 mm and a packing height of 800 mm. The heavy organic phase (1,2-dichlorobenzene) is dispersed at the top of the column while the aqueous continuous phase flows in counter-current mode. In order to reduce the droplet velocity to the temporal resolution of the tomographic setup, viscosity is increased by adding polyethylene glycol.

The influence of a perforated and unperforated corrugated sheet packing on the above-mentioned hydrodynamic parameters is compared. In both packing geometries, droplets pass only a fraction of the cross-sectional area and preferred droplet paths can be identified. In contrast to absorption/distillation columns, disperse phase holdup is distributed uniformly inside the structured packing without accumulation at the packing intersections.

Beside the hydrodynamic investigation, we present a suitable test-system for nonâ€'invasive tomographic measurements of mass-transfer in liquid-liquid systems. High atomic numbers and densities of metal ions allow the tomographic measurement of ion concentrations. Therefore, it is possible to measure non-invasively local concentration profiles in both phases during metal liquid-liquid extraction.

540384 Influence of Surfactants on Mass Transfer and Fluid Dynamics in Disperse Multiphase Systems

Abstract:

< style="padding-left: 24pt; text-align: center;">Influence of surfactants on mass transfer and fluid dynamics in disperse multiphase systems

Joschka M. Schulz, Matthias Kraume

Technische Universität Berlin, Chair of Chemical and Process Engineering Ackerstraße 76, 13355 Berlin, Germany

THE GRAPHIC ON THIS PAGE WAS NOT AVAILABLE ON THE E-MEDIA VERSION OF THIS PAPER.

< style="padding-left: 5pt; text-align: left;">Figure 1: Experimental transient drop rise velocity of 2 mm 1-octanol droplets in water for different surfactant concentrations. Calculated velocities for the movable and rigid interface are shown for comparison (Feng and Michaelides 2001, Martin 1980).

< style="padding-left: 5pt; text-align: left;">References

M. Wegener, J. Grünig, J. Stüber, A. R. Paschedag, M. Kraume, Chem. Eng. Sci. 2007, 62, 2067-2078.

D. Merker, L. Böhm, M. Oßberger, P. Klüfers, M. Kraume, Chem. Eng. Technol., 2017, 40, 1391-1399. Z.-G. Feng, E. E. Michaelides, J. Fluids. Eng., 2001, 123, 841-849.

H. Martin, Chem. Ing. Techn., 1980, 52, 199-200.

M. Wegener, N. Paul, M. Kraume, Int. J. Heat Mass Transfer, 2014, 71, 475-495.

533549 Separation Efficiency and Design Optimization of Gravity Settlers â€“ CFD Modeling and Experimental Investigation

Abstract:

Fig. 1: Velocity of water (m/s) in the DN150 settler

Separation efficiency and design optimization of gravity settlers – CFD modeling and experimental investigation

E. Charlafti[1], J. Steinhoff[2], D. Leleu[3], L. Reinecke[4], A. Pfennig[3], H. -J. Bart[2], M. Kraume[1]

[1]TU Berlin, Berlin, Germany, [2]TU Kaiserslautern, Kaiserslautern, Germany, [3]Université de Liège, Liège, Belgium, [4]Franken Filtertechnik KG, Hürth, Germany

The separation of two-phase liquid/liquid dispersions through gravity settlers is a fundamental and widely implemented process, thanks to its simplicity and its cost and energy efficiency. The joint research project ERICAA ('Energy and resources saving through innovative and CFD-based design of liquid/liquid gravity separators'), funded by BMWi (Federal Ministry for Economic Affairs and Energy), aims to optimize design standards for gravity settlers used for industrial liquid/liquid separation applications.

The experimental approach of the project is based on a continuous mixer-settler setup, where the separation of paraffin oil-in-water dispersions is monitored. The separation efficiency is determined through precise measurements of the dispersed phase at the inlet and outlet of a standard DN150 gravity settler. In parallel to the continuous separation process, batch settling experiments are carried out with the help of an inline calibration device, which provides complementary data on the optimal residence time in the continuous settler, as well as on the dispersion characteristics[1].

Numerical simulations of the DN150 gravity settler based on computational fluid dynamics (CFD) are performed, not only to predict the separation efficiency of the continuous process in a more cost-sustainable way, but also to generate additional information about parameters which are not easily measurable, such as the velocity field inside the vessel. The results of the CFD simulations are compared to the experimental results and their validation is discussed.

Fig. 1: Velocity of water (m/s) in the DN150 settler

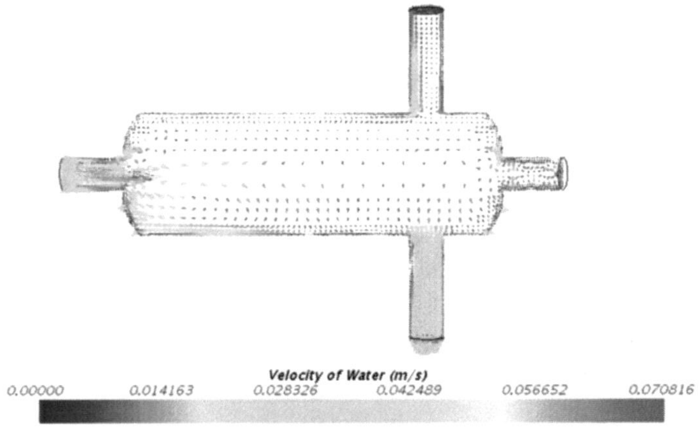

519168 Liquid-Liquid Equilibrium for Biodiesel-Glycerol-Methanol or Ethanol Systems Using Unifac Correlated Parameters

Abstract:

Liquid-liquid phase equilibrium (LLE) for the system of biodiesel, glycerol, methanol or ethanol is fundamental for modeling, simulation, design and economic evaluation of biodiesel production. This paper presents a correlated set of group interaction parameters (GIP) for the UNIFAC model that adequately describe the LLE of such biodiesel systems. To correlate and to verify the GIP, a database including 84 LLE experimental results reported in the literature was built. The database includes biodiesel produced using 23 different vegetable oils and 2 different alcohols (methanol and ethanol), as raw materials, at temperatures between 293.15 K and 353.15 K. The GIP were correlated using an evolutionary algorithm. The correlated set of GIP describes the LLE with an average deviation up to 0.7%, regardless of the type of biodiesel and the temperature (in the range studied). Model descriptions using the proposed set of GIP were compared with the description of others GIP reported in the literature. LLE descriptions using the new set of GIP proposed in this work exhibited the best fit between the experimental data and the calculated values for all the systems biodiesel-glycerol-methanol or ethanol evaluated.

515906 Optimization of the Extraction Parameters in a Millichannel-Based Packed and Unpacked Rectangular Serpentine Extraction Device

Abstract:

OPTIMIZATION OF THE EXTRACTION PARAMETERS IN A MILLICHANNEL-BASED PACKED AND UNPACKED RECTANGULAR SERPENTINE EXTRACTION DEVICE

Somen Mondal and Subrata Kumar Majumder*

Department of Chemical Engineering, Indian Institute of Technology Guwahati, Guwahati 781039, India, Email: somen.mondal@iitg.ernet.in; skmaju@iitg.ernet.in

Corresponding and presenting author's Email: skmaju@iitg.ernet.in

ABSTRACTS

The present study aims to investigate the effect of copper ion concentration in the solution (C_m), extractant concentration in the solvent (C_s), solution pH, mixture actual velocity of the phases (U_m) and temperature (T) on copper extraction process from a copper-laden solution. Copper is extracted from the aqueous phase to the organic phase using a commercially available extractant D2EHPA [Bis-(2-ethylhexyl) phosphate] dissolved in the organic phase (benzene). The present work introduced a rectangular serpentine packed millichannel as a novel extracting device. The main parameters for the copper extraction process were optimized for the flow through a rectangular serpentine channel and the effect of packing on the extraction efficiencies was explained in this present work. It is observed that pH is one of the most important parameters for the extraction process. At a pH = 9.26, C_m = 0.047M, C_e = 0.05M and the aqueous to organic phase ratio (A/O) = 1, the higher percentage of extraction (96.17%) as well higher extraction efficiency (96.32%) is obtained by the packed channel whereas the unpacked channel shows 86.81% extraction at an efficiency of 88.58%. After fixing the parameters of the extraction process at a particular value other than temperature, the percentage of extraction and efficiency for both the channels increases, but at a temperature 50^{0}C, the higher extraction (97.02%) for packed channel proved its benefit over the unpacked channel which showed 86.38% copper extraction with an efficiency of 88.28%. The equilibrium constant for the copper extraction by D2EHPA has also been incorporated in the study. It is seen that the K_e and the K_La values are quite satisfactory for the packed channel. Due to the large number of benefits like higher reaction rate, the higher mass transfer coefficient and also for the higher extraction efficiency

(98.89%), the rectangular packed channel device may be used as a highly efficient extraction or any other reaction devices. The performance study of the packed and unpacked channel reveals the fact that extraction efficiency of the packed channel is higher (85-99%) than the unpacked channel (80-91%). The present work may be helpful for further understanding about the separation and extraction of copper from a particle laden metal ion solution and also may be useful for the innovation of the new rectangular serpentine packed channel mass transfer devices in future.

Keywords: Serpentine millichannel; extraction parameters; packed and unpacked channel; equilibrium constant; volumetric mass transfer coefficient.

References

[1] Belkhouche NE, Didi MA, Villemin D. Separation of Nickel and Copper by solvent extraction using di 2 ethylhexylphosphoric acid based synergistic mixture. Solvent Extr. Ion Exch. 2005; 23: 677-693.

[2] Agarwal S, Ferreira AE, Santos SMC, Reis MTA, Ismael MRC, Correia MJN, Carvalho JMR. Separation and recovery of copper from zinc leach liquor by solvent extraction using Acorga M5640. Int. J. Miner. Process. 2010; 97: 85-91.

[3] Tang J, Steenari BM. Solvent extraction separation of copper and zinc from MSWI fly ash leachates. Waste Manage. 2015; 44: 147–154.

[4] Darekar M, Singh KK, Mukhopadhyay S, Shenoy KT. Single-stage micro-scale extraction: Studies with single microbore tubes and scale-up. Sep. Purif. Technol. 2016; 158: 160–170.

523951 My Membrane Research at National University of Singapore (NUS)

Abstract:

Clean water, clean air, renewable energy, global warming and affordable healthcare are five major concerns globally resulting from clean water shortages, severe haze in air, high fluctuations of oil prices, climate changes and high costs of healthcare. Clean water, clean air and public health are highly related, while renewable energy is essential for sustainable prosperity.

Among many potential solutions, advances in membrane technology are one of the most direct, effective and feasible approaches to solve these sophisticated issues. Membrane technology is a fully integrated science and engineering which consists of materials science and engineering, chemistry and chemical engineering, separation and purification phenomena, environmental science and sustainability, statistical mechanics-based molecular simulation, process and product design.

In this presentation, we will introduce our efforts on membrane development for clean water, clean air, clean energy and organic solvent nano-filtration at NUS.

527166 Novel Membranes, Membrane Processes and Membrane Devices

Abstract:

Neal Chung and his group have been extraordinarily productive in a number of areas of membrane research. These include a variety of membrane fabrication, membrane distillation, gas separation, mixed matrix membranes, pervaporation, organic solvent nanofiltration etc. I will describe some of our recent and earlier research in these and related areas. First, I will describe our very recent explorations on reverse osmosis separation of organic solvent mixtures using swelling-resistant hydrophobic membranes. I will also touch upon organic solvent nanofiltration. I will next describe our studies with a variety of novel membranes and membrane devices for highly productive performances in direct contact membrane distillation. These include both flat porous membranes and hollow fibers. I will next describe our earlier research on highly selective recovery of volatile biofuels from their aqueous solutions simulating dilute fermentation broths. I will conclude by describing our recent research on recovering various chemicals from their dilute solutions.

539858 Tuning the Interlayer Channels of GO Membranes for Molecule or Ion Transport

Abstract:

Graphene is a well-known two-dimensional (2D) material that exhibits preeminent electrical, mechanical and thermal properties owing to its unique one-atom-thick structure. Graphene and its derivatives (e.g., graphene oxide (GO)) have become emerging nano-building blocks for separation membranes featuring distinct laminar structures and tunable physicochemical properties. Extraordinary molecular separation properties for purifying water and gases have been demonstrated by graphene-based membranes, which have attracted a huge surge of interest during the past few years.[1, 2]

This talk will give an overview of our recent progresses on graphene-based membranes for molecular separation, with a focus on our attempts on the design and control of fast and selective channels within the membrane. We demonstrate a scalable fabrication of graphene oxide (GO) membranes on ceramic hollow fiber substrate and showed good pervaporation dehydration of aqueous organic solution.[3] Moreover, a novel bio-inspired strategy is proposed to utilize the synergistic effect of a hydrophilic polymer and GO laminates to realize fast water-transport channels for constructing high-efficiency membrane.[4] For gas separation, we design a novel type of membrane with fast and selective gas-transport channels of GO laminates enabled by polymer-GO hydrogen bonding,[5] showing excellent CO_2 permeation performance. Several strategies including the facile spray-evaporation and integrated external forces driven assembly approaches are proposed to precisely manipulate the GO membranes with highly ordered 2D microstructure for precise molecular gas separation.[6] Recent studies also show excellent ionic separation using GO-based membranes.[7] Future efforts should be taken to further develop high-quality graphene-based membranes with good structural stability for practical application.

References

- Liu, W. Jin, et al., *Chem. Soc. Rev.*, 44 (2015) 5016-5030.
- Liu, W. Jin, et al., *Angew. Chem. Int. Ed.*, 55(2016) 13384â€"13397.
- Huang, W. Jin, et al., *Angew. Chem. Int. Ed.*, 53 (2014) 6929-6932.
- Huang, W. Jin, et al., *Adv. Funct. Mater.*, 25 (2015) 5809â€"5815.
- Shen, W. Jin, et al., *Angew. Chem. Int. Ed.*, 54 (2015) 578-582.
- Shen, W. Jin, et al., *ACS Nano*, 10 (2016) 3398-3409.
- Chen, M. Wu., W. Jin, J. Li, H. Fang, et al., *Nature*, 550 (2017) 380-383.

538550 Catalytic Membranes for Biomass Hydrolysis and Dehydration

Abstract:

Biofuels and chemicals derived from lignocellulosic biomass are leading renewable energy candidates to replace fossil-based transportation fuels and chemicals. However, the economic viability of biproducts is significantly hindered by manufacturing costs. Our paradigm changing catalyst and membrane reactor is not only cheaper than cellulase enzyme (currently the workhorse for cellulose hydrolysis), but also, we combine three unit operations: thermochemical pretreatment, detoxification and enzymatic hydrolysis into a single unit operation.

A catalytic membrane reactor has been developed to simultaneously catalyze the hydrolysis of cellulosic biomass and separate the monomer sugars with improved sugar yields and efficiency. An enzyme mimic catalyst has been synthesized by growing poly(styrene sulfonic acid) (PSSA) chains as well as poly (vinyl imidazolium chloride) ionic liquid (PIL) chains from the surface of ceramic membranes. These polymeric solid acid catalysts are able to facilitate cellulose dissolution and catalyze its hydrolysis reaction. PSSA catalyzes biomass hydrolysis whereas PIL helps solubilize the biomass and enhance the catalytic activity. Our solid acid catalysts demonstrate over 97% and 32% TRS yields from crystalline Î±-cellulose hydrolysis in ionic liquid [EMIM]Cl and aqueous solutions respectively. Moreover, these catalysts are able to convert cellulose directly to 5-hydroxymethylfurfural (HMF) with over 60% yield. These catalysts are stable and maintain high catalytic activity after repeated runs.

Further we have shown that our solid acid catalyst can catalyze the production of sugars and platform chemicals (glucose, xylose, hydroxymethylfurfural (HMF) and levulinic acid) directly from the lignocellulosic biomass depending on the reaction temperature and solvent. Our catalyst is reusable, environmentally friendly and can be easily regenerated. Its catalytic activity can be easily tuned by altering the polymer chain length, chain density and the relative ratio of the two nanostructures. Moreover, it can be potentially used for many other acid catalyzed reactions.

536737 A Non-Thermal Process to Extract Aroma Compounds from Coffee Using Membranes

Abstract:

This work deals with a non-thermal process for extraction of aroma compounds from coffee solutions to prevent thermal degradation. Conventional soluble coffee processing involves extraction of soluble components from ground coffee, followed by thermal evaporation of water and drying of the concentrated extract. Unfortunately, the concentration step not only removes water, some coffee aromas are lost as well. While such techniques as gas stripping may be used to recover aromas prior to evaporation, but aroma components with high volatilities cannot be recovered, resulting in an undesired alteration in the flavor profile. Here we present a new processing concept to reduce thermal exposure of the coffee extract and its aromas based on an integrated process of nanofiltration (to concentrate coffee solubles without the use of heat) and low-temperature pervaporation (to recover aromas from the nanofiltration permeate). Nanofiltration is preferred over reverse osmosis for concentrating coffee extracts because of its higher flux, lower propensity for fouling, and its ability to attain a higher extract concentration. Since nanofiltration allows both water and aroma to permeate through the membrane, pervaporation is used to recover the aromas from the nanofiltrated coffee extract. Both continuous and batch pervaporation of multicomponent aroma solutions were carried out using organophilic membranes, and all the representative flavor compounds tested were shown to be preferentially permeable through the membranes. It was also shown that the membranes can be tailored to yield a balanced flavor profile in the aroma stream recovered.

526157 Advancing Water Remediation Technologies By Nanostructured Membranes (invited paper)

Abstract:

The contamination of aquifers by toxic metals and organic compounds is a widespread problem that prevents these potentially potable sources from being used for drinking and other water needs. Membranes provide flexible technologies, and various applications range from toxic pollutants removal to virus separations to desalination. Metal sorption with polypeptide (helix-coil transition) functionalized membranes is extremely high (for example, >1g metal/g for Pb) compared to conventional ion exchange (50 mg/g). The integration of pollutant degrading enzymes by layered assembly or metal/metal oxide (iron, or bimetallic) nano-particles in microfiltration type membrane pore domain allow catalytic detoxification (for ex, PCBs, TCE, TCP) of water through both reductive and oxidative pathways. Our research and technology development includes functionalization (green chemistry) of membrane pores with pH and temperature responsive polymers, and then subsequent ion exchange of metal ions (Fe (II)) followed by reduction allowed in-situ synthesis of non-aggregated, reactive metal particles in pore domain. We address the need for targeted remediation strategies by developing integrated, cost-effective technologies which incorporate both reductive and oxidative strategies in order to allow the complete remediation of chlorinated organic compounds without the production of toxic byproducts. This is accomplished through the development of a polymer/membrane platform in both lab-scale and full-scale for environmentally benign nanostructured iron synthesis and the establishment of individual and combined technology strategies to reduce the toxicity of chloro-organics. In addition, the presentation will include translational activities of functionalized membrane technology for trace metals separations from energy production related water. We acknowledge the funding by NIH-NIEHS-SRP, NSF EPSCoR program, and industries.

527837 Bioinspired Membranes and Membrane Processes

Abstract:

By mimicking the exceptional composition, structure, formation process and functions of natural materials, a variety of bioinspired membranes have been designed and fabricated using cell membrane, lotus, mussel, nacre, aquaporin as representative prototypes and biomineralization, bioadhesion, self-assembly as major methods. These membranes have exhibited fascinating properties and outstanding performances such as high permeability, multiple selectivity, superior stability. Meanwhile, these membranes have played critical roles in dealing with energy, environment and water stress. Bioinspiration, as the powerful platforms for sustainable innovation and exploitation of membrane science and technology, have great implications in developing novel membrane materials and intensifying diverse membrane processes. This presentation will give a brief overview about the preparations, applications as well as perspectives of bioinspired membranes.

530537 The Separation of Sulfide in Polluted Air from Molecular Simulation

Abstract:

The air pollution is serious problem to affect life and environment especial toxic gases. In this work, we employed molecular simulation method to study the separation capacity of nano microporous materials removing sulfide in polluted air. The studies show that the life-threatening concentration of hydrogen sulfide (H_2S) or sulfur dioxide (SO_2) is just higher than 100 ppm in air. There are 89 kinds of nano porous materials including MOFs and COFs were evaluated to capture H_2S and SO_2 from ratio of 79% nitrogen (N_2) and 21% oxygen (O_2) using grand canonical Monte Carlo (GCMC) simulation. The processes of pressure swing adsorption (PSA) and temperature swing adsorption (TSA) were simulated to obtain the selectivity, working capacity and selection parameter. The results show that UiO-66, ZIF-71, ZIF-69 and ZIF-97 are good adsorbents for separation of H_2S. Their selectivity and loading of H_2S are higher than 300 and 0.01 mmol/g, respectively; the working capacity of PSA is almost 0.06 mmol/g from 1 bar increasing to 10 bar; and the working capacity of TSA is about 0.008 mmol/g from 348 K decreasing to 298 K. For the separation of SO_2 the good candidates are UiO-66, ZIF-71, ZIF-69, MIL-100, Zn-DOBDC, ZnBDC, IRMOF-11and MIL-140B which have the higher selectivity of 500 and the loading of 0.06 mmol/g. Their working capacity of PSA and TSA is almost 0.2 mmol/g and 0.05 mmol/g at same condition as H_2S, respectively. At the same time the diffusivity of gases was investigated in good adsorbents with concentration changes of N_2 through molecular dynamics (MD) simulation. The diffusions of sulfide have no obviously change with decreasing of N_2. Our works provides useful computational insights for the separation of sulfide in governance of polluted air.

533704 Continuous Li-Mining from Secondary Resources Via Electrospun Nanofiber Membrane Adsorber with Lithium Ion Sieves

Abstract:

With unceasing technological developments, high demand for lithium (Li) has driven the need for its recovery from various sources such as mineral ores, brines, and seawater. Recent studies have shown great interests to recover Li from other sources such as Li-rich solid and liquid wastes. Thus, an efficient way of extracting lithium from these sources was developed for fast and high selective lithium recovery. Herein, lithium ion sieves (LIS) H_2TiO_3 embedded in electrospun polysulfone (PSf) nanofiber (NF) membrane was employed for continuous Li^+ mining from various samples, such as coal ash leachate and simulated wastewater. The LIS/PSf NF membrane was evaluated using different characterization techniques such as XRD, SEM-EDS, BET, capillary flow porometry, and mechanical testing. The performance of the membrane was thoroughly investigated through convective single- and multiple-pass permeation of the leachate via LIS/PSF NF adsorbent membrane filtration followed by Li^+ desorption and enrichment using mild acid solution. Adsorption isotherms, material recyclability and effects of competing ions were also examined.

This work suggests that the developed process is simple to operate and offers promising option for continuous Li^+ recovery from Li-enriched secondary resources. Overall, the results showed the ability of the prepared H_2TiO_3/PSf adsorbent nanofiber membrane to selectively, rapidly, and continuously mine Li in any type of Li-enriched sources and upgrade its potential for future industrial-scale application. This research was supported by the National Research Foundation of Korea (NRF) funded by the Ministry of Science and ICT (No. 2017R1A2B2002109 and No. 2017R1D1A1B03028102) and the Ministry of Education (No. 22A20130012051(BK21Plus)).

525014 Preparation of Amine Modified Bimodal Mesoporous Silica Particles for CO_2 Separation

Abstract:

Mesoporous silica has been extensively studied for many applications such as CO_2 adsorbents, drug delivery, catalysis, and so on. Among them, bimodal mesoporous silica (BMS), which had two different sizes of mesopores, have drawn much attention recently because it exhibited unique diffusion properties in the pores. For this reason, when BMS is modified to have affinity with CO_2, it is expected that high CO_2 adsorption capacity is obtained by small pores and fast adsorption rate is achieved by large pores.

Therefore, we prepared various kinds of amine modified BMS, and the effects of pore structure of BMS samples and the type of introduced amine on CO_2 adsorption/desorption properties were examined in this study.

In the experiments, BMS were prepared by co-condensation of two different silica precursors. Tetraethylorthosilicate (TEOS) and triblock copolymer Plulonic 123 (P123) were used as silica precursors and templating agent, respectively. Polymethylhydrosiloxane (PMHS) was not only used as silica precursors but also as template for large pores. First, various BMS samples were prepared with different PMHS concentration, and the pore structures of the resulting samples were investigated.

The results showed that three types of mesoporous silica with different pore structures (monomodal and bimodal) were successfully prepared and it was found that the pore diameter and the pore volume could be controlled by changing PMHS concentrations. To introduce amine groups to the BMS samples, 3-Aminopropyltriethoxysilane (APTES), 3-(2-Aminoethoxyl amino)propyltriethoxysilane (AEAPTES), 3-[2-(2-Aminoethylamino)ethylamino]propyltrimethoxysilane (AEAEAPTES) were used.

It is known that when conventional monomodal mesoporous silica is modified with AEAEAPTES, although introduced amount of amine is large, amine utilization efficiency is low because long chain length of the moiety plugs the pores and diffusion of CO_2 in the pores is inhibited. On the other hand, our experimental results showed that although the samples were modified with AEAEAPTES which had longest amino group, and large amount of amino groups were introduced, the amino group utilization efficiencies for CO_2 capture within short period adsorption (30 min.) were still high. This is probably due to better pore structure of our samples compared to conventional mesoporous sample.

In conclusion, in this study, preparation procedure of bimodal mesoporous silica (BMS) with different pore structure was established. Then, the effect of pore structure on CO_2 adsorption characteristics was examined using the resulting BMS samples after modifying with various amino groups. The results showed that by applying BMS for CO_2 adsorption, it was found that high CO_2 adsorption capacity was achieved by small pores and fast adsorption rate was achieved by large pores.

534050 Ash Modified with Surface Active Agents for the Adsorption of Chloro/Nitro Benzenes from Aqueous Phase

Abstract:

Boiler fly ash, collected from ICI soda ash plant, Pakistan, was modified by two different ionic surfactants namely hexadecyltrimethyl ammonium bromide (HDTMA) and sodium dodecyl sulfate (SDS), separately. Surfactant treated ash was characterized by SEM, TGA, and FTIR, to explore the surface morphology, thermal stability, and the presence of functional groups on the fly ash surface, respectively. The two modified forms of fly ash were also experimented to test their efficiency for the removal of two non-polar organic pollutants chloro- and nitrobenzene from aqueous phase. Different parameters like effect of adsorbent dose, pH of solution, temperature, adsorbent interaction time and initial conc. of solution were also studied. The analysis results showed SDS modified fly ash is superior than HDTMA modified fly ash. The maximum uptake capacity of chlorobenzene and nitrobenzene with SDS-modified fly ash (SFA) is 200 mg/g and 70mg/g respectively. The adsorption of chlorobenzene is exothermic over both adsorbents while nitrobenzene showed the endothermic behavior. The isotherm analysis depicted the multilayer adsorption of both pollutants onto the surface of surfactant modified ash.

538479 Synthesis of Zeolite X from Rice Husk Ash

Abstract:

Zeolites are materials of high industrial importance, derived from their molecular sieving and ion exchange capacity, and their catalytic properties. These materials have attracted great interest because they can be produced with different crystalline structures, with uniform pore sizes and channels, high surface area, high thermal stability and an excellent adsorption ability. In recent decades, the synthesis of zeolite has been investigated, trying to exploit more environmentally friendly raw materials under moderate processing conditions. Specifically, some biobased materials have been considered as alternative feedstock for the synthesis of zeolites, in particular the residues from biomass combustion processes.

Due to the high silica content in Rice husk ashes (nearly 90% amorphous silica) this material is suited as feedstock for zeolite production. This residue is obtained from rice drying and parboiling ovens, and it is commonly disposed in cement kilns or mixed directly into concretes as a partial cement replacement. However, production of zeolite materials would be a better value-added alternative, suitable for exploitation in Latin-America and Asia where rice is a major agricultural commodity.

Among the different possible zeolite structures that can be produced with rice husk ashes, zeolite X is of special interest given its molecular sieving capacity, and the high adsorptive selectivity in the separation of paraffin-olefin mixtures. In this regard, this work focused in the use of rice husk ashes (RHA) obtained from a Colombian milling facility (waste product of the parboiling process), as a raw material for the synthesis of zeolite X. Initially, the raw ashes with a SiO_2 content of 80% were calcined to remove the unburned carbonaceous material, reaching a 91% SiO2 content. Then, the calcined solids were processed under alkaline conditions to obtain sodium silicate. The resulting mixture was filtrated and characterized via Atomic absorption spectroscopy to determine the sodium and silicon content. In order to obtain a high content of silicon in solution, the effect of the reaction temperature and the NaOH/SiO2 ratio were assessed. Once sodium silicate was obtained, different formulations for the zeolite synthesis were evaluated to optimize yield and crystallinity degree (determined with X ray diffraction). Under the optimized conditions, a zeolite x material with a surface area of 870 ±4.6 m^2/g was obtained. Finally, the adsorption capacity of the material was evaluated using methane, nitrogen and carbon dioxide at 25 °C and 1 atm.

539036 Synthesis and Adsorption Kinetics of Hierarchical 5A Zeolites

Abstract:

Based on the molecular scale management, double object optimization of ethylene and aromatics yield can be achieved through separating normal paraffins from naphtha using shape-selective absorption 5A zeolites. In order to improve the diffusivity and adsorption separation efficiency, the micro/mesopore structure hierarchical 5A zeolites was synthesized, and the performance was studied.

The hierarchical 5A zeolites was synthesized by hydrothermal method with the dimethyl octadecyl [3-(trimethoxysilyl)propyl] ammonium chloride (TPOAC) as the soft template. Micropores with the diameter of 0.5 nm were connected with 7-13nm mesopores. The amount of mesopores could be controlled by the template dosage. The optimal synthesis conditions wereSiO_2 : Na_2O : Al_2O_3 : H_2O: TPOAC = 100 : 333 : 67: 20000 :0.1(mol), crystallization at 100â„ƒ for 6h, washed by methanol to pH=8, exchanged by calcium ion. The mesopore volume was 0.0674 cm^3/g, which was 26.7 % of total pore volume. The adsorption capacity for normal hexane was 0.1183 g/g.

The liquid phase adsorption kinetics of normal paraffins in the hierarchical 5A zeolites was investigated. When the template dosage is 0.1%, the liquid phase diffusion coefficients of n-paraffins in the hierarchical 5A zeolites were 1.8-4.8 times larger than those in the micropore 5A zeolites. Diffusion coefficients of n-paraffins in the hierarchical 5A zeolites increased with the template dosage. The diffusion activation energies of n-pentane, n-hexane and n-heptane increased with carbon numbers in the chains.

The molding of hierarchical 5A zeolites was studied. The adsorption capacity of 1mm bar-type hierarchical 5A zeolites for n-hexane was 0.1048g/g, and diffusion coefficient was 2.34 times larger than that in 5A micropore zeolites.

545959 Cr-, Fe-, and Ga-Doped CaO Adsorbents for High Temperature CO_2 Capture: An Adsorption and In-Situ XRD Study

Abstract:

Calcium oxide is an efficient adsorbent for high temperature CO_2 capture process, however, it suffers from rapid deactivation and capacity loss after a few cycles as a result of particles sintering sintering. Metal doping is an effective strategy to address the durability issue of CaO. In this investigation, we report development of novel metal-doped CaO adsorbents with high capture capacity, fast kinetics, and long-term stability. In particular, Cr, Fe, and Ga with carried concentration were used as promoters to improve the adsorption performance of CaO adsorbent. The doped adsorbents comprising of 1% Cr@CaO, 10% Fe@CaO, and 10% Ga@CaO exhibited the highest adsorption capacities of 12.3, 13.7, and 14.2 mmol/g, respectively at 650 °C which were at least two folds higher than that of the bare CaO. Moreover, the doped-CaO sorbents showed reversible performance by desorbing almost all of the adsorbed CO_2 during desorption step at the same temperature. Through *in-situ* XRD measurements, it was shown that the carbonation under CO_2 and desorption under N_2 flow take place resulting in efficient CO_2 capture and adsorbent regeneration. The cyclic adsorption-desorption runs demonstrated the excellent stability of the materials by retaining 95% of their initial capacity after 10 cycles. Moreover, adsorption temperature was found to have a favorable impact on CO_2 uptake over the doped adsorbents. The findings of this study highlight the importance of metal-doped CaO adsorbents as promising candidates for use in calcium looping or combined capture-utilization processes.

530682 Development of Polymeric Ionic Liquid Thin Films for Ion-Selective Anion Exchange Membranes in Electrodialysis Separations

Abstract:

The onset of climate change and rising global population has caused greater water scarcity, a considerable problem which threatens our world. Ion exchange membranes have been used for water purification since 1950s in a process called Electrodialysis (ED). ED is an electrically-driven separation process aimed at water treatment through ion removal, which is achieved through the selective control and transport of ionic species. ED is a popular water treatment method; however, ED is limited by membrane scaling, lack of ion selectivity, and high energy consumption for high salinity feeds. Recent studies suggest that ion-selective ion exchange membranes can enhance selective ion separations in ED systems, resulting improved productivity for water treatment. This study is focused on the synthesis of unique ionic liquid-based anion exchange membranes in order to obtain ion-selectivity in ED systems through addition of a divalent ion repulsion layer. Thin films of polymerizable ionic liquids 1-ethyl-3-methylimidazolium bis(trifluoromethane)sulfonimide and trioctyl-(4-vinylbenzyl)phosphonium bis(trifluoromethane)sulfonimide were coated and cured on the anion exchange membrane in order to enhance ion selectivity. The membrane characteristics were studied through FTIR, SEM, AFM, contact angle measurement, and ED performance. Results suggest that while imidazolium coatings were incompatible with ED systems due to rapid degradation, phosphonium coatings exhibited enhanced ion selectivity with negligible dissolution of selective layer and comparable separation efficiency to uncoated membranes. However, as current selective layers are thick (~150 Î¼m), ED productivity decreased, indicating that thinner films would be more beneficial.

539062 Zwitterionic Interactions with Charge Mosaic Membranes Prepared Via Electrohydrodynamic Jet Printing

Abstract:

Charge mosaic membranes, which possess an array of anionic and cationic charged functionalities, can more readily transport ions compared to neutral solutes. Using inkjet printing of reactive solutions, a patterned charge mosaic membrane can be produced in a scalable manner. These membranes have an overall neutral surface charge, but within each patterned stripe exhibit a positive or negative charge, which are susceptible to fouling. Attractive dispersive, electrostatic, and hydrodynamic interactions between dissolved solutes and the membrane leads to the attachment and accumulation of material on the surface of the membrane, which reduces the throughput and performance. One of many emerging techniques for reducing fouling is the use of polyzwitterions. Zwitterions contain an equal number of anionic and cationic groups on the molecular chain, with a net neutral charge. These units form a strong hydration layer, thereby increasing the free energy of adsorption for foulants. This study evaluates novel multi-functionalized charge mosaic membranes that contain a zwitterionic anti-fouling layer.

Membranes cast from a poly(trifluoroethyl methacrylate-*co*-oligo-(ethylene glycol) methyl ether methacrylate-*co*-(3-azido-2-hydroxypropyl methacrylate) P(TFEMA-OEGMA-AHPMA) copolymer are the parent substrates utilized in this study. The azido groups lining the pore wall of the parent substrate lead to multiple functionalization avenues. A polyzwitterionic nature is attained through the controlled deposition of propiolic acid onto the membrane via electrohydrodynamic jet printing, in which an applied voltage drives fluid flow from the nozzle onto the substrate. The penetration depth of the zwitterionic anti-fouling layer, which is deposited first, is a function of the volume of reactive ink dispensed on the membrane. Reactions with 3-(dimethylamino)-1-propylamine and 1,2-propane-sultone form the zwitterionic functionality. Charge patterning was accomplished through the selective deposition of alkynyl-terminated reactants on the membrane, penetrating past the zwitterionic functionalities, creating cationic and anionic domains traversing the remaining membrane thickness. Energy dispersive x-ray spectroscopy (EDS) was used to develop elemental maps locating the presence of functional elements within the zwitterionic layer and absent from the charge-mosaic layer. Assessing the performance of these membranes was done through measuring the permeance, ionic transport, and fouling performance. Bovine serum albumin was used as the model foulant and was tested in both aqueous and salt solutions to measure permeability drop over time, along with ion permeance, to quantify the performance of these membranes.

The development of multi-functionalized membranes offers the ability to combine the attractive performance characteristics of multiple chemistries into a single membrane system. Zwitterions have been shown to possess high resistance to irreversible fouling but have shown limited ionic interactions. The incorporation of zwitterions into charge-mosaic membranes possess the ability to reduce the extent of fouling while maintaining the inherent ionic transport properties of charge mosaic membranes.

535125 Formation of Activated Carbon/Polymer Bilayer Membranes By Solution Electrospraying for Water Purification

Abstract:

Hybrid membrane processes utilizing activated carbon treatment and membrane filtration have recently attracted attention from researchers due to their overall improved performance over each individual step. However, hybrid processes still possess weaknesses associated with both activated carbon adsorption and membrane separation, such as fouling and attrition. We present a novel method where activated carbon/polymer bilayer membranes are fabricated, inspired by hybrid membrane processes, as a single filtration unit by solution electrospraying. This method is expected to potentially increase throughput by forming surface pores on otherwise dense membranes, while adding adsorption capacity with a layer of embedded activated carbon. The fabrication process, membranesâ€™ properties, and their water purification performance are described.

536653 Integrated Electrocoagulation-Ultrafiltration System for Treating Poultry Processing Wastewater

Abstract:

Industrial water reuse is considered as necessary for substantial reduction in water supply needs and savings in related costs. In 2016, the poultry industry within the United States processed approximately 8.9 billion broilers, using an average of 26.5 L of water per bird and produced over 60 billion gallons of poultry processing wastewater (PPW) requiring treatment. PPW contains proteins, fats, carbohydrates, feather, blood and skin, resulting in much higher biochemical oxygen demand (BOD) and chemical oxygen demand (COD) compared to municipal wastewater. Dissolved air floatation (DAF) is the most popular treatment method currently utilized in US poultry processors with approximately 80% of slaughter plants employing the technology. While in most cases, the DAF-treated PPW is sent to a biological wastewater treatment plant for further removal of contaminants and discharge into environment, at least a portion of PPW could be recycled and reused.

Ultrafiltration (UF) has been widely used in the food processing industry for the past 20 years. UF is a low-pressure and cost-effective option in terms of higher permeate flux compared to competing technologies (e.g. NF and RO) for treating highly impaired wastewaters. Similar to all membrane separation processes, UF suffers from membrane fouling, which causes a dramatic drop in the efficiency of the water recovery, due to accumulation of separated particles both on the membrane surface and inside the membrane pores. Here, we focus on electrocoagulation (EC) as a pretreatment step prior to UF regarding fouling mitigation. In EC, sacrificial electrodes are utilized to release coagulant counter ions into solution using electricity and aid in physical separation of suspended solids and organic macromolecules from the feed solution.

We report significant membrane fouling of the UF membrane due to high levels of oil and grease (O&G), proteins and suspended solids in the feed water that result in rapid flux decline. However, we show that the membrane fouling can be greatly mitigated using EC. We show that for an EC reaction time of 5 min, equivalent to 0.15 kWh m^{-3} energy consumption for EC, followed by 6 h sedimentation, over 85% reduction of O&G, BOD and suspended solids results. This reduction leads to higher initial permeate flux as well as lower flux decline during UF. In addition, higher contaminate rejection is obtained in EC-UF process compared to individual UF. Finally, we show that intermittent cleaning can increase the water recovery to a great extent. Here, the recovered water volume was increased by over 30% in a week of experimental time when employing membrane cleaning after each 24 h.

534676 Determination of Binodal Curves and Tie Lines for Aqueous Two-Phase Systems with Osmolytes for Bioseparations

Abstract:

Aqueous two-phase system (ATPS) is considered a potential alternative to currently used expensive methods in the biopharmaceutical industry such as chromatography and filtration. ATPS, formed by two incompatible aqueous solutions, provide inexpensive, biocompatible, mild, and environment-friendly advantages over the conventional methods, and can be easily adapted to continuous processing. However, the lack of understanding of the driving mechanisms of biomolecules poses a major challenge for industrial implementation. Our recent investigation for the recovery of porcine parvovirus (PPV) in a PEG 12kDa-citrate system indicates that hydrophobicity of the viral particles and the system play a vital role in the virus partitioning. The partitioning was affected by the salting-out effect from the citrate-rich phase and hydrophobic interaction from the PEG-rich phase.To build on the hydrophobicity theory, we incorporated osmolytes in the system. Osmolytes are naturally occuring compounds that have a high affinity for water. Common osmolytes include sugars and amino acids. Glycine has shown to preferentially flocculate relatively hydrophobic viruses in a complex lysis mixture. Using this property of osmolytes, we hypothesized that osmolytes would increase the hydrophobic driving force in ATPS. The driving force can be examined by generating the binodal curves and tie lines. The binodal curve separates the homogeneous region from the two-phase region and the tie line represents the systems separating into two-phases having the same PEG-rich and citrate-rich phase compositions. We have determined the binodal curves for PEG 12kDa-citrate by including 0.5M glycine or betaine (trimethyl glycine) at pH 7. The binodal curves shifted towards the origin as compared to the system without osmolytes, suggesting an elevated driving force to form two phases.Tie lines were determined by conductivity measurements of the citrate-rich phase. The osmolyte concentrations were determined in the citrate-rich phase by LC-MS. The tie line slope of the glycine system was found to be greater than the betaine system. This demonstrated that any two-phase system formed with glycine will have higher concentration of citrate and lower concentration of PEG in their respective phases as compared to betaine. The osmolytes were found to be in the citrate-rich phase. The occurrence of the osmolytes in the citrate-rich phase will help in salting-out the virus from the citrate-rich phase, providing more separation driving force within the ATPS.

536856 Imidazolium Based Poly(ionic liquids), the Tunable Membranes Having Antimicrobial Activity

Abstract:

Due to the favorable properties like low vapour pressure, large electric potential window, high degree of chemical and thermal stability, wide liquidus range, high solubilizing capability of inorganic, organic and polymeric substances; room temperature ionic liquid finds a wide range application in the field of separation, catalysis, synthesis, electrochemistry and even in energy applications and hence they are world-wide considered as potential 'green' alternatives to highly volatile organic solvents. The most attractive property of ionic liquid is the high degree of tenability. A very small structural modification can bring significant changes in the physico-chemical properties of the ionic liquid and hence the suitable ionic liquid can be designed based on desired properties and applications, provided the structure-activity relationship is understood properly. Introduction of functional groups onto the ionic liquid has also been reported to be used for task specific applications. The presence of ionic liquid was also demonstrated to show some unusual chemistry, which otherwise is not observed in presence of conventional solvents. For example; the extraordinarily high extraction efficiency of metal ion in presence of ionic liquid was attributed to the predominance of 'ion exchange' mechanism, which otherwise proceeds through 'solvation' mechanism for conventional molecular diluents. During 'ion exchange' mechanism, the separation takes place at the cost of transfer of cationic or anionic part of ionic liquid. Hence, with time we are always going to loose the amount of ionic liquid. Ionic liquid being expensive, loosing of significant amount during large scale industrial applications is not worth. Moreover, usually the high viscosity coefficient of the ionic liquid makes the mass transfer kinetics slower during the biphasic separation.

In view of these, poly(ionic liquid) membranes are the solutions to address such issues. The poly(ionic liquid) (PIL) membranes are the polymeric membranes containing repeated units of ionic liquid and is a nice blend of membrane properties as well as ionic liquid properties. Being membrane, it is a solid polymeric boundary layer used for separation, hence the problems associated with ionic liquid loss and high viscosity become irrelevant. The PIL membranes may be obtained from polymerizable cationic part of ionic liquid or anionic part of ionic liquid or both. The combination of the unique properties of ILs with the macromolecular architecture in PIL membranes is not only fascinating but also provide way to creating new properties and functions in these novel materials in the field of membrane technology and advanced materials for energy applications. The enhanced mechanical stability, durability, spatial controllability over the IL species and improved processability are the major advantages of PIL membranes over ionic liquids.

A simple two step synthetic methodology was optimized for the preparation of three polyionic liquid membranes from imidazolium family. These PIL membranes were found to be hydrophilic, while with increase in polymerization time was found to reduce the hydrophilicity of the membranes. The FTIR spectroscopic analysis indicated the presence of polyvinyl moieties, polystyrene moieties, polyvinyl imidazolium moieties and nitrile group in these PIL membranes. The detailed structural characterization of these membranes were also done by ^{1}H and ^{13}C NMR spectroscopy in terms of magnetically nonequivalent protons and carbons. At acidic pH (below pH 7), the surface charge of these membrane was found to be positive and it appeared zero near pH ~ 7, while alkaline pH showed negative charge on the membrane surface. The AFM and SEM imaging revealed that 15 min of polymerization led to fabrication of microfiltration as well as ultrafiltration membranes. These PIL membranes were found to have appreciable stability in presence of polar-protic; polar-aprotic and apolar-aprotic solvents, indicating its suitability in a wide variation of solvent applications. The plate counting method revealed that, in presence of PIL membranes, the number of colonies of *staphylococcus aureus* and *pseudomonas aeruginosa* reduced significantly and also with time. In case of plate reader, the ratio of live to dead bacterial cell was also found to reduce significantly in presence of PIL membranes. This antibacterial activity of PIL membranes were found to depend on the substituents present in the imidazolium ring. Finally, these PIL membranes were used for water flux measurement. After precompaction; membranes from allybromide and bromohexane showed water flux of ~982 LMH and 948 LMH (0.5 Psi), respectively; while that for chlorobutane PIL membrane was 315 LMH (1.5 Psi). Rejection in BSA and dextran revealed the tenability of the membrane's property by different substituents on imidazolium ring and polymerization time.

The direct plate counting method was employed to count the colony forming unit of bacteria per mL at different time of contact with PIL membranes. In this present investigation, two gram positive bacteria; *staphylococcus aureus* and *pseudomonas aeruginosa* were used. The initial CFU/ mL was kept sufficiently high to observe significant reduction in antibacterial activity. The blank did not contain any PIL membranes (but same amount of CFU/mL). In blank, there is no appreciable change in the CFU/mL value upto 24 hours, while for PIL membranes, 5 h of contact led to significant reduction in the number of bacterial colony. This clearly indicated the anti-bacterial activity of the PIL membranes. It was also to be noted, the number of colonies after 5 hours of contact with different PILs are different, indicating difference in the antimicrobial activity of the PIL membranes. Though, all these PIL membranes are from imidazolium family, the nature of substituents on imidazolium ion played significant role in their antibacterial activity. After 24 hours, CFU/mL was found to decrease further. Though the PIL membranes showed similar trend in antibacterial activity for both the gram positive bacteria; *staphylococcus aureus* and *pseudomonas aeruginosa,* yet the extent of antibacterial activity was found to be slightly different. The PIL membrane derived from chlorobutane was found to have maximum antibacterial activity compared to that of bromohexane followed by the allylbromide derived PIL membrane.

The direct plate reading method is based on the preferential tagging of two different dyes on dead or live cells of the bacteria. Since these two dyes have their own characteristic photo luminescence emission, monitoring the intensity of their characteristic wavelengths gives a quantitative measurement of the ratio of live bacterial cell to dead bacterial cell. Both the gram positive bacteria *staphylococcus aureus* and *pseudomonas aeruginosa* were investigated in present study to confirm the antibacterial activity of these PIL membranes. The live to dead cell ratio of both the bacteria was monitored at each hour interval up to 25 hours. The initial L/D ratio values for *staphylococcus aureus* and *pseudomonas aeruginosa* were kept 65 and 58, respectively. For blank, the L/D ratio values for both the bacteria were not found to change drastically, although a minor variation was still seen. For PIL membranes, the L/D ratio values were found to decrease drastically with time. This investigation confirmed the antibacterial activity of these PIL membranes. This investigation also revealed that, the antibacterial activity of the PIL membranes from allylbromide was lesser compared to the other two, while other two PIL membranes showed almost similar antibacterial activity. Though both plate counting and plate reading method almost revealed the similar antibacterial activity, yet some variation was still present.

535042 Synthesis and Characterization of Novel Sulfonated Amine Block Copolymers for Direct Methanol Fuel Cells

Abstract:

Direct methanol fuel cells (DMFC) have continued to gain traction over past years as a promising alternative source of energy. The efficiency of the DMFC depends on the composition of the proton exchange membrane (PEM) used. Although Nafion is currently the state of the art PEM, it exhibits some disadvantages such as permeability to fuels and high cost. In order to overcome these limitations, there is a considerable interest in synthesizing PEMs based on aromatic polymers because of their availability and easy modification.

In this work, atom transfer radical polymerization (ATRP) was used to synthesize styrene-based amine block copolymers for DMFC applications. The block copolymer was then sulfonated by chemical grafting with pendants sulfobutyl groups onto the polymer structure. This method has been shown to generate PEMs with better proton conductivity and water absorption, than PEMs with sulfonic groups bonded on aromatic rings. Properties of the resulting membranes were determined as a function of amine block composition and sulfonation level. A series of materials characterization techniques (*e.g.,* FT-IR, AFM, SAXS) was utilized to understand the interaction between sulfonic and amine groups in the polymer matrix and their effect on the morphology and transport properties of the membranes for fuel cells.

526718 Elucidating the Effects of Asymmetric Charge Patterning on Ion Transport through Charge Mosaic Membranes

Abstract:

Charge mosaic membranes consist of cationic and anionic domains that traverse the membrane thickness. Due to their charge-patterned structure, these membranes are capable of transporting ionic solutes more rapidly than neutral particles of comparable or smaller sizes. To date, the majority of the understanding that has been developed regarding the novel transport properties of charge mosaic membranes has been generated using 1:1 monovalent salts (e.g., KCl, NaCl) permeating through mosaic membranes with equal coverage of cationic and anionic domains. The ability of mosaic membranes to separate asymmetric, multivalent salts (e.g., $MgCl_2$, Na_2SO_4), however, has not been investigated in detail due to limited control over membrane properties provided by existing patterning processes. Here, using a recently-developed inkjet printing process that enables the ready fabrication of charge mosaic membranes with controlled patterns, we investigate the throughput of asymmetric salts by adjusting the areal coverage of the charge-functionalized domains. First, parent nanofiltration membranes based on a poly(acrylonitrile-*co*-[oligo(ethylene glycol) methyl ether methacrylate]-*co*-[3-azido-2-hydroxypropylmethacrylate]) [P(AN-OEGMA-AHPMA)] copolymer were prepared. These parent membranes possess pore walls lined by reactive azido moieties that made them amenable to post-synthetic modification via a printing device. Next, charge patterning was accomplished through the selective deposition of alkynyl-terminated reactants on the surface of the azido-functionalized parent membranes. In this way, cationic and anionic domains in patterns of alternating stripes were generated via the copper(I)-catalyzed alkyne-azide cycloaddition (CuAAC) reaction mechanism. The use of the striped pattern allowed asymmetric charge-patterns to be generated by implementing different values of the stripe thickness. Single solute rejection experiments using magnesium chloride and sodium sulfate as model solute were executed using these membranes with asymmetric patterns. The fundamental knowledge developed by studying charge patterned membranes with controlled areal coverages of the charge-functionalized domains will enable further development of charge mosaic membranes that can be deployed in the many established and emerging technologies where the selective transport of ionic solutes is of critical importance such as water softening and protein purification.

527830 Fundamental Pure and Mixed Liquid Sorption Properties of Osn Membranes Based on Polybenzimidazoles

Abstract:

Organic solvent nanofiltration is attracting increasing interest as it could potentially replace distillation. As reported by Marchetti et al.[1], for a given separation, OSN would help save up to 3 orders of magnitude of energy compared to distillation. Moreover, since most of chemical synthesis occur in solution, solvent recovery and solute concentration is a common issue in the chemical, petrochemical and pharmaceutical industry. Although flux and rejection data are available in the literature for several materials, fundamental structure-transport property correlations for OSN membranes are still missing, with the specific role of solvent and solute sorption and diffusion coefficients being fairly unknown. Moreover, most of available experimental data refer to composite membranes, where the presence of a fabric backing makes it difficult to provide a fundamental description of solvent and solute sorption and transport in the active layer. This lack of fundamental information has hampered the development of rational methods for the design of better materials for OSN.

In this study, free standing, mechanically robust flat membranes based on Celazole® were cast from DMAc[2]. Pure and mixed solvent and solute sorption experiments were carried out at room temperature. Among solvents, hydrocarbons, alcohols, acetone, toluene, ethyl acetate and water were considered. N-decane and PEG400 were used as mimic standard solutes. Interestingly, molar solubility decreases by an order of magnitude with increasing penetrant molar volume, which points out that liquids sorption in Celazole® is controlled essentially by entropic factors. Indeed, as long as penetrant size increases, it is less likely to accommodate penetrant molecules into the polymer matrix[3]. However, this rapid decrease is followed by a plateau, where sorption changes little with penetrant chemical nature and size.

When considering sorption of isomers, solubility starts to be affected by enthalpic effects. Specifically, 1-propanol solubility is 5 times higher than 2-propanol solubility. Such behavior can be justified by considering that, while 1-propanol can easily form hydrogen bonds with the amine groups on the polymer backbone, 2-propanol is less prone to do so, since the â€“OH group is much more sterically screened.

Finally, the effect of the second component on solvent and solute solubility in CelazoleÂ® was also investigated using model mixtures, such as methanol-PEG400.

References

1. P. Marchetti et al., Chem. Rev. 2014 114 10735-10806
2. H. Borjigin et al., Polymer 2015 71 135-142
3. M. Galizia et al., J. Membr. Sci. 2012 405-406 201-211

530540 Effects of Regiochemistry on the Properties and Gas Separation Performances of Ionic Polyimides

Abstract:

A bis(imidazole) monomer synthesized from 6-FDA and 1-(3-aminopropyl) imidazole can be used to form "ionic polyimides" through condensation polymerizations with benzylic dihalides such as dichloroxylenes, and subsequent ion-exchange with lithium bistriflimide ($LiTf_2N$). Our synthetic techniques have yielded ionenes with M_N values typically > 75,000 and as high as 100,000 amu. Ionic polyimides can also absorb and incorporates ionic liquids (ILs) within their structures which can have distinct effects on ordering the ionic polyimide. This work explores how the regiochemistry of the xylene linkage (i.e., para, meta, ortho) affects polymer properties and their performances as gas separation membranes for the separation of CO_2 from species such as H_2, CH_4 and N_2.

532525 Improved Gas Separation Performance of Mixed-Linker Zeolitic Imidazolate Framework ZIF Membranes Via Post Synthetic Ligand Exchange

Abstract:

Zeolitic-imidazolate frameworks (ZIFs), a subclass of MOFs, consisting of transition metals (Zn, Co, etc.) and imidazole-derived ligands, have drawn attention mainly due to their ultra-micropores (< 5 Å) and relatively high thermal/chemical stabilities.[1] A prototypical ZIF, ZIF-8, consisting of Zn and 2-methylimidazole with sodalite (SOD) zeolitic topology, is one of the most widely studied ZIFs mainly due to its robust synthesis and potential for gas separations due to its well-defined pore size. ZIF-7, consisting of Zn and benzimidazole with SOD, exhibits smaller pore size (ca. 3.2 Å)[2] than ZIF-8 (ca. 4.0 Å)[3], therefore promising for helium or hydrogen separation.[4]

Properties of MOFs including gas transport properties are affected by structures of MOFs. To obtain MOFs with new properties, one can obtain via 1) direct (*de novo*) synthesis of new MOFs and 2) post-synthetic modifications (PSMs) of existing MOFs. PSMs include post-synthetic ligand and metal exchange[5]. Compared to the complex and sensitive nature of MOF formation process via *de novo* synthesis,[6] PSMs are a simple and facile approach to impart new properties since daughter MOFs with new properties can be obtained without compromising the structures of mother MOFs. Recent studies have shown that PSMs can modify pore size and surface properties, thereby separation performances.[7-9]

In this talk, we would like to present improved gas separation performances of ZIF-8 membranes via post-synthetic ligand exchange by systematically reducing the effective aperture size of ZIF-8. Mother ZIF-8 membranes prepared by a secondary growth were subjected to post-synthetic linker exchange with ZIF-7 linkers, i.e., benzimidazole. The effects of the PSM on the gas separation performance and water stability of the modified membranes will be discussed.

References

1 Park KS, Ni Z, Côté AP, et al., (2006) "Exceptional chemical and thermal stability of zeolitic imidazolate frameworks", *Proceedings of the National Academy of Sciences,* 103 (27), pp 10186-10191

2 Thompson, J. A., Blad, C. R., Brunelli, N. A., Lydon, M. E., Lively, R. P., Jones, C. W., Nair, S., "Hybrid Zeolitic Imidazolate Frameworks: Controlling Framework Porosity and Functionality by Mixed-Linker Synthesis", *Chemistry of Materials*, 2012, 24 (10), pp 1930–1936

3 Zhang C, Lively RP, Zhang K, Johnson JR, Karvan O, Koros WJ., (2012) "Unexpected Molecular Sieving Properties of Zeolitic Imidazolate Framework-8", *The Journal of Physical Chemistry Letters*., 3 (16), pp 2130-2134.

4 Yan Shuo Li., YS., Liang, FY., Bux, H., Feldhoff, A., Yang, WS., Caro, J., "Molecular Sieve Membrane: Supported Metal–Organic Framework with High Hydrogen Selectivity", *Angewandte Chemie International Edition*., 2010, 122, pp 558-561

5 Tanabe, KK. and Cohen, SM., (2011), "Postsynthetic modification of metal-organic frameworks-a progress report", *Chemical Society Reviews,* 40 (2), pp 498-519.

6 Kim, M., Cahill, JF., Fei, H., Prather, KA., and Cohen, SM., (2012) "Postsynthetic Ligand and Cation Exchange in Robust Metal–Organic Frameworks", *Journal of the American Chemical Society,* 134 (43), pp 18082-18088

7 Lee, M. J., Kwon, H. T., Jeong, HK., "High Flux Zeolitic Imidazolate Framework Membranes for Propylene/Propane Separation by Postsynthetic Linker Exchange", *Angewandte Chemie International Edition*, 2018, 57, pp 156-161

8 Karagiaridi, O., Lalonde, M. B., Bury, W., Sarjeant, A. A., Farha, O. M., Hupp, J. T., "Opening ZIF-8: A Catalytically Active Zeolitic Imidazolate Framework of Sodalite Topology with Unsubstituted Linkers", *Journal of the American Chemical Society*, 2012, 134 (45), pp 18790-18796

9 Zhang, H., James, J., Zhao, M., Yao, Y., Zhang, Y., Zhang, B., Lin, Y.S., "Improving hydrostability of ZIF-8 membranes via surface ligand exchange", *Journal of Membran*

533258 Evaluation of Desorption and Diffusion in Zeolite Membrane with Nano-Perm Porometry

Abstract:

Zeolite is an aluminosilicate crystal having a pore structure of molecular size. It is expected that zeolite membranes can be applied to separation of small molecule mixtures based on the differences of molecular size and/or adsorption. It is important to evaluate the adsorption, diffusion and desorption of molecules to understand the mechanism of permeation through zeolite membrane. Though there have been a lot of reports on adsorption in zeolite, methods for evaluating diffusion and desorption in zeolite membrane have not yet been established. In this study, we aimed to evaluate diffusion and desorption of hydrocarbon adsorbed on silicalite-1 membrane by using Nano-Perm Porometry.

Silicalite-1 membrane was synthesized by a hydrothermal synthesis method after application of seed crystals by dip coating on Î±-alumina tubular support (outer diameter 10 mm, length 30 mm, and average pore diameter 150 nm). Silicalite-1 membrane was set in a module and heated to a given temperature while flowing He in a Nano-Perm Porometer. Hydrocarbon, which was condensable in zeolite pore, was supplied at different relative pressures with He. Then, after stopping hydrocarbon supply, we measured the He permeance every 1 second. Hydrocarbons used were *n*-hexane and 3-methylpentane. The measurement temperature was 373 - 473 K.

He permeance through silicalite-1 membrane was about 3.0×10^{-7} mol m^{-2} s^{-1} Pa^{-1} after pretreatment. The He permeance with *n*-hexane or 3-methylpentane substantially decreased to less than 4.0×10^{-8} mol m^{-2} s^{-1} Pa^{-1}. Clearly, hydrocarbon adsorbed in the pore of silicalite-1 membrane and inhibited the permeation of He. After the supply of hydrocarbon stopped, the He permeance recovered to about 3.0×10^{-7} mol m^{-2} s^{-1} Pa^{-1}. This result showed that hydrocarbon diffused and desorbed from silicalite-1 membrane and thereby He was able to permeate through silicalite-1 membrane. We found that the recovery rate of the He permeance became faster as the membrane temperature increased. With increasing membrane temperature, the adsorption equilibrium of hydrocarbon was shifted to desorption and its diffusion in the micropore increased. These results indicated that this method can evaluate differences of hydrocarbon diffusion and desorption rate in zeolite membrane. Moreover, the recovery rate of He permeance was extremely slow when we used 3-methylpentane at 373 K, suggesting that diffusion and/or desorption of 3-methylpentane in silicalite-1 membrane was extremely slow. In summary, this method can evaluate the rates of diffusion and desorption, possibly leading to the understanding of diffusion of hydrocarbon and its desorption in zeolite membrane.

533624 Membrane Surface Modification Using Acrylate- and Thiol-Containing Zwitterionic Materials Via Polydopamine

Abstract:

Fouling is one of the major challenges for the use of membrane technology for water purification. One effective way to mitigate the fouling is to chemically modify membrane surface to reduce its favorable interactions with foulants, such as grafting with superhydrophilic zwitterions. However, zwitterionic materials are water-soluble, and it is challenging to graft or coat zwitterions for long-term underwater operation. In this work, we demonstrate facile covalent grafting of zwitterionic materials (functionalized by acrylate and thiol) on the surface of ultrafiltration (UF) membranes using dopamine. Specifically, dopamine is deposited with sulfobetaine methacrylate (SBMA) or a thiol-containing zwitterionic polymer (p(MPC_{160}-*co*-$DTMAEL_{42}$, or PMD). In the presence of oxygen, dopamine forms polydopamine (PDA) adhering onto the membrane surface and covalently grafts SBMA or PMD via Michael addition to form a robust thin superhydrophilic layer, as confirmed by contact angle measurement and X-ray photoelectron spectroscopy (XPS). Interestingly, the coating does not completely cover the pores (as evidenced by the SEM) and thus, the modified membranes still exhibit reasonably high water permeance. The antifouling properties of the membranes were determined using a crossflow system and bovine serum albumin (BSA) as a model foulant. The modified UF membrane with SBMA and dopamine demonstrates up to 30% higher water flux than the uncoated one when tested with 1 g/L BSA. Moreover, the membrane modified with PMD and dopamine exhibits less flux decline (38% reduction) than the unmodified membrane (53% reduction). The facile approach of membrane modification employing SBMA/dopamine is also adapted for postâ€'modification of a commercial nanofiltration (NF) membrane module, which demonstrates enhanced antifouling properties when tested with real wastewater at a wastewater treatment plant.

533833 Effect of Sulfonated Graphene Oxide Nanofiller on the Performance and Properties of Poly(vinyl alcohol) Thin Film Composite Forward Osmosis Membrane

Abstract:

Forward osmosis (FO) is a very promising technology for desalination due to its noteworthy energy efficiency. However, the lack of effective draw solutes and highly efficient membranes impede FO industrialization. In FO membrane fabrication, thin film composites (TFC) with asymmetric structures have been commonly used as the support layer can significantly enhance the mechanical stability of the membrane. On the support, formation of an extremely thin selectively polyamide layer can be formed resulting in TFC FO membranes with high water flux and reverse salt selectivity. To further enhance the FO performance of TFC membranes, this study examines the feasibility of poly(vinyl alcohol) (PVA) as a support layer prepared via electrospinning. Meanwhile, the selective polyamide layer was formed on PVA through interfacial polymerization. To further enhance the mechanical stability and hydrophilicity of PVA, hydrophilic fillers such as sulfonated graphene oxide (sGO) was used. The sGO filler has high specific surface area and is rich with oxygenous and sulfonate groups which can contribute to the enhancement of water permeability and mechanical stability of the TFC membrane. The sGO was initially dispersed via sonication in the PVA dope solution prior electrospinning.

Initial results reveal enhanced hydrophilicity of PVA support in the presence of sGO (i.e. sGO/PVA) as indicated by its lower contact angle relative to the pure PVA support layer (control). Furthermore, tensile strength and Youngâ€™s modulus of sGO/PVA support were improved in comparison with the control sample. The successful polymerization of thin polyamide layer at the surface of sGO/PVA was confirmed through Scanning Electron Microscopy (SEM-EDS) and Fourier Transform Infrared spectroscopy (FTIR) analyses. Meanwhile, membrane performance of sGO/PVATFC FO membrane is currently being explored.

This research was supported by the National Research Foundation of Korea (NRF) under the Ministry of Science and ICT(2016R1A2B1009221) and Ministry of Education (No. 22A20130012051(BK21Plus)).

534045 Thermo-Responsive Ionic Liquids with LCST-Type Phase Transition Property As Draw Solutes in Forward Osmosis for Seawater Desalination

Abstract:

Desalination technologies have attracted significant attention owing to their immense potential to solve water crisis. Forward osmosis (FO) has emerged as one of the most promising technology in desalination. FO is an innovative membrane separation technology that relies solely on concentration gradient differences between two solutions separated by a semi-permeable membrane. However, the full implementation of FO remains highly constrained by the lack of ideal draw solutes imparting substantial water permeation, low reverse solute leakage, and convenient recyclability for long-term use.

Ionic liquids (ILs) are coined as green substances due to their low volatility and high thermal stability. In addition, ILs have high ionicity, moderate to high molecular weight, and moderate diffusivity, potentially permitting them to be suitable drawing agents in FO system. ILs having suitable hydrophobicity can undergo low critical solution temperature (LCST)-type phase transition in water. Herein, ILs exhibiting LCST-type phase transition in IL/water mixtures were synthesized. The synthesized ILs were thoroughly characterized by FT-NMR, FTIR, and TGA. The LCST property of the ILs was tested by UV-Vis absorption spectrum. Osmotic pressure, ionic strength, and diffusion coefficient measurement were conducted and later correlated with the ILs FO performance. Moreover, the capability of LCST-type ILs as osmotic pressure generators were compared with NaCl as a control in FO system using commercially available HTI-CTA membranes. The performance of the LCST-type ILs was evaluated in terms of water flux (J_v), reverse solute flux (J_s) and specific solute selectivity (J_s/J_v) during FO runs against deionized (DI) water and simulated sea water (0.6M) in both pressure retarded osmosis (PRO) and FO modes. The effect of membrane orientation, concentration of draws solutes and the respective FO performances were thoroughly investigated.

The investigated ILs generated considerable water flux with negligible solute leakage. This was attributed to their ionic strength, low diffusivity, and low affinity towards the FO membrane. All draw solutes had performed better in PRO mode attributed to lower concentration polarization effects relative to the FO mode. Over all, results demonstrated that the performance of LCST-type ILs is superior over the extensively studied NaCl attributed to their low specific solute selectivity during FO runs. These ILs can undergo phase transition upon mild heating leading to an IL-rich phase that can be recycled and re-used. The ability of these LCST-type ILs to phase separate from water would be beneficial for its convenient recovery and recyclability for long-term use in FO desalination technology.

This research was supported by the National Research Foundation of Korea (NRF) under the Ministry of Science and ICT (No. 2016R1A2B1009221) and the Ministry of Education (No. 22A20130012051(BK21Plus)).

534114 Hybrid Zeolitic-Imidazolate Frameworks (ZIFs) Membranes with Tunable Gas Separations

Abstract:

Metal organic frameworks (MOFs) are emerging organic-inorganic crystalline microporous solids for gas separation application owing to their tunable pore structure and functionality with diverse metal nodes and organic linkers.[1] Zeolitic-imidazolate frameworks (ZIFs),[2] a subclass of MOFs, in particular have attracted many attentions due to their chemical/thermal stabilities, their ultra-microporosities, and high surface area when compared to other MOFs material.[2] ZIFs possess zeolite-like topologies in which the tetrahedral Si or Al and the bridging O in zeolite structure are replaced by transition metals (such as Zn,[2] Co,[3] and Cd[4]) and imidazolate-derived ligands. A common drawback for gas separation is the limited availability of aperture pore size and functionalities of the porous materials. Furthermore, the common approach to fabricate ZIF polycrystalline membranes are lengthy requiring up to weeks of synthesis.[5-11] Studies have shown that through mixing metals[12,13] and linkers,[14,15] one can continuously tune the MOFs/ZIFs (termed hybrid ZIFs) properties to match with the characteristic of specific gas mixture. However, no study have shown the tuning effect of the hybrid ZIFs grown as polycrystalline membrane.

Here, we plan to present a rapid microwave-assisted in situ synthesis of well-intergrown mixed linker ZIF-7-8 membranes in under â^¼90 s.[16] ZIF-7-8 consists of Zn^{2+} metal nodes bridged by a mixture of benzimidazolate (bIm, ZIF-7 linker) and 2-methylimidazolate (mIm, ZIF-8 linker) linkers.[15] To the best of our knowledge, this is the fastest synthesis of any polycrystalline MOF membranes reported up to now. Furthermore, the gas separation performances (separation factor and permeance) of hybrid ZIF-7-8 membranes were systematically tuned by varying the bIm-to-mIm ratio incorporated into the framework. This is attributed to the fact that an increase in bulky benzimidazolate linker incorporation reduces the effective aperture size of the hybrid framework, consequently improving the separation factor at the expense of reduction in permeance. The unprecedentedly rapid synthesis of ZIF-7-8 membranes with tunable molecular sieving properties is an important step forward for their practical applications in the separation of gas mixtures of interest.

References:

1. Li, J.-R.; Kuppler, R. J.; Zhou, H.-C. Selective gas adsorption and separation in metal-organic frameworks. *Chem. Soc. Rev.* **2009**, *38*, 1477-1504.
2. Park, K. S.; Ni, Z.; Cote, A. P.; Choi, J. Y.; Huang, R.; Uribe-Romo, F. J.; Chae, H. K.; O'Keeffe, M.; Yaghi, O. M. Exceptional chemical and thermal stability of zeolitic imidazolate frameworks. *Proc. Natl. Acad. Sci. U. S. A.* **2006**, *103*, 10186-10191.
3. Banerjee, R.; Phan, A.; Wang, B.; Knobler, C.; Furukawa, H.; O'Keeffe, M.; Yaghi, O. M. High-Throughput Synthesis of Zeolitic Imidazolate Frameworks and Application to CO2 Capture. *Science* **2008**, *319*, 939-943.
4. Karagiaridi, O.; Bury, W.; Sarjeant, A. A.; Stern, C. L.; Farha, O. K.; Hupp, J. T. Synthesis and characterization of isostructural cadmium zeolitic imidazolate frameworks via solvent-assisted linker exchange. *Chem. Sci.* **2012**, *3*, 3256-3260.
5. Pan, Y.; Lai, Z. Sharp separation of C2/C3 hydrocarbon mixtures by zeolitic imidazolate framework-8 (ZIF-8) membranes synthesized in aqueous solutions. *Chem. Comm.* **2011**, *47*, 10275-10277.
6. Bux, H.; Chmelik, C.; Krishna, R.; Caro, J. Ethene/ethane separation by the MOF membrane ZIF-8: Molecular correlation of permeation, adsorption, diffusion. *Journal of Membrane Science* **2011**, *369*, 284-289.
7. Bux, H.; Feldhoff, A.; Cravillon, J.; Wiebcke, M.; Li, Y.-S.; Caro, J. Oriented Zeolitic Imidazolate Framework-8 Membrane with Sharp H2/C3H8 Molecular Sieve Separation. *Chem. Mater.* **2011**, *23*, 2262-2269.
8. Bux, H.; Liang, F.; Li, Y.; Cravillon, J.; Wiebcke, M.; Caro, J. Zeolitic Imidazolate Framework Membrane with Molecular Sieving Properties by Microwave-Assisted Solvothermal Synthesis. *J. Am. Chem. Soc.* **2009**, *131*, 16000-16001.
9. Kwon, H. T.; Jeong, H.-K. Highly propylene-selective supported zeolite-imidazolate framework (ZIF-8) membranes synthesized by rapid microwave-assisted seeding and secondary growth. *Chem. Comm.* **2013**, *49*, 3854-3856.
10. Kwon, H. T.; Jeong, H.-K. In Situ Synthesis of Thin Zeoliticâ€"Imidazolate Framework ZIF-8 Membranes Exhibiting Exceptionally High Propylene/Propane Separation. *J. Am. Chem. Soc.* **2013**, *135*, 10763-10768.
11. Kwon, H. T.; Jeong, H.-K.; Lee, A. S.; An, H. S.; Lee, J. S. Heteroepitaxially Grown Zeolitic Imidazolate Framework Membranes with Unprecedented Propylene/Propane Separation Performances. *J. Am. Chem. Soc.* **2015**, *137*, 12304-12311.
12. Wang, C.; Yang, F.; Sheng, L.; Yu, J.; Yao, k.; Zhang, L.; Pan, Y. Zinc-Substituted ZIF-67 Nanocrystals and Polycrystalline Membranes for Propylene/Propane Separation. *Chem. Comm.* **2016**, *52*, 12578-12581.
13. Kaur, G.; Rai, R. K.; Tyagi, D.; Yao, X.; Li, P.-Z.; Yang, X.-C.; Zhao, Y.; Xu, Q.; Singh, S. K. Room-temperature synthesis of bimetallic Co-Zn based zeolitic imidazolate frameworks in water for enhanced CO2 and H2 uptakes. *J. Mater. Chem. A* **2016**, *4*, 14932-14938.
14. Eum, K.; Jayachandrababu, K. C.; Rashidi, F.; Zhang, K.; Leisen, J.; Graham, S.; Lively, R. P.; Chance, R. R.; Sholl, D. S.; Jones, C. W.; Nair, S. Highly Tunable Molecular Sieving and Adsorption Properties of Mixed-Linker Zeolitic Imidazolate Frameworks. *J. Am. Chem. Soc.* **2015**, *137*, 4191-4197.
15. Thompson, J. A.; Blad, C. R.; Brunelli, N. A.; Lydon, M. E.; Lively, R. P.; Jones, C. W.; Nair, S. Hybrid Zeolitic Imidazolate Frameworks: Controlling Framework Porosity and Functionality by Mixed-Linker Synthesis. *Chem. Mater.* **2012**, *24*, 1930-1936.
16. Hillman, F.; Brito, J.; Jeong, H.-K. Rapid One-Pot Microwave Synthesis of Mixed-Linker Hybrid Zeolitic-Imidazolate Framework Membranes for Tunable Gas Separations. *ACS Applied Materials & Interfaces* **2018**, *10*, 5586-5593.

534364 A Scalable Method to Prepare Zeolitic-Imidazolate Framework ZIF-8 Membranes on Polymer Hollow Fibers for Propylene/Propane Separation

Abstract:

Separation of olefin/paraffin mixtures (e.g., ethylene/ethane and propylene/propane), is one of the most challenging separations due to the similarity of their physicochemical properties. Currently, propylene/propane separation is typically achieved through energy-intensive cryogenic distillation process. Although energy-efficient membrane-based separation processes have received significant attention, however, there are no commercially available membranes for olefin/paraffin separations mainly due to the limitations of current membrane materials and lack of cost-effective and simple processing methods to fabricate membrane with large area[1-3].

A sub-class of metal-organic frameworks (MOFs), zeolitic-imidazolate frameworks (ZIFs) with zeolite topologies are crystalline nanoporous materials with well-defined aperture sizes (< 5 Å). ZIFs, which are composed of divalent metal ions (typically Zn^{2+} and Co^{2+}) and imidazole-based ligands[4], have been extensively studied especially for gas separation applications. ZIF-8 in particular is a promising membrane material for high resolution propylene/propane separation because the effective pore size of ZIF-8 (~ 4 Å) [5] lies in between the size of propylene and propane molecules. Although ZIF-8 membranes (mostly planar configurations) synthesized either by *in-situ* or seeded growth method showed excellent propylene/propane separation performances, scaling-up and commercialization of the membranes requires them to be fabricated on a cheap and scalable support (i.e., hollow fiber or tubular support). Polymeric hollow fibers have been utilized as substrates, but only a few of the resulting ZIF-8 membranes showed promising propylene/propane separation performances.[6-9]

Herein, we would like to present a simple and scalable method to synthesize well intergrown ZIF-8 membranes on polymeric hollow fibers. The method is based on substrates modification followed by solvothermal reaction in linker solution. High-quality ZIF-8 seed layers prepared from the solvothermal reaction are secondarily grown to form high quality ZIF-8 membranes. The resulting ZIF-8 membranes are tested for propylene/propane separation. In this work, the effects of different synthesis parameters on the microstructures of ZIF-8 membranes will be discussed. Finally, the gas separation performances of ZIF-8 membranes especially propylene/propane separation will be presented.

References and Notes

1 Koros, William J. and Fleming, G.K. (1993) "Membrane-based gas separation", *Journal of Membrane Science*, 83, pp l-39

2 Baker RW., (2002) "Future Directions of Membrane Gas Separation Technology" *Industrial & Engineering Chemistry Research.*, 41 (6), pp 1393-1411

3 Tanaka, K., Taguchi, A., Hao, JQ, Kita, H., and Okamoto, K., (1996) "Permeation and separation properties of polyimide membranes to olefins and paraffins", *Journal of Membrane Science*, 121, pp 197-207

4 Park KS, Ni Z, Côté AP, et al., (2006) "Exceptional chemical and thermal stability of zeolitic imidazolate frameworks", *Proceedings of the National Academy of Sciences,* 103 (27), pp 10186-10191

5 Zhang C, Lively RP, Zhang K, Johnson JR, Karvan O, Koros WJ., (2012) "Unexpected Molecular Sieving Properties of Zeolitic Imidazolate Framework-8", *The Journal of Physical Chemistry Letters.*, 3 (16), pp 2130-2134.

6 Andrew J. Brown, Nicholas A. Brunelli, Kiwon Eum, Fereshteh Rashidi, J. R. Johnson, William J. Koros, Christopher W. Jones, Sankar Nair, (2014) "Interfacial microfluidic processing of metal-organic framework hollow fiber membranes", *Science*, 345, 6192, pp 72-75

7 FernandoCacho-Bailo, SilviaCatalán-Aguirre, MirenEtxeberría-Benavides, Oğuz Karvan, VictorSebastian, CarlosTéllez, JoaquínCoronas (2015) "Metal-organic framework membranes on the inner-side of a polymeric hollow fiber by microfluidic synthesis", *Journal of membrane science*, 476, pp 277-285

8 Kiwon Eum, Chen Ma, Ali Rownaghi, Christopher W. Jones, and Sankar Nair (2016) "ZIF 8 Membranes via Interfacial Microfluidic Processing in Polymeric Hollow Fibers: Efficient Propylene Separation at Elevated Pressures", *ACS Applied materials and interfaces*, 8, pp 25337 – 25342

9 Anne M. Marti, Wasala Wickramanayake, Ganpat Dahe, Ali Sekizkardes, Tracy L. Bank, David P. Hopkinson, and Surendar R. Venna (2017) "Continuous Flow Processing of ZIF 8 Membranes on Polymeric Porous Hollow Fiber Supports for CO2 Capture", *ACS Applied materials and interfaces*, 9, pp 5678-5682

534726 Membrane Synthesis and Process Design for Hydrogen Purification from Coal-Derived Syngas

Abstract:

Membrane Synthesis and Process Design for Hydrogen Purification from Coal-derived Syngas

Yang Han and W.S. Winston Ho

William G. Lowrie Department of Chemical and Biomolecular Engineering, The Ohio State University, Columbus, OH

In this presentation, three amine-containing facilitated transport membranes were tailored for a single-stage membrane process to decarbonize coal-derived syngas. In these membranes, water-swellable crosslinked polyvinyl alcohol was used as polymer matrix, along with high MW polyvinylamine as fixed-site carrier. For Membrane 1, an amino acid salt of 2-(1-piperazinyl)ethylamine and sarcosine was used as the mobile carrier; sarcosine has a mild degree of steric hindrance. In Membrane 2, the 2-(1-piperazinyl)ethylamine salt of 2-aminoisobutyric acid (PZEA-AIBA) was used as the mobile carrier; AIBA is a sterically hindered amine. Membrane 3 utilized 1-(2-hydroxyethyl)piperazine as the mobile carrier, which contains 2 amino groups and a CO_2-phile hydroxyl group. In addition, poly(ethylene glycol) dimethyl ether was incorporated in Membrane 3 as a CO_2-phile moiety. Nanoporous graphene oxide was dispersed in all three membranes to avoid membrane compaction upon high feed pressure. The membranes were tested at 107 °C and 31.7 bar feed pressure with a CO_2 partial pressure of 12.5 to 0.5 bar, which corresponded to the feed CO_2 partial pressures before and after the bulk CO_2 removal, respectively. The separation performances of Membranes 1 and 2 depended strongly on the CO_2 partial pressure, yielding a CO_2 permeance of 72 – 435 GPU (1 GPU = 10^{-6} cm^3 (STP)·cm^{-2}·s^{-1}·$cmHg^{-1}$) with a CO_2/H_2 selectivity of 93 – 558. The CO_2 permeance reduced significantly with increasing CO_2 partial pressure due to the carrier saturation phenomenon. On the other hand, the presence of amino groups and the abundant CO_2-phile moieties in Membrane 3 enhanced the CO_2 solubility at high CO_2 partial pressure, thereby a CO_2 permeance of 206 – 246 GPU and a decent CO_2/H_2 selectivity of 103 – 123 were obtained for a CO_2 partial pressure of 12.5 – 3.5 bar. The different degrees of dependence on the feed CO_2 partial pressure provide a design opportunity for a single-stage membrane process. The more permeable but less selective Membrane 3 can be implemented to the proximity of feed inlet, where the CO_2 partial pressure is high but the H_2 partial pressure is low. Upon CO_2 removal, the feed CO_2 partial pressure sharply reduces whereas the H_2 transmembrane driving force increases. In this case, Membrane 1 or 2 can be used to elevate the CO_2 permeance and CO_2/H_2 selectivity. Initial techno-economic analysis shows that the single-stage membrane process can achieve 90% CO_2 removal with >99% H_2 recovery. A 15.7% increase in cost of electricity is estimated, which is much cost-efficient then Selexol.

534760 Synthesis and Techno-Economic Analysis of Novel Facilitated Transport Membrane for Post-Combustion Carbon Capture

Abstract:

Large-scale application of membrane in post-combustion carbon capture has been limited by the trade-off between CO_2 permeance and CO_2/N_2 selectivity of most polymeric membrane materials. In order to overcome this limitation, research efforts on facilitated transport membrane (FTM) have been devised with the objectives of (1) developing carriers with high CO_2 loading capacity and reactive diffusivity, and (2) designing membrane processes that can capitalize on the outstanding selectivities shown by FTM. In this presentation, novel FTM was synthesized in a composite membrane configuration with an 170 nm selective layer coated on a polyethersulfone nanoporous substrate. In the selective layer, polyvinylamine with amine-sites covalently bound to the polymer backbone was used as fixed-site carrier and an amino acid salt, synthesized by deprotonating sarcosine with 2-(1-piperazinyl)ethylamine, was blended as mobile carrier. The membrane demonstrated a CO_2 permeance of 1500 GPU (1 GPU = 10^{-6} cm^3 (STP)·cm^{-2}·s^{-1}·$cmHg^{-1}$) and a CO_2/N_2 selectivity >140 at 67 °C with 1 atm feed and permeate pressures. Two membrane processes were designed for the FTM to decarbonizing coal derived flue gas with 90% CO_2 recovery. In the first membrane process, i.e., air sweep process, the flue gas flowed to a vacuum membrane stage to produce a >95% pure CO_2 permeate; the remaining CO_2 was removed by an air-sweep membrane stage. The CO_2-laden sweep air was fed to a power plant boiler as combustion air. In the second membrane process, i.e., retentate recycle process, the flue gas was passed to two membrane stages in an enriching cascade. A portion of the CO_2-depleted retentate of the first membrane stage was recycled as its own sweep gas, while a vacuum was pulled for the second membrane stage. Both membrane processes yielded a capture cost <40/tonne CO_2 in the 2007 dollars, which met the target set by the Department of Energy (DOE) for 2025. A common element in these two processes was a vacuum membrane stage with a permeate pressure as low as 0.3 atm. Under this testing condition, however, the elastic selective layer of the synthesized FTM sunk into the nanoporous substrate, which resulted in a drastically reduced CO_2 permeance. To address this issue, multi-walled carbon nanotubes (MWNTs) wrapped by a copolymer poly(vinylpyrrolidone-*co*-vinyl acetate) were dispersed in the selective layer as reinforcement fillers. The gas permeation measurements showed that the incorporation of MWNTs strengthened the polymer matrix and the selective layer penetration was refrained by a 3 wt.% MWNTs loading. The presence of MWNTs also mitigated the polymer compaction if the feed gas was compressed. The reinforced membrane demonstrated stable separation performance for a feed pressure up to 4 atm with a vacuum of 0.3 atm pull on the permeate side.

534766 Fabrication of Spiral-Wound Membrane Modules for CO_2 Capture from Flue Gas

Abstract:

A spiral-wound membrane module, including a spiral-wound membrane element, Plexiglas fiber-reinforced plastic tube, and membrane housing with face compression â€œOâ€ rings for effective seal of gases was developed. The spiral-wound membrane modules showed essentially no leakage and demonstrated an efficient gas separation. At a dry feed gas flow rate of less than 1000 sccm before humidification, a concentration polarization phenomenon was observed. Therefore, both dry feed and sweep gas flow rates each at 1000 sccm before humidification were used to evaluate the performances of the spiral-wound membrane modules at the typical flue gas temperature of 57°C. The feed and sweep gas pressures were at 1.5 and 1 psig, respectively. The spiral-wound membrane modules showed similar performance results with a CO_2 permeance of greater than 800 GPU and a CO_2/N_2 selectivity of more than 140 when tested with either a simulated flue gas or an actual flue gas. The results were similar to the results obtained from the flat-sheet membrane used for the module fabrication. The pressure drops of the modules were less than 1.5 psi/m and met the requirements for CO_2 capture from flue gas. The performances of the modules revealed some unexpected issues during the field test using the actual flue gas at the National Carbon Capture Center in Wilsonville, Alabama, including the feed spacer indentations on the membrane selective layer and the leakage at glue lines. Improvement of the spiral-wound membrane module fabrication was made by using a smoother feed spacer and a longer glue curing time. The spiral-wound membrane module was scaled up to a 5 inches diameter to accommodate a membrane area of around 14000 cm^2. A suitable spacer was selected to accommodate a spiral-wound membrane module test with vacuum at the permeate side. The field test results of the modules have shown their good potential for the post-combustion CO_2 capture from coal-fired power plants.

535840 A Combined Seeding Approach for High-FluxÂ Zeolitic-Imidazolate Framework ZIF-67 Membranes for Olefin/Paraffin Separation

Abstract:

A Combined Seeding Approach for High-flux Zeolitic-Imidazolate Framework ZIF-67 Membranes for Olefin/Paraffin Separations

Jingze Sun[1], Chen Yu[1], Kumar Varoon Agrawal[2], and Hae-Kwon Jeong*[1,3]

[1]Artie McFerrin Department of Chemical Engineering and [3]Department of Materials Science and Engineering, Texas A&M University, College Station, TX 77843-3122

[2]Laboratory of Advanced Separations (LAS), École Polytechnique Fédérale de Lausanne (EPFL) CH-1951, Sion, Swizerland

* Corresponding author: hjeong7@tamu.edu

Having been extensively researched as potential membrane materials, zeolite-imidazole frameworks (ZIFs) are famous for their robust synthesis protocols, chemical and thermal stabilities, and well-defined ultra-micropores (~ 2 to 5 Å).[1, 2] Composed of zinc nodes and 2-methylimidazole linkers, ZIF-8 has been shown very promising for propylene/propane separation, because of its effective aperture size lying between that of propylene and propane..[3, 4] Due to its slightly smaller aperture, ZIF-67 (Co-substituted ZIF-8)[5, 6] seems promising for improving propylene/propane separation performances as compared with ZIF-8. Previous works from our group for ZIF-8/ZIF-67/ZIF-8 multi-layer membranes has exhibited propylene/propane separation factors as high as 209.[7] Unfortunately, no ZIF-67 membranes have been reported primarily due to the transformation of ZIF-67 seed crystals into a double hydroxide phase upon the secondary growth.

Here, we are glad to present the first phase-pure ZIF-67 membranes as thin as ~ 300 nm. The resulting ultrathin ZIF-67 membranes resulted from microwave-assisted (MW) seeding followed by Electrophoretic Nuclei Assembly (ENACT), by which high-quality densely-packed ZIF-8 seed layers prepared.[8] Much increased propylene permeances were observed from these ultra-thin ZIF-67 membranes as compared to ZIF-8 membranes. To obtain phase-pure ZIF-67 membranes, it was found critical 1) to have densely-packed ZIF-67 seed layers by combining MW and ENACT seeding and 2) to lower secondary growth temperature. This MW/ENACT seeding technique was demonstrated effective for ZIF-8 and ZIF-90 membranes on alumina supports as well.

References and Notes:

1. S. Park, W. R. Kang, H. T. Kwon, S. Kim, M. Seo, J. Bang, S. hyup Lee, H. K. Jeong and J. S. Lee, *Journal of Membrane Science*, 2015, **486**, 29-39.
2. C. Zhang, R. P. Lively, K. Zhang, J. R. Johnson, O. Karvan and W. J. Koros, *The journal of physical chemistry letters*, 2012, **3**, 2130-2134.
3. K. Li, D. H. Olson, J. Seidel, T. J. Emge, H. Gong, H. Zeng and J. Li, *Journal of the American Chemical Society*, 2009, **131**, 10368-10369.
4. H. Bux, F. Liang, Y. Li, J. Cravillon, M. Wiebcke and J. r. Caro, *Journal of the American Chemical Society*, 2009, **131**, 16000-16001.
5. F. Hillman, J. M. Zimmerman, S.-M. Paek, M. R. Hamid, W. T. Lim and H.-K. Jeong, *Journal of Materials Chemistry A*, 2017, **5**, 6090-6099.
6. P. Krokidas, M. Castier, S. Moncho, D. N. Sredojevic, E. N. Brothers, H. T. Kwon, H.-K. Jeong, J. S. Lee and I. G. Economou, *The Journal of Physical Chemistry C*, 2016, **120**, 8116-8124.
7. H. T. Kwon, H.-K. Jeong, A. S. Lee, H. S. An and J. S. Lee, *Journal of the American Chemical Society*, 2015, **137**, 12304-12311.
8. G. He, M. Dakhchoune, J. Zhao, S. Huang and K. V. Agrawal, *Advanced Functional Materials*, 2018.

537059 Microscopic Diffusion of Ethylene in ZIF-11 Based Mixed Matrix Membranes (MMMs) By Pulsed Field Gradient (PFG) NMR

Abstract:

Mixed matrix membranes (MMMs) have attracted attention of the scientific community due to their easy and scalable fabrication, as well as superior gas separation performance compared to pure polymeric membranes. MMMs are formed by dispersing permeable filler particles in a polymer matrix. Recently, zeolitic imidazole frameworks (ZIFs) have been considered for filler particles due to their great thermal and chemical stability, high surface area and tunable frameworks which have made them more compatible with polymers compared to other types of filler particles.

Detailed understanding of molecular diffusion inside ZIF particles dispersed in MMMs is important for their potential applications in gas separations. Here we report microscopic studies of diffusion of ethylene inside different ZIF-11-based MMMs at equivalent loading pressures. Diffusion measurements were performed using ^{13}C pulsed field gradient (PFG) NMR utilizing a high field (17.6 T) and high magnetic field gradients (up to 25 T/m) to obtain diffusivities for the length scales of displacements smaller than the mean size of the ZIF-11 crystals. It was found that the diffusivity of ethylene inside ZIF-11 particles for displacements much smaller than the particle size is influenced by the particle confinement in the polymer. This observation is attributed to the ability of the polymer to restrict framework flexibility of the dispersed ZIF-11 particles in the polymer.

538080 Novel High-Performance Hollow Fiber Membrane Modules for Water Desalination through Direct Contact Membrane Distillation

Abstract:

High-performance (high flux and energy efficient) hollow fiber membrane modules were fabricated for water desalination by the membrane distillation (MD) process. We introduce a new approach to pack hollow fiber membranes into the modules using 3D spacers designed and fabricated through mathematical modeling and an additive manufacturing method, respectively. Functional Polyvinylidene difluoride (PVDF) hollow fiber membranes were fabricated through a dry-jet wet-spinning method. The membranes were characterized in the fabricated modules and direct contact membrane distillation (DCMD) process. We observed a close agreement between the experimental and simulation results. The experimental results showed that the flux decreases when the interspacing parameter--the ratio of fiber radius to center-to-center distance of fibers--approaches the value of 0.5. Moreover, the permeate flux increased with the feed temperature and flow rate. The unique design of the module reduces temperature polarization, improves flow pattern, and enhances the moduleâ€™s thermal efficiency, leading to a high permeate flux. The results confirmed that the high-performance hollow fiber membrane modules can be fabricated by properly designing the modules, arranging hollow fiber membranes, and adjusting the interspacing distance between the fibers.

535712 Multilayer Composite Membranes with Superior CO_2 Separation Properties

Abstract:

Graphene oxide (GO)-based membranes have shown high potential for gas separation[1]. Numerous membranes containing different forms of GO have been fabricated thanks to its appealing properties[2]. Different functional groups of GO allow for its functionalization and preparation of functionalized GO membranes[3]. Also, ionic liquids have been found to be promising materials for CO_2 capture[4].

This paper deals with the fabrication and gas-transport-performance evaluation of multilayer composite membranes. These membranes were made from a porous polyacrylonitrile (PAN) substrate, GO-melamine (GO-M), an imidazolium-based ionic liquid (IL), and a polydimethylsiloxane (PDMS) coating layer. The doped GO-melamine (GO-M) nanomaterials were prepared by modifying the surfaces of GO nanosheets with melamine and then mixing the modified nanosheets with the IL. The GO-M-IL suspension was dip-coated on the PAN substrate, which was then coated with a layer of PDMS. Results from a thorough investigation of the influence of the GO concentration on the gas-transport properties and long-term stability of the multilayer membranes are presented. Compared to the membrane containing just the IL, the GO-M-IL-based membranes have better CO_2/N_2 and CO_2/CH_4 selectivities and slightly lower CO_2 permeances.

References:

1. Wang, S.; Wu, Y.; Zhang, N.; He, G.; Xin, Q.; Wu, X.; Wu, H.; Cao, X.; Guiver, M. D.; Jiang, Z., A highly permeable graphene oxide membrane with fast and selective transport nanochannels for efficient carbon capture. *Energy & Environmental Science* **2016,** *9* (10), 3107-3112.
2. Shen, J.; Liu, G.; Huang, K.; Chu, Z.; Jin, W.; Xu, N., Subnanometer two-dimensional graphene oxide channels for ultrafast gas sieving. *ACS nano* **2016,** *10* (3), 3398-3409.
3. Karunakaran, M.; Villalobos, L. F.; Kumar, M.; Shevate, R.; Akhtar, F. H.; Peinemann, K.-V., Graphene oxide doped ionic liquid ultrathin composite membranes for efficient CO 2 capture. *Journal of Materials Chemistry A* **2017,** *5* (2), 649-656.
4. Zeng, S.; Zhang, X.; Bai, L.; Zhang, X.; Wang, H.; Wang, J.; Bao, D.; Li, M.; Liu, X.; Zhang, S., Ionic-liquid-based CO2 capture systems: structure, interaction and process. *Chemical reviews* **2017,** *117* (14), 9625-9673.

529556 Properties, Processing and Performance of Aromatic Ionic Polyimides and Polyamides As Gas Separation Membranes

Abstract:

We have synthesized a variety of ionenes (main-chain polycations) via condensation reactions which results in complex, highly aromatic ionic polyimides and ionic polyamides. These imidazolium-based ionenes are designed from aromatic precursors derived from imidazoles and analogous dihalides. Furthermore, these ionenes are also able to interact with and â€œholdâ€ ionic liquids (ILs) within their structures. The addition of IL creates unique ordering effects and often increases the processability of the ionic polyimides and polyamides. This presentation will discuss our characterization and processing of aromatic ionenes, the impact of adding IL to the material, and their use as gas separation membranes.

534119 Rapid Synthesis of Hybrid Zeolitic-Imidazolate Frameworks (ZIFs) Membranes with Tunable Gas Separations

Abstract:

Metal organic frameworks (MOFs) are emerging organic-inorganic crystalline microporous solids for gas separation application owing to their tunable pore structure and functionality with diverse metal nodes and organic linkers.[1] Zeolitic-imidazolate frameworks (ZIFs),[2] a subclass of MOFs, in particular have attracted many attentions due to their chemical/thermal stabilities, their ultra-microporosities, and high surface area when compared to other MOFs material.[2] ZIFs possess zeolite-like topologies in which the tetrahedral Si or Al and the bridging O in zeolite structure are replaced by transition metals (such as Zn,[2] Co,[3] and Cd[4]) and imidazolate-derived ligands. A common drawback for gas separation is the limited availability of aperture pore size and functionalities of the porous materials. Furthermore, the common approach to fabricate ZIF polycrystalline membranes are lengthy requiring up to weeks of synthesis.[5-11] Studies have shown that through mixing metals[12,13] and linkers,[14,15] one can continuously tune the MOFs/ZIFs (termed hybrid ZIFs) properties to match with the characteristic of specific gas mixture. However, no study have shown the tuning effect of the hybrid ZIFs grown as polycrystalline membrane.

Here, we plan to present a rapid microwave-assisted in situ synthesis of well-intergrown mixed linker ZIF-7-8 membranes in under â¨90 s.[16] ZIF-7-8 consists of Zn^{2+} metal nodes bridged by a mixture of benzimidazolate (bIm, ZIF-7 linker) and 2-methylimidazolate (mIm, ZIF-8 linker) linkers.[15] To the best of our knowledge, this is the fastest synthesis of any polycrystalline MOF membranes reported up to now. Furthermore, the gas separation performances (separation factor and permeance) of hybrid ZIF-7-8 membranes were systematically tuned by varying the bIm-to-mIm ratio incorporated into the framework. This is attributed to the fact that an increase in bulky benzimidazolate linker incorporation reduces the effective aperture size of the hybrid framework, consequently improving the separation factor at the expense of reduction in permeance. The unprecedentedly rapid synthesis of ZIF-7-8 membranes with tunable molecular sieving properties is an important step forward for their practical applications in the separation of gas mixtures of interest.

References:

1. Li, J.-R.; Kuppler, R. J.; Zhou, H.-C. Selective gas adsorption and separation in metal-organic frameworks. *Chem. Soc. Rev.* **2009**, *38*, 1477-1504.
2. Park, K. S.; Ni, Z.; Cote, A. P.; Choi, J. Y.; Huang, R.; Uribe-Romo, F. J.; Chae, H. K.; O'Keeffe, M.; Yaghi, O. M. Exceptional chemical and thermal stability of zeolitic imidazolate frameworks. *Proc. Natl. Acad. Sci. U. S. A.* **2006**, *103*, 10186-10191.
3. Banerjee, R.; Phan, A.; Wang, B.; Knobler, C.; Furukawa, H.; O'Keeffe, M.; Yaghi, O. M. High-Throughput Synthesis of Zeolitic Imidazolate Frameworks and Application to CO2 Capture. *Science* **2008**, *319*, 939-943.
4. Karagiaridi, O.; Bury, W.; Sarjeant, A. A.; Stern, C. L.; Farha, O. K.; Hupp, J. T. Synthesis and characterization of isostructural cadmium zeolitic imidazolate frameworks via solvent-assisted linker exchange. *Chem. Sci.* **2012**, *3*, 3256-3260.
5. Pan, Y.; Lai, Z. Sharp separation of C2/C3 hydrocarbon mixtures by zeolitic imidazolate framework-8 (ZIF-8) membranes synthesized in aqueous solutions. *Chem. Comm.* **2011**, *47*, 10275-10277.
6. Bux, H.; Chmelik, C.; Krishna, R.; Caro, J. Ethene/ethane separation by the MOF membrane ZIF-8: Molecular correlation of permeation, adsorption, diffusion. *Journal of Membrane Science* **2011**, *369*, 284-289.
7. Bux, H.; Feldhoff, A.; Cravillon, J.; Wiebcke, M.; Li, Y.-S.; Caro, J. Oriented Zeolitic Imidazolate Framework-8 Membrane with Sharp H2/C3H8 Molecular Sieve Separation. *Chem. Mater.* **2011**, *23*, 2262-2269.
8. Bux, H.; Liang, F.; Li, Y.; Cravillon, J.; Wiebcke, M.; Caro, J. Zeolitic Imidazolate Framework Membrane with Molecular Sieving Properties by Microwave-Assisted Solvothermal Synthesis. *J. Am. Chem. Soc.* **2009**, *131*, 16000-16001.
9. Kwon, H. T.; Jeong, H.-K. Highly propylene-selective supported zeolite-imidazolate framework (ZIF-8) membranes synthesized by rapid microwave-assisted seeding and secondary growth. *Chem. Comm.* **2013**, *49*, 3854-3856.
10. Kwon, H. T.; Jeong, H.-K. In Situ Synthesis of Thin Zeoliticâ€"Imidazolate Framework ZIF-8 Membranes Exhibiting Exceptionally High Propylene/Propane Separation. *J. Am. Chem. Soc.* **2013**, *135*, 10763-10768.
11. Kwon, H. T.; Jeong, H.-K.; Lee, A. S.; An, H. S.; Lee, J. S. Heteroepitaxially Grown Zeolitic Imidazolate Framework Membranes with Unprecedented Propylene/Propane Separation Performances. *J. Am. Chem. Soc.* **2015**, *137*, 12304-12311.
12. Wang, C.; Yang, F.; Sheng, L.; Yu, J.; Yao, k.; Zhang, L.; Pan, Y. Zinc-Substituted ZIF-67 Nanocrystals and Polycrystalline Membranes for Propylene/Propane Separation. *Chem. Comm.* **2016**, *52*, 12578-12581.
13. Kaur, G.; Rai, R. K.; Tyagi, D.; Yao, X.; Li, P.-Z.; Yang, X.-C.; Zhao, Y.; Xu, Q.; Singh, S. K. Room-temperature synthesis of bimetallic Co-Zn based zeolitic imidazolate frameworks in water for enhanced CO2 and H2 uptakes. *J. Mater. Chem. A* **2016**, *4*, 14932-14938.
14. Eum, K.; Jayachandrababu, K. C.; Rashidi, F.; Zhang, K.; Leisen, J.; Graham, S.; Lively, R. P.; Chance, R. R.; Sholl, D. S.; Jones, C. W.; Nair, S. Highly Tunable Molecular Sieving and Adsorption Properties of Mixed-Linker Zeolitic Imidazolate Frameworks. *J. Am. Chem. Soc.* **2015**, *137*, 4191-4197.
15. Thompson, J. A.; Blad, C. R.; Brunelli, N. A.; Lydon, M. E.; Lively, R. P.; Jones, C. W.; Nair, S. Hybrid Zeolitic Imidazolate Frameworks: Controlling Framework Porosity and Functionality by Mixed-Linker Synthesis. *Chem. Mater.* **2012**, *24*, 1930-1936.
16. Hillman, F.; Brito, J.; Jeong, H.-K. Rapid One-Pot Microwave Synthesis of Mixed-Linker Hybrid Zeolitic-Imidazolate Framework Membranes for Tunable Gas Separations. *ACS Applied Materials & Interfaces* **2018**, *10*, 5586-5593.

522347 Organic Solvent Nanofiltration Via Zeolitic-Imidazolate Framework Membranes: Insights from Molecular Simulation

Abstract:

Organic solvents are intensively used in chemical and pharmaceutical industries, their separation and recovery account for a significant portion of energy consumption and capital cost in many processes. Toward this end, switching from traditional distillation to membrane nanofiltration could reduce the energy cost and carbon emission to a large extent. In this study, three zeolitic-imidazolate framework (ZIF-25, -71 and -96) membranes were examined for their application in organic solvent nanofiltration (OSN) by molecular dynamics (MD) simulations. ZIF-25 membrane exhibits the highest flux and permeability for hydrophilic solvents due to its hydrophobic nature, despite the smallest window size among the three ZIFs; whereas ZIF-96 membrane shows the lowest flux due the strong interaction between its highly hydrophilic surface and solvent molecules. Good correlation between the combined solvent properties and solvent permeance is established. This study reveals that the chemical properties of the inner membrane surface, rather than the channel size, plays a more important role in determining the solvent flux during permeation process; and suggests that ZIFs are potential materials for organic solvent recovery.

522386 Separation of Zinc and Nickel from Industrial Wastewater through Supported Liquid Membrane Using Environmentally Benign Solvent

Abstract:

Effectiveness of supported liquid membrane (SLM) process for separation of Zn (II) and Ni (II) has been studied in this work. Nowadays SLM has received increased attention towards separation of heavy metals such as Hg, Pb, Cd, As, Zn and Ni *etc.*[1-3]. SLM technology has great potential as an alternative for separation of heavy metal as opposed to reverse osmosis, chromatography, adsorption, ion exchange, solvent extraction *etc*. The effluents emanating from various industries such as metal plating, electroplating, battery *etc.* are the main source of zinc related pollution whereas nickel is found in wastewater released from industries such as electroplating, battery, rubber processing, mining *etc.* Zn (II) and Ni (II) above concentrations 3 ppm and 0.02 ppm, respectively lead to environmental pollution. Traditionally the solvents used in the SLM are organic in nature which are inflammable and toxic to environment as well as to human health. Nevertheless the solvent used in this study, *viz.* sunflower oil, is environmentally as well as physiologically benign in nature. Addition of an extractant (*a.k.a* carrier agent) into the solvent seldom enhances efficiency of extraction and recovery of solute(s). Various extractants, such as tri-n-octylamine (abbreviated as TOA), di-(2-ethylhexyl) phosphoric acid (abbreviated as D2EHPA) and n-methyl-n,n,n-trioctylammonium chloride (*a.k.a* Aliquat 336), were employed in order to choose the best extractant for the SLM. D2EHPA has been found to be the best carrier agent for transport of Zn (II) whereas TOA is the best choice in favour of Ni (II) for the same purpose. The combination of solvent and carrier is termed as membrane phase. Both the membrane phases in this study, *viz.* "D2EHPA-sunflower oil" for Zn (II) and "TOA-sunflower oil" for Ni (II), are environmentally and physiologically "green" in nature. The polymeric support of SLM is made of polyvinylidene fluoride (or abbreviated as PVDF) with pore size of 0.22 μm. Among various strippants employed in this study for recovery of Zn ion, such as sulphuric acid (H_2SO_4 leading to 33.32% recovery of zinc and 35.46% recovery of nickel), ammonium chloride (NH_4Cl leading to 23.58% recovery of zinc and 21.87% recovery of nickel), sodium chloride (NaCl leading to 10.07% recovery of zinc and 11.96% recovery of nickel) and ethylene di-amine ($C_2H_4(NH_2)_2$ leading to 6.20% recovery of zinc and 7.52% recovery of nickel), it has been observed that H_2SO_4 is the best stripping agent. The performance of SLM in transport of solute(s) has been optimized by tuning various parameters, *viz.* concentration of solute in source phase (effluent), pH of source phase, concentration of strippant in stripping phase, concentration of carrier in the membrane phase, stirring speed and run time. It has been observed that the optimum concentration of Zn (II) and Ni (II) in source effluent should be 100 mg L^{-1} (for both solutes) in order to achieve best transport performance while the studied range in this category was 50 to 400 mg L^{-1}. Further 3% (vol/vol) D2EHPA in the membrane phase, within studied range of 0 to 5%, was ideal for transportation of Zn (II) whereas 5% (vol/vol) TOA was ideal for transportation of Ni (II) within the studied range 0 to 8%. Similarly 0.1 M H_2SO_4 in stripping phase (with studied range of 0.01 to 1 M), pH 5 in source phase (with studied range 2 to 8), 120 rpm of stirring speed in source and stripping phase (with studied range 60 to 250 rpm), 48 hours of run time (with studied range of 24 to 72 hrs) have been found to be most suitable for transport of Zn (II) in the most efficient manner. On the other hand 0.1 M H_2SO_4 in stripping phase (with studied range of 0.01 to 1 M), pH 4 in source phase (with studied range 2 to 8), 120 rpm of stirring speed in source and stripping phase (with studied range 60 to 250 rpm), 48 hours of run time (with studied range of 24 to 48 hrs) have been found to be most suitable for transport of Ni (II) in the most efficient manner. The efficiencies of extraction and recovery with the aforementioned optimized operating conditions are 42% and 53%, respectively for Zn (II) whereas the same for Ni (II) are 45% and 53%, respectively. Combinatorial transport of both Zn (II) and Ni (II) in a single SLM setup is being investigated at present. The results of SLM process can further be used for the separation of solute(s) through hollow fibre membrane (HFM).

References:

[1] E. Jean, D. Villemin, M. Hlaibi, L. Lebrun, Heavy metal ions extraction using new supported liquid membranes containing ionic liquid as carrier, Sep. Purif. Technol. 201 (2018) 1–9.

[2] B. Swain, K. Sarangi, R. P. Das, Effect of different anions on separation of cadmium and zinc by supported liquid membrane using TOPS-99 as mobile carrier, J. Membr. Sci. 277 (2006) 240–248.

[3] P.K. Parhi, K. Sarangi, Separation of copper, zinc, cobalt and nickel ions by supported liquid membrane technique using LIX 84I, TOPS-99 and Cyanex, Sep. Purif. Technol. 59 (2008) 169–174.

532387 A Novel Cationic Guanidine Compound Grafted Polyvinylidene Fluoride Membrane for Biofouling Mitigation

Abstract:

A novel cationic guanidine compound grafted polyvinylidene fluoride membrane for biofouling mitigation

Shanshan Zhao [a, b], Guimei Liu [a, b], Fangang Meng [a, b, *]

[a] School of Environmental Science and Engineering, Sun Yat-sen University, Guangzhou 510275, PR China

[b] Guangdong Provincial Key Laboratory of Environmental Pollution Control and Remediation Technology, Sun Yat-sen University, Guangzhou 510275, PR China

Abstract

Membrane separation technologies have gained wide application in water and wastewater treatment. However, membrane fouling, particularly biofouling, remains a critical issue in the real applications. The undesired bacterial adhesion and subsequent biofilm formation on the membrane surface severely declines the separation efficiency and increases operation cost. Membrane surface modification is an emerging method to mitigate biofouling. In this study, a novel antibiofouling polyvinylidene fluoride (PVDF) membrane was fabricated via coating polydopamine (PDA)-silane (3-aminopropyltrimethoxysilane, APTMS) hybrid layer as a building block layer, followed by alkoxysilane hydrolysis induced cationic guanidine compounds (CGC) grafting. The CGC modified membrane showed comparable pure water permeability with the pristine membrane, but presented efficient antibacterial activity against both *S. aureus* and *E. Coli*. The grafting of CGC inhibited cell attachment on the membrane surface and significantly alleviated membrane flux decline during the filtration experiment, i.e. ~20 % flux decline of CGC modified membrane compared to ~50% flux decline of the unmodified membrane. The biofilm formation on the membrane surfaces was further investigated using confocal laser scanning microscope (CLSM). The result revealed that less live bacteria were observed on the CGC modified membrane in comparison to the pristine membrane, indicating the excellent biofouling mitigation function of CGC. Moreover, the PDA/APTMS building block layer enhanced the stability of the CGC layer during harsh environmental conditions. This novel CGC grafted membrane shows great potential for water and wastewater treatment application for biofouling mitigation.

Keywords: Antibiofouling, cationic guanidine compounds, polydopamine/silane hybrid layer, membrane filtration, wastewater treatment

533025 Star Polymers As a New Building Block for the Fabrication of Reverse Osmosis and Nanofiltration Membranes

Abstract:

Star polymers with a globular structure are attractive materials since the chemistry of their core and arms can be tailored for the suitable applications. In this work, star polymers with the high density of amine groups were synthesized *via* atomic transfer radical polymerization (ATRP), followed by amine end-functionalization for their application to the membranes for water treatment and desalination. First, the well-defined poly (methyl acrylate) (PMA) star polymers consisting of PMA linear polymer arms and a cross-linked poly (ethylene glycol dimethacrylate) core were synthesized *via* ATRP based on an arm-first approach. The methyl oxy groups of PMA arm polymers were then substituted with hydrazine groups to produce the amine-functionalized poly (acryloyl hydrazide) (PAH) star polymer. The synthesized PAH star polymer was employed to fabricate thin film composite membranes for water desalination *via* the interfacial polymerization (IP) of the star polymer and trimesoyl chloride. High density reactive functional groups and compact globular structure of the star polymer enable to form a unique selective layer structure, consisting of a highly cross-linked, selective topmost layer on a less dense, permeable sublayer containing the large number of interstitial voids. Importantly, the star polymer-assembled membrane exhibited a significant improvement both in the permeselectivity and antifouling performance compared to commercial membranes. In addition, reverse osmosis and nanofiltration grade performance can be achieved by simply adjusting the IP process parameters, which is not feasible with conventional monomers, demonstrating the excellent versatility of the star polymer.

537030 Effect of Pressure and Spacer Configuration on Assisted Reverse Osmosis Performance

Abstract:

Water supply crisis is a reality in many places and it became a problem also in Brazil, typically known as a water-abundant country. Lack of suitable public policies and harsh droughts led to agricultural and industrial water shortage in the past few years. In this scenario, liquid effluents are a potential water source mainly in Northeast and Southeast regions. Particularly, oil produced water is abundant in these regions, although it can contains high levels of contaminants and salt content. Aiming to apply an economical and suitable technology for desalinate high salinity streams, assisted reverse osmosis (ARO) has been proposed as a potential low-energy technology, mainly for effluents out of conventional reverse osmosis (RO) range. This process is very similar to RO, but it uses a sweep or draw solution to diminish osmotic pressure difference across the membrane. With this feature, the same hydraulic pressure for RO can lead to a greater flux in ARO, although it is usually necessary to recover the sweep solution with a conventional RO process. As it is a novel technology, there is a lack of studies on its real performance, mainly for sweep solution flow and spacer configuration. In this paper, the effect of feed hydraulic pressure and sweep solution spacer configuration were studied in an adapted plate-and-frame ARO module. Three commercial spacers were tested over feed pressures until 60 bar with a commercial seawater reverse osmosis membrane. Head loss was also measured for draw solution module path as a function of feed pressure and number of spacers. It was verified that spacer design has a strong influence on membrane deformation and consequently its salt rejection. These results are similar to those obtained in literature for pressure retarded osmosis (PRO) technology. Besides, the feed pressure increases the compaction of spacers, also increasing the head loss significantly. This could be a major issue for plate and frame ARO modules design and flow optimization.

537431 Crystal Morphology and Process Control of Multiple High-Salinity Wastewater Treatment Via Membrane Distillation Crystallization

Abstract:

Crystal Morphology and Process Control of Multiple High-salinity Wastewater via Membrane Distillation Crystallization

Guannan Li, Gaohong He, Xiaobin Jiang*

State Key Laboratory of Fine Chemicals, Research and Development Center of Membrane Science and Technology, School of Chemical Engineering, Dalian University of Technology, Dalian, P. R. China

*Corresponding author: Email: xbjiang@dlut.edu.cn

Abstract

Nowadays, the separation of water recovery and salinity is mainly carried out by evaporative crystallization (ECr) [1]. However, ECr has the disadvantages of high operation temperature and uncontrolled supersaturation. Therefore, it is necessary to introduce a new separation technology [2,3].

This study is based on existing research work, and propose treatment of multiple high-saline wastewater with evaporation membrane distillation crystallization (VMDC). Coal chemical industry wastewater is a typical multiple high-saline wastewater, which is mainly composed of NaCl and $MgCl_2$. The influence of different mass fraction of NaCl and $MgCl_2$ in the feed fluid on the permeation flux of membrane has been studied. Besides, the effect of impurity $MgCl_2$ on the NaCl crystal in the process of MDC has also been investigated. The results indicated that, as the mass fraction of $MgCl_2$ in the feed increases from 0% to 100%, the viscosity of the solution gradually increases, and permeation of the water of the membrane distillation gradually decreases(shown in Fig. 1),. In addition, as the mass fraction of $MgCl_2$ in the feed increases continuously, crystal surface appearance defects has been obviously (shown in Fig. 2),and average crystal size has been gradually increased(shown in Fig. 3).

The study has been demonstrated the possibility of VMDC in the treatment of multiple high-saline wastewater. Compared with traditional evaporative crystallization, it provides a theoretical basis for the practical application of VMDC.

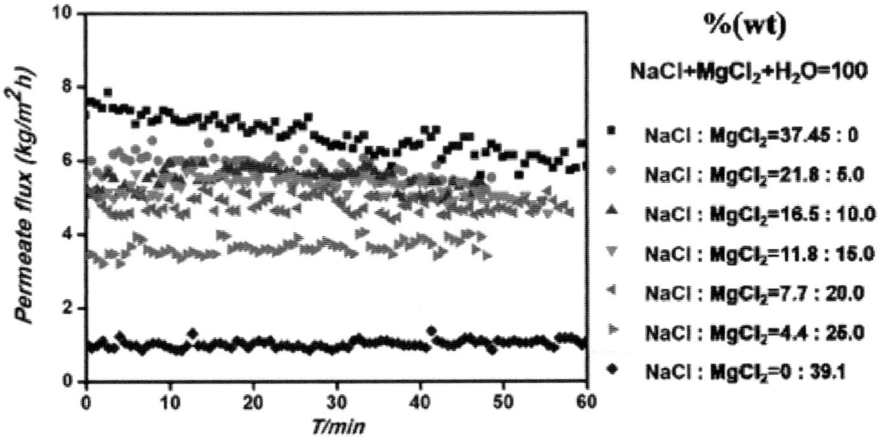

Fig. 1 The permeate flux changing with different NaCl and $MgCl_2$ mass fraction

Fig.2 The crystal surface morphology with different NaCl and $MgCl_2$ mass fraction

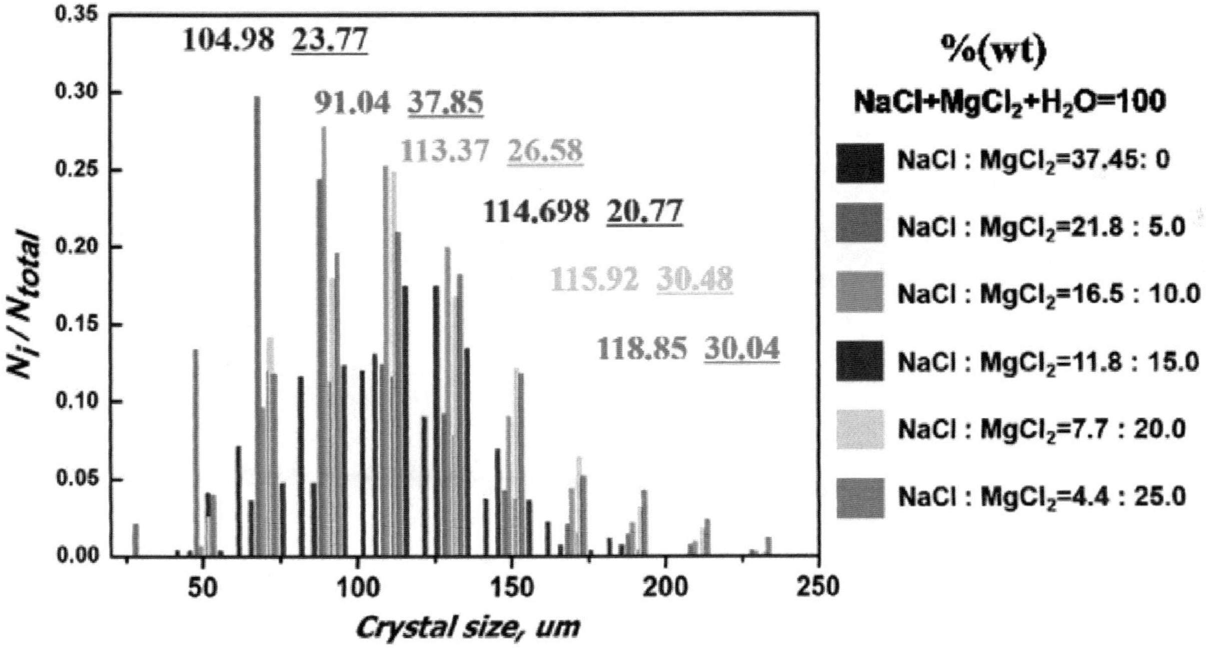

Fig. 3 The crystal size distribution with different NaCl and $MgCl_2$ mass fraction

References

[1] Lu H, Wang J, Wang T, et al. Crystallization techniques in wastewater treatment: An overview of applications[J]. *Chemosphere*, 173: 474-484 (2017).

[2] Zhang Y, Peng Y, Ji S, et al. Review of thermal efficiency and heat recycling in membrane distillation processes[J]. *Desalination*, 367: 223-239(2015).

[3] Chabanon E, Mangin D, Charcosset C. Membranes and crystallization processes: State of the art and prospects[J]. *J. Mem. Sci.*, 509: 57-67(2016).

527018 Membrane-Based Controlled Release: A Useful Tool in Oilfield Operations

Abstract:

Oilfield operations involve the use of expensive chemicals to enhance the reservoir performance and improve hydrocarbon recovery. These chemicals are often injected into hostile downhole environment existing at high temperature, pressure, salinity, and pH. There is, therefore, a need to develop advanced techniques to protect and control the release of these chemicals so that they can be successfully delivered to targeted zones underground. It is also important to improve the efficiency of these chemicals by minimizing their consumption in order to save cost and protect the environment. A very few number of publications have appeared on this topic employing technologies in the pharmaceutical, cosmetics, and agrochemical industries. Controlled release of oilfield chemicals has been tested in oilfield operations such hydraulic fracturing, and surfactant delivery in enhanced oil recovery by using the chemicals in the form of microparticles and complex nanoparticles. None of these works have explored the use of membrane technology in the designing, preparation, and testing of these chemicals. Traditional types of membrane-based controlled release used in the pharmaceutical industry include diffusion-controlled, osmotically controlled, swelling-controlled and chemically controlled membrane system. This work will evaluate possible applications of these systems in oilfield applications. It will delve into the release kinetics and how various parameters that are associated with the membrane can be harnessed for applications in controlled release and delivery of oilfield chemicals. Finally, self-assembled polymers with potential for controlled-release applications will be reviewed.

521211 Modeling and Optimization of Membrane Based Process for CO_2 Separation from Flue Gas

Abstract:

A two-stage membrane process is proposed to efficiently separate high purity CO_2 from flue gas via statistical analysis and simulation study. To collect data representing CO_2 separating performance such as CO_2 recovery rate and CO_2 concentration at the permeate side, experiments at various operating conditions have been conducted with a membrane devised for CO_2 separation from flue gas. Preliminary analysis for the data makes it possible to understand features of the membrane including relationship between the operating conditions and the above performance indices. Then, a statistical model to predict the performance indices with the operating conditions has been built and inserted into a simulation program. With the simulation program, we can perform various case studies, so that the optimal operating condition can be found out while both minimizing the number of real experiments and considering constraints of equipment.

522350 Molecular Simulation Study of Polymers of Intrinsic Microporosity Nanofilms for Organic Solvent Nanofiltration

Abstract:

Chemical separations account for 40-70% of capital and operating costs in chemical production. Existing separation technologies such as distillation are energy-intensive. As an alternative, membrane separation has been proven to be more energy-efficient and easier to operate. While membrane filtration in aqueous systems is well-established, its application in organic media has been on the rise recently. In particular, organic solvent nanofiltration (OSN) shows great potential for the molecular separation of high value solutes from organic solvents. Microporous polymer membranes are drawing considerable attention owing to their ease of processing and capability for molecular discrimination. From a molecular perspective, the study aims to explore the performance of polymers of intrinsic microporosity (PIM-1) nanofilms for OSN applications. Molecular simulations are performed for the permeation through PIM-1 nanofilms by a series of organic solvents such as acetone, acetonitrile, ethanol and methanol. Solvent permeabilities obtained are in good agreement with experimental data, thereby validating the simulation results. The permeabilities are found to correlate well with combined solvent property and comparable to those of existing polyamide nanofilms. Subsequently, molecular simulations of OSN on the rejections of hesperetin solute from organic solvents are investigated. While the solute rejection is 100%, the solvent fluxes and permeabilities decrease. For all solvents, the adsorption of hesperetin on PIM-1 is unraveled to be solute rejection mechanism. In the case of acetone, solvent permeability is the highest due to high polymer swelling and hesperetin solubility. Further detailed analyses of the molecular simulations provide the underlying insights on the separation performance. Therefore, this simulation study illustrates the performance of PIM-1 nanofilms for OSN based on the manifestation of molecular interactions among solute, solvent and membrane.

528429 A Phase Field Method for Mesoscopic Modeling of Porous Polymer Membrane Formation Via Phase Inversion

Abstract:

Phase inversion, a process that occurs when a polymer collapses from solution into a thin film upon quenching or exposure to another phase, is the most widely used process for producing porous polymer membranes. Modeling of this process has been mostly limited to the continuum level and empirically-derived parameters, with little theoretical basis to connect the process conditions and important characteristics of membrane microstructure. We have developed a large scale three-dimensional mesoscopic simulation model to examine the dynamics of the phase inversion process to predict the morphology of pore structures that develop within polymer membranes. The computational approach is a novel phase-field method based on the Cahn Hilliard equation, which incorporates the thermodynamic and transport parameters relevant to phase separation in polymer solutions. The thermodynamics of mixing for phase equilibrium is described by the Flory-Huggins free energy of mixing for binary solutions. The transport processes are assumed to be diffusive in nature and are dependent on the local temperature and polymer concentration. The proposed large scale three-dimensional model explores parameters relevant to industry in order to capture mesoscopic pore formation. A range of polymer volume fractions and quench rates were investigated to verify the modelâ€™s capability of forming bi-continuous to discrete morphologies. Analysis of pore size and continuity quantify the resulting porous structures and allow comparison to experimental results. The modeling tools developed will enable significant speed-up in membrane development. This talk will focus on initial results of the Cahn Hilliard model and compare simulated structures and morphologies to current manufactured polymer membrane filters.

529847 Molecular Simulation on Separation of CO_2/CH_4 Mixture By Carbon Membrane with Zigzag Pore Structure

Abstract:

Combustion of natural gas (CH_4) can reduce carbon dioxide (CO_2) emission by 40% compared with coals and crude oils used widely. CH_4 is therefore accepted as a relatively clean and environmentally friendly energy. Separation of CO_2/CH_4 mixture is an essential process in the natural gas industry. Membrane technique is one of the most effective options for gas separation. As a promising material, carbon molecular sieve membrane (CMSM) yields well-developed pore structures and high permselectivity. Compared with experimental study, modelling can provide reproduction and prediction of a process of interest, allowing ones to gain insight into the process on the microscopic scale and even to reveal some new information. However, classical transport theory could not be suitable for nano-scale pore structure like CMSM. Molecular simulation is an appropriate choice to express permeation and separation behaviours of molecules. Previous modelling of gas separation in CMSM assumed a uniform and defect-free pore structure. Recent experimental investigations showed that some pore imperfections like malposition and defect were actually existent inside CMSM. Such imperfections may be the reason why simulation results did not agree with experimental data. It is therefore important to develop a new pore structure to represent realistic pores in CMSM concisely.

Zigzag-type pore structure was proposed for the separation of CO_2/CH_4 binary mixture by CMSM. Defect-free pore and three kinds of defect pores, i.e., random, uniform and partial defects, were considered to improve the separation performance. Adsorption and diffusion behaviours of pure gases and gas mixture were simulated by the Grand-Canonical Monte-Carlo (GCMC) method and the Non-Equilibrium Molecular Dynamics (NEMD) method, respectively. Operating pressure ranged from 10 to 100 kPa and temperature was between 273 and 348 K during simulation. The simulative isotherms of CO_2 and CH_4 showed an acceptable agreement with the experimental data for the defect-free pore of 0.67 nm in size at 298 K. Examination on the effect of pore sizes showed that 0.67 nm was the appropriate option for separation. The total selectivity was 20.1 at 298 K and 100 kPa with the pore of 0.67 nm, which is consistent with the experimental value. The adsorption was determined as the dominant separation mechanism between adsorption and diffusion. Compared with the defect-free pores, the introduction to the random and uniform defect pores can improve the total selectivities and the random defect provided the superior separation performance. Appropriately low temperature and small pore size were beneficial to the separation.

Acknowledgement

The authors express their sincere appreciation to the financial supports from the Fundamental Research Funds for the Central Universities of China (DUT17JC07) and the National Natural Science Foundations of China (21376037, 21436009).

References

[1] Baker R W. Future directions of membrane gas separation technology. Membrane Technol. 2001, 138, 5–10.

[2] Li L, Song C W, Jiang D W, Wang T H. Preparation and enhanced gas separation performance of Carbon/Carbon nanotubes (C/CNTs) hybrid membranes. Sep. Purif. Technol. 2017, 188, 73–80.

[3] Xu L F, Sedigh M G, Tsotsis T T, Sahimi M. Non-equilibrium molecular dynamics simulation of transport and separation of gases in carbon nanopores. II. Binary and ternary mixtures and comparison with the experimental data. J. Chem. Phys. 2000, 112, 910–922.

[4] Pan Y Q, Wang T T, Sun H M, Wang W. Preparation and application of titanium dioxide dynamic membranes in microfiltration of oil-in-water emusions. Sep. Purif. Technol. 2012, 89, 78–83.

[5] Pan Y Q, Wang W J, Wang W, Wang T H. Prediction of particle deposition and layer growth in the preparation of a dynamic membrane with cross-flow microfiltration. RSC Adv. 2015, 5, 89015–89024.

[6] Wang T H, Zhang B, Qiu J S, Wu Y H, Zhang S H, Cao Y M. Effects of sulfone/ketone in poly (phthalazinone ether sulfone ketone) on the gas permeation of their derived carbon membranes. J. Membr. Sci. 2009, 330, 319–325.

[7] Furukawa S, Sugahara T, Nitta T. Non-equilibrium MD studies on gas permeation through carbon membranes with belt-like heterogeneous surfaces. J. Chem. Eng. Jpn. 1999, 32, 223–228.

[8] Song C W, Wang T H, Jiang H W, Wang X Y, Cao Y M, Qiu J S. Gas separation performance of C/CMS membranes derived from poly (furfuryl alcohol) (PFA) with different chemical structure. J. Membr. Sci. 2010, 361, 22–27.

520616 Combined Concentration Polarization and Pore-Flow Modeling to Predict the Performance of a Nano Filtration Membrane for NaCl Rejection

Abstract:

Abstract:

Membrane based filtration techniques are gaining importance day by day for its effectiveness and simplicity [1]. Different pressure driven techniques are exercised for membrane separation; nano filtration is one of them. Nano filtration membranes stand handy for different applications because of their thermal, chemical and mechanical stability [2]. High surface charge of these membranes help in separation of charged species from a solution. This particular property can be used to segregate any type of salt from water. Hence, it can be a useful technique for desalination of water which is a challenge for present day's world [3,4].

Modeling and simulation always provides useful lead to understand the underlying physical phenomenon behind a real life system. It also proves to be useful in scaling up the system and predict its performance in remote scenarios. Transport phenomena based membrane modeling is not a new field of study; a number of works have been reported till date which deals with it. Different studies have been done starting from prediction of different fouling mechanism [5,6] of membranes to prediction of flux for different processes [7,8]. In order to predict the flux and permeate concentration across the membrane a number of theoretical models have been proposed such as (a) osmotic pressure model [9–11]; (b) film theory [12–14]; (c) solution diffusion model; (d) Kedem-Katchalsky equation etc. Each of these models has their advantages and limitations also. In film theory the mass transfer boundary layer is assumed which under predicts the permeate flux of the membranes. Prediction of exact relation between osmotic pressure and solute concentration is very hard and often leads to erroneous flux prediction. If these models are used in tandem with similarity solution then gives further erroneous result. Similarity solution technique uses the order of magnitude analysis method which is an approximation method to club different coordinates into a uniform one. On the other hand integral analysis gives much accurate result than its similarity counterpart. Pore flow model on the other hand gives extremely satisfactory result in case of transport of charged species through membranes which possesses effective surface charge [15–17]. This model uses the Nernst-Planck equation within the membrane pores to predict actual amount of species transport through a membrane which other models fail to predict to that accuracy.

However, to the best of the knowledge of the authors there is no attempt has been made where integral method is coupled to the pore flow model in order to predict the flux and amount of species transported across a membrane with realizable surface charge. In this present work attempt has been made to bridge this gap. Here integral method has been used to calculate the concentration within the mass transfer boundary layer and on the membrane surface and then pore flow technique is used to predict actual flux and concentration profile at permeate. Simulation data are then compared with the experimental findings and it can be observed from the results that they are in excellent agreement with each other. This study will be extremely helpful in predicting the performance of nano filtration membranes and also to scale up the process.

Key words: Membrane; pore flow, nano filtration,

Result:

Some results are given below which shows the parity between experimentally determined permeate flux and concentration (with respect to the transmembrane pressure) with that of the theoretical one.

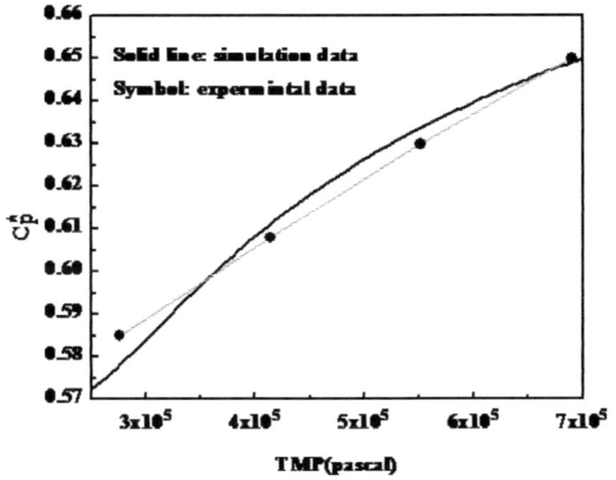

Variation of non dimensional permeate concentration with transmembrane pressure.

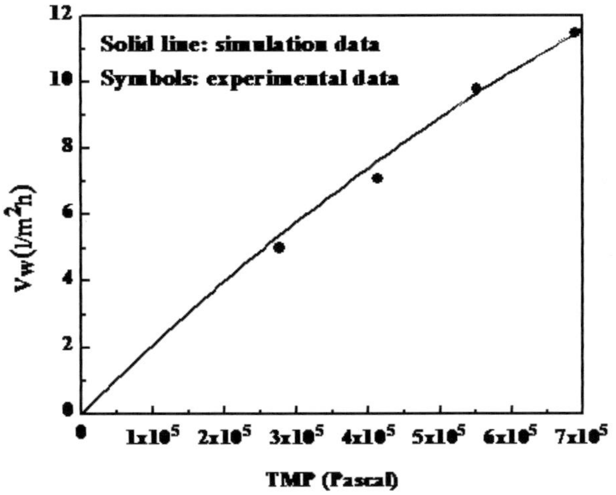

Variation of permeate flux with transmembrane pressure.

References:

[1] G. Hagmeyer, R. Gimbel, Modelling the rejection of nanofiltration membranes using zeta potential measurements, Sep. Purif. Technol. 15 (1999) 19–30. doi:10.1016/S1383-5866(98)00050-1.

[2] S.R. Panda, S. De, Preparation, characterization and performance of ZnCl2 incorporated polysulfone (PSF)/polyethylene glycol (PEG) blend low pressure nanofiltration membranes, Desalination. 347 (2014) 52–65. doi:10.1016/j.desal.2014.05.030.

[3] M. Talaeipour, J. Nouri, A.H. Hassani, A.H. Mahvi, An investigation of desalination by nanofiltration, reverse osmosis and integrated (hybrid NF/RO) membranes employed in brackish water treatment, J. Environ. Heal. Sci. Eng. 15 (2017) 18. doi:10.1186/s40201-017-0279-x.

[4] C.K. Diawara, Nanofiltration Process Efficiency in Water Desalination, Sep. Purif. Rev. 37 (2008) 302–324. doi:10.1080/15422110802228770.

[5] S. Mondal, S. De, A fouling model for steady state crossflow membrane filtration considering sequential intermediate pore blocking and cake formation, Sep. Purif. Technol. 75 (2010) 222–228. doi:10.1016/j.seppur.2010.07.016.

[6] S. Mondal, S. De, Generalized criteria for identification of fouling mechanism under steady state membrane filtration, J. Memb. Sci. 344 (2009) 6–13. doi:10.1016/j.memsci.2009.08.015.

[7] S. Mondal, S. De, A. Cassano, F. Tasselli, Modeling of turbulent cross flow microfiltration of pomegranate juice using hollow fiber membranes, AIChE J. 60 (2014) 4279–4291. doi:10.1002/aic.14594.

[8] R. Prabhakar, S. DasGupta, S. De, Simultaneous prediction of flux and retention for osmotic pressure controlled turbulent cross flow ultrafiltration, Sep. Purif. Technol. 18 (1999) 13–24. doi:10.1016/S1383-5866(99)00043-X.

[9] M. Elimelech, S. Bhattacharjee, A novel approach for modeling concentration polarization in crossflow membrane filtration based on the equivalence of osmotic pressure model and filtration theory, J. Memb. Sci. 145 (1998) 223–241. doi:10.1016/S0376-7388(98)00078-7.

[10] J.G. Wijmans, S. Nakao, C.A. Smolders, Flux limitation in ultrafiltration: Osmotic pressure model and gel layer model, J. Memb. Sci. 20 (1984) 115–124. doi:10.1016/S0376-7388(00)81327-7.

[11] H. Nabetani, M. Nakajima, A. Watanabe, S.-I. Nakao, S. Kimura, Effects of osmotic pressure and adsorption on ultrafiltration of ovalbumin, AIChE J. 36 (1990) 907–915. doi:10.1002/aic.690360612.

[12] Z.V.P. Murthy, S.K. Gupta, Estimation of mass transfer coefficient using a combined nonlinear membrane transport and film theory model, Desalination. 109 (1997) 39–49. doi:10.1016/S0011-9164(97)00051-9.

[13] J.D. Rogers, R.L. Long, Modeling hollow fiber membrane contactors using film theory, Voronoi tessellations, and facilitation factors for systems with interface reactions, J. Memb. Sci. 134 (1997) 1–17. doi:10.1016/S0376-7388(97)00074-4.

[14] J. Miranda, Mass transfer in the vicinity of a separation membrane—the applicability of the stagnant film theory, J. Memb. Sci. 202 (2002) 137–150. doi:10.1016/S0376-7388(01)00747-5.

[15] P. Banerjee, S. De, Modeling of Nanofiltration of Dye Using a Coupled Concentration Polarization and Pore Flow Model, Sep. Sci. Technol. 46 (2011) 561–570. doi:10.1080/01496395.2010.534525.

[16] P. Banerjee, S. De, Coupled concentration polarization and pore flow modeling of nanofiltration of an industrial textile effluent, Sep. Purif. Technol. 73 (2010) 355–362. doi:10.1016/j.seppur.2010.04.023.

[17] S. Bhattacharjee, J.C. Chen, M. Elimelech, Coupled model of concentration polarization and pore transport in crossflow nanofiltration, AIChE J. 47 (2001) 2733–2745. doi:10.1002/aic.690471213.

530725 Investigation of Oceanic Microfiber Pollution and Development of Inexpensive Filtration Units to Reduce That from Residential and Commercial Washing Machines

Abstract:

Polymer-based clothing, such those composed of nylon and polyester, contribute to a significant percentage of all micron-scale oceanic plastic pollution in the form of microfibers. These microfibers fall off clothes during the high RPM and vigorous shaking of a typical washing machine unit, and exit in the wastewater expelled from the units during the washing cycle. Too small to be filtered out by municipal septic systems, these microfibers enter natural bodies of water and act like micro- sponges for organic compounds. Furthermore, these fibers are also ingested or inhaled by marine life that may also be consumed by humans. This research seeks to investigate the effects of polymer distribution in clothing material, load size, clothing age, and cycle type on the number of microfibers that leave washing machines in an average cycle. Preliminary research obtained by microscopic analysis has estimated one 80% polyester, 20% nylon jacket to contribute between .02 - .07 ± .016 g/L of mircofibers to the wastewater of a typical washing machine cycle. In addition to quantifying the mass of microfiber pollution, simultaneous research is also being performed on developing inexpensive membrane separators that can reduce the amount of microfiber pollution. Current research, also using microscopic analysis, has found both cellulose and sand as microfiltration membranes to be sufficient at removing 95 â€“ 99% of all microfibers in washing machine wastewater. Further studies to scale these filters are being performed, with heavy consideration towards their durability, longevity, and ease to clean and repair.

534786 Electrospun Polyvinylidene Fluoride Membranes for Direct Contact Membrane Distillation

Abstract:

An estimated 1.1 billion people worldwide lack access to water and 2.7 billion people will find water scarce for at least a month of the year. With this, 2.4 billion people are exposed to water-borne diseases due to inadequate sanitation processes. The driving forces behind water scarcity are the growing uses of fresh water and the depletion of natural freshwater sources. About two-thirds of the worldâ€™s population are expected to experience water shortages by the year 2025. Addressing this problem, several methods of water purification including reverse osmosis, multistage flash distillation, and direct contact membrane distillation have been explored. Of these options, Direct Contact Membrane Distillation (DCMD) has the lowest manufacturing and operating costs and seems the most promising technology. Current DCMD membranes are expensive and inefficient due to the lack of optimization of the membraneâ€™s desirable properties. To solve this, the fabrication method of free surface electrospinning was used, it allowed for the control of properties such as fiber diameter, pore size, and membrane thickness. Membranes were electrospun from a 22 wt% polyvinylidene fluoride (PVDF) solution with 78 wt% dimethylacetamide (DMAC), each being specifically chosen for their combined properties. Free surface electrospinning is achieved by using a rotating wire spindle partially submerged in the PVDF/DMAC solution. Over the 6.5 inch working distance, a 40 kV voltage was applied while the humidity and spin times were varied, 60-80% and 30-60 minutes respectively. A grounded rotating drum collected all the gelled fibers which solidified there. The produced membranes were then characterized by their thickness readings and scanning electron micrograph (SEM) images. Finally, a DCMD apparatus was setup to test the productivity of the membranes by distilling salt water through them. These productivity values were compared to those of commercially produced membranes, produced by phase-inversion, through the same DCMD setup.

525637 Modeling the Effects of Mass Transfer on Microstructure Formation in Polymer Membranes

Abstract:

Widely used in industry, phase-inversion membranes represent the most common instance of a self-assembled polymer membrane. However, despite their ubiquity, quantitatively predicting and controlling their microstructure remains a difficult challenge. This challenge persists because phase-inversion membranes are produced by a non-equilibrium processes that is challenging to model due to the need to capture phase separation kinetics and complex hydrodynamics over a wide range of length and time scales. We have recently developed and validated a "multi-fluid" model that is capable of spanning these scales by incorporating continuum fluid dynamics and a field-theoretic description of polymer thermodynamics. We proceed to use this model to explore the kinetics of the mass-transfer driven phase separation that drives microstructure formation in these membranes. We will show that our model predicts the existence of different kinetic regimes, which qualitatively matches experimentally observed behaviors in the phase-inversion process.

525280 Enhancing Ionic Conductivity of Anion Exchange Membrane Via Incorporating Tetra-Quaternized Calix[4]Arene

Abstract:

Anion exchange membranes (AEMs) have been widely applied in flow redox battery, fuel cells and electrodialysis. However, low ionic conductivity is difficult to satisfy the requirement of such electrochemical systems. To address this issue, trimethylamine tetra-quaternized calix[4]arene (TQCA) was successfully synthesized and incorporated into imidazolium-functionalized polysulfone AEMs (PSF-ImOH/TQCA). The unique conical architecture of TQCA, i.e. cationic groups at the upper rim and alkyl chains at the lower, makes it easy to aggregate into a globular micelle. The cationic groups at the outer edge of the globular micelle could involve imidazolium groups to form big ionic clusters, which are confirmed by TEM images and SAXS. The cavity of calix[4]arene functions as a reservoir for water storage due to the high free volume, increasing water uptake for effective OH^- transport and controlling dimensional change at the same time. Consequently, the PSF-ImOH/TQCA shows a remarkable 1.7-fold enhancement of ionic conductivity at 25 °C over that of the membrane without TQCA. Additionally, the upper or lower rims of calixarenes can be readily functionalized, which provides more possibilities for preparation of high-performance AEMs.

We gratefully acknowledge the financial support of the National Science Foundation of China (21476044, 21406031 and 21776034) and the Changjiang Scholars Program (T2012049).

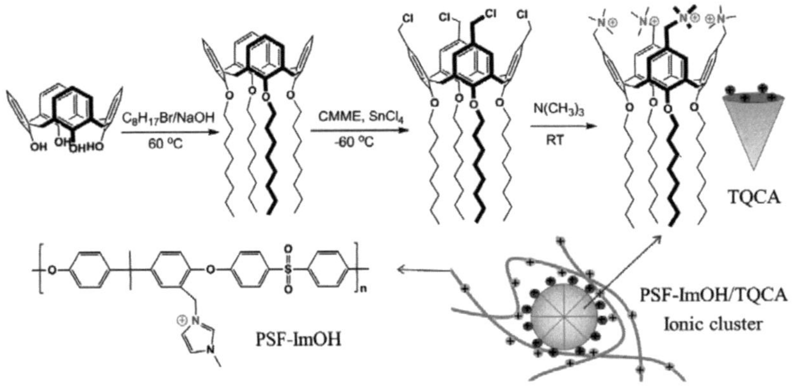

524954 Highly Hydroxide Conductive Quaternized Polybenzimidazole Anion Exchange Membranes

Abstract:

Polybenzimidazole (PBI) is a polymer with excellent mechanical, thermal and chemical stabilities. However, most anion exchange membranes (AEMs) based on quaternized PBI reported in the literature suffer from extremely poor hydroxide conductivity compared to other AEM categories. In this study, it is found that the poor hydroxide conductivity might be caused by the electrostatic interaction between imidazole anion and quaternary ammonium cation, which makes quaternization of PBI difficult and limits the formation of ion clusters in PBI AEM. Based on this analysis, highly hydroxide conductive quaternized PBI AEM are prepared by a two-step method to avoid the electrostatic interaction: First, PBI was grafted non-ionic side chains via N-substitution reaction, and then reacted with tri-quaternization reagent. The designed membrane exhibits high hydroxide conductivity (154 mS cm^{-1}) and low swelling ratio (23%) at 80°C. After immersion in 60 °C, 1 M KOH solution for 168 h, this hydroxide conductivity of the membrane still remains constant. The two-step method mentioned in this work provides a universal promising method to prepare high hydroxide conductivity and alkaline stability quaternized PBI anion exchange membranes.

The authors thank the National Science Foundation of China (21476044, 21406031 and 21776034), the National Key Research and Development Program of China (2016YFB0101203) for financial support of this work.

527573 One-Step Formation of Polyethersulfone Inner-Selective Hollow Fiber Membranes for Dye Removal

Abstract:

The development of a loose inner-selective polyethersulfone (PES) hollow fiber by one-step spinning process for nanofiltration (NF) has been demonstrated in this study. It is found that the addition of a small amount of sulfonated polysulfone (SPSf) and the use of a specially designed bore fluid made of hydrophilic additives such as polyethylene glycol (PEG) and polyvinylpyrrolidone (PVP) are of great importance to fabricate this type of NF membranes. Not only are they able to reduce the pore size but also suppress the formation of finger-like macrovoids. After optimizing both dope solution and spinning conditions, the as-fabricated membrane can have a 70 % rejection to PEG 1000 and a pure water permeability (PWP) of 12.8 $Lm^{-2}h^{-1}bar^{-1}$. Its rejection of indigo carmine (MW: 466 $gmol^{-1}$) can reach 96.64 %. In addition, it shows good stability in a continuous 76-hour operation with good efficiency of dye removal. The newly developed membrane may have great potential in the treatment of textile wastewater.

520601 Zero Valent Iron Nps Impregnated UF Membrane for Nitrobenzene Reduction and Fluoride Rejection

Abstract:

Zero valent iron NPs impregnated UF membrane for nitrobenzene reduction and fluoride rejection

Piyal Mondal, Mihir Kumar Purkait*

Iron nanoparticle (NPs) impregnated PVDF-co-HFP membrane was prepared by a novel method of phase inversion technique where instead of water, 0.5 (M) $FeSO_4$ solution was used as non-solvent. The phase inversed casted membrane was then dipped into a solution (70 °C) of clove extract (optimized with respect to total flavonoid and phenolic content). Green synthesis of iron nanoparticles on the membrane surface and within the pores occurs where iron salt solution remains during phase inversion process. The surface and cross-sectional FESEM morphology images confirms the formation and well homogenous distribution of iron NPs over the membrane and within the pores.

Clove extract acts as reducing agent for the iron salt. Freshly prepared zero valent iron NPs impregnated membrane was then used for nitrobenzene reduction (initial concentration 100 ppm) and fluoride removal study. After each run of nitrobenzene reduction study zero valent iron gets oxidized. In order to reuse and retain the reduction capacity of the membrane 0.1N HCl and H_2SO_4 mixture was passed through the membrane and again it was dipped in the prepared clove extract solution.

It was observed that aniline formation was 19.5 ppm during 1st run at first 10 min of operation and gradually the value decreases to 14.4 ppm and 13.7 ppm at the end of 2nd and 3rd run respectively. Moreover the iron oxide formed on the membrane surface and membrane pores contributed well towards the fluoride rejection study due to increased surface area of iron NPs. The maximum fluoride removal was observed 72.4% during 1st run for 20 ppm stock solution and it gradually decreases with lower initial fluoride concentration (69.4% for 10 ppm solution and 55.2% for 5 ppm solution). Similarly with run order the values gradually decreases . Hence, through this investigation a reusable membrane for nitrobenzene reduction and fluoride rejection was fabricated by a new technique where $FeSO_4$ solution used as non solvent additive during phase inversion process.

525149 Organic Solvent Nanofiltration (OSN) Membranes Made from Plasma Grafting of Polyethylene Glycol on Cross-Linked Polyimide Ultrafiltration Substrates

Abstract:

A novel organic solvent resistant nanofiltration (OSN) membrane was developed via plasma grafting of a high-density polyethylene glycol (PEG) on a cross-linked polyimide (cPI) ultrafiltration (UF) substrate in this work. The argon plasma treatment successfully formed an ultrathin dense layer on the substrate surface and directly converted the UF membrane to a nanofiltration (NF) membrane. The PEG-modified composite membrane has an isopropanol (IPA) permeance of 5.91 Lm$^{\hat{a}``2}$h$^{\hat{a}``1}$bar$^{\hat{a}``1}$ at 10 bar because of its relatively high polarity. Comparing to the cPI membrane, the PEG-modified membrane shows a significant increase in rejection of Rose Bengal (a molecular weight of 1017.64 gmol^{-1}) from 70.10% to 99.64% in IPA. The newly developed OSN membranes also demonstrate good stability in alcohols for 400-hour tests, especially in IPA and butanol. To our best knowledge, this is the first endeavor of utilizing PEG to directly develop OSN membranes by means of plasma grafting. The work may provide useful insights to design green, cost-effective and high-performance composite OSN membranes.

537774 Hydrophilic ZSM-5 Zeolite Membrane for Forward Osmosis

Abstract:

Forward osmosis (FO) is a promising membrane separation process in water treatment, osmotic membrane bioreactor, food processing and power generation. In these applications, FO has advantages with low energy consumption and less membrane fouling compared with reverse osmosis, nanofiltration and ultrafiltration. Although various types of polymeric membranes and mixed matrix membranes has been studied for FO operations, pure inorganic membranes for FO were rarely reported.

ZSM-5 is one of the most popular synthetic zeolite which is attracted attention as inorganic membrane material. ZSM-5, aluminum-containing MFI-type zeolite have been studied as membrane materials to separate water from alcohols and organic acids.â€€This type of zeolite possesses hydrophilicity and molecular sieving property. In this study, we demonstrated the potential of a hydrophilic ZSM-5 membrane for FO operations. This is the first report about ZSM-5 membrane for FO operations in our best knowledge.

ZSM-5 membrane was prepared on the outer surface of tubular a-alumina support (outer diameter, 10 mm; inner diameter, 7 mm; length, 30 mm; average pore size, 700 nm) by a secondary growth method. The support was seeded by using dip-coating method. The seeded support was set in a PTFE-lined autoclave with synthesis solution having the molar composition of Al_2O_3:240SiO_2:53.3Na_2O:8000H_2O, and then hydrothermal treatment was carried out at 453 K for 12 h under stationary condition. After the crystallization, obtained membrane was washed with boiling water for several hours and dried at 383 K overnight.

FO performance of the ZEM-5 membrane was investigated as follows. NaCl aqueous solution and distilled water were used as the draw solution (DS) and the feed solution (FS), respectively. The DS and FS were fed to the outside and inside of the tubular ZSM-5 membrane. Each flow rate was adjusted to be 20 mL min^{-1}. The fluxes of water and salt were calculated based on the changes of the volume of DS and of Na concentration in FS.

Dependencies of the of water and salt fluxes on the NaCl concentration in DS were investigated. The water flux increased from 0.96 to 2.2 L m^{-2} h^{-1} with increasing NaCl concentration from 2.0 to 9.0 wt% at 313 K. In contrast, the salt flux was almost constant ca. 0.6 g m^{-2} h^{-1} despite increasing DS concentration. The salt flux was very low even in the case of using 9.0 wt% DS. In addition, the water flux increased with rising membrane temperature from 298 to 333 K, and the salt flux was almost constant despite changing temperature.

The FO performance under acidic condition was also investigated. pH value of DS was adjusted to be 4.0 by adding hydrochloric acid. As a result, the water and salt fluxes hardly changed under acidic condition compared with those obtained under neutral conditions, suggesting that the ZSM-5 membrane kept its performance under such acidic condition. Consequently, we found the possibility for FO operation with zeolite membrane for the first time.

526110 Gas Separation Performance of Polymer-Ionene Hybrids

Abstract:

Ionic liquids (ILs) have been explored as media for membrane-based gas separations, both in supported liquid membrane (SLMs) and polymerized ionic liquid (poly(ILs)) forms. Ionenes are another class of ionic polymers wherein the ionic groups are located directly within the polymer backbone, yet there is very little research focusing on the use of ionenes as gas separation membranes. While many ionenes are relatively simple ammonium salts tethered by long alkyl chains, there are ample opportunities to design much more complex ionenes which are inspired by the other classes of polymer materials such as polyimides, thermally-rearranged (TR) polymers, and polymers of intrinsic microporosity (PIMs). We are investigating the design and synthesis of novel polymer-ionene hybrids derived from structural and functional components typically associated with these high-performance and ultra-high-performance polymer materials. These sophisticated material classes possess distinct advantages and disadvantages, yet there are a number of unique properties and behaviors which may arise from a precise spacing and alternating sequence of multiple types of linkages (i.e. amide-imide) and imidazolium cations along the main chain. Here, we present the design philosophy and development of polymer-ionene hybrids resulting from complex monomer units and the impact of design on gas separation performance.

532855 Mitigation of Bidirectional Solute Flux Via Membrane Surface Coating of Zwitterion Functionalized Carbon Nanotubes in Forward Osmosis Process

Abstract:

Osmotically driven membrane processes are being developed to address global freshwater shortage, wastewater reuse, and water-energy nexus. Among those processes, forward osmosis (FO) utilizes osmotic pressure gradient across a semi-permeable membrane to reclaim high-quality water for versatile applications, such as irrigation, food processing and life support. The major advantages of FO-based water treatment systems include reduced operating pressure, high rejection of undesired compounds, low fouling intensity of FO membranes, and less energy demand if energy-intensive draw regeneration can be properly addressed [1, 2].

A key challenge and also impediment for FO applications is the bidirectional solute flux, including reverse solute flux (RSF) and forward solute flux (FSF). RSF is defined as the cross-membrane diffusion of draw solutes to the concentrated feed solution, and can result in severe loss of draw solutes and gradual salinity build-up at the feed side, leading to reduced osmotic driving force, increased fouling propensity, and elevated operation cost due to periodical replenishment of draw solutes [3, 4]. Accumulation of the reversed draw solutes will require further treatment of the feed solution before it can be discharged to a natural water body. Thus, mitigation of solute accumulation at the feed side is important, and several approaches have been proposed and studied. However, most of these approaches, including integrated microfiltration or electrodialysis, can only reduce salinity accumulation in the feed solution, they cannot mitigate or reduce RSF effectively.

FSF (also known as membrane rejection) is defined as the cross-membrane diffusion of feed solute to the diluted draw, and can lead to undesirable contamination of final product water and enhanced fouling propensity in downstream draw solute regeneration process (e.g. reverse osmosis, RO). Previous study indicated that ~20% of the phosphorus in wastewater (the feed) could penetrate through FO membrane and accumulate in the final draw [5]. Similar to RSF, FSF is resulted from FO membrane failing to be a perfect barrier, and has yet to be effectively addressed by advanced membrane fabrication or modification.

Herein, mitigation of bidirectional solute flux in FO process has been comprehensively investigated via surface coating of zwitterion-functionalized carbon nanotubes (Z-CNT). If embedded in membrane structure, the Z-CNT was confirmed to enhance RO desalination performance, leading to elevated salt rejection and enhanced biofouling resistance [6]. Hence, the Z-CNT was further introduced into FO process to evaluate potential mitigation of bidirectional solute flux via surface deposition. Specific RSF and FSF with a unit of gram diffused solute per liter recovered water (g L^{-1}) were selected to enable better comparison with previous studies. With simple evaporation coating of 0.1 mg cm^{-2} Z-CNT on active layer, the coated commercial FO membrane presented a consistent RSF of 0.28 ± 0.02 g L^{-1} for four consecutive cycles (12 hours each with 1 mol L^{-1} NaCl as the draw and deionized water as the feed), rendering a 56.3% reduction comparing to that of the original membrane (0.64 ± 0.01 g L^{-1}). Comparable water flux ($p>0.10$) was observed between the coated (12.17 ± 0.26 LMH) and the original (12.46 ± 0.73 LMH) FO membranes. Subsequently, various initial draw concentrations (0.25, 0.50, 0.75, and 1.00 mol L^{-1} NaCl) were successively tested, rendering a very stable and mitigated specific RSF (~0.30 ± 0.01 g L^{-1}). The Z-CNT coated FO membrane also exhibited effective RSF mitigation over a wide selection of commonly used draw solutes, including $(NH_4)_2HPO_4$ (74.5% reduction), $NH_4H_2PO_4$ (83.8% reduction), NH_4Cl (70.8% reduction), and NH_4HCO_3 (61.9% reduction). Further investigation focused on assessing potential mitigation of specific FSF by replacing deionized water with synthetic wastewater as the feed and 1 mol L^{-1} NaCl as the draw. The penetration efficiencies of contaminants across the Z-CNT coated FO membrane were significant reduced from 87.8% to 50.2% for NH_4^+-N, 26.1% to 16.4% for NO_3^--N, 38.5% to 27.5% for $NO2^-$-N, 60.9% to 26.4% for K^+, 4.6% to 0% for Mg^{2+}, 0% to 0% for Ca^{2+}, 4.3% to 2.6% for SO_4^{2-}-S, and 2.4% to 0% for PO_4^{3-}-P. These results have collectively demonstrated that consistent mitigation of bidirectional solute flux in commercial FO membrane could be effective achieved with simple surface coating of Z-CNT and warrants further investigation on membrane fouling resistance by using real wastewater.

References

[1] Liu, Z., Bai, H., Lee, J. and Sun, D.D. (2011) A low-energy forward osmosis process to produce drinking water. Energ. Environ. Sci. 4(7), 2582-2585.

[2] Su, J., Zhang, S., Ling, M.M. and Chung, T.S. (2012) Forward osmosis: an emerging technology for sustainable supply of clean water. Clean Technol. Environ., 1-5.

[3] Achilli, A., Cath, T.Y. and Childress, A.E. (2010) Selection of inorganic-based draw solutions for forward osmosis applications. J. Membr. Sci. 364(1), 233-241.

[4] Boo, C., Lee, S., Elimelech, M., Meng, Z. and Hong, S. (2012) Colloidal fouling in forward osmosis: Role of reverse salt diffusion. J. Membr. Sci. 390â€“391, 277-284.

[5] Zou, S., Qin, M., Moreau, Y., He, Z. Nutrient-Energy-Water Recovery from Synthetic Sidestream Centrate Using a Microbial Electrolysis Cell - Forward Osmosis Hybrid System. J. Clean. Prod. 154, 16-25.

[6] Chan, W.F., Marand. E., Martin, S.M. Novel zwitterion functionalized carbon nanotube nanocomposite membranes for improved RO performance and surface anti-biofouling resistance, J. Memb. Sci. 509 (2016) 125â€“137.

525792 Hydrosilylation-Based UV-Curable PDMS Pervaporation Membranes for *N*-Butanol Recovery

Abstract:

Pervaporation (PV) using a membrane is the most attractive technology for the effective recovery or separation of organic compounds due to its energy and cost effectiveness compared to other conventional separation methods. Polydimethylsiloxane (PDMS) has been the most widely used PV membrane material for the alcohol recovery. The PDMS membrane is typically prepared using casting method followed by a heat-curing process. However, the heat-curing process is time-consuming and its batch-wise nature requires a large-scale equipment for the commercialization. In this regard, a UV-curing process is beneficial since its curing rate is fast even at room temperature and it can be processed in a continuous manner. In this study, for the first time, we fabricated the PDMS-based PV membranes (pristine and mixed matrix membranes (MMM) containing silicalite-1) using a new class of hydrosilylation-based UV-curable PDMS for *n*-butanol recovery. The structures and separation performance of the prepared UV-cured PDMS membranes (UV-PDMS) were compared with those of conventional heat-cured PDMS membranes (Heat-PDMS). While there was no difference in PV performance between UV-PDMS and Heat-PDMS for the pristine membranes, the difference in PV performance was remarkable for the MMMs. Especially, for the silicalite-1 loading of 20 %, the UV-cured MMM exhibited an 58 % increase in flux with similar separation factor compared to the heat-cured MMM. This significant increase in flux was attributed to the unique structure of the UV-cured MMM having the increased surface roughness and loosely packed morphology.

530193 Removal of Antibiotics Using Polyethylenimine Cross-Linked Nanofiltration Membranes: Relating Membrane Performance to Surface Charge Characteristics

Abstract:

In this study, surface positive charge intensity of a polyethylenimine (PEI) modified nanofiltration (NF) membrane was characterized by adsorption of counter ions WO_4^{2-} on the membrane surface at pHs 2-10 followed by quantitative analysis using X-ray photoelectron spectrometry (XPS). Retentions of NaCl and two representative antibiotics (cefadroxil and enrofloxacin) by this membrane were systematically evaluated at pHs 2-10. It was found that the PEI (25000 $g{\cdot}mol^{-1}$) modified NF membrane exhibited positive charge and the charge intensity was strong, moderate and weak in the pH range of < 4, 4 - 7, and 8 - 10, respectively. Such surface charge characteristics at different pH had significant effect on separation performance of ionic species such as NaCl and antibiotics. Donnan exclusion mechanism played a critical role in NaCl retention across different pH. The electrostatic interaction between the dissociated species of antibiotics and the membrane surface governed the retention behavior at various pH conditions. The orientation of the polar antibiotic molecules may be induced by the membrane surface charge, which could decrease retentions. This study may provide fundamental understanding of positively charged NF membrane application in removal of trace organic contaminants.

532810 A Thin Film Composite Membrane Prepared from Monomers of Guaiacol and Trimesoyl Chloride for Organic Solvent Nanofiltration

Abstract:

Organic solvent nanofiltration (OSN) membranes have shown promising applications in the recovery of antibiotic residues in organic solutions and the separation of homogeneous catalysts in organic mixtures. For the strong polar organic solvents including tetrahydrofuran (THF), N,N-dimethylformamide (DMF), N-methyl pyrrolidone (NMP), dimethyl sulfoxide (DMSO), etc., there still need rational approaches to explore solvent resistant polymeric NF membranes with high separation performance. Guaiacol (GA) and trimesoyl chloride (TMC) were adopted as the monomers to synthesize a hydrophobic polyarylester OSN membrane via interfacial polymerization on the ethylenediamine cross-linked poly(ether imide) substrate. The optimal NF-3GA membrane exhibits the higher stability in THF and the superior separation performance with the rejection of 96% for Rose Bengal (RB) and a THF permeance of 76 L m^{-2} h^{-1} MPa^{-1}. The relationship between the membrane permeance and the membrane-solvent solubility parameter distance were analyzed using Hansen solubility parameter methods. The results provide a promising route for the selection of monomers to synthesize OSN membranes with high separation performance.

536192 Preparation and Characterization of Graphene Oxide-Based Nanofiltration Membranes for Water Desalination

Abstract:

To address the pressing global issue of clean water scarcity, water desalination and purification technologies are being used widely to turn seawater to drinkable water, yet we face challenges with the design of sustainable, energy-efficient technologies. Membrane-based separation methods for desalination can provide adequate clean water resources. Desalination technologies used today have high energy consumption and capital costs. Preparation of novel membranes which can reduce the energy consumption together with high water permeability and high salt rejection capacity is highly demanding. Graphene-oxide (GO) based nanofiltration membranes have high stability and high ion rejection for seawater desalination to surpass the present upper bounds for better membrane performance. Recently, GO materials have emerged as potential candidates with excellent desalination characteristics. In this research, graphene oxide is used to fabricate membranes for desalination. Subsequently, the nematic liquid crystalline phases of GO are produced and then exposed to shear-alignment using a film applicator to form large-area GO membranes and membrane modules are fabricated. These modules are characterized using a variety of tests such as AFM, SEM, UV-VIS, FTIR, Raman Spectroscopy, and water permeability and bacterial adhesion tests are also carried out to test the efficiency of the membranes.

536727 Gas Transport Properties of Polysulfone Mixed-Matrix Membranes Embedded with Hexamethylenetetramine Dicyanamide Cadmium Nanoparticles

Abstract:

Polymer membranes are used wieldy in gas separation [1,2]. A major challenge in polymer membranes is the trade-off between gas permeability and selectivity [3]. To address this challenge, mixed-matrix membranes (MMMs) have been introduced and studied. These membranes consist of organic or inorganic nanoparticles dispersed in a polymeric matrix [4]. MMMs benefit from the high flexibility and processability of polymers and the outstanding thermal resistance and gas selectivity of nanoparticles. Polysulfones (PSUs) are a class of glassy polymers with high thermal, hydrolytic, and oxidative stability. They are widely used in manufacturing gas-separation and water-purification membranes [5-6]. The polymer contains sulfonyl groups, which have high affinity for CO_2 rather than CH_4, making it suitable for manufacturing CO_2/CH_4 separation membranes. However, PSU suffers from low free volume and consequently low gas permeability. This problem can be overcome by adding nanoparticles to the PSU matrix and manufacturing a MMM [7].

This paper presents a study of embedding hexamethylenetetramine dicyanamide cadmium (HDC) nanoparticles into the matrix of PSU to fabricate a MMM using solvent evaporation. The effect of HDC nanoparticles on gas transport properties was evaluated through N_2, O_2, CH_4 and CO_2 permeation tests. It was found that membranes with 2.5 wt% of the nanoparticles have the best gas transport properties. At this nanoparticle loading level, the permeability of the membrane was lower than that of the pure PSU. However, the CO_2/CH_4, CO_2/N_2 and O_2/N_2 selectivities increased sharply. For example, at the optimum loading level, CO_2/CH_4 and CO_2/N_2 gas selectivities increased by 42.6% and 61.6%, respectively, compared with the pure PSU. The optimal membrane showed CO_2/CH_4, CO_2/N_2 and O_2/N_2 selectivities of 41.66, 20.08 and 5.09, respectively. Analyses of the chemical structures of the MMMs using SEM and FTIR spectroscopy further confirmed excellent gas sieving properties of the membranes. The results indicate that the incorporation of HDC nanoparticles into polymers allows for addressing the high free volume and poor sieving ability problems in polymer membranes.

References

1. Aitken, W. Koros, and D. Paul, *Macromolecules*, **25**, 3424 (1992).
2. A. Shamsabadi, A. Kargari, F. Farshadpour, and S. Laki, *J. Membrane Sep. Technol.*, **1**, 19 (2012)
3. M. Robeson, *J. Membrane Sci.*, **320**, 390, (2008).
4. P. Isfahani, M. Sadeghi, A.H.S. Dehaghani, and M.A. Aravand, *J. Ind. Eng. Chem*, **44**, 67 (2016).
5. C. Bidsorkhi, H. Riazi, D. Emadzadeh, M. Ghanbari, T. Matsuura, W. Lau, and A. Ismail, *Nanotechnology*, **27**, 415706 (2016).
6. Mokhtari, A. Rahimpour, A.A. Shamsabadi, S. Habibzadeh, and M. Soroush, *App. Surf. Sci.*, **393**, 93 (2017).
7. Ahn, W.-J. Chung, I. Pinnau, and M.D. Guiver, J. Membrane Sci., **314**, 123 (2008).

533664 Bimetallic Nanoparticles Composite Poly(acrylic acid) Membrane for Water Remediation: Synthesis, Advance Characterization and Reactive Properties

Abstract:

The detoxification of chlorinated organics in groundwater has been achieved by using bimetallic (Fe/Pd) nanoparticle methods [1,2]. However, the agglomeration as well as the difficulties to recycle the particles prevent the application of the methods. The integration of catalytic particles and functionalized commercial membranes was achieved (in both commercial flat-sheet membranes and spiral membrane modules) to address these issues. The Fe/Pd nanoparticle was formed inside the pores of membrane through in situ reduction after chelating by introduced carboxylate groups (acrylic acid (AA) or methacylic acid (MAA) was polymerized inside the membrane). The functionalized AA/MAA could prevent nanoparticle aggregation and metal ion loss (recapture after dechlorination treatment). The metal nanoparticle-based membrane system showed great performance on the treatment of lab synergic polychlorinated biphenyls and field water samples (from a Superfund site, which includes trichloroethylene, tetrachloroethylene and carbon tetrachloride). At 2 s residence time, 95% of chlorinated organic species (ppm level) in the field sample were dechlorinated in the convective flow mode. The pilot studies with functionalized commercial spiral membrane modules are in progress.

Advanced characterization methods, such as FIB, TEM, XPS were applied to understand the correlation between nanoparticle properties and depth inside membrane pores, leading to the optimization of membrane design and treatment performance. The depth profile of functionalized membrane indicated that particle size was uniform inside membrane pores (particle size: 24±6 nm) but slightly smaller than those nanoparticles located on the surface (39±9 nm). Besides, the distribution of particles inside the membrane as well as the composition of single palladized-iron nanoparticles were analyzed by using EDS and EELS. These investigations illustrated the advantages of the Fe/Pd particle based membrane system method, compared to the conventional applications of the metal nanoparticles.

In XRD analysis, the Fe/Pd particle samples (which were deliberately oxidized and then reduced) exhibited the same crystalline patterns as the original samples. The membrane maintained reactivity after four degradation cycles with regeneration between each cycle. The increase of surface particle size of 22% resulted in a decrease of 9.7% PCB conversion for the 4 hr reaction time [3].

This research is supported by the NIEHS-SRP grant P42ES007380. Partial support is also provided by NSF KY EPSCoR grant (Grant no: 1355438).

537035 Brackish Water Desalination with a Novel Polymer Nanocomposite Membrane

Abstract:

BRACKISH WATER DESALINATION WITH A NOVEL POLYMER NANOCOMPOSITE MEMBRANE

Liliana R. Villanueva López and David Suleiman, Ph.D

Chemical Engineering Department, University of Puerto Rico, Mayaguez

The rapid growth of global population, the industrialization and environmental problems caused by global warming has made it necessary to implement advanced water purification techniques capable of fulfilling the increasing demand for drinking water. According to the UNESCO, the world population surpasses 7.0 billion people and groundwater resources across the world are limited to supply the basic daily water needs for the world.

Salty water is the most abundant on earth, which represents approximately 97% of the water available. However, it is the less recovered source of water, because of the high costs required by the ion exchange technologies used. These processes are still the most commonly used for desalination processes due to their high efficiency and environmentally benign nature.

In this research, sulfonated poly(styrene-isobutylene-styrene) (SIBS) has been used to validate its efficiency as a filtration media in brackish water desalination. The effect of counter-ion substitution in the ionic domains (*e.g.,* Ba +2, Ca +2, and Mg +2) was used as they changed the nanostructure and the resulting permeability of water and salt.

Characterization techniques employed to evaluate physicochemical properties of the polymer membranes include: infrared spectroscopy, energy dispersive X-ray spectroscopy thermogravimetric analysis, elemental analysis, water uptake, ion exchange capacity, water and salt permeability. Sulfonated membranes with counter-ions showed higher selectivities in the nanofiltration system for brackish water desalination.

523278 CO_2 Adsorption Performance of Functionalized Metal-Organic Frameworks with Different Topologies By Molecular Simulations

Abstract:

Anthropogenic carbon dioxide (CO_2) is the major greenhouse in the atmosphere that leads to global climate change. Metal-organic frameworks (MOFs) are regarded as a promising means to efficiently capture CO_2. The Multivariate MOFs (MTV-MOFs) consisting of a variety of organic linker functionalized by multiple functional groups could exhibit better performance in carbon capture and separation (CCS). In order to understand the effect of the framework topology and mixed functionalities on the CCS performance, Zr-MOFs in three topologies (*csq, ftw, scu*) with each composed by one of the three types of linkers of different lengths were constructed, and then functionalized by three types of functional groups (–F, $-NH_2$, $-OCH_3$), respectively. Grand canonical Monte Carlo (GCMC) simulations were used to investigate CO_2 adsorption performance of all created MOFs. We revealed that among the parent MOFs consisted of identical building blocks but in different topologies, *ftw*-MOFs exhibit the highest CO_2 working capacity and CO_2/N_2 selectivity due to the strong affinity towards CO_2. Moreover, the CO_2 adsorption performance of *ftw*-MOFs shows obvious dependence on the pore size, consistent with their host-adsorbate interaction energy, which is not the fact for *csq*-MOFs and *scu*-MOFs. The enhanced CO_2 adsorption performance upon functionalization of MOFs displays pore-size dependence as well, especially for $-NH_2$ functionalized frameworks in *csq* and *scu*, which is mostly attributed to the enhanced host-adsorbate Coulombic interaction. The proposed effects of the topology, the pore size and the functional group of MOFs on CO_2 adsorption performance obtained in this study may provide meaningful insight into the rational design of high-performing MOFs for carbon capture by tuning the topology, pore size and functionality of frameworks.

528220 Fibrous Carbon Molecular Sieve with 3-5 a Tunable Pores for Many Industrial Gas Separations (Poster)

Abstract:

Traditional adsorbents in the form of pellet/bead used in Pressure Swing Adsorption have significant constraint to balance the needs of high mass transfer rate and low pressure drop. Structured adsorbent, in the form of fiber, film, and honeycomb are desirable, which have large external contact area, small diffusion length, and low-resistance channels for gas flow. This talk introduces a new fibrous carbon molecular sieve adsorbent made by a two-step pyrolysis of melt extruded polyvinylidene chloride copolymer (PVDC) fiber. The PVDC fiber was first pretreated in air at low temperature (< 180 C) to form conjugated double bond and crosslinking via dehydrochlorination, then pyrolyzed in N_2 at high temperatures 1000-1400 C. A high throughput kinetic adsorption tool with 48 parallel testing cells using 10 probe molecules (from CO_2, σ_K=3.3 Å to SF_6, σ_K=5.5 Å) was used to 1) characterize the effective micropore size and its changes; and 2) fast screen the adsorbents for many industrial gas separations. The new fibrous CMS adsorbents have effective micropore size in the range of 3-5 Å and show very high separation factors of 20, 26, and 88 for the industrial gas pairs of N_2/CH_4, C_2H_4/C_2H_6, and C_3H_6/C_3H_8, respectively.

533908 Selective Adsorbents Based on Thia-Crown Ether Functionalized Composite Mesoporous Silica for Selective Recovery of Silver Ions from Aqueous Sources

Abstract:

Selective recovery of valuable metals in great demand like silver (Ag) from aqueous sources has been big challenge due to the large concentration difference between the target metal ions and the competing base metal ions in the medium.

Herein, highly selective and recyclable composite adsorbents were successfully developed for selective recovery of Ag(I) ions (Ag^+). The adsorbents are composed of thiacrown ethers (thia-CEs) as Ag^+-specific ionophores functionalized on mesoporous silica with well-defined pores as support material.The adsorbents (thia-CE@SBA-15) were prepared through immobilization of thia-CEs onto the surface of 3-(chloropropyl)triethoxysilane (CPTES) functionalized mesoporous silica (SBA-15).Various thia-CEs were synthesized by intermolecular cyclization of bulky epoxides with 1,2-benzenedithiol. Bulky epoxides with different sizes were synthesized via etherification of corresponding alkene halides with 2,2-diethylpropandiol followed by epoxidation of alkenes. Moreover, syntheses of thia-CEs were optimized by varying the metal templates, reactant concentrations, reaction solvents, time, and temperature. The synthesized thia-CE intermediates and final thia-CE products, SBA-15, and thia-CE@SBA-15 were thoroughly characterized by high-resolution spectroscopic methods such as ^1H and ^{13}C NMR, FTIR, HRMS, SEM, XRD, TGA, XPS and BET. On-going works are focused on the adsorption properties of thia-CE@SBA-15 for the selective recovery of Ag^+. Current results suggest the strong potential of thia-CE@SBA-15 as selective Ag^+adsorption.

This research was supported by the National Research Foundation of Korea (NRF) under the Ministry of Science and ICT (No. 2017R1A2B2002109) and the Ministry of Education (No. 22A20130012051(BK21Plus)).

537909 Study of Functional Groups of Ligands in Cu^{2+} MOFs in the Efficiency and Selectivity of Gas Adsorption

Abstract:

Metal Organic Frameworks (MOFs), are coordination compounds with crystalline structure formed by two sections organic ligands, comprising carboxylic acids, imidazoles, and other molecules with active heteroatoms, and metal compounds (ions or clusters). MOFs form reticular structures with big cavities and large surface area.

MOFs are extensively used for gas adsorption and separation [1-4], which are directly related to the pore size of the material and the affinity of the host molecules with the metals or the ligands of the MOF.

In this work we synthesize and analyze a family of copper derived MOFs as adsorbents of gases such as NH_3 and SO_2. The objective is to relate the chemical structure with the efficiency and selectivity of the adsorption process. MOFs are synthesized by mixing the ligands (terephthalic acid, phthalic acid and succinic acid) with the copper source ($Cu(NO_3)_2$ $6H_2O$) in a molar ratio of 1:4 in dimethyl formamide (DMF). The mixtures were homogenized and left at 80°C for 48 hours. The solid formed is filtered and washed with DI water and let dry at 120°C for 3 hours. MOFs are then characterized by solubility tests, melting point determination, FT-IR and X-ray diffraction.

For the adsorption tests, filters with 0.1 g of MOF were built, and two different tests were carried on. For the first test, NH_3 detection tubes are connected to a source of NH_3 (a solution of NH_4OH), the adsorption efficiency is determined in one step by correlating the concentration of gas detected by the color tubes in the system with and without filter. MOFs were compared to active carbon. MOFs showed better efficiency than carbon; terephthalic acid MOF adsorbed twice than the succinic acid MOF.

In a second test, a solution containing 3% of NH_4OH v/v is heated. The liberated gas is separated in two currents, one of them is connected to the MOF filter and then both currents are bubbled in corresponding vapor traps containing a solution of 0.1 M H_2SO_4. After 10 minutes, the solution in the vapor trap is titrated with NaOH 0.01 M previously validated with a reference. Terephthalic acid MOF registered a 64.7% of efficiency of NH_3 while succinic acid MOF registered 35.2%. Phthalic acid MOF showed similar efficiency than terephthalic acid MOF.

Test was repeated for SO_2. For generating SO_2 the reaction of $NaSO_3$ and H_2SO_4 was carried instead of the heating of the solution and using water in the vapor traps. Terephthalic acid MOFs showed bigger efficiency than succinic acid MOF but lower than activated carbon. A specific efficiency could not be registered because degradation of the MOF was observed due to the acidity of the gas.

Aromatic ligands in MOFs have a better efficiency in adsorbing basic and acid gases, however the stability of the material gets compromised in strong acid environments. The work in our laboratory continues in the synthesis of new MOF ligands and the design of systems for gas separations.

[1] Han, S.; Huang, Y.; Watanabe, T.; Nair, S.; Walton, K.; Sholl, D.S.; Meredith, C. *Microporous and Mesoporous Materials* **2013,** *173,* 86-91 "MOF stability and gas adsorption as a function of exposure to water, humid air, SO_2 and NO_2"

[2] Jacoby, M. *C&EN Chicago, Science and Technology* **2014,** *September,* 26-28 "New sorbents for greener cooling"

[3] Zhao, Z.; Ma, X.; Kasik, A.; Li, Z.; Lin, Y.S. *Ind. Eng. Chem. Res.* **2013,** *52,* 1102-1108 "Gas Separation Properties of Metal Organic Framework (MOF-5) Membranes"

[4] Hartlieb, K.J.; Holcroft, J.M.; Moghadam, P.Z.; Vermeulen, N.A.; Algaradah, M.M.; Nassar, M.S.; Botros, Y.Y.; Snurr, R.Q.; Stoddart, F. *J. Am. Chem. Soc.* **2016,** *138(7),* 2292-2301 "CD-MOF: A versatile separation method"

536779 Task-Specific Ionic Liquids Functionalized with Cobalt(II)Salen for Biomimetic Reversible Dioxygen Binding

Abstract:

Task-specific ionic liquids functionalized with Cobalt(II)salen for biomimetic reversible dioxygen binding

Qinghe Zheng, Marty Lail, Shaojun Zhou*, John Thompson, Kelly Amato

Discovery of Science and Technology Division, RTI International, Durham, NC 27709, United States.

*szhou@rti.org

Introduction

In biological systems, the respiratory pigments (e.g. hemoglobins and myoglobins) are able to reversibly bind dioxygen from the atmosphere without appreciable loss of activity. It is common knowledge that transition metals with certain organic coordination environment plays an important role in the oxygen binding in these natural oxygen carriers. In synthetic chemistry, low-molecular biomimetic substances containing a transition metal center and being able to reversibly bind oxygen have attracted considerable interest for air separation application. By far the greatest number of synthetic oxygen complexes are based on cobalt, and one of the most thoroughly investigated systems is a Schiff base complex *N,N_*-Bis(salicylidene)ethylenediaminocobalt(II), i.e. (Co(II)salen) also labeled as ÒSalcomineÓ, and its ring-substituted derivatives. The Co(II)salen molecule is a tetradentate ligand coordinating to the central Co^{2+} cation through the hydroxyl and amine groups. This tetradentately chelated compound has been found with oxygen uptake at the ratios of Co:O=2:1 (binuclear) in solid state, and at 2:1 or 1:1 (mononuclear) in different solvation environment.

Room temperature ionic liquids (RTILs) have various applications in a large variety of disciplines. These liquids have desirable properties including easy preparation, tunable structure, negligible vapor pressure, non-flammability, high thermal stability, environmentally nontoxicity and recyclability. Due to the Coulombic attraction between the ions of these liquids, they exhibit no measurable vapor pressure up to their thermal decomposition point, generally above 350 ^{o}C. Because of this feature, they would not add any contamination to the gas stream when used for gas sorption process, especially during regeneration, which minimizes the associated energy consumption. Much attention has been devoted to functionalized ILs (often termed task-specific ionic liquids, TSILs) as solvents for various gas separation applications. Matsuoka and colleagues published the first research (2017) on novel TSILs with reversible and stable O_2 absorbability. The TSILs were synthesized by complexing Co(II)salen complex and ionic liquid-based ligands. The as-prepared TSILs could chemically and selectively absorb a large amount of O_2 owing to the liquidity of the TSILs, and could function as an O_2 carrier in supported TSIL membranes.

The present study explored biomimetic dioxygen binding by various types of TSILs functionalized with Co(II)salen for air separation application. IL ligands composed of different combinations of phosphonium-type cations including triethylpentylphosphonium (P_{2225}^{+}), triethyloctylphosphonium (P_{2228}^{+}), tributylpentylphosphonium (P_{4445}^{+}), and trihexyltetradecylphosphonium (P_{66614}^{+}), and anions including N-methylglycinate ($NmGly^{-}$) or/and bis(trifluoromethanesulfonyl)imide ($NTf2^{-}$) were synthesized, and complexed with Co(II)salen successfully. The dioxygen sorption capacity and reversibility of the as-synthesized TSILs were investigated at various operating conditions. The study provides important results for room-temperature oxygen-selective sorbent and membrane development for modular air separation application.

Experimental

The ionic liquids were synthesized following literature procedure using air-free technique. Typical synthesis procedure for Co(II)salen functionalized a TSIL is as the following. In a dried Schlenk flask under continuous N_2 flow, 100 mL of ethanol was added, followed by 1 eq. of Co(II)salen. The mixture was continuously stirred at room temperature for 1 hr. A solution containing 2 eq. of IL compound in ethanol was added dropwise to the above suspension. It was worth noting that Co(salen) gradually dissolved upon addition of IL. The mixture was under agitated stirring at 25 °C for 12 hrs to form an IL-Co(II)salen complex. The solvent was then removed on a rotary evaporator, and the residual product was further dried in a vacuum line at 50 °C for overnight to yield a viscous liquid product. The as-synthesized IL and IL-Co(II)salen complex products were characterized by CHNS, NMR (1H, 31P, and 13C), IR, and UV-Vis analyses. The gas sorption was studied by TGA and ASAP-isotherm measurements at various conditions. The viscosity of the complex materials were studied between 25 °C and 80 °C.

Results and Discussion

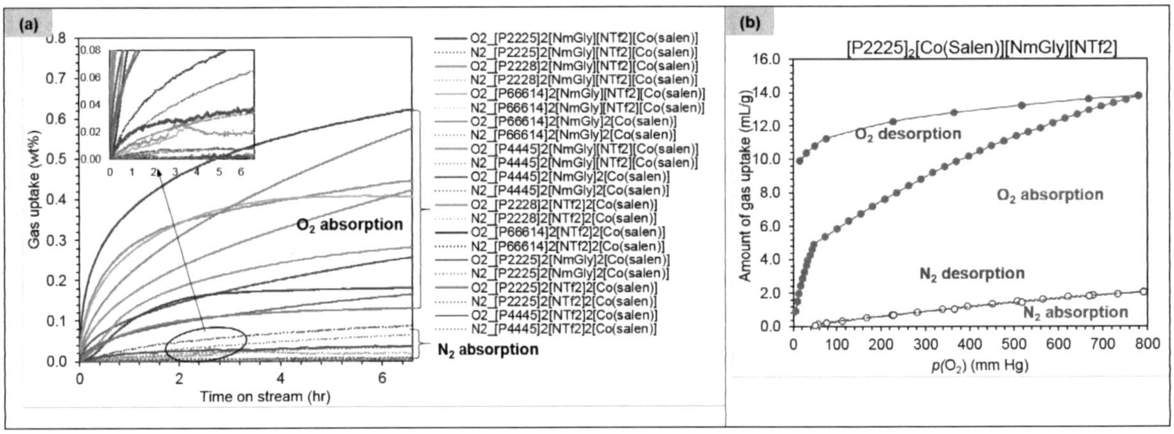

Figure. (a) TGA screening of O_2 and N_2 uptakes by as-synthesized IL-Co(salen) complexes. (b) O_2 and N_2 sorption isotherms for selected $[P_{2225}]_2[NmGly][NTf_2][Co(salen)]$-type TSIL at 25 °C.

A series of IL-Co(salen) complex-type O_2 sorbent materials were successfully synthesized, characterized, and screened for selective O_2 absorption over N_2. Types and molecular weights of IL ligands, and IL-Co(salen) coordination stoichiometry all have significant influences on the gas sorption capacity and kinetics. Among the studied samples, $[P_{2225}]_2[NmGly][NTf_2][Co(salen)]$ showed the highest O_2 absorption capacity and kinetics, with maximum capacity of 4368 uL/g sample and nearly half of the capacity reached (2065 uL/g sample) within 30 min of TOS. Reaction mechanisms during O_2 absorption in IL-Co(salen) complexes were proposed.

Reference

[1] A. Matsuoka, E. Kamio, T. Mochida and H. Matsuyama, *J. Membr. Sci.*, 2017, **541**, 393Ð402.

[2] Y. Kohno, M. G. Cowan, A. Okafuji, et al. *Ind. Eng. Chem. Res.*, 2015, **54**, 12214Ð12216.

525748 Separation of Ammonium Iodide and 1,4-Phenylenediamine from Their Mixture

Abstract:

Since 1960s, 1,4-phenylenediamine (PPD) has been surfaced to a raw material for para-aramid fiber having superior mechanical strength and thermal resistance under harsh condition. Recently, aromatic amination of 1,4-diiodobenzene using Cu(II) compounds as a catalyst has been started for commercial development of PPD process. Substitution of two iodine through amination of NH_3 have sparked the cheap and efficient production of high purity 1,4-phenylenediamine. However, to the best knowledge of known processes, the incomplete removal of mono-substituted benzene derivatives and iodide salts has determined the mechanical strength, molecular weight, viscosity and polymerization rate of synthesized polymers. Therefore, separation of iodine salts and side products is one of key operations in improving the economic feasibilities of developing process.

In the present work, drowning-out crystallization was applied to separate highly pure NH_4I from NH_4I/1,4-phenylenediamine mixture. Solubilities of NH_4I and 1,4-phenylenediamine were measured isothermally in various organic solvents. Yield and solubility of NH_4I was the primary criterion for selecting appropriate solvent/anti-solvent pairs in drowning-out crystallization. Thus, we tested the reduction of iodine salts into the recovery of elemental iodine by reducing agents. Purity and yield were achieved to 99% and 93% at 65 °C, respectively. XRD pattern reveals that pure NH_4I was precipitated without containing other crystalline structures.

525763 Amorphization of Azilsartan By Drowning-out Crystallization Combined with Freeze-Drying

Abstract:

Amorphization has been extensively studied to enhance the solubility and dissolution ability of poorly water soluble drug candidates. There are a number of different ways to generate the amorphous drug substances, but some of these applications may be difficult to control the product quality due to its high energy input which leads to contamination or degradation of drug substances. In the present study, amorphous solid of azilsartan (treatment of hypertension) was prepared by drowning-out crystallization which can exclude the thermal energy or high pressure. During drowning-out, azilsartan solid was appeared as amorphous phase at first and gradually transformed into crystalline type I. Amorphous suspension of azilsartan was collected and dried in two different ways (freeze-drying, vacuum filtration and drying). Duration time in amorphous state of azilsartan was significantly influenced by operating conditions such as temperature, stirring speed and additives. To obtain the well dispersed and amorphous azilsartan with nano-size, a freeze-drying method was adapted. When amorphous intermediate of azilsartan was freeze-dried, the powder obtained was shown to be well dispersed contrary to those prepared by vacuum filtration and drying.

520799 Purification of L-Menthol Enantiomers from the Racemic Mixture By Stripping Crystallization

Abstract:

Stripping crystallization (SC) is a new separation technique which combines distillation and crystallization. Basically, SC is operated at a triple-point condition, in which the liquid is simultaneously vaporized and crystallized due to the three-phase equilibrium. In essence, the process is continued until the liquid phase is completely eliminated and only the pure solid crystals remain in the feed. Thus, crystal washing is not required since only pure solid crystals remain in the feed and no impurities are adhered on the crystal surfaces at the end of SC. In the previous research, SC has been successfully applied to separate p-xylene from the mixed xylenes by Shiau et al. (2005, 2008, 2013).

Menthol is widely used in perfumery, cigarette, cough drops, and nasal inhalers for its cooling effect and refreshing flavor. Only L-menthol has been reported to have analgesic effect whereas D-menthol is lack of analgesic and flavor properties. Although natural L-menthol is obtained by distillation from the leaves of various subspecies of mint, menthol can also be synthesized via a number of routes, by which a racemic mixture is obtained. Thus, separation and purification of L-menthol from the racemic mixture has been an important research issue. A thermodynamic model is developed in this work to simulate the three-phase equilibrium during the SC operation and to direct the batch SC experiments. The experiments show that no crystallization occurs during SC without stirring due to subcooling of the liquid enantiomers. Stirring can eliminate subcooling during SC, leading to the L-menthol crystals gradually formed. The experimental results, including the final enantiomeric purity and recovery ratio of L-menthol, are consistent with the simulation results predicted by the model.

Keywords: crystallization; vaporization; purification; menthol; thermodynamics process

References:

1. Shiau, L. D.*, Wen, C. C. and Lin, B. S. "Separation and purification of p-xylene from the mixture of m-xylene and p-xylene by distillative freezing", *Ind. Eng. Chem. Res*, **44(7)**, p2258-2265, 2005.

2. Shiau, L. D.*, Wen, C. C. and Lin, B. S. "Separation of p-xylene from the multicomponent xylene system by stripping crystallization", *AIChE Journal*, **54(1)**, p337-342, 2008.

3. Shiau, L. D.*; Liu, K. F. "Investigations into the effects of the cooling rate on stripping crystallization", *Ind. Eng. Chem. Res*, **52**, p1716-1722, 2013.

529906 Continuous High-Purity Recovery of Xylobiose from the Output of *Bacillus Pumilus* β-Xylosidase Reaction Using a Well-Designed Simulated Moving Bed Process

Abstract:

Xylobiose (X2) is recognized to possess great prebiotic function and to be highly favorable for application in food and prebiotic industries. In this study, we developed an efficient simulated moving bed (SMB) process for continuous recovery of xylobiose from the output of xylose(X1) → xylobiose(X2) reaction based on the cloned β-xylosidase of *Bacillus pumilus* IPO. From the SMB experimental results, it was confirmed that X2 could be obtained with nearly 100% purity and 92% yield on a continuous-separation mode. The developed SMB process could also recover the unreacted X1 almost completely, which leaves room for a further increase in the overall X2 reaction yield by reusing the recovered X1 from the SMB as the reactant of the upstream processing (i.e., *Bacillus pumilus* IPO β-xylosidase X1 → X2 reaction). The results of this study will enable a highly economical and environmentally-friendly production of high-purity X2 from X1.

532624 Effect of Rotating Elements on HETP of a Horizontal Distillation Column

Abstract:

Small scale continuous separation technology needs to be developed to realize continuous pharmaceutical production systems. There are several studies on the development of small scale unit operations such as extraction, crystallization and absorption. However, the number of reports on the small scale distillation is limited. Thus, this study aims at development of a new compact distillation apparatus which can be operated in a continuous mode. To realize small scale continuous distillation apparatus, it is essential to intensify the mass transfer rate between vapor and liquid phases in the column. Usually this has been achieved by improving packing materials and trays. Recently Katayama et al. [1] proposed a distillation apparatus that incorporates rotating elements at the center of a horizontal column to promote mass transfer. By the motion of the rotating elements, a thin liquid film is formed and mass transfer between vapor and liquid is expected to be promoted.

In this study, a horizontal type distillation apparatus equipped with rotating elements was constructed and the separation performance was evaluated by performing a total reflux distillation experiment. A horizontal column with an inner diameter of 3.0 cm and a length of 40 cm was used. Distillation experiments were carried out under total reflux conditions using cyclohexane / n-heptane system. Several different types of rotating elements were tested in the study, since the performance of the column is expected to be dependent on the shape of them. For fabrication of some rotating elements, a 3D printer was used.

Experiments were conducted to investigate the effect of rotating elements on the HETP and pressure drop. It was confirmed that HETP decreases greatly by increasing the rotation speed, and it reduced to 4 cm when the speed was 1000 rpm. However, pressure drop increased with increasing the rotation speed.

HETP and The magnitude of pressure drop depended on the shape of the rotating elements. Further optimization of the rotating element would lead to an improvement in both of them.

[1] H.Katayama et al., Journal of Chemical Engineering of Japan, 33, (5), pp. 689-695, 2000

532730 Numerical Investigation of the Effect of Bend on the Gas Absorption Rate in Microchannels

Abstract:

The internal volume of microreactor is much smaller than the conventional reactors, and thus microreactors are capable of achieving high mass transfer and energy transfer rates. Thus, microreaction technology is expected to help us realize highly efficient chemical processes and bring about dramatic reduction in the consumption of resources and energies. When microreactor is applied to a multiphase system, the fluid shows various flow patterns. We focused on the gas-liquid slug flow which is one of the typical flow patterns, and studied the relationship between the fluid behavior and mass transfer in this flow using computational fluid dynamics. In this flow, gas and liquid slugs flow alternately and the gas-liquid interfacial area increases. It is known that the fluid in each liquid slug exhibits strong circulating motion. The circulating motion of the liquid enhances the mass transfer from the gas-liquid interface to the liquid bulk. This study focused on the gas absorption behavior in a gas-liquid slug flow in a bent microchannel. The relationship between fluid behavior and the volumetric mass transfer coefficient was investigated by means of computational fluid dynamics.

Before conducting a simulation study, a numerical model to express gas absorption rate was developed. The rate equation used in theoretical studies of gas absorption usually expresses the rate per unit interfacial area. This rate was converted to show the value per unit volume so that it can be used in the discretized equation in the finite volume framework. This enabled us to numerically simulate gas absorption behavior through a moving interface.

A series of numerical simulation was carried out to investigate the effects of fluid behavior on the volumetric mass transfer coefficient. A commercial CFD software Fluent was used in numerical simulation. The linear velocity of the slug flow was changed from 0.4 to 0.8. It was assumed that the slug length was 2.0 mm and a single liquid slug flowed through the channel. The distribution of the mass transfer coefficient along the channel was examined.

It was confirmed that the mass transfer rate increased at a bend in a microchannel. This is because the gas-liquid interfacial area became larger due to deformation of the slug. When the flow velocity was 0.4 m/s, the shape of the slug recovered quickly after bend, but did not recover when the velocity was 0.8 m/s. It was also observed that the volumetric mass transfer coefficient in the latter case was greater than in the former case. The presentation covers the details of the results including the relationship between the fluid motion and the gas absorption rate.

534289 Methyl Palmitate Separation from the Reaction Mixture of the Solvent-Free Transesterification to Produce Sucrose Esters

Abstract:

Sucrose esters are biobased emulsifiers used in high value niche markets such as cosmetics and food [1]. They are commonly produced by transesterification of sucrose and fatty acid methyl esters (FAMEs). With the aim of improve reactants incompatibility and avoid the use of toxic solvents, a solvent-free process using emulsifiers as contacting agents has been developed [2]. However, this reaction system is characterized by incomplete conversions so that the remaining reactants should be separated from the final product and recovered for recycling. However, the purification of product streams is a complex task owing to the complexity of the reactive mixture.

Different purification processes to obtain commercial sucrose ester have been reported [3–8]. Mainly, an organic non-polar solvent is used to extract residual FAMEs. Then, water or other polar solvent is used to wash the precipitated sucrose esters and to separate them from residual sucrose and catalyst. The effectiveness of this recovery process is highly dependent of the selectivity of the organic solvent for the FAMEs, its ability to enhance the color of the product, and how easily it can be removed and recycled. In the open literature there is lack on information regarding the impact of these key parameters in the process, which are required for further modeling and design of the separation process.

This work systematically studied the separation and recovery of methyl palmitate from a solvent-free sucroester production media, using different solvents. The solvents were selected among those included in the "*Green Solvents Guide Selection*" developed by *Glaxo Smith & Kline*. From this list, a screening was done using the Hansen Solubility Parameters to choose the solvents with highest affinity for the methyl ester. The resulting list was assessed considering the price, normal boiling point and solvent availability. The three chosen solvents were ethyl acetate, *iso*-amyl acetate and methyl isobutyl ketone. The separation performance of each solvent was evaluated by doing solid-liquid extractions using a specific solvent/solid mass ratio. After proper contact between the solvent and the solid sucroester mixture, the remaining solids were separated from the solvent by centrifugation. Quantification of FAME was done by high performance liquid chromatography (HPLC). The remaining solids were mixed with water, citric acid, sucrose and sodium chloride to precipitate the sucrose esters. Humid sucrose esters were dried in a vacuum oven at 50°C. Additionally, the recovery of the methyl palmitate extracted was performed by evaporation of the organic solvent. The yield of the recovery of the methyl palmitate and the organic solvent was calculated from the overall mass balance.

According to results ethyl acetate exhibited the best performance, in terms of FAME recovery, color enhancement, and solvent recovery straightforwardness. Then, further separation experiments were carried out with the selected solvent, under different separation temperature (5, 20 y 35°C), mass ratio of solvent to solids (2, 3, 4 and 6) and number of extraction stages (1, 2 and 3). The optimum operation conditions during the extraction process were identified, and the conceptual design of the separation process was done. Results of this work will be used in a further process design for solvent-free sucroester production.

References

[1] B.A.P. Nelen, J.M. Cooper, Sucrose esters, in: R.J. Whitehurst (Ed.), Emuls. Food Technol., Backwell Publishing Ltd, 2004: pp. 131–162.

[2] J. Fitremann, Y. Queneau, J. Maitre, A. Bouchu, Co-melting of solid sucrose and multivalent cation soaps for solvent-free synthesis of sucrose esters, Tetrahedron Lett. 48 (2007) 4111–4114. doi:10.1016/j.tetlet.2007.04.015.

[3] K. James, Purification of Sucrose Esters, US 4104464, 1978.

[4] S. Kea, C.E. Walker, Separation and purification of sugar esters synthesized from both aqueous and nonaqueous systems, US 4710567, 1987.

[5] S. Matsumoto, Y. Hatakawa, A. Nakajima, Process for purifying sucrose fatty acid esters, US 5008387, 1991.

[6] S. Matsumoto, Y. Hatakawa, A. Nakajima, Process for recovering unreacted sucrose from reaction mixture synthesis of sucrose fatty acid esters, US 4995911, 1991.

[7] S. Matsumoto, Y. Hatakawa, A. Nakajima, Method of producing powdery high hlb sugar fatty acid ester, 5144022, 1992.

[8] Y. Koyama, N. Kawase, H. Yamamoto, S. Kawata, Y. Kasori, Process for producing sucrose fatty acid ester, EP 0659760A2, 1994.

534696 Ultrasound-Mediated Nonequilibrium Separation of Ethanol-Water Solutions, Including Avoidance of the Azeotropic Bottleneck

Abstract:

Title: Ultrasound-mediated nonequilibrium separation of ethanol-water solutions, including avoidance of the azeotropic bottleneck

Ozan Kahraman*

Department of Food Science and Human Nutrition

University of Illinois at Urbana-Champaign

kahramn2@illinois.edu

Arne J. Pearlstein

Professor

Department of Mechanical Science and Engineering

University of Illinois at Urbana-Champaign

ajp@illinois.edu

Hao Feng

Professor

Department of Food Science and Human Nutrition

University of Illinois at Urbana-Champaign

haofeng@illinois.edu

Introduction

Separation of liquid mixtures, frequently by distillation, is ubiquitous in the chemical and process industries

(CPI). Distillation accounts for ~95% of the energy used in liquid separations, ~25–40% of overall energy used in CPI, and ~3% of global energy consumption.[1-2] The low efficiency of distillation is largely due to two issues. First, there are large irreversible losses due to heat transfer.[3] Second, a significant fraction of energy used in liquid separations is used to separate azeotropic mixtures in azeotrope-forming systems (e.g., ethanol/water). While a number of conventional distillation technologies[4-5] (e.g., pressure-swing, extractive distillation, and azeotropic distillation[6]) and new separation approaches[5] (e.g., dividing-wall columns, membranes, molecular sieves, and bio-absorbance) have been developed for azeotropic systems, these approaches largely rely on thermal separation via phase equilibrium or involve large capital and/or operational costs.

Ultrasound passing through a liquid layer with a free surface above produces droplets that form a mist. Current thinking is that this occurs at crests of capillary waves at the gas/liquid interface, by bursting of cavitating micro-bubbles near the interface, or possibly by both mechanisms.[5] Regardless of the mechanism, it is known that mist formation is accompanied by favorable partitioning of solute between the mist and the mother solution,[7] with studies in EtOH/water solutions showing significant EtOH enrichment in the mist[8-10], including enrichment of EtOH from 0.07 mass fraction in the mother solution to nearly 1.0 in the mist.[9] Current theories attribute enrichment to an excess of ethanol in a thin layer on the liquid side of the interface[10-11], enhanced evaporation due to enlargement of surface area, or to EtOH-rich clusters in the bulk solution, directly brought into the gas phase by bursting cavitation bubbles.

In this study, we propose to develop, test, and demonstrate scalable nonequilibrium liquid separations (including for azeotrope-forming systems), that avoid the heat transfer losses of distillation, as well as the azeotropic bottleneck. Success will result in significant savings in energy and capital costs, for a broad class of systems.

Method

In the first ultrasonic separation prototype, a 2.4 MHz ultrasonic transducer with a diameter of 18 mm and rated power of 13.2 W was installed in the center of the bottom of a custom-designed cylindrical separation unit. The unit (see Figure 1) was made from transparent polyvinyl chloride resin (10 mm thickness), with a height of 305 mm and an inside diameter of 100 mm. The temperature of the bulk liquid (at the bottom of the unit) was kept nearly constant by circulating controlled-temperature water through a volume laterally surrounding the bottom section of the vessel. Mist generated by ultrasonic actuation was carried out of the cylindrical volume by an air flow.

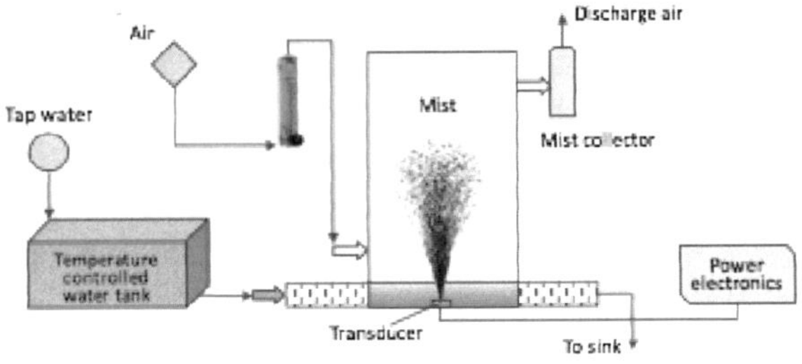

Figure 1. The ultrasonic separation prototype.

In experiments, the entire set-up was pre-conditioned to a target separation temperature, e.g., 30°C, by water circulation, followed by adding approximately 200 ml of EtOH solution with mass fraction $0.04 \leq w \leq 0.76$ into the separation unit, to a liquid depth of about 25 mm. The ultrasonic transducer was then turned on to produce a mist from the bulk liquid. Then the unit was opened to a flow of carrier air, set at a flowrate between 10 to 30 L/min, to carry the mist to a collector. The collected mist was filtered through a syringe filter into a 1.5 ml capped vial for HPLC analysis. Duplicate samples were analyzed using a Waters e2695 HPLC series system (Waters Associates, Milford, MA, USA) equipped with a Prevail Carbohydrate ES 5 µm column, automatic injector, and a Waters 2414 refractive index detector at 40°C. The mobile phase was

160 μL H_2SO_4/L HPLC grade water, with a flow rate of 1 mL/min. Distilled water and 200 proof ethanol (Decon Labs, King of Prussia, PA) were used.

Results

Enrichment of ethanol was observed in the mist for mother solutions at all initial compositions (see Figure 2). For initial ethanol mass fractions exceeding about $w_{initial} = 0.60$, the ethanol mass fraction in the collected mist exceeded the azeotropic vaalue of 0.956. When $w_{initial}$ exceeded about 0.65, the collected mist was nearly pure ethanol.

The volumetric flow rate of the carrier air had a significant effect on enrichment. For instance, when the air flowrate was increased from 10 L/min to 30 L/min, the ethanol mass fraction in the mist increased from 0.147 to 0.203 and 0.813 to 0.969, for $w_{initial} = 0.040$ and 0.648, respectively (see Figure 3a-b). It can be seen that the air flowrate is even more critical for mixtures with an initial mass fraction $w > 0.25$. We found that the optimal air flowrate for separation at 30°C was between 25 and 30 L/min. Another important parameter for this separation is the collection time. At an air flowrate of 25 L/m, when the sampling time increases, water droplets in the mist tend to reach equilibrium with the surrounding air, which may result in a decrease in the efficiency of separation. As shown in Figure 3c, increasing the sampling time from 2 min to 10 min for samples with $w_{initial} = 0.165$, reduced the EtOH mass fraction in the mist from 0.326 to 0.193. For $w_{initial} = 0.542$, the same change in carrier air flow rate reduced the EtOH mass fraction in the collected mist from 0.871 to 0.675.

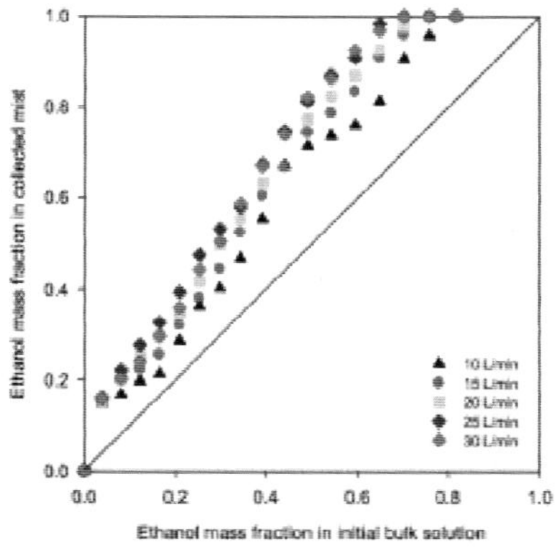

Figure 2. Ethanol mass fraction in the bulk solution and mist at different carrier air flow rates.

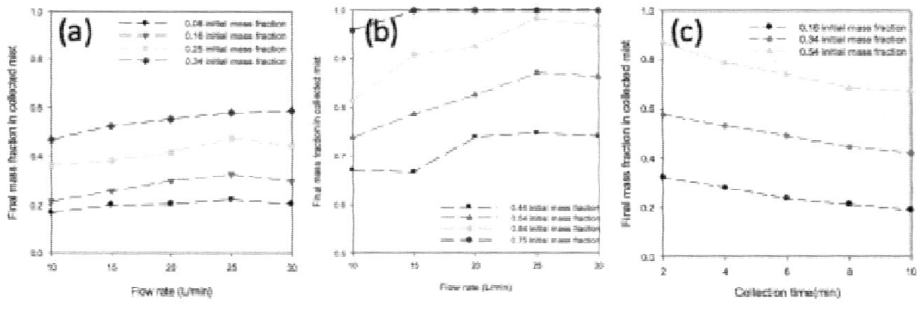

Figure 3. Effect of carrier air flowrate (a & b) and collection time (c) on ethanol mass fraction in the collected mist.

Conclusion

Ultrasound-mediated ethanol separation at low temperature is shown to be effective in enriching ethanol in the ultrasound-generated mist compared to the bulk ethanol-water solution. This nonthermal, nonequilibrium separation method is able to break the ethanol-water azeotrope, showing promise to reduce energy consumption in ethanol production.

References

1. Rush, F.E. 1980. Energy-saving alternatives to distillation. *Chem. Eng. Prog.* **76**, 44–49.

2. Hewitt, G., Quarini, J., Morell, M. 1999. More efficient distillation. *The Chemical Engineer* (London), **690**, 16–18.

3. Kazemi, A., Faizi, V., Mehrabani-Zeinabad, A., Hosseini, M. 2017. *Sep. Sci Technol.* **52**, 1387–1396.

4. Widagdo, S., Seider, W.D. 1996. Azeotropic distillation, *AIChE J.* **42**, 96–130.

5. Mahdi, T., Ahmad, A., Nasef, M.M., Ripin, A. 2015. *Sep. Purif. Rev.* **44**, 308–330.

6. Huang, H-J., Ramaswamy, S., Tschirner, U.W., Ramarao, B.V. 2008. *Sep. Purif. Technol.* **62**. 1–21.

7. Kirpalani, D.M., Toll, F. 2002. *J. Chem. Phys.* **117**, 3874–3877.

8. Yasuda, K., Tanaka, N., Rong, L., Nakamura, M., Li, L., Oda, A., Kawase, Y. 2003. *Jap. J. Appl. Phys.* **42**, 2956–2957.

9. Sato, M., Matsuura, K., Fujii, T. 2001. *J. Chem. Phys.* **114**, 2382–2386.

10. Li, Z.X., Lu, J.R., Styrkas, D.A., Thomas, R.K., Rennie, A.R., Penfol, J. 1993. *Mol. Phys.* **80**, 925–939.

11. Yano, Y.F. 2005. *J. Colloid Interface Sci.* **284**, 255–259.

535021 Techno-Economic Analysis of Deep Eutectic Solvent Based so_2-CO_2 Co-Capture Process for Flue Gas

Abstract:

Over the last few decades, coal-fired power plants emission system has been upgraded in comply with baghouse, selective catalytic reducer (SCR), and flue gas desulphurization (FGD) unit to remove fine particulates (PM 2.5), NO_x, and SO_2, respectively. Each individual unit increase the parasitic loads (or, decrease overall efficiencies) of the coal-fired power plant. In fact, state-of-the-art wet limestone based FGD unit from CH2M reduces the overall coal-fired power plant efficiency around 2.0-2.5% depends of the throughput of the power plant. Another eminent technology is carbon capture and storage (CCS) to reduce the greenhouse gas emission from the coal-fired power plant. Unfortunately, this additional technology will further reduce the efficiency (~28% from ~34%) from the current practice. As a result, the power plant will have to consume more coal to deliver the current electricity demand. Therefore, a common solvent, which is not harmful to the environment but capture SO_2 and CO_2 simultaneously, is of great need to sustain the clean electricity production from coal-fired power plant.

Deep eutectic solvents (DES) are environmentally friendly and relatively less expansive solvent than ionic liquids and amine-based solvents. DESs are eutectic mixtures of salts or complexing agents with a certain stoichiometric ratio. Wide range of metal salts, metal oxides, and hydrogen-bond donors (HBD) can be the salts or complexing agents to form DES. Despite being a relatively newer technology, the wide applications of DES including CO_2 adsorption makes it very attractive separation technologies. However, the applicability of DES in CO_2-SO_2 co-sequestration has never been explored before.

In this project, experimentally and computationally determined parameters were evaluated and were incorporated into a full-scale techno-economic analysis for flue gas treatment. This evaluation includes several different solvents with sensitivity analysis and comparison to an NETL baseline capture process for a 550 MW coal fired power plant.

521618 Optimization of Distillation Processes

Abstract:

Distillation is the primary technology used in industry for liquid separation and purification. This is because it has a proven track record and is often the most cost effective method, particularly in crude separations. Even though it is the economical choice, distillation is very costly both in energy and capital. With new column designs a significant amount of time and effort is consumed with optimizing the tradeoffs between capital and the long term energy costs to get the column positioned for the lowest economic investment for the life of the asset. In existing processes we continue to make empirical adjustments to further push columns to reduce costs and help the company improve the bottom line.

In most cases, column installations are complex and have multiple variables that can be manipulated to optimize the column and each parameter needs to be optimized. Traditional methods of optimization involve performing parametric studies. It is difficult to arrive at the optimal condition of all process parameters that will minimize/maximize the desired objective(s) while fully meeting the design and operational constraints. This makes it a challenge to get to the most optimal set of conditions for the separation. However with the implementation of a rigorous optimization algorithm it is possible to explore all the possible scenarios required to get to the true optimum. The optimization of a distillation column involves both continuous and discrete variables. In this work, we used the Mixed-Integer Non Linear Programming (MINLP) optimization formulation and systematically optimized the performance of different distillation processes using the non-dominated sorting genetic algorithm II (NSGA-II) available in the MATLAB® global optimization toolbox.

To demonstrate the improvements the MINLP optimization approach can have over parametric methods, several case studies were completed pitting an experienced column designer against the computer. In each case the results indicated that the MINLP would have resulted in a more economical design, especially as the columns became more and more complex. This shows the long term advantages implementing this optimization strategy can have over traditional methods and improvements to the company cost position.

525098 Multi-Objective Optimization of a Batch Distillation Column

Abstract:

Over the years, intuition and experience based heuristic approaches are used for the development and operation of batch processing in a distillation column. An optimal operating condition of a batch distillation may lead to run the column most efficiently in terms of both utility consumption and cost. Robust optimization like genetic algorithm has a significant advantage over gradient-based optimization to find the global optimal point due to non-convexity of design space. This work aims at formulating a multi-objective optimization (MOO) strategy to optimize the design and operating parameters of a conventional batch distillation (CBD), probably for the first time in the literature. Elitist non-dominated sorting genetic algorithm (Deb et al., 2002) is employed for the development of the optimization strategy along with the selection of the optimal point by using the technique for order of preference by similarity to ideal solution method (TOPSIS) along with entropy information for weighting method. Two objective functions in the aspect of performance criteria, i.e. total annual cost (TAC), and total annual production (TAP) are considered, which are often conflicting in nature. The number of trays, reboiler heat duty and reflux ratio are considered as decision variables to be optimized. The Pareto optimal front is constructed from the stated MOO studies, and then the optimal point is chosen by the TOPSIS-entropy weighting method. A CBD column separating a mixture of acetone and water (Miladi and Mujtaba, 2004) is considered for the demonstration of the proposed optimization strategy. The optimum value for the TAC and TAP are found to be 94689 $/year and 65610 kmol/year, respectively; and the corresponding decision variables are 22 trays (excluding the reboiler and total condenser), constant reboiler heat duty of 8204 kJ/min and a constant reflux ratio of 0.72.

References

Deb K., Pratap A., Agarwal S., Meyarivan T., A fast and elitist multi-objective genetic algorithm: NSGA-II, IEEE Tran. Evol. Comput. 6 (2002) 182-197.

Miladi M.M., Mujtaba I.M.. Optimization of design and operation policies of binary batch distillation with fixed product demand. Comput. Chem. Eng. 28 (2004) 2377â€“90.

A Novel FO-MED Hybrid System for MED Brine Further Concentration

Ye Yang (speaker) , Yuzhu Sun*, Jianguo Yu*

State Key Laboratory of Chemical Engineering, College of Chemical Engineering, East China University of Science and Technology, Shanghai 200237, China

Abstract

Multi-effect distillation (MED) is one of the most common thermal desalination processes. However, MED brine disposal is still an inevitable problem. In this paper, forward osmosis (FO) process was integrated into multi-effect distillation (MED) process for brine further concentration, as shown in Figure 1.The feasibility of FO for LT-MED brine was studied in experiments, and the result showed that $CaCO_3$ scaling not only caused flux decline, but also reduced the forward salt rejection remarkably at lower temperature due to cake-enhanced concentration polarization (CECP) as figure 2 shows. The membrane scaling was almost reversible under simple physical cleaning, which was explained by the interaction of scaling ions with membrane interface using radial distribution function (RDF) via Materials Studio software. Besides, the effects of operating condition on flux and rejection were investigated to improve FO performance, and it was found that the flux was linearly increased with temperature in premise of thermal stability. The influence of flow rate was lower than that of temperature. Based on above operational characteristics, a FO-MED hybrid system was proposed to improve the performance of existing traditional LT-MED system as shown in Figure 3. The Aspen Plus simulation results showed that this hybrid system had higher concentration factor (CF) of 2.67, greater gained output ratio (GOR) of 7.68, as well as lower specific area (SA) of 198, which was a bridge linking thermal seawater desalination and resource utilization of brine.

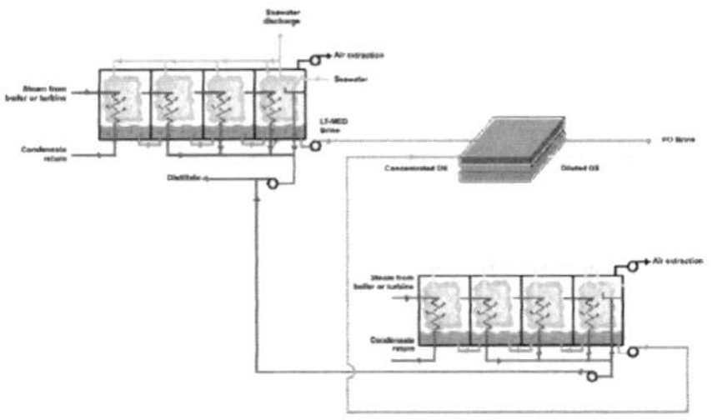

Figure 1 Schematic diagram of MED-FO hybrid system

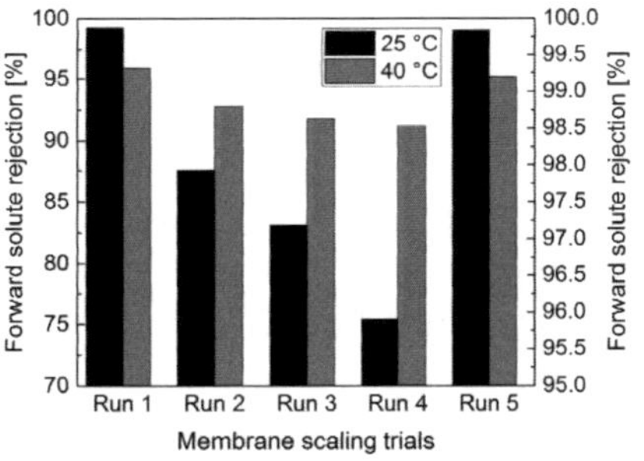

Figure 2 Forward salt rejection during membrane scaling at 25 degree and 40 degree

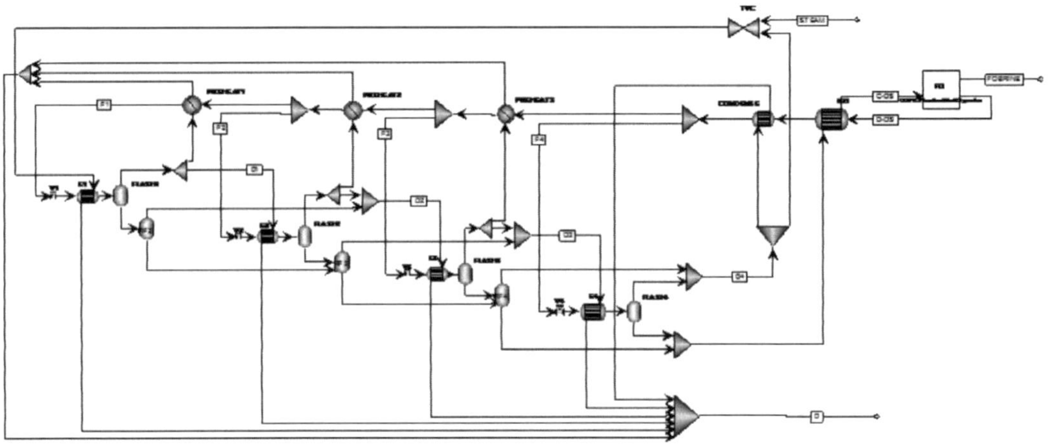

Figure 3 Aspen Plus flow sheet of FO-HT-MED system for LT-MED brine further concentration

534779 Predicting the Productivity of Chromatography Processes By Repeated Cyclic Operations or By Continuous Column Switching Operations

Abstract:

Chromatography processes are commonly optimized in terms of the productivity P defined as P = (mass recovered)/[(bed volume)(cycle time)]. P can be increased by either pseudo-continuous multi column operations such as simulated moving bed (SMB) chromatography or a single column repeated cyclic operation (RCO). It is first needed to optimize a RCO process in order to evaluate the benefit of continuous chromatography. We have developed a method for predicting the P in RCO isocratic elution processes by considering the distribution coefficient K and HETP as a function of mobile phase velocity. A model system chosen was reversed phase chromatography of two polyphenols with an ethanol-water mixture mobile phase. In this system the retention and the selectivity (the difference between the two distribution coefficients, Î”K) can be changed by the ethanol concentration markedly. The calculation showed the highest P value (considering mobile phase consumption) at 24-26% ethanol concentration. Based on these optimized conditions, we performed numerical simulation of a standard 4-zone SMB separation process. The column height of SMB was one-fourth of the RCO. The sample and the solvent feed rates were calculated from the optimized RCO conditions. At 24% ethanol concentration where Î”K was large(ca.2.5), the performance of RCO was better than that of SMB. At 26% ethanol concentration (Î”K=1.5), the performance of SMB was similar to that of RCO. At 28% ethanol concentration (Î”K=0.9), SMB productivity was higher by 10-20%. These results show that RCO is suited as a pseudo continuous chromatography operation when the selectivity is high (large Î”K).

541565 25 by 25: Chemical Engineering in the Next 25 Years

Abstract:

In 2008, the centennial year for AIChE, we asked 25 chemical engineers for their visions of the profession from around the US and the world: Nine faculty, eight from industry, and eight post-docs and graduate students. We posed four questions related to each individual's industrial sector or area of research, asking for extrapolation, new impacts on existing sectors, new sectors, and an optional, more open-ended comment on the future of the profession as a whole.

In this 110th year of AIChE, we have revisited these questions with many of the same people. Ten years has seen many developments in our profession, in society, and in our thought-leaders' experiences. Here we report on their perspectives, comparing them to the 2008 visions.

541542 The Future of Chemical Engineering Itself

Abstract:

Chemical engineeringâ€™s future is bright because it is vital to so much of society and industry -- and its future is complicated by its rapidly expanding universe of activities. It must represent depth in core skills centered on the chemical sciences including biology and materials. At the same time, it must be transdisciplinary; it must exemplify the need and value of reaching across disciplines in borderless ways.

One way is to create and disseminate a compelling vision and strategy for chemical engineering, research, and education for the future. AIChE and the National Academies are working together to that end. In 1988, the National Academies published "Frontiers in Chemical Engineering: Research Needs and Opportunities.â€ Its committee of authors outlined a wide-ranging roadmap to turn promising research opportunities into reality, while calling for researchers to embrace new frontiers.

Since 1988, that has happened -- along with many other transformative changes like the Web, fracking, personalized medicine, and social media. A new nationwide effort, aided by international contributions, will address both the identity of chemical engineering and its future directions. Its goal is to ensure that the people in chemical engineering are primed to meet society's needs and rapid changes.

544708 Accelerating Innovation through Academic-Industrial Partnerships

Abstract:

In 2011, The Dow Chemical Company launched the University Partnership Initiative (UPI) to sponsor collaborative research at leading U.S. Universities. This 10-year commitment aims to address fundamental questions in industrial chemistry and chemical engineering with alignment to Dowâ€™s next-generation process technology and product development efforts. To date, this portfolio of projects has generated an impressive array of intellectual property, publications, and numerous examples of the collaborative developments implemented at Dow. This presentation will introduce UPI program and highlight several contributions from our collaborators and leaders in the chemical engineering profession.

540559 Maximizing Uptime, Efficiency, and Safety of Industrial Operations through Early Risk Detection

Abstract:

Achieving zero incidents while maximizing process uptime and efficiency is a key vision for operating companies. An important building block towards this vision is establishment of a proactive risk mitigation culture, supported by effective systems and a strong management. Studies have shown that most of the incidents and unexpected process failures can be avoided if the operating teams get timely information about developing risks and take preventive actions early on. While advances have been made in process monitoring and automation in the past decades, there still remains significant technological and behavioral gaps that prevent plant operations from proactively mitigating process risks.

A modern industrial plant monitors thousands of parameters, generating upwards of 50-100 million data points every day. Thanks to recent breakthroughs in machine learning and artificial intelligence approaches, today there are autonomous systems that can sift through this data and point out meaningful and timely insight. This can help operating teams ascertain process issues that are hidden in the data, long before process variables reach critical levels.

In this presentation, a new autonomous system, Dynamic Risk Analyzer™ (DRA), will be introduced that points out early indicators of process issues using proprietary machine learning. The risk indicators enable operations team to respond and make required process changes, days and sometimes weeks ahead of any alarm, avoiding late stage (expensive) fixes. To realize these objectives fully, a proactive management workflow must also be in place. The presentation will discuss new initiatives and workflows developed by plant operations to implement a proactive culture. Real-life case studies from plants using DRA technology will be presented on how this culture change resulted in increased uptime, efficiency and safety of plant operations.

531309 Gaussian Processes for Hybridizing Analytical & Data-Driven Decision-Making

Abstract:

Surrogate models are widely appreciated in process systems engineering [1]. The typical setting focuses on expensive-to-evaluate, possibly uncertain functions. Examples include: modular process simulators [2], integrated gasification combined cycle processes [3], a carbon capture absorber [4], and many other applications, e.g. [5-8]. Resources are typically limited, so effective decision making requires data-efficient learning.

The data science and statistical machine learning communities typically focus on models learned solely from observed data. But chemical engineering applications may also require explicit, parametric models, e.g. modeling known process constraints, operations constraints, and cost objectives [9]. So, work has integrated semi-algebraic functions with those learned from data [10] or developed semi-physical modeling techniques [11, 12].

This presentation surveys the state-of-the-art in hybridizing analytical and data-driven decision making [13]. We consider three probabilistic modeling applications to these hybrid situations:

Design of experiments for model discrimination [14]. We bridge the gap between classical, analytical methods [15] and Monte Carlo-based approaches [16]. Classical methods may have difficulty managing non-analytical model functions and data-driven Monte Carlo approaches come at a high computational cost. We replace the original, parametric models with probabilistic, non-parametric Gaussian process surrogates learned from model evaluations. The surrogates are flexible regression tools that extend classical analytical results to non-analytical models, while providing us with model prediction confidence bounds and avoiding the computational complexity of Monte-Carlo approaches.

Multi-objective optimization [17, 18]. We make novel extensions to Bayesian multi-objective optimization in the case of one analytical objective function and one black-box, i.e. simulation-based, objective function. The resulting method has been applied to a bone neotissue application [19] and a more general test suite.

Scheduling plant operations under uncertainty. For processes with equipment degradation, we use Gaussian processes to approximate large-scale, mixed-integer optimization problems.

We close by offering a broad outlook on applying probabilistic surrogate models to chemical engineering. Statistical machine learning has recently attracted significant interest in process systems engineering [20]. Here we show that state-of-the-art research in Gaussian processes [21, 22] and probabilistic modeling more generally [23] can have a big impact on chemical engineering.

References

[1] Bhosekar A, Ierapetritou M. Advances in surrogate based modeling, feasibility analysis and and optimization: A review. *Comput Chem Eng*. 108: 250-267, **2018**.

[2] Caballero JA, Grossmann IE. An algorithm for the use of surrogate models in modular flowsheet optimization. *AIChE J*. 54(10):2633-2650, **2008**.

[3] Lang Y, Zitney SE, Biegler LT. Optimization of IGCC processes with reduced order CFD models. *Comput Chem Eng*. 35(9):1705-1717, **2011**.

[4] Cozad A, Sahinidis NV, Miller DC. Learning surrogate models for simulation-based optim-ization. *AIChE J*. 60(6):2211-2227, **2017**.

[5] Boukouvala F, Ierapetritou M. Surrogate-based optimization of expensive flowsheet modeling for continuous pharmaceutical manufacturing. *J Pharm Innov*. 8: 131, **2013**.

[6] Soepyan FB, Cremaschi S, Sarica C, and others. Estimation of percentiles using the Kriging method for uncertainty propagation. *Comput Chem Eng*. 93:143-59, **2016**.

[7] Shokry A, Ardakani MH, Escudero G, Graells M, Espuña A. Dynamic Kriging-based fault detection and diagnosis approach for nonlinear noisy dynamic processes. *Comput Chem Eng*. 106:758-76, **2017**.

[8] Tran AP, Georgakis C. On the estimation of high-dimensional surrogate models of steady-state of plant-wide processes characteristics. *Comput Chem Eng*. DOI 10.1016/j.comp-chemeng.2018.02.014, **2018**.

[9] Boukouvala F, Hasan MF, Floudas CA. Global optimization of general constrained grey-box models: new method and its application to constrained PDEs for pressure swing adsorption. *J Glob Optim*. 67(1-2): 3-42, **2017**.

[10] Boukouvala F, Floudas CA. ARGONAUT: AlgoRithms for Global Optimization of coNstrAined grey-box compUTational problems. *Optim Lett*. 11(5): 895-913, **2017**.

[11] Pearson RK, Pottmann M. Gray-box identification of block-oriented nonlinear models. *J Process Contr*. 10(4):301-315, **2000**.

[12] Cozad A, Sahinidis NV, Miller DC. A combined first-principles and data-driven approach to model building. *Comput Chem Eng*. 73: 116-27, **2015**.

[13] Boukouvala F, Misener R, Floudas CA. Global optimization advances in mixed-integer nonlinear programming, MINLP, and constrained derivative-free optimization, CDFO. *Eur J Oper Res*. 252: 701-727, **2016**.

[14] Olofsson S, Deisenroth MP, Misener R. Design of Experiments for Model Discrimination Hybridising Analytical and Data-Driven Approaches. *arXiv:1802.04170*. **2018**.

[15] Buzzi-Ferraris G, Forzatti P, Emig G, Hofmann H. Sequential experimental design for model discrimination in the case of multiple responses. *Chem Eng Sci*. 39(1): 81-85, **1984**.

[16] Vanlier J, Tiemann CA, Hilbers PAJ, van Riel NAW. Optimal experiment design for model selection in biochemical networks. *BMC Syst Biol*, 8(20), **2014**.

[17] Olofsson S, Mehrian M, Geris L, Calandra R, Deisenroth MP, Misener R. Bayesian Multi-Objective Optimisation of Neotissue Growth in a Perfusion Bioreactor Set-Up. *Computer Aided Chemical Engineering*. 40: 2155-2160, **2017**.

[18] Beykal B, Boukouvala F, Floudas CA, Pistikopoulos EN. Optimal design of energy systems using constrained grey-box multi-objective optimization. *Comput Chem Eng*. DOI 10.1016/j.compchemeng.2018.02.017, **2018**.

[19] Mehrian M, Guyot Y, Papantoniou I, Olofsson S, Sonnaert M, Misener R, Geris L. Maximizing neotissue growth

kinetics in a perfusion bioreactor: An *in silico* strategy using model reduction and Bayesian optimization. *Biotechnol Bioeng*. 115(3):617-29, **2018**.

[20] Lee JH, Shin J, Realff MJ. Machine learning: Overview of the recent progresses and implications for the process systems engineering field. *Comput Chem Eng*. DOI 10.1016/j.compchemeng.2017.10.008, **2017**.

[21] Damianou A, Lawrence N. Deep Gaussian processes. In *Artificial Intelligence and Statistics*. 207-215, **2013**.

[22] Deisenroth MP, Ng JW. Distributed Gaussian processes. In *32nd International Conference on International Conference on Machine Learning*. 37: 1481-1490, **2015**.

[23] Ghahramani Z. Bayesian non-parametrics and the probabilistic approach to modelling. *Phil Trans R Soc A*. 371(1984):20110553, **2013**.

529978 Competitive and Cooperative Adsorption of Ethanol on ZIF-8 in the Presence of 1-Butanol

Abstract:

The Metal Organic Framework ZIF-8 has been studied extensively in the context of the recovery of renewable alcohols produced via fermentation. The vapor and liquid phase adsorption isotherms of these linear alcohols on ZIF-8 do not have a classical Type I shape, but show a distinctive S-shape, containing an inflection point. In this work, the effect of the isotherm shape on competitive adsorption of ethanol and 1-butanol was investigated. The Ideal Adsorbed Solution Theory (IAST) was used to predict the binary adsorption equilibria on ZIF-8 for ethanol and 1-butanol. IAST predictions showed that the presence of 1-butanol at low partial pressures, increases the equilibrium capacity of ethanol. These predictions were further investigated via Monte Carlo simulations, to study the molecular interactions governing this cooperative effect. Breakthrough experiments with binary ethanol/1-butanol mixtures at different ratios were performed to confirm the cooperative adsorption of ethanol. The experimental breakthrough profiles were modeled, employing IAST to predict the equilibrium capacities during the dynamic simulations. The developed dynamic model using IAST was able to accurately describe the influence of the isotherm shape on the experimental breakthrough profiles.

522603 Modeling of Multicomponent Sorption of Acetone-Butanol-Ethanol in a Fixed-Bed Adsorber Using the General Rate Model

Abstract:

Butanol production via fermentation is gaining renewed interest in recent years due to its various applications as raw material in the production of butyl carboxylates, as an industrial solvent, and as an alternative biofuel due its high energy density and heat of vaporization closer to that of gasoline. Biobutanol is being produced on a large scale using fermentation processes based on various renewable feedstocks such as corn, sugar beets, and agro-wastes. Compared to bioethanol, biobutanol is less volatile, less corrosive, and has higher energy content. It can be blended with gasoline without any modifications to the automotive engine. Biobutanol can become economically viable if the upstream feedstock costs are reduced, yields increased, and downstream separation costs are reduced. In the fermentation processes, the end product concentration of butanol is only about 20 g/L due to product inhibition at high concentrations. The high normal boiling point of butanol relative to water makes distillation process complex and energy intensive. Separation and recovery strategies using various resin sorbents in fixed-bed columns have been studied as more efficient alternatives. This is due to the high product selectivity of the resins and the feasibility of continuous mode of operation in fixed-beds. Mathematical modeling of sorption dynamics provides the ability to understand the effects of different operational parameters on the separation process. The aim of this research is to implement a general rate model of the multicomponent sorption process using COMSOL Multiphysics® software and to evaluate the impact of process variables on multicomponent separations. The PDEs representing mass balance in the liquid phase and the solid phase along with axial dispersion, external mass transfer, and intraparticle mass transfer effects for multicomponent sorption of acetone, butanol, and ethanol in the product stream were developed and solved in the coefficient form PDE interface available in COMSOL Multiphysics® software. The breakthrough profiles of weakly adsorbed components show roll over effect and their outlet concentrations are about 50% greater than inlet feed concentrations. The results show that the operational conditions need to be selected appropriately to achieve good separation between the three components. The effects of feed flow rate, initial concentration, and bed length on fixed-bed breakthrough curves including roll-over effects will be presented.

529464 Rare Earth Elements Extraction from Geothermal Brine Using Nanofluids

Abstract:

Rare earth elements (REE) are critical materials in a wide variety of applications such as generating and storing renewable energy. Extracting rare earth metals from geothermal brines is a very challenging problem due to the low concentrations of these elements and engineering challenges with traditional chemical separations methods involving packed sorbent beds or membranes that would impede large volumetric flow rates of geothermal fluids transitioning through the plant. We are demonstrating a simple and highly cost-effective nanofluid-based method for extracting rare earth metals from geothermal brines. Core-shell composite nanoparticles are produced that contain a magnetic iron oxide core surrounded by a shell made of metal-organic framework (MOF) sorbent functionalized with chelating ligands selective for the rare earth elements. By introducing the nanoparticles at low concentration (â‰ˆ0.05 wt%) into the geothermal brine after it passes through the plant heat exchanger, the brine is exposed to a very high concentration of chelating sites on the nanoparticles without the need to pass through a large and costly traditional packed bed. Instead, after a short residence time flowing with the brine, the particles are effectively separated out with an electromagnet and standard extraction methods are then applied to strip the rare earth metals from the nanoparticles, which are then recycled back to the geothermal plant. This process was investigated in a magnetic separation loop system and parameters such as magnetic power, flow rate, particle dispersion and recycling lifetime were studied. We have shown that it is possible to retain the magnetic particles and recover the REE adsorption capacity of the magnetic particles after over 500 h continuous cycling. A detailed preliminary techno-economic performance analysis of extraction systems showed potential to generate a promising internal rate of return (IRR) up to 16%.

530605 Microporous Coating Modified 5A Zeolite for Propylene/Propane Separation

Abstract:

Separation of propylene/propane mixtures is a crucial processing step in the chemical industry. Compared with traditional cryogenic distillation, adsorptive separation, for example, by Pressure Swing Adsorption (PSA), may serve as a more energy-efficient separation alternative. So far, this process has yet been demonstrated industrially, and lack of high performance adsorbents is probably an important restricting factor. It is very challenging to prepare selective adsorbents for separating propylene/propane mixture because of their small size difference and similar chemical/physical properties.

In this study, Molecular Layer Deposition (MLD) and Atomic Layer Deposition (ALD) were combined to deposit uniform microporous coatings on 5A zeolite surface to tune the effective pore size for propylene/propane separation. The pore entrance of 5A zeolite was fine-tuned at the Angstrom scale by changing the MLD and ALD cycle numbers. In this process, MLD was first applied to form a conformal and uniform coating on the external surface of 5A zeolite, which was converted into microporous coating by calcination to remove the organic compound. We have shown that MLD coating effectively reduced the pore size of 5A zeolite due to pore misalignment. Then, ALD was applied to further fine tune the MLD coating pores and thus further reduce the pore size of 5A zeolite. The results showed that the propylene/propane adsorption selectivity increased to ~21 after optimizing MLD/ALD cycles. Meanwhile, diffusivity ratio of propylene/propane increased drastically from ~1 of bare 5A to ~75 for MLD/ALD modified 5A. Our research work suggests that MLD/ALD modified zeolites may have great potential for the adsorptive separation of propylene/propane mixtures.

536714 Development of MOF-74@Zeolite-5A Composite Adsorbents for H_2 Purification

Abstract:

Hydrogen is considered as one of the most important clean and renewable energy sources for a sustainable energy future. However, its efficient and cost-effective purification still remains challenging. In this work, we report the development of novel MOF@Zeolite composites comprised of MOF-74 and Zeolite-5A with core-shell structure for efficient purification of H_2. The composites were synthesized hydrothermally through the addition of Zeolite particles with and without carboxyl functional groups to the MOF synthesis solution. The MOF:Zeolite weight ratio was varied systematically to find the optimum composition based on the adsorption performance. The formation of MOF@Zeolite composites was confirmed by various characterization techniques. Single-component adsorption isotherms of CO_2, CH_4, H_2, CO, and N_2 were measured at 298 K and in the pressure range of 0 to 25 bar. The dynamic adsorption measurements were also performed using binary gas mixtures of CO_2/H_2, CH_4/H_2 and N_2/H_2. The selectivity values were estimated using ideal adsorbed solution theory (IAST) and also from breakthrough profiles. The MOF-74@Zeolite-5A with weight ratio of 95:5 exhibited a similar morphology to that of pristine MOF-74 but with a more ordered crystalline structure and higher surface area. Moreover, this composite showed 20-25% increase in CO_2, CH_4, CO, and N_2 uptake and higher selectivity values in comparison to the bare MOF and Zeolite which could be attributed to the formation of new micropores at the MOF-Zeolite interface and also to the enhanced surface dispersive forces of the composite. The experimental adsorption isotherms were well fitted with single-site Langmuir-Freundlich model. Overall, the findings of this study suggest that the MOF-74@Zeolite-5A composites with core-shell structure are promising candidates for industrial H_2 purification processes.

529856 Microstructure Design of Carbon Materials and Matched Adsorption Process for the Low-Grade Methane Separation from Its Mixture with Nitrogen

Abstract:

Microstructure design of carbon materials and matched adsorption process for the low-grade methane separation from its mixture with nitrogen

Kai Lu **(speaker)**, Donglei Qu, Ying Yang, Li Ping*, Yu Jianguo

State Key Laboratory of Chemical Engineering, College of Chemical Engineering, East China University of Science and Technology, Shanghai 200237, China

*Correspondence author email: 13671654719@163.com, liping_2007@ecust.edu.cn

Abstract

Methane is not only a clean and high caloric fossil fuel, but also the second largest greenhouse gas after carbon dioxide with dramatic influence on climate change. Besides natural gas, unconventional nature gases, such as coalbed gas, shale gas, oil-field gas, landfill gas, biogas, etc, also have CH_4 as the main component. Due to the high impurity content in these feedstocks, the CH_4 separation and enrichment processes are always required before unconventional nature gases can be used for power generation, chemical raw material and other processes. Due to the weak polarities of CH_4 and N_2 and their similar physicochemical properties, the microporous adsorbents have a little amount of adsorption capacity for the low-concentration methane, and the separation of CH_4/N_2 is also very difficult to accomplish by the adsorption-based process.

In this work, we shall investigate the microstructure effect of carbon materials on the low-concentration methane adsorption separation from its mixture with nitrogen, including raw material or precursor for carbon skeleton formation, effect of the specific surface area of carbon material on the low-concentration methane adsorption amount, relationship between micropore size distribution of carbon material and CH_4/N_2 adsorption separation mechanism, and effect of different activator on the weak polar methane adsorption capacity, as shown in Fig. 1.

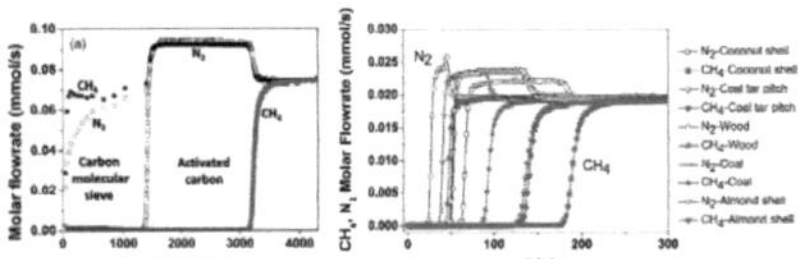

Fig. 1 microstructure effect of carbon materials on the low-concentration methane adsorption separation from its mixture with nitrogen. (a) Comparison between carbon molecular sieve and activated carbon, (b) comparison among activated carbons with different precursor.

After microstructure analysis, the potential carbon materials are recommended for the low-concentration methane adsorption separation from its mixture with nitrogen. Then, two adsorption processes, vacuum pressure swing adsorption (VPSA) process and CO_2 displacement VPSA process, as shown in Fig. 2, are adopted to separate CH_4/N_2 and recover the low-concentration methane. According to the experimental results and simulated data, the matched adsorption process to the potential carbon material is discussed in order to recover effectively the low-grade methane from its mixture with nitrogen.

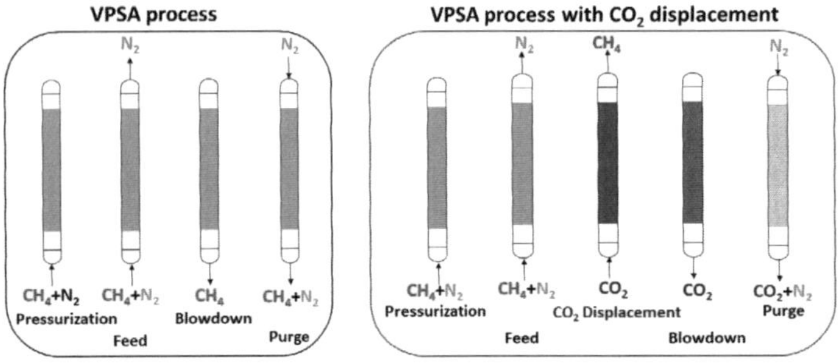

Fig. 2 VPSA processes with and without CO_2 displacement for the low-concentration methane adsorption separation from its mixture with nitrogen.

Keywords: Methane enrichment; CH_4/N_2 separation; microstructure of carbon material; VPSA process; unconventional nature gas

535295 Lithium Ion Sieves Vs. Li^+ Intercalation Electrodes As Effective and Energy-Efficient Materials for Li^+ Mining from Aqueous Resources

Abstract:

Lithium is becoming an increasingly important element due to its expanding application in energy related industries. This indicates inevitable Li supply risks considering that > 80% of global Li production depends on lengthy operations of conventional lime-soda evaporation. Thus, more reliable high throughput Li production technology must be developed. This study presents a comprehensive analysis on the benefits and drawbacks of two of the most promising Li^+ recovery processes: (1) adsorption using lithium ion sieves (LIS) and (2) electrochemical Li^+ recovery systems using selective Li^+ intercalating electrodes.

For adsorption, highly selective LIS such as $Li_{1.6}Mn_{1.6}O_4$, $Li_{1.33}Mn_{1.67}O_4$, $LiMn_2O_4$, $Li_4Ti_5O_{12}$ and Li_2TiO_3 were employed due to their high theoretical Li^+ capacities (38 – 127 mg g^{-1}) and acceptable Li^+ selectivities. For electrochemical Li^+ recovery, Li^+ intercalating materials such as l-MnO_2, $LiNi_{1/3}Co_{1/3}Mn_{1/3}O_2$ and $LiNi_{0.5}Mn_{1.5}O_4$ were prepared as electrodes and paired with Ag as Cl^--selective counter electrode. These active materials (i.e. Li^+ insertion capacity of 39 – 74 mg g^{-1}) were chosen for their stability in water environment and selectivity towards Li^+.

The LIS were electrospun into nanofiber (NF) with polymer support matrix (i.e. LIS NF) and were employed as membrane adsorption filters. Upon saturation of the LIS NF, Li^+ was desorbed and recovered in a small volume of mild acid solution. This step also regenerated the membrane for the next cycle of Li^+ adsorption. Using the same mild acid solution repeatedly for Li^+ recovery eventually produced an LiCl-rich solution. Meanwhile, Li^+-capturing electrodes (i.e. the active materials were coated on Pt foil, which also served as current collector) paired with Ag, captured Li^+ from the feed source by applying a positive current in the cell. At reverse current, LiCl is released and recovered in a receiving electrolyte solution. Similar with membrane adsorption, the electrochemical process was repeatedly carried by "discharging" and "charging" the cell to capture and release the LiCl, respectively, in the same receiving solution until an enriched LiCl stream was produced.

Both processes were evaluated in simulated streams containing low to high Li^+ concentrations which are typical to the following aqueous sources: < 10 mg Li^+ L^{-1} in desalination retentate and coal ash leachates, 100 – 1000 mg Li^+ L^{-1} in brine pools and > 1000 mg Li^+ L^{-1} in certain industrial wastewaters.

Overall results indicate that LIS NF in membrane configuration features the following benefits: (1) Simplicity of operation and provision for full automation considering the adaptable state-of-the-art technology of existing membrane separation processes at industrial scale, (2) convective delivery of the feed allows faster mass transport of Li^+ from the feed to the LIS surface (3) small dimensional property of the nanofibers significantly reduces the diffusion path lengths for Li^+ to reach the LIS adsorption sites thereby shortening the down-time for LIS saturation and (4) convenient control of feed residence time either by appropriate membrane design (i.e. membrane thickness control) or changing the operating conditions (5) well-controlled cycling adsorption/desorption time (i.e. 24 h/cycle) depending on desired Li^+ production rate and (6) the highly porous LIS NF membrane required minimal applied pressure, even lower than those applied in microfilters hence the process is highly energy-efficient and (7) superior LiCl quality due to high Li^+ selectivity and Li^+ capacity of the LIS, particularly the Li_2TiO_3 or H_2TiO_3 (active LIS form).

On the other hand, electrochemical Li^+ recovery system offers the benefit of fast Li^+ capture and release rates, as short as 1 h/cycle. However, the system suffers from the following drawbacks: (1) Even at high Li^+ feed concentration, the Li^+ capturing electrodes cannot be induced (i.e. charge rate increase, total charge flow increase) to achieve their maximum Li^+ intercalation capacity due to current efficiency reduction and severe co-intercalation of competing ions, (2) aside from the energy requirement of the electrode system, existing batch design needs additional mechanical steps for back-and-forth transfer of the electrodes between the feed source and receiving solution, making the process more tedious to operate, (3) long-term stability of the electrodes is difficult to predict but surely will not exceed the lifetime of a typical aqueous Li^+ battery system.

Both processes are not the best options to recover Li^+ from low concentration streams. But between the two, LIS NF membrane adsorption is more competitive over electrochemical due to the high energy consumption and relatively lower Li^+ capacity of the latter system. For feed streams with high Li^+ concentrations, electrochemical system requires much lower energy to capture and release Li^+. Thus, it can somehow compete with the LIS NF membrane adsorption in terms of Li^+ production rate. Thus, for high throughput production of low quality Li, electrochemical system is the completive option. However, LIS have superior selectivity over the Li^+ capturing electrode materials. Thus, if high product purity is desired, LIS NF membrane adsorption remains superior over the electrochemical process albeit at lower production rates.

This research was supported by the National Research Foundation of Korea (NRF) under the Ministry of Science and ICT (No. 2016R1A2B1009221) and the Ministry of Education (No. 22A20130012051(BK21Plus)).

535100 Development of a Novel Emission Control Technology for Onshore-Offshore Applications

Abstract:

Condensate storage tank emissions have been identified as potential sources of emission release into the atmosphere. They are usually accounted for as working, breathing and flash losses. With the rapid expansion in unconventional crude oil and natural gas production in the United States and with propose regulations by USEPA underway, emission control technology would proffer a sustainable research path to effectively maximized production in the oil & gas and petrochemical industry. While the most widely used technology is the vapor recovery units or adsorption, this may not be economical in some cases or may lead to creation of another waste stream.

This study focuses on the design, modeling and testing of a field scale emission monitoring, control technology (EMCT) to capture, and recovers vapors from point source. Volatile organic compounds and natural gas are the focus of this study.

The field test design involved the use of novel carbon adsorbent. The operation of the field scale EMCT unit occurs in continuous cyclic steps including. Online-mode: The unit operations presented received the feed streams of a mixture of gases emitted from the point source, which then comes into the EMCT vessel. The EMCT vessels are connected in parallel and operated in a continuous cyclic manner independent of either each other (online or offline mode). The mixture of gases is exposed to layers of a chemically impregnated novel adsorbent of various forms used to enhance molecular attraction of the gases emitted for it to be accumulated on the adsorbent surface area to capture the VOCs, hydrogen sulfide and other inert gases separated from natural during treatment. The capturing phase by adsorption is monitored until a specified breakthrough point is observed, by which the automated control system switched from adsorption to bed regeneration, which occurred in about 15 minutes. As the online EMCT vessel is undergoing selective adsorption, the other vessel is offline. When breakthrough is evident in the online vessel, the system proceeds to regeneration, while the second vessel is open for adsorption.

Regeneration of the bed is automatically initiated as a result of the state of the art control systems design implemented, which in turn controls the in-vessel heat source to enhance the desorption of the VOCs and other gases, subsequently followed by liquid recovery via condensation.

The combinations of approach involved for the developed methodologies and apparatus in this inventions to efficiently controls the emitted vapor through adsorption, desorption and condensation provide a pathway for a reduce system cycle time, which consequently leads to less power requirement.

The study is supported by a built in mathematical model based on derived governing equations, embedded in the control systems to allow comparisons between the field scale treatment efficiency and the model prediction in real time.

The competitive edge of this invention over the existing technologies includes, but not limited to its ability to make new product in liquid form (Liquid VOCs), recovery of natural gas, shorter cycle times , 100% automation during operations, reduce footprint, 50-75% reduction in adsorbent requirement due to bed regeneration, treatment efficiency stands at 99.8% compared to 95% required by regulations. Preliminary mathematical results from this study at temperature of 1 atm at well production rate of 1000 bbl/day indicate adsorption, desorption and recovery of the VOC emissions of 97%, and 98% respectively.

This project proffers an ability to make new viable petro-chemical product in liquid form by ultimately eliminating flares or vent during well completions.

521377 High-Throughput Process Development of Recombinant Human Serum Albumin Separation with Mixed-Mode Chromatography

Abstract:

The recombinant human serum albumin (rHSA) expressed by microorganisms, such as *P. pastoris*, is a potential pathogen-free and economic alternative for the plasma-derived HSA (pHSA), which could avoid virus infection and resolve blood shortage issues. However, high purity of rHSA is essential for therapeutic use because of high dosage requirement, which leads to a great challenge to purification processes.

In this work, four mixed-mode resins (Capto adhere, Capto MMC, MX-Trp-650M and Nuvia cPrime) were tested to capture rHSA from *P. pastoris* broth and compared with two ion-exchangers (Q Sepharose FF and SP Sepharose FF). The high-throughput screening method with microtiter filter plate was used to screen the optimal pH and dilution ratio for rHSA adsorption. The adsorption capacity and selectivity were investigated. It was found that the cation-exchange resins with the negatively-charged group (such as carboxyl or sulfonic group) were more suitable for the binding of rHSA, and mixed-mode resins showed high salt-tolerant property and the pretreatment of dilution was unnecessary. For MX-Trp-650M, 200 mM NaCl could even promote the adsorption of rHSA. In addition, four resins with cation-exchange group were used to screen the elution conditions for rHSA separation with microtiter filter plate. For Capto MMC and MX-Trp-650M, high recovery and purity could be obtained due to high adsorption selectivity. Finally, the separation conditions based on the high-throughput studies were verified by the column chromatography. The results indicated that high binding capacity and elution recovery could be obtained under the optimized loading and elution conditions. In general, Capto MMC and MX-Trp-650M showed an efficient practicability for rHSA separation with high capacity (~50 mg/ml dynamic binding capacity at 10% breakthrough), suitable salt-tolerance (about 20 mS/cm), high recovery (above 95%) and high purity (above 96%), as well as excellent removal of pigments and HCPs. The results indicated that high-throughput process development with microtiter filter plate is a powerful tool to determine a wide range of variables and then optimize the separation conditions. Mixed-mode chromatography showed high process efficiency for rHSA separation from cell broth with high capacity and purity, suitable salt-tolerance and high elution recovery.

This work was supported by the International Science & Technology Cooperation Program of China and National Natural Science Foundation of China.

538301 Process Development of a Continuous Precipitation and Filtration Unit Operation for the Capture of Biotherapeutics

Abstract:

The biopharmaceutical industry is currently facing several challenges in the downstream purification of recombinant proteins of diverse and significant therapeutic values. The advancement in genetic engineering has enabled a ten-fold increase of target protein titers in bacteria or mammalian cell cultures. However, this improvement in productivity has brought a bottleneck to the subsequent purification steps. This bottleneck refers to the decline in performance of the first crude purification step that relies heavily on chromatography, where the desired proteins are selectively separated by passing through a packed column. One appealing alternative option is target protein precipitation by the addition of reagents. It is an efficient and selective bulk purification technique that takes advantage of the higher titer because it enhances supersaturation which is the driving force for precipitation. Therefore, we propose a continuous precipitation-filtration coupled unit operation to serve as the capture step in the protein purification train. We have identified that the combination of polyethylene glycol (PEG) and $ZnCl_2$ is efficient and effective in inducing target protein precipitation. Equilibrium and kinetic information on this synergistic precipitation was obtained with human immunoglobulin (hIgG) and bovine serum albumin (BSA) in a batch setting. In addition, study of the critical flux during tangential flow microfiltration of BSA showed that the critical flux increases with increasing $ZnCl_2$ concentration but reaches a maximum at an intermediate PEG concentration. In this work, we couple a tubular reactor with static mixer for precipitation and tangential flow filtration for the subsequent solid-liquid separation to construct a continuous unit operation for the capture step in downstream processing. Using this laboratory-scale system, we test and demonstrate its applicability to industrially relevant clarified cell culture fluid (CCCF) and show process optimization based on fundamental solubility, kinetics and precipitate particle filtration behavior to fulfill purity and yield requirements for the target therapeutic.

533134 Crystallisation of Short Peptides

Abstract:

Bio-active peptides can be used as highly effective pharmaceutical compounds to treat a wide range of diseases. The purification of peptide currently relies on chromatography which contributes significantly to the high cost of downstream processing. Although crystallisation is a more economical and scalable option for the purification of peptide, there is a knowledge gap in peptide crystallisation which the current study aims to fill by investigating the crystallisation of short peptides that have less than 10 amino acids in a sequence. Since the measurement of solubility is the first step in the development of a crystallisation strategy, the solubility of short peptides such as aspartame and polyglycines between 5 and 40 degrees Celcius is determined before performing the crystallisations. The experimental data of this study not only confirms the feasibility of peptide crystallisation over a range of conditions, but are also useful for constructing thermodynamic models for predicting the solubility of short peptides in the future studies. The knowledge of this work can aid the downstream processing of bio-active peptides in general.

Acknowledgement

We acknowledge the EPSRC (EP/N015916/1) for funding.

Reference

1. Kastin, A. *Handbook of Biologically Active Peptides. Academic press* (2013).
2. IUPACâ€IUB Joint Commission on Biochemical Nomenclature (JCBN) Nomenclature and Symbolism for Amino Acids and Peptides Recommendations 1983. *Int. J. Pept. Protein Res.* **24,** 9â€“37 (1984).

528635 Molecular Modeling to Efficiently Screen Chromatographic Separation of Challenging Enantiomer Separations

Abstract:

More than half of all pharmaceuticals are chiral compounds. Although the enantiomers of chiral compounds have the same chemical structure, they can exhibit marked differences in physiological activity; therefore, it is important to remove the enantiomer that exhibits undesirable physiological activity. Chromatographic separation of chiral enantiomers is one of the best available methods to obtain enantio-pure substances, but the optimization of the experimental conditions can be very time-consuming. Generally, there are several chiral stationary phases to choose from; the most commonly used ones are polymers that have an amylose or cellulose backbone, with various sidechains. These could be coated on the solid support or covalently bonded to it. Then there is the choice of the solvent system that constitutes the mobile phase, dictated by whether the CSP is coated or covalently bonded to the support, and also by the solubility of the drug. Furthermore, modifiers (acids or bases or buffers) may be added to the solvent in order to change or protect some functional groups on the drug to realize a better resolution between the enantiomers. One could try different CSPs and mobile phases using educated guesses based on experience, but the number of possible combinations is huge, and could easily run into hundreds of combinations. To reduce possible combinations of experimental approaches to chiral separations, we proposed a computational method that would provide the largest separation between the enantiomers of the drug in question. Chromatographic separation is a dynamic process, with the interactions between the drug and the chiral stationary phase mediated by the solvent, so no single property, such as the minimum interaction energy, could possibly describe and account for the ratio of residence times in the chromatographic column for the enantiomeric pair. We chose to use explicit-solvent fully atomistic molecular dynamic simulations, permitting all the interactions between the atoms constituting the polymeric chiral stationary phase, the solvent molecules and the drug molecule.

Using molecular dynamics simulations, we have developed a predictive molecular model to determine the experimental elution order of enantiomers and the experimental value of the separation factor. This predictive capability is beyond what HPLC separation experiments can provide. The latter can only provide separation factors and additional experiments are required to find which absolute configuration enantiomer was eluted first. Our model predicts which of the enantiomers elutes first. At the present stage of this work, the best correlation provides separation factors with a correlation coefficient of **0.80** and the elution order is predicted correctly for **90%** of the cases. The metric which provides this level of prediction is the maximum value of the average lifetime of hydrogen bonds between the drug and the CSP, including all donor-acceptor pairs between the drug and the CSP in this overall measure.

In our next phase of work, we are investigating several other CSPs, not only different side chains on the amylose backbone, but also cellulose-based CSPs, investigate the effect of using modifiers (small concentrations of a compound that is added to the solvent system in order to improve the baseline separation and improve the separation factor for a given drug/CSP/solvent system combination), and investigate the currently widely-used supercritical media, such as supercritical CO_2 .

537986 Advances in Centrifugal Separation in Biotechnology

Abstract:

Advances in centrifugal separation in biotechnology

Wallace Woon-Fong Leung

Mechanical Engineering, The Hong Kong Polytechnic University, Hung Hom, Hong Kong

Abstract

Centrifugal separation has been used to recover recombinant protein after fermentation using, respectively, bacteria, mammalian cells and yeast as host cells. The separation is being tuned to the special requirements for each process. For example, for mammalian cells are highly sensitive to shear due to absence of a cell wall as with yeast or bacteria. High shear in the feed acceleration can lead to lysing of the cell thereby spoiling the entire batch. We will discuss possible provision whereby this can be minimized. For bacterial cells, the protein is produced inside the cell and the bacteria have to be lysed before the separation by centrifugation. Separation followed by reslurrying/washing are required. The issue is to how to effectively separate and purify the protein before downstream purification by the column. For yeast processing, the requirement is to extract to high yield with minimal processing. Typically, a two-stage separation is required to warrant high yield. In addition to protein harvest, there are numerous biotech separation processes that require high yield, high purity, high concentration of the solid phase, solid-free centrate liquid, etc. Technological advancement in the continuous centrifuges, such as the disc nozzle centrifuges, and semi-continuous centrifuges, such as the intermittent disk and the tubular centrifuges are also discussed. The decanter is also used to process bioproducts, especially the waste stream. Flocculation can be used to enhance separation. The optimal floc size and minimum polymer dosing are of great interest. Various challenges and opportunities will be discussed in the presentation.

Biography of Wallace Woon-Fong LEUNG

Prof. Wallace W-F Leung received his MS and ScD both in Mechanical Engineering from MIT. For the past 40 years, he has worked on filtration and separation from ultrafiltration of protein solution, centrifugation of process stream, vacuum filtration of minerals, to air filtration. For 18 years, he was with Bird/Baker Hughes as Senior Research Scientist and Director of Process Technologies directing centrifugal separation/filtration. Earlier, he has worked for Gulf Oil and Schlumberger on flow in petroleum reservoirs. For the past 14 years, he is Chair Professor of Innovative Products and Technologies at The Hong Kong Polytechnic University. Wallace's research interest is in centrifugation of bioprocesses. Besides, he is also interested in applications of nanofiber technologies in separating nano-aerosols, renewable energy (photovoltaics, hydrogen production using water spliting), clean air and water (photocatalysis), and health technologies (wound dressing).

He is well known in the world for his work in centrifugation. He is author of the 1998 McGraw-Hill book "Industrial Centrifugation Technology" and 2007 Elsevier/AP book "Centrifugal Separation for Biotechnology". He has 36 US patents on centrifugation and 13 US patents on other technologies.

Dr. Leung is a fellow of American Institute of Chemical Engineers, American Society of Mechanical Engineers, American Filtration and Separations Society, Hong Kong Institute of Engineers, and Hong Kong Academy of Engineering Sciences. He is the Chairman for the World Filtration Congress in 2004 (New Orleans) and 2020 (San Diego). He is also Chairman for the International Delegation on Filtration 2016-2020 with 12 member countries, overseeing the world Filtration Congress and promoting filtration and separation activities worldwide.

528148 Separation of Intrinsically Magnetic Red Blood Cells Using Combination of Numerical Modelling and Microfluidic Magnetic Deposition System

Abstract:

Annually, an average of 21 million units of human blood components are transfused in the US for various medical purposes. Moreover, as more and more researchers study the human blood components for various medical research purposes, demand for efficient red blood cell (RBCs) separation technique increases as well. The traditional method for the separation of RBCs is centrifugation, which requires a lot of resources (equipment/energy) and time. To overcome this issue, various approach to develop novel RBC separation system has been developed over the years, but opportunities still exist for improvement, and recent development in analysis technology led to focus on the study of red blood cell themselves.

Based on the observation made by Pauling and Coryell back in 1936, the blood community has concluded that RBCs in different chemical state (oxygenated, deoxygenated and oxidized) express different magnetic moment due to different number of unpaired electrons in the iron (stored in hemoglobin). This difference in magnetic moment has led many researchers to develop novel separation devices that could exploit this magnetic susceptibility difference. The magnetic deposition microscopy (MDM) is a system that uses permanent magnets and pole pieces in special configuration to generate magnetic gradient field. This magnetic gradient drives the magnetic entities from a mixture towards the deposition slides and eventually separate them from the non-magnetic population.

For this study, the magnetic separation of the RBCs was conducted using combination of experiment and numerical modelling approach. To model the behavior of RBCs in the MDM system, the magnetic mobility (a parameter that describes the responsiveness of individual entity to a magnetic energy gradient) of the oxidized hemoglobin (metHb) and oxygenated hemoglobin (oxyHb) was measured using the cell tracking velocimetry system (CTV). The CTV system uses charge coupled device camera connected to microscope to record/calculate the magnetic mobility of entities in buffer under the influence of constant magnetic energy gradient in steady state flow. After their magnetic mobility values were recorded, the data was fed into the MDM model developed through combination of the two software, FEMM and ANSYS Fluent. Both software uses finite element and volume analysis method to solve electromagnetic and flow problems. The discrete phase modelling approach connected to custom user defined function, which incorporates the magnetic force balance and the magnetic mobility distribution from CTV, was used to model the behavior of the oxyHbs and metHbs in the system, as well as calculating the theoretical deposition/separation rate of the magnetic RBCs.

Based upon the simulation results, the optimum flow condition was determined to be 0.2 ml/min for the MDM system. Also, the experimental result showed a close match in the deposition location compared to the simulation results.

However, this low volumetric flow rate is not yet realistic enough for clinical applications. To overcome this issue, we mean to utilize this study method more vigorously, where current magnetic separation system can be modified and simulated through FEMM, determine the optimal flow setting for RBC lysis using ANSYS Fluent, and finally carry out magnetic separation experiment in the newly designed system. Furthermore, we are now exploring the possibility of using polymer mixed with iron oxide particles to 3-D print a novel disposable magnetic separation system. Using this combinatorial approach, we aim to increase the throughput of the device to separate 5 ml of RBCs unit in 10 minutes.

539098 Ionicity and Hydrogen–Bonding As Critical Factors for Protic Ionic Liquids (PILs) Dissolution of Polysaccharides and Lignin

Abstract:

Effectively partitioning lignocellulosic biomass into its various fractions—cellulose, hemicellulose and lignin—is essential for the implementation of a biofuel/biorefinery-based economy. In particular, an efficient, low-cost technique for the selective **removal and recovery of lignin**, the component that largely renders biomass intractable, is necessary to facilitate easier access to the polysaccharides and the production of valuable side-product streams based on lignin. Current separation techniques for lignin removal suffer significant drawbacks such as a high energy intensity and a harmful environmental impact, and need to be optimized to minimize waste generation and resource (lignin) under-utilization.

A highly effective method has been developed for the facile extraction of lignin from lignocellulosic biomass using potentially inexpensive protic ionic liquids (PILs). With systematic variations in the cation of acetate-based PILs, several PILs were evaluated to reveal the important structural features that pertain to lignin extraction. Solubility tests (with commercially available biomass components), in conjunction with the physical and chemical properties of the PILs, and force field energy minimizations (MM2), were correlated with the PILs' lignin extraction efficiency. These results indicate that the PIL's ability to partially dissolve xylan (i.e., hemicellulose) and disperse the cellulose in a stable suspension results in greater fiber disruption/penetration, and consequently, enhances the effectiveness of the lignin extraction. This observation was also directly linked to two significant PIL properties: **ionicity and hydrogen bonding interactions.**

PILs from cyclic amines (e.g pyrrolidinium acetate – [Pyrr][Ac]) and PILs from alkanolamines (e.g. ethanolammonium acetate – [Eth][Ac]) are two noteworthy PIL groups that were identified with a superior performance for this process (i.e lignin extraction). Further analyses revealed that these PILs are associated with ions that are strongly interacting, which is linked to key structural features of the PIL, the ease of accessibility of the PIL ions, and the ability of the PIL ions to form hydrogen bonds with ideal bond lengths. In particular, the PIL, [Eth][Ac], is able to selectively extract up to 85% of the lignin found in cornstover (a biomass resource). After the lignin-extraction step, the PILs are easily recovered using distillation leaving the separated lignin and cellulose-rich residues available for further processing.

Careful selection of the ions present in PILs is essential for their utilization in many applications especially in the biomass/bioenergy industry. The strategies employed in this research project can be easily applied as an approach for designing task-specific PILs for targeted applications (such as biomass processing), which will enable improved process efficiencies and product yield/purity.

531239 How Proton Transfer Equilibria Influence Ionic Liquid Properties: Molecular Simulations of Alkylammonium Acetates

Abstract:

Protic ionic liquids (PILs) form through proton transfer from a Brönsted acid to a Brönsted base. The thermodynamics of this process are studied by molecular simulation in two ways. In the first, we use molecular dynamics to determine how PIL properties vary with χ, the extent of the proton transfer reaction. Three PILs are considered: *N*-propylammonium acetate, [N3][Ac], *N*-butylammonium acetate, [N4][Ac], and *N,N*-dimethylbutylammonium acetate, [N114][Ac]. In all cases density and viscosity increase with increasing χ, while diffusivities of all species decrease with increasing χ. In each PIL the ionic conductivity exhibits a maximum at intermediate χ due to competition between increasing ion concentration and decreasing ion mobility. Ionicity analysis suggests that strongly correlated behavior is present at all χ. Finally, we determine the χ for which the properties of each simulated PIL best agree with experimental data; these are χ = 0.95, 0.91 and 0.14 for [N3][Ac], [N4][Ac] and [N114][Ac], respectively. These results suggest that proton transfer is nearly complete in the primary ammonium PILs but not in the tertiary ammonium PIL, consistent with recent experimental observations. We propose that this difference is due to cooperative production of hydrogen bonds with increasing χ in the primary ammonium PILs, which does not occur in the tertiary ammonium PIL. In a second study, free energy calculations using aqueous media as a reference state are used to determine the proton-transfer reaction free energy in the PIL environment. Since the ionic strength and other properties of the PIL vary with χ, this reaction free energy also depends on χ. This dependence is determined for selected PILs and used to predict *a priori* whether the PILs are completely or incompletely proton-transferred at equilibrium.

536887 Investigating Diffusivity of Solvated Ionic Liquids through Molecular Dynamics Screening

Abstract:

Supercapacitors are electrochemical devices that exhibit high power density but are currently limited in their role of energy storage due to low energy density. A potential solution for improving energy density is the use of room temperature ionic liquids (RTILs), a class of organic solvents that are liquid at room temperature[1]. RTILs have certain physiochemical properties, including a wide operational voltage window which theoretically allows for a higher energy density of supercapacitor devices[2],[3]. Despite this attractive property, RTILs exhibit low diffusivities which can negatively affect the charge/discharge rates of supercapacitors. To overcome this issue, RTILs are often solvated in organic solvents to improve transport properties without a substantial loss in operational voltage window[4]. Thus in order to optimize the performance of supercapacitor devices, it is desirable to fully understand the interactions between RTILs and solvents.

While RTIL interaction with a limited number of organic solvents has been explored with molecular dynamics (MD), difficulties in predicting behavior remain due to the large design space and complex interactions of these systems[5]. With this as motivation, we take a screening approach to expand our knowledge of RTIL-solvent interactions. As a result, we have implemented the necessary algorithms and force fields within MoSDeF[6] to enable screening of [BMIM+][Tf2N-] solvated in 21 distinct solvents over a range of 18 different concentrations. Self-diffusivities of RTILs in each system are obtained by mean squared displacement (MSD) calculations in order to determine trends between RTIL transport, solvent selection, and solvent concentration. Additionally, we compute the ion pairs and ion cages in each system to provide insight into how the addition of each solvent affects RTIL structure. From the ability to sample a larger portion of the RTIL-solvent design space, we find that RTIL diffusivity increases with the bulk diffusivity of the pure solvent and increases with greater solvent concentration. The trends presented can further guide the selection of solvents and concentration for electrochemical applications of RTILs.

References

[1] J. F. Brennecke and E. J. Maginn, "Ionic Liquids: Innovative Fluids for Chemical Processing",AIChE

Journal, **2001**, 47, 2384-2389.

[2] L. P. Rebelo, J. N. Lopes, J. M. Esperanca, E. Filipe, "On the Critical Temperature, Normal Boiling Point, and Vapor Pressure of Ionic Liquids", Journal of Physical Chemistry B, **2005**, 109, 6040-6043

[3] T. Sato, G. Masuda, K. Takagi, "Electrochemical Properties of Novel Ionic Liquids for Electric Double Layer Capacitor Applications", Electrochimica Acta, **2004**, 49, 3603-3611.

[4] R. Lin, P. Huang, J. Segalini, C. Largeot, P.L. Taberna, J. Chmiola, Y. Gogotsi, P. Simon, "Solvent Effect on the Ion Adsorption from Ionic Liquid Electrolyte Into Sub-nanometer Carbon Pores", Electrochimica Acta, **2009**, 54, 7025-7032.

[5] N. Osti, K. Van Aken, M. Thompson, F. Tiet, D. Jiang, P.T. Cummings, Y. Gogotsi, E. Mamontov, "Solvent Polarity Governs Ion Interactions and Transport in a Solvated Room-Temperature Ionic Liquid", The Journal of Physical Chemistry Letters, **2016**, 8, 167-171.

[6] "MoSDeF" [Online]. Available: https://github.com/mosdef-hub.

535129 Thermodynamics & Thermophysical Properties of Thermally Robust Ionic Liquids and Their Mixtures

Abstract:

Although ionic liquids are often described as having high thermal stability, evidence shows that many traditional ionic liquids begin decomposing at temperatures much lower than those reported from TGA decomposition experiments, often losing a few percent mass in just a few hours at temperatures 50-100C lower their reported decomposition temperatures. Functional groups appended to the ions to impart a particular physical or chemical property often dramatically decrease the thermal stability of compound. Aryl groups provide a great deal of thermal stability; however, they typically increase the melting point of the compound significantly. Thus, there is an inherent tradeoff between thermal stability, properties of interest (high heat capacity, solubility, etc.) and low melting point, and the vast structural diversity available in designing ionic liquids is highly constrained when thermal stability is a desired characteristic.

For the past few years, our groups have studied and characterized a number of classes of thermally robust ionic compounds. In this work, we study several classes of thermally robust ionic liquids, including per-aryl phosphonium and sulfonium salts, from a cation perspective, as well as benzesulfonate salts from an anion perspective. In addition to pure component thermodynamic and thermophysical properties (melting points, heat capacities, phase equilibria, etc.) we demonstrate that desired properties for these fluids (low melting point, high heat capacity) can be achieved through mixtures of these salts with each other and with molecular aromatics.

Traditionally, ionic liquids are distinguished from molten salts by having melting points lower than 100C, however, the term â€œionic liquidâ€ is evolving to encompass ionic compounds that are liquids in the temperature range in which they used. Here we embrace the expanded term as the classes of compounds that we are examining have structurally similar representatives some of which melt above 100C and some below, highlighting the arbitrary nature of the distinction.

531069 Development of an Ionic Liquid Based Low-Temperature Electrolyte System for Sensing Applications of Planetary Exploration

Abstract:

It is of great importance to understand the interior structure and the composition of other planets via seismological studies for a foreign planet. The molecular electronic transducer (MET) technology has been utilized in designing planetary seismometers that use liquid electrolyte to enable the higher shock tolerance and independence of installation angle during deployment, comparing to conventional spring-mass based systems. However, it is challenging to adapt MET sensors using any aqueous-based electrolyte to the extreme low-temperature environments on planets such as Europa.

Ionic liquids have a wide liquid temperature range and are highly ionic conductive, non-volatile, and electrochemically stable, making them a class of materials as the excellent candidate for electrolytes for a variety of applications. Despite the advantages, the drawbacks of using ionic liquids as electrolyte materials for space applications include their relatively high viscosity and undesired thermal events, which can both contribute to depressing the charge transport in the electrolyte, especially when operating at low temperatures. Herein, we report a liquid electrolyte system based on a mixture of the ionic liquid 1-butyl-3-methylimidazolium iodide ([BMIM][I]) and water with incorporation of an alkylammonium-based ionic liquid for the planetary application of MET sensors at extremely low temperatures. The properties of thermal, mass transport, and conductivity of developed electrolyte solutions were examined at varying temperatures down to â€“75 Â°C. The effect of incorporating alkylammonium-based ionic liquid into the [BMIM][I]/water mixture at the optimized concentration can effectively prevent the formation of crystallization within the solution, which is critical in employing ionic liquids as electrolytes for any low-temperature applications, and also further tune the properties of the electrolyte by lowering the glass transition temperature, reducing viscosity, and enhancing the conductivity. The electrochemical stability of the selected candidate formulations was studied via cyclic voltammetry, showing that the incorporation of the alkylammonium-based ionic liquid did not appear to introduce undesired redox reactions and the ILs-water-salt system didnâ€™t suffer from additional electrochemical reactions other than iodide/triiodide redox within a proper potential window. The results of this presented work will not only extend the MET sensing technology to the field of space exploration but also allow the potential applications of other electrochemical devices with iodide/triiodide redox pairs at low temperatures.

525897 Solubility and Diffusivity of Ammonia in Aprotic and Protic Ionic Liquids

Abstract:

Ionic liquids (ILs) are salts composed of discrete cations and anions with melting points below 100 °C and many are liquid at ambient temperature. Due to their negligible vapor pressure, nonflammability, relatively high thermal stability and excellent solvation potential, ILs have drawn broad attention in various applications including catalysis, gas separation, and heating/cooling cycles. Our motivation in this study is to replace water (high vapor pressure absorbent) with a feasible IL (negligible vapor pressure absorbent) in the ammonia-water absorption refrigeration cycle to reduce or eliminate separation costs. Therefore, we are thoroughly investigating the thermodynamic and kinetic properties of ammonia in aprotic ionic liquids (AILs) and protic ionic liquids (PILs). The major difference between AILs and PILs is that PILs have a proton available for hydrogen bonding due to the proton transfer from the acid to the base during formation. Therefore, PILs can be used to create hydrogen bonded network for ammonia which can result in exceptionally high ammonia sorption. The Vapor-Liquid-Equilibria (VLE) measurements were performed using advanced, fully automated and highly accurate magnetically coupled gravimetric microbalance (XEMIS, Hiden,Inc). Pressure-temperature-composition (*PTx*) data for NH_3 in ILs at isothermal conditions up to 7 bar was correlated using the Peng-Robinson equation of state. In addition to thermodynamic analysis, to understand time dependent behavior of ammonia with ionic liquids, we applied a simplified diffusion model to ammonia absorption in ILs and ammonia desorption from NH_3-IL mixtures. Findings on thermodynamic and kinetic behavior of NH_3 and ILs in this study will contribute to the understanding of the interaction of ammonia and ionic liquids.

527454 Efficient and Reversible Separation of Ammonia with Ionic Liquid-Based Materials

Abstract:

Efficient and Reversible Separation of Ammonia with Ionic Liquid-based Materials

Shaojuan Zeng [a], Dawei Shang [a], Haifeng Dong [a], Xiangping Zhang [a,*], Suojiang Zhang [a,*]

[a] *Institute of Process Engineering, Chinese Academy of Sciences, 100190, Beijing, China*

**Corresponding author: xpzhang@ipe.ac.cn; sjzhang@ipe.ac.cn*

As one typical alkaline and poisonous gas, NH_3 emission inevitably causes serious air pollution problems, such as fog and haze, which has raised extensive attentions worldwide. The commercial methods, like water scrubbing and acid scrubbing, can purify the NH_3-containing gas mixtures, but they substantially transfer the NH_3 pollute from gas phase to liquid or solid phase, which doesnot mean total elimination[1]. Therefore, how to efficiently separate NH_3 and simultaneously recover high purity of NH_3 easily is the most fundamental way but also a great challenge. The emerging of Ionic liquids (ILs) provides a promising way to efficient separation of NH_3 owing to their peculiar properties, like nonvolatility, designability and high stability[2].

In this work, the interaction between NH_3 and ILs and the structure-property relationship was firstly revealed by combining simulation calculations and experimental characterizations. The results indicated that the H-bond and cations play an important role in NH_3 absorption, and the stronger H-bond means higher NH_3 capacity. Meanwhile, the introduction of the metal center into ILs greatly improves NH_3 absorption through chemical complexion. Based on these results, a series of functionalized Lewis-acid ILs with the metal center on anions and Brønsted-acid ILs with the acidic H groups on cations were designed for efficient and reversible absorption of NH_3, respectively. For example, the Brønsted-acid IL [Bim][Tf_2N] can absorb 2.69 molNH_3·molIL^{-1} at 40 and 0.10 MPa through the strong hydrogen bonding between NH_3 and cations[3]. The cobalt ILs $[C_nmim]_2[Co(NCS)_4]$ not only have exceptional NH_3 capacity and high NH_3/CO_2 selectivity, but also excellent recyclability. The maximal NH_3 capacity is up to 6.09 molNH_3·molIL^{-1} at 30 and 0.10 MPa, which is over 30 times higher than the conventional ILs[4]. On the other hand, considering the higher viscosity of these ILs, the ILs and porous materials were combined to form new adsorbents for NH_3 separation. Compared with pure silica gel, NH_3 adsorption capacity of IL-based materials increases by 70-100%, and the adsorption and desorption process is also very fast. These findings will provide significant theoretical basis to develop efficient and reversible IL-based technologies for NH_3separation.

1. J. W. Erisman, A. Bleeker, J. Galloway and M. S. Sutton, *Environ Pollut*, 2007, **150**, 140-149.
2. S. J. Zeng, X. P. Zhang, L. Bai, X. C. Zhang, H. Wang, J. J. Wang, D. Bao, M. D. Li, X. Y. Liu and S. J. Zhang, *Chem Rev*, 2017, **117**, 9625-9673.
3. D. W. Shang, X. P. Zhang, S. J. Zeng, K. Jiang, H. S. Gao, H. F. Dong, Q. Y. Yang and S. J. Zhang, *Green Chem*, 2017, **19**, 937-945.
4. S. J. Zeng, L. Liu, D. W. Shang, J. P. Feng, H. F. Dong, Q. X. Xu, X. P. Zhang and S. J. Zhang, *Green Chem*, 2018, DOI: 10.1039/c8gc00215k.

532755 Electrical Conductivities of Binary and Ternary Deep Eutectic Solvents Via Molecular Simulation

Abstract:

Solvent properties, such as electrical conductivity, are crucial to the successful design and operation of batteries. In several instances, the types of conventional solvents used in batteries are both potentially toxic to an individual, as well as to the environment. As an alternative, deep eutectic solvents (DES) are biologically based ionic liquid eutectic mixtures which possess tunable properties according to the selected cation, anion, and hydrogen bond donating co-solvent partner. These tunable properties include density, dielectric constant, and solvating power. DES possess extremely low vapor pressures, and therefore do not volatilize as easily as some conventional solvents used in the battery industry. Moreover, because DES are comprised of biologically based ions (i.e. choline) rather than synthetic ions (i.e. 1-ethyl-3-methyl-imidazolium), they are less toxic to organisms and the environment, and less expensive.

This work seeks to characterize environmentally benign, non-toxic, tunable DES through molecular simulation. In particular, molecular dynamics (MD) simulations are used to compute electrical conductivities for various binary and ternary DES, comprised of mixtures of choline chloride, urea, and glycerol. Electrical conductivities are calculated by two approaches: 1.) via post-processing of mean-squared displacement data from conventional MD simulations and 2.) on-the-fly during the course of an MD simulation. This latter on-the-fly approach involves the development of a new computational method within the popular open-source MD code, LAMMPS, developed by Sandia National Laboratories. Comparisons of computed electrical conductivities between the two approaches for $NaCl/H_2O$ mixtures are presented for benchmarking against experimental data, along with new predictions for various binary and ternary DES. Comparison against electrical conductivities of conventional battery solvents are made, too.

535804 Solvatochromic Evaluation of Hydrophobic Deep Eutectic Solvents

Abstract:

Deep eutectic solvents (DES) are attractive for separation based applications due to the ability to design a solvent with a particular set of properties. Type III DES are formed by a hydrogen bond donor (HBD), typically an acid, and a hydrogen bond acceptor (HBA) such a quaternary ammonium halide salt. The hydrogen bond sharing results in a deep melting point depression, or eutectic point. Even if both HBA and HBD are solids at room temperature, the correct ratio of the two can form a liquid. By varying HBA and HBD species and ratio it is possible to adjust properties such as melting point, density, conductivity, hydrophobicity, viscosity, etc. Numerous studies have investigated a small selection (typically 2 to 8 solvent combinations) of DES and a relatively fewer number of studies have taken a broader take on how DES work. Thus far, a predictive model to estimate DES properties that does not use computationally intensive methods has not been described.

Solvatochromic methods of analysis are useful for measurements of a solvent's physical properties such as dipolizability, hydrogen bond donating ability, and hydrogen bond accepting ability. H-bond characteristics will be measured using betaine dyes in a UV-VIS. The solvatochromic method was developed in the 1980s and has been applied for IL and DES. Reichardt's dye or betaine dye 30 (2,6-diphenyl-4-(2,4,6-triphenylpyridinium- 1-yl)phenolate) is one of the most widely used solvatochromic probes, however, acidic conditions will strongly affect its solvatochromic nature. Since, the goal of this project is to characterize HDES, where HBAs are long chain acids, Reichardt's dye 33 (2,6-dichloro-4-(2,4,6-triphenylpyridinium1-yl)phenolate) will be chosen for this project instead of dye 30. Kamlet-Taft (KT) parameters will be evaluated for hydrophobic DES. The KT β parameter provides H-bond accepting ability of HDES. This parameter will be obtained by solvatochromic method by comparing solvent-induced shifts of the absorption bands of probes. Kamlet and Taft proposed probes, 4-nitroaniline/N,N-diethyl-4-nitroaniline will be used in this project, since they are the most commonly used KT probes for ILs and DES. A lower value of β is expected as the formation of HDES through establishment of H-bonds increases the capacity of the HDES to engage in further H-bonds network with HBD. In this paper, effects of anion, aliphatic chain length, and acidic hydrogen bond donor will be evaluated for β parameter. Solvatochromic parameters will then be used to compare solvent physical properties with that of the HBA and HBD components.

529928 Robust Polymeric Thin Film Composite Membranes for Organic Solvent Nanofiltration

Abstract:

Membrane-based filtration systems have been taken up for use in various applications over the past five decades as a key green process engineering platform for process intensification. Particularly, membrane technology in liquid filtration has been successful in desalination applications. Though the recent research progress in membrane technology exhibits further potential of state-of-the-art membranes in a pharmaceutical, petroleum, and energy-related industries beyond wastewater purification and desalination applications, however, it is yet challenging to expand the limited operating conditions of the currently available membranes and to solve the exacerbating environmental burden during the fabrication of polymeric membranes. Therefore, in this study, we had developed new kinds of robust polymeric membranes in terms of support and active layers for both (i) fabricating a robust membrane that can function beyond the typical operating conditions of conventional membranes and (ii) minimizing waste creating and production time. The developed robust thin film composite membranes incorporating new support and active layers had remarkable stability and performance for the separation of solutes in *N,N*-dimethylformamide, even at elevated temperatures not feasible with conventional polymeric membranes. Furthermore, the developed support layer can reduce the mass transfer resistance inside of the support layer due to its highly porous and interconnected pore structure with thin thickness and consequently can generate high power density through pressure retarded osmosis system. Also, the proposed fabrication method for the new active layer can reduce the usage of polymers and solvents and eliminate the time-consuming coagulation and washing steps, both of which generate an enormous amount of solvent-contaminated wastewater. Therefore, in this presentation, we will introduce the potential of developed robust support and active layers for organic solvent nanofiltration and also present the distinct significances of both developed layers comparing with conventional systems.

522222 Evidence for Entropic Selection of Xylene Isomers in Carbon Molecular Sieve Membranes

Abstract:

The purification of benzene derivatives, particularly xylene isomers, is one of the most important organic mixture separations practiced in industry. The separation of xylene isomers is especially challenging due to the similarity of their physical properties, such as kinetic diameter, molecular weight and boiling point. Carbon molecular sieve (CMS) membranes are promising materials capable of performing such challenging solvent separations. Indeed, CMS materials possess excellent thermal and chemical stability, even in aggressive solvent environments. Xylene isomer transport and sorption properties in polymer of intrinsic microporosity-1 (PIM-1) derived CMS materials are examined over a temperature ranging from 35 to 55 ^{o}C. The diffusivity, sorption coefficient and permeability under different operating conditions are reported. Diffusion and permeation activation energies, as well as apparent heats of sorption for the p-xylene and o-xylene in the CMS, are compared. The results reveal that diffusive selectivity was the dominant factor in contributing to the permselectivity. Moreover, the contributions of "enthalpic" and "entropic" selectivity to the diffusive selectivity are studied in detail and reveal that the entropic factors dominate the xylene selection mechanism. Overall, this study shows that the rigid PIM-1 derived carbon membrane provide a strong molecular sieving effect that distinguishes between p-xylene and o-xylene based on molecular shape.

536149 Design of Affinity and Size-Based Membranes for Organic Solvent Nanofiltration: Experiments and Simulations

Abstract:

Mixed-matrix or polymeric organic solvent resistant nanofiltration (OSN) membranes with high solvent permeability and selectivity will have broad applications in chemical and pharmaceutical industries. Nano-porous metal-organic frameworks (MOFs) with tunable pore sizes and ligand properties have generated tremendous interest due to their many possible applications. Here MOFs are embedded in polymer matrices to make mixed matrix membranes (MMM) with enhanced selectivity and permeability. Crystalline Cu(I) and Cu(II) MOFs that can selectively interact with targeted aromatic compounds are synthesized with particle sizes ranging from nanometers (nm) to micrometers (mm). MMMs containing these MOFs were explored for affinity based hydrocarbon separations. Quantum mechanical calculations coupled with classical simulations were conducted to elucidate its selectivity based on the specific solvent-MOF interactions. Besides affinity based MMMs, size based OSN membranes using high free volume glassy perfluoropolymer were investigated. The fractional free volumes (FFV) and pore size distributions of various perfluoropolymeric membranes with different ratios of cyclic to linear monomer units were determined using classical molecular dynamics (MD) simulations.

527034 Polyethylene Glycol Grafting of Ultrafiltration Cross-Linked Polyimide Membranes Via Plasma Modification to Fabricate Organic Solvent Nanofiltration (OSN) Membranes

Abstract:

A two-step process of organic solvents resistant nanofiltration membrane fabrication via high-density polyethylene glycol (PEG) plasma grafting on cross-linked polyimide (cPI) ultrafiltration (UF) substrate was developed in this work. An ultrathin PEG dense layer was successfully formed on cPI substrate surface that converted the UF membrane to a nanofiltration (NF) membrane by plasma treatment. In addition, the presence of PEG effectively prevents pore collapse of the UF support and maintained relatively high solvent permeance of substrate. Moreover, comparing to the cPI membrane, the PEG-modified membranes offer extremely high hydrophilicity resulting in the significant enhancement of isopropanol (IPA) permeance. The optimized PEG-grafted composite membrane shows a 99.64 % rejection of Rose Bengal (a molecular weight of 1017.64 g mol^{-1}) and a 5.91 L m$^{\hat{a}\text{ˆ'}2}$ h$^{\hat{a}\text{ˆ'}1}$ bar$^{\hat{a}\text{ˆ'}1}$ IPA permeance at 10 bar. The newly developed organic solvents nanofiltration (OSN) membranes also demonstrate good stability in alcohols over 400 hours. To our best knowledge, this is the first endeavor of utilizing PEG to directly develop OSN membranes by mean of plasma grafting. The study may not only widen the application of PEG grafting for OSN but also provide useful insights to design a green and cost-effective scale-up membranes fabrication process for industrial applications.

523016 Membranes with Porous Organic Cages for Organic Solvent Nanofiltration

Abstract:

Porous organic cages (POCs) are individual molecules that are porous and soluble in certain organic solvents. Compared to their extended framework counterparts, such as zeolites and metal-organic frameworks (MOFs), POCs offer the advantage of solution processability. Moreover, when fabricated into mixed matrix membranes, the soluble POC molecules have the potential to exhibit molecular-level intimate mixing with matrix polymer. In this work, we fabricated mixed matrix membranes with vertex functionalized amorphous scrambled porous organic cages (ASPOCs). The distribution of ASPOCs bearing different crystallization tendencies within a polymer matrix are probed using Raman imaging and energy dispersive X-ray (EDX) mapping. The performance of the mixed matrix membranes in organic solvent nanofiltration were tested using poly(styrene) oligomer standards. Rejections of over 90 % were observed for all poly(styrene) oligomers starting from poly(styrene) dimer. This high rejection was attributed to the effective molecular sieving provided by the individual pores distributed throughout the polymer membrane.

533513 Organic Solvent Nanofiltration Membranes Developed By Mussel-Inspired Strategy

Abstract:

Most of the reported nanocomposite solvent nanofiltration (OSN) membranes were fabricated by interfacial polymerization or direct coating method. Interfacial polymerization suffers sophisticated multistep fabricating steps, while direct coating usually yields separation layer with thickness up to tens of micrometers. Herein, a mussel-inspired polydopamine (PDA) coating layer has been first explored as a separating layer for organic solvent nanofiltration (OSN). A PDA based separating layer was constructed on polyimide (PI) support via dopamine coating. The subsequent membrane was then treated with 1,6-hexanediamine for cross-linking on both the PDA layer and PI support. After an optimized coating time of 4 h, the resultant membrane showed an EtOH permeance of 0.91 L m−2 h−1 bar−1 and a RB rejection of 99%. More importantly, the composite membrane also exhibited good performance for dyes separation from a wide range of solvents including challenging polar aprotic and strongly swelling solvents, such as dimethylformamide and acetone. Furthermore, to make OSN more economically and efficiently via deposition approach, we also develop a mussel-inspired co-deposition strategy to fabricate nanocomposite OSN membranes, which possess the optimized ethanol (EtOH) permeance of 1.26 L m−2 h−1 bar−1 with a rejection of 99% to Rose Bengal (RB). Particularly, the nanocomposite membrane demonstrated stable performances during a two-day long term test in DMF for RB concentration. In addition to providing a highly-efficient way to high-performance OSN membrane, this work may stimulate the bio-inspired design of advanced nanocomposite membranes for environmental applications.

535864 Cermet-Based Hydrogen Transport Membrane Reactors for Conversion of Methane to Value-Added Chemicals

Abstract:

An abundance of natural gas presents new opportunities for use in the chemical industry, particularly gas-to-liquids (GTL) applications. The catalytic conversion of methane to benzene in the absence of oxygen is a reaction equilibrium-limited process (~12% at 700 °C, 1 atm) that may be enhanced by selective hydrogen removal. Cermet membranes consisting of a metal plated on a ceramic support provide the best combination of thermal stability, mechanical strength while enhancing the $H2$ flux. Nickel and palladium were electroless plated on porous alumina supports. The hydrogen separation performance of these membranes was evaluated for use in the methane-to-benzene process.

526882 Experimental and Numerical Study of an Intensified Water-Gas Shift (WGS) Reaction Process Using a Membrane Reactor (MR)/Adsorptive Reactor (AR) Sequence

Abstract:

Reactive separation processes, such as membrane reactors (MRs) and adsorptive reactors (ARs), have been attracting increased interest for industrial applications. Such a hybrid system combining a membrane reactor (MR) and an adsorptive reactor (AR) in tandem (with the AR following the MR, and the MRâ€™s reject stream serving as the ARâ€™s feed) can produce an ultra-pure H_2 product (without the need for using a post-processing step), and can deliver WGS reaction conversion exceeding equlibrium. The proposal configuration (AR following MR) provides significant flexibility in addition to high-purity hydrogen and carbon dioxide production.

The primary objective of this work is to first develop and simulate an advanced, and detailed mathematical model for the MR/AR sequence and then validate this model using experimental data obtained at the bench-scale. The MR system is composed of a reaction zone packed with catalyst pellets and glass beads, and a water/hydrogen permeation zone. In the AR, both catalyst and adsorbent materials are used for simultaneous reaction and separation. The AR is a dynamically operated process. Both steady-state and dynamic experiment are carried out in a laboratory setting to capture the individual MR/AR performances. Experiments are carried out for a broad range of temperatures (up to 300 ^{o}C) and pressures (up to 25 bar) relevant to realistic application conditions. The velocity, temperature and species concentration profiles in the MR reaction and permeation zones are captured by momentum/species transport models accounting for convection/reaction /diffusion/conduction mechanisms. The rigorous Maxwell-Stefan and dusty gas models are applied to describe mass diffusion fluxes. Finally, the performances of the combined system (the membrane reactor (MR) followed by an adsorptive reactor (AR) is simulated for a broad range of the operating conditions and design parameters. The simulation results are compared with the experimental findings.

532647 Hydrogen Production in Pd-Based Membrane Reactor Via Reforming Reactions

Abstract:

Abstract

In the last decades, the excessive use of fossil fuels to produce energy, as well as the growth of environmental pollution due to green house gases (GHGs) and other harmful emissions has led to the development of new technologies and to the utilization of renewable sources. In this context, hydrogen is recognized as one of the most promising energy carriers of the future. Unfortunately, hydrogen is not available in the Earth in elementary form and is currently produced primarily from steam reforming of natural gas, an energy-intensive process.

Nowadays, different methods, such as alcohols reforming reactions, and alternative devices, such as membrane reactors (MRs), are covering an important role for hydrogen production/separation. Specifically, alcohols as ethanol, methanol, glycerol etc. are largely available from renewable sources and they could be used as â€œbio-feedstocksâ€ for hydrogen generation via reforming reactions, while MR technology seems to represent a valid choice for replacing the conventional process due to its ability to provide the generation and separation of pure hydrogen in a single stage [1]. Among several renewable sources, bioethanol has received considerable attention due to its potential environmental benefits [2].

In the last years, Pd-based composite membranes, i.e. thin metallic layer supported on such porous materials as ceramics or stainless steel, have been considered owing to their lower cost (thin Pd layer) and higher mechanical resistant (porous support) than dense Pd-based ones [3]. Moreover, the use of Pd-alloys, in particular, with Ag, Au, Pt, is grown in the last years given their ability to enhance the hydrogen permeability and to resist sulfur poisoning.

Hence, the current study reviews the most relevant scientific results on hydrogen production from bio-ethanol reforming processes performed through membrane reactor technology. The benefits and the main challenges of inorganic membrane reactors, when applied to ethanol-reforming processes, are examined and the performances in terms of hydrogen yield, hydrogen recovery, and conversion are compared to those of conventional systems.

References

[1] A. Iulianelli, S. Liguori, A. Vita, C. Italiano, C. Fabiano, Y. Huang, A. Basile, The oncoming energy vector: hydrogen produced in Pd-composite membrane reactor via bioethanol reforming over Ni/CeO_2 catalyst, *Catal. Today*, 259 (2015) 368-375

[2] T. Yamazaki, et al., Behavior of steam reforming reaction for bio-ethanol over Pt/ZrO_2 catalysts, *Appl. Catal. B: Environ.* 99 (2010) 81-88.

[3] S. Liguori, et al., Performance of a Pd/PSS membrane reactor to produce high purity hydrogen via WGS reaction. *Catal. Today*, 193 (2012) 87-94.

531555 Na-LTA Membranes with High Water Selectivity for Dimethyl Ether Production in a Catalytic Membrane Reactor

Abstract:

As an important industrial intermediate, Dimethyl Ether (DME), can be used as a promising clean fuel in various ways. Traditionally, DME is produced either through methanol dehydration or a couple of consecutive reactions for DME directly from syngas. However, the yield of DME is low, limited by the equilibrium in both processes. In order to increase the DME yield, a novel catalytic membrane reactor to remove the produced water in-situ from the reaction system will be beneficial.

In this study, a highly water permeable NaA membrane, prepared on ceramic hollow fibers with secondary growth method, was investigated to selectively remove water from the simulated gas mixtures of $CO_2/H_2/CO$/Methanol/H_2O with representative compositions from the reactions producing DME from syngas route at very challenging conditions (temperatures up to 250 °C and pressures up to 600 psig). Current results showed that, when the feed water concentration was 2.57 mol.%, the membrane selectivity of H_2O/CO_2 was higher than 200, and no CO, H_2, Methanol were detected in the permeate side of the membrane at 250 °C and 300 psig. Stability testing over 11 h showed the NaA membrane had a good stability with <10% performance degradation. These results suggest that NaA membranes are capable of removing water in-situ in a membrane reactor, and thus increase the DME yield in syngas route at elevated temperatures and high pressures. We also expect NaA membranes be used to increase the product yield of reactions with water as a by-product in a catalytic membrane reactor at harsh conditions.

539731 PDMS/Ceramic Composite Membrane in Glycerol Fermentationâ €“PV Coupled Process for Biobutanol Production

Abstract:

Facing the depletion of fossil energy and the pollution of the environment, the development and utilization of biomass fuels have brought dawn to the sustainable development of human beings. Biomass fuel, as a renewable energy, has unparalleled advantages. With the development of biodiesel industry, the production of crude glycerol, as an inevitable by-product in the production of biodiesel, is huge. As we all know, bioconversion of crude glycerol without pretreatment to biofuels is the way to disposal of the surplus crude glycerol. In addition, biomass fuel butanol, which is produced by bio-fermentation, can be used as a clean renewable energy biofuel. At the same time, pervaporation is considered to be the greatest potential separation technology because of its energy-saving and efficiency, as well as no harmful effects on the microorganisms. Therefore, fermentation-PV coupled process is a membrane-based process so as to in-situ separate bio-product from fermentation broth by the alcohol-permselective membranes. PDMS membrane has been attracting more and more attention for the separation of biobutanol by coupling bio-fermentation process especially.

In this work, the PDMS/ceramic composite membrane separated the biobutanol directly from the crude glycerol fermentation broth, which not only relieves the inhibition effect of butanol, but also improves the yield of butanol. The pervaporation performance and the stability of the membrane in batch fermentation and fed-batch fermentation were systematically researched. During 250 h of continuous fed-batch fermentation-PV coupled process, the membrane exhibited high flux of 550 g/ (m^2 h) and separation factor of 24.1. Furthermore, compared with fermentation, the yield of butanol was increased by 55.6%. In this process, membrane fouling leaded to the degradation of membrane pervaporation performance after 200 h. The SEM, AFM, FT-IR and XPS characterizations demonstrated that the membrane fouling is due to the adsorption of active cells on the surface of PDMS membrane. Nevertheless, the total flux and separation factor of the water-washed membrane were the same as those of the fresh membrane, implying that the effect of membrane fouling on the membrane was reversible.

526892 Thermochemial Stability of ZIF Membranes for Membrane Reactor Applications

Abstract:

Thermochemical stability of the supported Zeolitic Imidazolate Framework (ZIF) membranes is studied for membrane reactor applications. It is found that the membrane/support interface acid-base chemistry has paramount inducing effects on the hydrothermal stability of ZIF membranes, resulting in the fatal acid-catalyzed decomposition of the ZIF frameworks. The ZIF-7/MgO-Al_2O_3 of marked thermochemical stability has permitted the first significant application of MOF membranes for catalytic membrane reactor in severe and practical process conditions, performing water-gas shift reaction ($CO + H_2O \leftrightarrow CO_2 + H_2$) at considerably high temperatures (473–673 K) and steam concentrations (20–40%).

517820 Enhancing CO_2/CH_4 Separation Performance and Mechanical Strength of Mixed-Matrix Membrane Via Combined Use of Graphene Oxide and ZIF-8

Abstract:

Gas separation using high-performance membranes have demonstrated itself as an energy-efficient and environment-benign process that is capable of replacing the conventional energy-intensive processes. In general, polymeric membranes have been widely incorporated in view of their good mechanical stability and processability. However, the trade-off effect between gas permeability and selectivity severely limits their future development of gas separation. Thus, mixed-matrix membranes (MMMs) that utilized synergistic effect of both polymer matrix and nanoporous fillers have been a focus of attention to researchers. This is because the addition of nanoporous materials acts as an efficient aid to favor selective gas transport. Nevertheless, the utilization of one type of filler tends to optimize either gas permeability or selectivity in gas mixtures. Therefore, in this context, an effective and feasible strategy in improving both gas permeability and selectivity was proposed by utilizing binary fillers.

High-performance MMMs that comprise both zeolitic imidazolate framework-8 (ZIF-8) and graphene oxide (GO) were synthesized to realize excellent CO_2/CH_4 separation. The incorporation of ZIF-8 nanocrystals alone in ODPA-TMPDA polyimide can be used to significantly enhance CO_2 permeability compared with that of pure ODPA-TMPDA. Meanwhile, the addition of a GO nanostack alone in ODPA-TMPDA contributes to improved CO_2/CH_4 selectivity. Hence, a composite membrane that contains both fillers displays significant enhancements in CO_2 permeability (up to 60%) and CO_2/CH_4 selectivity (up to 28%) compared with those of pure polymeric membrane. Furthermore, in contrast to the ZIF-8 mixed-matrix membrane, which showed decreased mechanical stability, it was found that the incorporation of GO could improve the mechanical strength of mixed-matrix membranes. Overall, the synergistic effects of the use of both fillers together are successfully demonstrated in this paper. Such significant improvements in the mixed-matrix membraneâ€™s CO_2/CH_4 separation performance and mechanical strength suggest a feasible and effective approach for potential biogas upgrading and natural gas purification.

534745 Identifying Nucleation and Growth Kernels of Crystallization

Abstract:

Crystallization, a separation process, is simply attachment and detachment of molecules across time scales. Minor changes in the operating conditions have proven to change the outcome of crystallization and controlling the process of crystallization remains a challenge. Nucleation and crystal growth are the two processes of the crystallization and are modeled using the thermodynamic approaches. Mechanism of both processes is the same, but the time scales of the two processes are different. The challenge in modeling the process of crystallization is to derive nucleation rate of crystal structure and growth rates of energetically stable faces on the crystal. The theories used to find these rates rely on calculating interfacial free energies which is often difficult and lacks generality. With the help of a novel approach based molecular simulations, we were able to find the nucleation and growth rates. Once the rates are known, it is necessary to connect the operating conditions to outcome of crystallization. A set of operating conditions which lead to a specific outcome of crystallization will be called as kernel of crystallization. To effectively capture the time scales and the mechanism of crystallization without lack of generality, we propose the use of Population Balance Modeling (PBM). PBM finds the rate of change of internal coordinate and for crystallization, the internal coordinates are nucleation and growth rates. The rates obtained from different operating conditions can be connected to the outcomes of crystallization and hence PBM successfully finds the kernel of crystallization. Knowledge of these kernels can successfully control the process of crystallization and also shows potential to solve the problem of polymorphism.

538036 A Novel Mode of Supersaturation Feedback Control: Semi-Batch Cooling Crystallization By Feeding Flow Rate Profiles

Abstract:

Concentration Feedback Control (CFC) approaches, based on ATR-FTIR, ATR-UV/Vis or Raman spectroscopy coupled with chemometrics, have been widely applied in cooling and antisolvent crystallization systems at laboratory as well as industrial scales, by which the improved product quality have been extensively demonstrated [1]. In the feedback loop, traditionally the supersaturation is controlled to an optimal set point by manipulating the solution or cooling jacket temperature. However, the traditional batch cooling crystallization is limited by the temperature dependency and has limited applicability in some specific, but often encountering situations like crystallization of heat sensitive materials or polymorphic crystallization systems.

In this work, we propose a novel CFC implementation of semi-batch mode (feed hot saturated solution into lower temperature saturated solution), which can achieve the feedback control of supersaturation by manipulating the feeding flow rate rather than the temperature. In such, the crystallization temperature can be fixed throughout the process and thus eliminate the temperature dependence of the system. The semi-batch mode complements the classical mode in CFC implementation to some crystallization systems, such as enantiotropic polymorphic system, heat sensitive system, temperature dependent reversible reaction-crystallization and also could be designed as a start-up strategy for continuous MSMPR crystallizers [2].

Using the proposed operational method for enantiotropic polymorphic system, in the feeding step the process temperature is fixed or varied in a narrow range. This is advantageous over the cooling crystallization of much more stable polymorphic form, because the polymorph stability order is a function of temperature for enantiotropic polymorphic system. For instance, the transition temperature of α and β form of para-amino benzoic acid (pABA) in pure ethanol [3], is 13.8 (β form is stable under 13.8). Therefore, it is difficult to obtain pure β form with CFC in batch cooling mode because the operating temperature range was limited by this transition point. Whereas the crystallization temperature is fixed in the proposed CFC with semi-batch mode, which can be flexible for crystallization of easily design for each polymorphic form. In this context, the semi-batch operation to generate different polymorphic forms combines the advantages of batch and continuous crystallization: applicable to small scale production and stably produces one or the other pure polymorphic forms.

We modelled, simulated and evaluated three supersaturation control procedures for cooling crystallization processes: batch crystallization (T-SSC), semi-batch crystallization (F-SSC) and combined cooling and semi-batch crystallization (TF-SSC). For this, the monovariate population balance equation was used, involving nucleation and growth, solved by the method of moments and extended with the solute mass balance

equation [4]. In this work, the comparison of these three scenarios will be presented through numerical simulations, which enables the identification of optimal operation modes of cooling crystallizers for various types of crystallization systems. Proof of concept experimental demonstrations for cooling and semi-bath crystallization will be also presented to support the numerical results followed by the implementation of T-SSC and F-SSC for enantiotropic polymorphic system (o-ABA) to achieve the efficient crystallization of both polymorphic forms.

References:

[1] Nagy Z K, Braatz R D. Advances and new directions in crystallization control[J]. *Annual Review of Chemical & Biomolecular Engineering, 2012, 3(5):55.*

[2] Su Q, Nagy Z K, Rielly C D. Pharmaceutical crystallisation processes from batch to continuous operation using MSMPR stages: Modelling, design, and control[J]. *Chemical Engineering & Processing Process Intensification, 2015, 89:41-53.*

[3] Hao H, Barrett M, Hu Y, et al. The Use of in Situ Tools To Monitor the Enantiotropic Transformation of p-Aminobenzoic Acid Polymorphs[J]. *Organic Process Research & Development, 2012, 16(16):35-41*

[4] Poehlein G W, Wenzel L A. Theory of particulate processes-analysis and techniques of continuous crystallization: By Alan D. Randolph and Maurice A. Larson, Academic Press, New York, *Journal of Colloid & Interface Science, 1972, 40(1):130-130.*

524824 Parameter Range Optimization and Modeling for a Reactive Crystallization Process

Abstract:

Chemical purity control is an essential component of quality risk management in active pharmaceutical ingredient (API) manufacturing. Impurity control typically relies on optimization of reaction parameters to minimize impurity formation, with downstream unit operations such as extractions and crystallizations serving as additional means of purity upgrade. In cases where crystallization occurs during the reaction, process performance can be more difficult to predict as the nucleation and uncontrolled crystallization of the product introduces stochastic variability with respect to impurity entrapment in the crystal lattice. This presentation details an industrial case study focused on process analysis and design of a robust impurity control strategy for an API step. In this step, due to a large solubility difference between API and penultimate starting material, high supersaturation conditions are achieved in the reaction, resulting in API self-nucleation and crystallization. Variability in nucleation and crystallization dynamics observed during process parameter space development resulted in variable levels of starting material entrapped in the isolated product. Extensive experimental studies were conducted to determine factors that influenced residual starting material impurity levels in isolated API, including reagent charge rate, solvent composition, and mixing conditions. A model was produced to assess process robustness over wide parameter space and quantify risks associated with operation on edges of or outside of the recommended ranges.

524971 *Enhanced Mass Transfer Process Control Via Hollow Fiber Membrane Assisted Antisolvent Crystallization*

Abstract:

Enhanced Mass Transfer Process Control via Hollow Fiber Membrane Assisted Antisolvent Crystallization

Xiaobin Jiang[a*], Linghan Tuo[a], Xuehua Ruan[b], Wu Xiao[a], Gaohong He[a,b*]

a State Key Laboratory of Fine Chemicals, Engineering Laboratory for Petrochemical Energy-efficient Separation Technology of Liaoning Province, School of Chemical Engineering, Dalian University of Technology, Dalian, Liaoning 116024, China

b School of Petroleum and Chemical Engineering, State Key Laboratory of Fine Chemicals, Dalian University of Technology at Panjin, Panjin 124221, China

Corresponding author: xbjiang@dlut.edu.cn; hgaohong@dlut.edu.cn

Abstract

Antisolvent crystallization plays an important role in various fields as an effective separation[1-4] and crystal manufacture technology[5-10]. Due to the operation under ambient temperature and atmospheric pressure, antisolvent crystallization is environment friendly and low energy consumption [11,12].

In this work, a novel hollow fiber membrane assisted antisolvent crystallization (MAAC) was proposed to enhance the antisolvent crystallization process control. PES membrane module was introduced as the key device for antisolvent mass transfer and solution mixing. An antisolvent liquid film formed on the membrane surface was renewed by the crystallization solution. The liquid film can additionally prevent the membrane from contacting with crystallization solution. By controlling the shell side flow velocity and the antisolvent mass transfer process, the antisolvent permeate rate can achieve sensitive, stable and accurate control during long term utilization.

Benefited from the nano and sub-nano size pores on the membrane surface and high packing density of the membrane module, MAAC greatly improved the micromixing between antisolvent and crystallization solution and reducing the local supersaturation degree. The calculated interfacial mass transfer rate of MAAC was approximately 1/50 of the one of the conventional antisolvent crystallization. On the basis of this effective and accurate antisolvent mass transfer control, MAAC provided the erythritol crystal products with better morphology, narrower CSD and smaller C.V. compared to the products manufactured by conventional antisolvent crystallization under the same antisolvent adding rate and operation duration.

Additionally, due to the the intrinsic feature of antifouling of the proposed MAAC operation, the permeate rate of MAAC could keep a stable value in a long running and the permeate performance of the membrane module was repeatable after more than twenty four times utilization. With well-controlled nucleation and crystal growth in the membrane module reducing the required retention time and the size of the tank for suspension liquid in MAAC, the production capacity *P.C.* (kg products¡¤m^{-3}¡¤h^{-1}) can be also enhanced via MAAC.

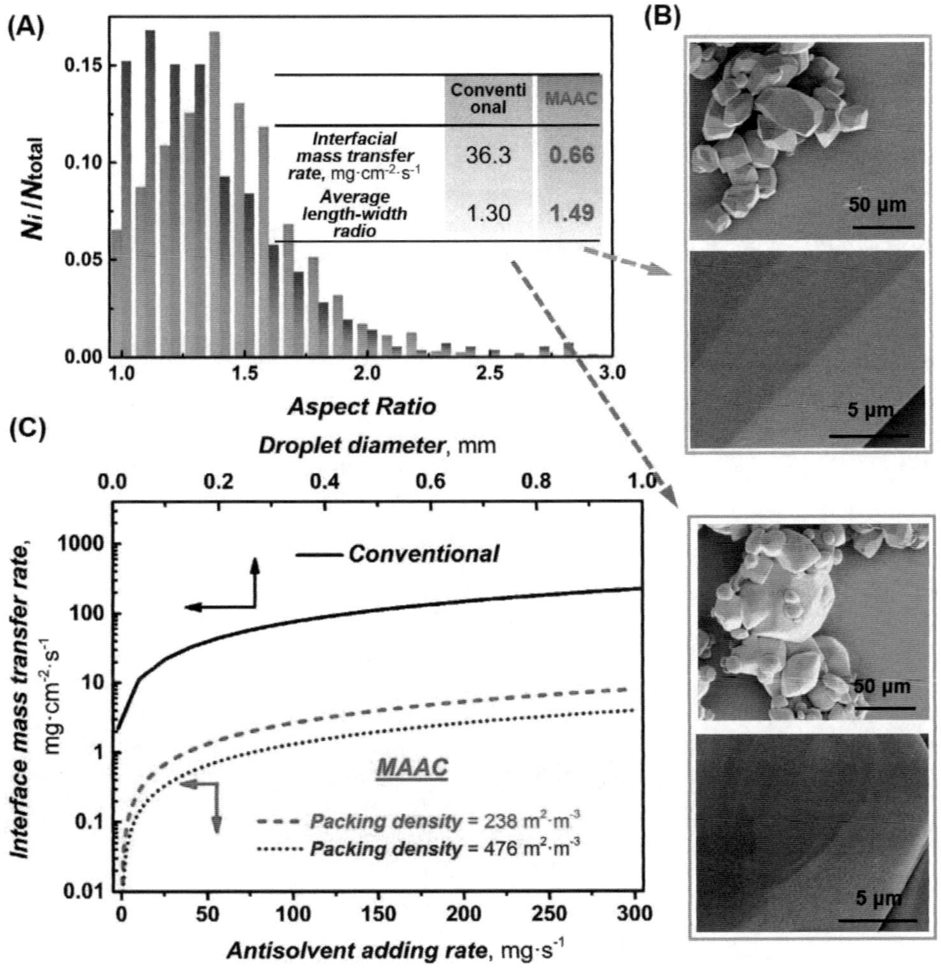

Figure. 1 Comparison of size distributions of aspect ratio (A), crystal morphology (B) and interfacial mass transfer rate (C) between erythritol produced by MAAC and conventional antisolvent crystallization after 1 hour operation.

Acknowledgements

We acknowledge financial contribution from National Natural Science Foundation of China (Grant No. 21527812, 21676043, U1663223, 21606035), Changjiang Scholars Program (T2012049), the Fundamental Research Funds for the Central Universities (DUT16TD19, DUT17ZD203) and Education Department of the Liaoning Province of China (No. LT2015007).

References

1. Biradha K, Su CY, Vittal JJ. Recent Developments in Crystal Engineering. *Crystal Growth & Design.* 2017;11(11):875-886.

2. Cisternas LA, V¨¢squez CM, Swaney RE. On the design of crystallization-based separation processes: Review and extension. *Aiche Journal.* 2010;52(5):1754-1769.

3. Simon L, Allan MS. Continuous antisolvent plug-flow crystallization of a fast growing API. Paper presented at: International Symposium on Industrial Crystallization - Isic2011.

4. Ulrich J, Frohberg P. Problems, potentials and future of industrial crystallization. *Frontiers of Chemical Science and Engineering.* 2013;7(1):1-8.

5. Lu H, Wang J, Wang T, Wang N, Bao Y, Hao H. Crystallization techniques in wastewater treatment: An overview of applications. *Chemosphere.* 4// 2017;173:474-484.

6. Park M-W, Yeo S-D. Antisolvent crystallization of carbamazepine from organic solutions. *Chemical Engineering Research and Design.* 2012;90(12):2202-2208.

7. Zhang C-t, Wang H-r, Wang Y-l. Internally generated seeding policy in anti-solvent crystallization of ceftriaxone sodium. *Chemical Engineering and Processing: Process Intensification.* 2010;49(4):396-401.

8. Aditya NP, Yang H, Kim S, Ko S. Fabrication of amorphous curcumin nanosuspensions using ¦Â-lactoglobulin to enhance solubility, stability, and bioavailability. *Colloids and Surfaces B: Biointerfaces.* 2015;127:114-121.

9. Zijlema TG, Geertman RM, Witkamp GJ, Rosmalen GMV, Graauw JD. Antisolvent Crystallization as an Alternative to Evaporative Crystallization for the Production of Sodium Chloride. *Industrial & Engineering Chemistry Research.* 2000;39(5):1330-1337.

10. † GXZ, Mitsuko Fujiwara, Yi Woo, ¡ì Xing, et al. Direct Design of Pharmaceutical Antisolvent Crystallization through Concentration Control. *Crystal Growth & Design.* 2006;6(4):892-898.

11. Lindenberg C, Krättli M, Cornel J, Mazzotti M, Brozio J. Design and Optimization of a Combined Cooling/Antisolvent Crystallization Process. *Crystal Growth & Design.* 2009;9(2):1124-1136.

12. Nowee SM, Abbas A, Romagnoli JA. Antisolvent crystallization: Model identification, experimental validation and dynamic simulation. *Chemical Engineering Science.* 2008;63(22):5457-5467.

525202 Experimental Implementation of a Model-Free Feedback Controller for the Size and Shape of Needle-like Crystals Growing in Suspension

Abstract:

Particle size and shape have a major impact on downstream processes in a crystallization process. The size and shape influences the physical characteristics of the final product obtained at the end of a crystallization step. Feedback control-based approaches have been proposed to control the size and shape of a population of crystals. Application of feedback control in a process can be particularly powerful as they have the ability to guide a population towards a desired target despite the presence of un-modeled phenomena and process disturbances. The prerequisite for the development and implementation of such feedback controllers are the availability of online monitoring and characterization tools for both the solid and liquid phase.

Studies aimed at experimental application of feedback control for size and shape control are scarce in literature [1]. This could be partly attributed to the unavailability of tools that could characterize the size and shape of a population of crystals. Recent advancements in imaging based tools have nevertheless addressed the above-mentioned issue of solid phase characterization [2]. Independent simulation studies performed using model-based and model-free feedback controllers in a growth-dominated seeded batch cooling crystallization process have led us to conclude that a model-free controller can robustly reach final target dimensions with a comparatively low level of complexity. The model-free approach requires only knowledge of the thermodynamics eliminating the need to develop multi-dimensional kinetic models for the system-of-interest [3].

In this work, a feedback controller was implemented in a lab scale batch crystallizer to guide two different seed populations of β L-glutamic acid in water towards desired target dimensions under growth-dominated conditions. The size and shape of the population of crystals was characterized using the online monitoring tool, μ-DISCO [2]. The experimental campaign was a two-step process, involving namely constant supersaturation controller (CSC) and model-free path following controller (PFC) experiments. The CSC experiments were performed at four different constant supersaturation levels to serve two purposes. First, to verify the monotonic behavior of the ratio of the rate of change of the average particle dimensions in the population with temperature, which is essential for the PFC. Second, to estimate an attainable region in the particle size and shape space where the controller could act on the population of crystals.

Upon the completion of the first step, PFC experiments were performed on both population of crystals, to reach two different target dimensions that falls within the attainable region. Repeatability experiments for all the PFC experiments were also performed. The controller was able to successfully reach target dimensions with only the thermodynamic knowledge of the system despite the controller acting on different batches of seed crystals, and operating conditions. The robustness provided by the controller used and experimental verification of the same leads us to conclude that model-free approach towards size and shape control could be a promising alternative to model-based control strategies.

References

[1] Eisenschmidt, H.; Bajcinca, N.; Sundmacher, K. Optimal Control of Crystal Shapes in Batch Crystallization Experiments by Growth-Dissolution Cycles. *Cryst. Growth Des.* **2016**, *16* (6), 3297–3306.

[2] Rajagopalan, A. K.; Schneeberger, J.; Salvatori, F.; Bötschi, S.; Ochsenbein, D. R.; Oswald, M. R.; Pollefeys, M.; Mazzotti, M. A Comprehensive Shape Analysis Pipeline for Stereoscopic Measurements of Particulate Populations in Suspension. *Powder Technol.* **2017**, *321*, 479–493.

[3] Ochsenbein, D. R.; Schorsch, S.; Vetter, T.; Mazzotti, M.; Morari, M. Growth Rate Estimation of β L-Glutamic Acid from Online Measurements of Multidimensional Particle Size Distributions and Concentration. *Ind. Eng. Chem. Res.* **2014**, *53* (22), 9136–9148.

537109 Morphology Based Adsorption Kinetics of a Selectively Modulated Metal-Organic Framework

Abstract:

Metal organic frameworks (MOFs) have gained interest in recent decades due to their high porosity, tunable pore aperture, and easily synthesized crystalline structures for applications such as catalysis, separations, sensing, gas storage, medicine, and energy. MOFs are composed of metal ions or oxo-metallic secondary building units (SBUs) coordinated with organic linkers to form a porous, open framework. The rational selection of these components in addition to modulating species and structure directing agents allows for the control of pore geometry and size at the surface of MOF crystals. By tuning the crystal size, morphology, and the dominant surface pore as determined by the crystal orientation, it is possible to tune the selectivity of a MOF to enhance the adsorption and separation of a given species of gas molecule. Previous results have shown that a copper based MOF with anisotropic pore distribution in the unit cell, $[Cu_2(ndc)_2(dabco)]_n$, can show controlled crystalline morphology through the addition of modulators during the synthesis. In this work, we leverage the modulation of crystallization growth processes to synthesize MOF surfaces with selective pore sizes and orientations. We demonstrate how changing the morphology of MOFs can affect the selectivity and adsorption kinetics of select species. The relationship between particle morphology, surface pore size and the adsorbed species is discussed. Additionally, tuning MOF morphology and orientation for higher selectivity and separations is discussed.

525827 Effects of Scale-up on the Mechanism and Kinetics of Crystal Nucleation

Abstract:

Crystal nucleation is an important step in solution crystallization of pharmaceutical compounds as it determines the crystal shape, polymorphic form, chiral form and the crystal size distribution of the product. Understanding this fundamental phase separation of molecules from the liquid to the solid phase is essential to exhibit process control. Nucleation is a stochastic process for unseeded crystallization processes involving low supersaturation levels or for small crystallizer volumes. Stirred small volume solution methods can be utilized to measure the stochastic nature of nucleation and to estimate nucleation kinetics. This way, probability distributions of induction times can be obtained from isothermal induction time experiments which can be used in conjunction with the classical nucleation theory (CNT) to estimate the nucleation rate and other important nucleation parameters. However, it remains unclear how scale-up affects the nucleation mechanism and nucleation kinetics in solution crystallization processes as the isothermal induction time method is currently limited to small scale experiments.

In this work, the effects of scale-up on the mechanism and kinetics of crystal nucleation will be presented. Four different crystallization setups were investigated, ranging from small magnetically-stirred 10 mL solutions to overhead-stirred solutions of 680 mL. The onset of crystallization was characterized using a Mettler Toledo Focussed Beam Reflectance Measurement (FBRM) probe. The crystallizer was programmed to start a new experiment after the FBRM probe detected crystallization. This way, many induction times were recorded in an automated fashion. The nucleation kinetics were estimated from the probability distributions of the induction times.

Nucleation kinetics were estimated for paracetamol in 2-propanol in the different crystallization platforms and at different supersaturation ratios. The nucleation rate was an order of magnitude faster in the magnetically-stirred crystallizer as compared to the crystallizers involving overhead stirring. The thermodynamic part of nucleation did not significantly change the nucleation rate whereas the kinetic nucleation parameter was found to be the rate-determining process when the crystallization process was scaled-up. In particular, the shear rate was rationalized to be the part of the kinetic parameter that changes most significantly when the crystallization process was scaled-up. The shear rate increases with volume and this effect becomes stable after the experiment has sufficiently increased in volume. In addition to these kinetic results, experiments involving the chiral sodium chlorate showed that the single nucleus mechanism is the underlying nucleation mechanism in all four tested crystallization setups. On the other hand it was found that cooling crystallization experiments were driven by a multi-nucleus mechanism. The presence of the single nucleus mechanism provides support for the use of probability distributions of induction times to estimate nucleation kinetics across all four tested crystallization setups.

In conclusion, a robust methodology was used to automatically record induction times using an FBRM probe. This approach is easily transferable and allows to capture induction times in different crystallizer types and volumes. The results presented herein on paracetamol and sodium chlorate contribute to a better understanding of the scale-up effects on the mechanism and kinetics of crystal nucleation of fine chemicals.

539974 Direct Feed Low Activity Waste Test Platform â€“ Alternative Demonstrations

Abstract:

The Low-Activity Waste Pretreatment System (LAWPS) facility is planned to pretreat Hanford tank waste supernate by filtering solids and processing through ion exchange columns to remove cesium. The filtration process is achieved through dead end filtration. The effluent from ion exchange will be vitrified to produce LAW glass.

The goal of this effort was to demonstrate alternative processing approaches for filtration, ion exchange and melter off-gas treatment. To that end, dead end filtration and alternative ion exchange media were tested. Actual waste samples were retrieved from Hanford Site Tank AP-107 and were composited for testing. These composited samples were first tested with cross flow filtration and dead end filtration to determine the filter flux under typical LAWPS conditions. The material was subsequently treated for cesium removal with CST followed by vitrification using a flow through melter system. These results of these tests will be described.

516311 Improved Foam Control during High Level Radioactive Waste Processing

Abstract:

Antifoam is needed during processing of HLW due to the high gas generation from steam and chemical reaction offgas products to prevent foam from contaminating the condensate. Antifoam is used to minimize foam[1] during chemical processing and HLW evaporation at SRS[2] and Hanford[3]. The antifoam used in the Defense Waste Processing Facility (DWPF) impacts flammability during chemical processing (generates three flammable degradation products) and while feeding the melter[4-5] (can decompose to CO/hydrogen). It also is the likely source of methyl for the methyl mercury present in the tank farm and excessive mercury in Saltstone[6]. The planned startup of The Salt Waste Processing Facility (SWPF), with much higher throughput, will challenge DWPF to process more efficiently.

DWPF employs Antifoam 747[7], a superspreader produced by Momentive Performance Materials, as an antifoaming agent during waste processing. Despite its ability to control foam, processing issues have risen from use, which has led to two â€œPotential Inadequacy in the Safety Analysesâ€ (PISAs) and has limited DWPF facility throughput.

During DWPF chemical processing, antifoam must be effective up to 103ËšC between pH of 3-13. Antifoam 747 is most effective at pH 6-8[8] and degrades as pH deviates. SRNL identified three flammable antifoam degradation[9] products using mass spec and FTIR offgas analyzers during experiments[10]. Equipment has not been used in identifying decomposition products from antifoam agents used elsewhere.

A new Antifoam or a new method to control foam is needed to minimize DWPF processing time. The progress in developing the new foam control method will be discussed.

1. Bindal, S. K.; Nikolov, A. D.; Wasan, D. T.; Lambert, D. P.; Koopman, D. C., Foaming in Simulated Radioactive Waste. *Environ. Sci. Technol.* **2001,** *35* (19), 3941-3947.
2. Wasan, D. T.; Lambert, D. P. *Foaming and Antifoaming in Radioactive Waste Pretreatment and Immobilization*; Illinois Institute of Technology: Chicago, IL, 2001.
3. Calloway, J. T. B.; Martino, C. J.; Jantzen, C. M.; Wilmarth, W. R.; Stone, M. E.; Pierce, R. A.; Josephs, J. E.; Barnes, C. D.; Daniel, W. E.; Eibling, R. E.; Choi, A. S.; White, T. L.; Crowley, D. A.; Baich, M. A.; Johnson, J. D.; Vijayaraghavan, K.; Nikolov, A. P.; Wasan, D. T., Radioactive Waste Evaporation: Current Methodologies Employed for the Development, Design and Operation of Waste Evaporators at the Savannah River Site and Hanford Waste Treatment Plant. **2003,** (37327), 157-170.
4. Choi, A. S.; Smith III, F. G.; McCabe, D. J. *Preliminary Analysis of Species Partitioning in the DWPF Melter. Sludge batch 7A*; SRNL-STI--2016-00540; Savannah River National Laboratory: Aiken, SC, 2017.
5. *Evaluation of the Safety of the Situation (ESS): Melter Feed Rate Temperature Correlation Basis (PISA PI-2014-0009)*; Savannah River Remediation LLC: Aiken, SC, 2016.
6. Meraw, H. J. *Volatilization and Flammability Characteristics of Elemental and Organic Mercury*; X-ESR-G-00048, Revision 2; Savannah River Remediation LLC: Aiken, SC, June 2015, 2015.
7. Koopman, D. C. *Comparison of Dow Corning 544 Antifoam to IIT747 Antifoam in the 1/240 SRAT*; WSRC-TR-99-00377; Savannah River Technology Center: Aiken, SC, 2000.
8. Lambert, D. P.; Koopman, D. C.; Newell, J. D.; Wasan, D. T.; Nikolov, A. P.; Weinheimer, E. K., Improved Antifoam Agent Study End of Year Report, EM Project 3.2.3. 2011.
9. McCord, J. B. *Evaluation of the Safety of the Situation (ESS): Melter Feed Rate Temperature Correlation Basis (PISA PI-2014-0009)*; Savannah River Remediation LLC: Aiken, SC, 2016.
10. Lambert, D. P.; Zamecnik, J. R.; Newell, J. D.; Martino, C. J. *Impact of Scaling on the Glycolic-Nitric Acid Flowsheet*; Savannah River National Laboratory: Aiken, SC, 2016.

536472 The Nature and Quantity of Solids Suspended in Hanford Nuclear Waste Supernatants

Abstract:

The Hanford Site contractors are currently designing and constructing a waste treatment process to treat highly concentrated sodium-bearing supernatants. The treatment process is expected to require suspended solids removal. In order to design the solids removal process, basic properties of suspended solids (size, density, chemical composition, shape, abrasively, quantity, etc.) are required. In order to provide insight in the solids characterization, a three pronged approach was taken: 1) Evaluate historic records on the solids concentration in supernatant grab samples; 2) characterize two samples that contained a floating solids layer at the surface; and 3) perform filtration studies of waste supernatants to characterize those solids that are not visible. The historic records indicated that the overwhelming majority of grab samples had no visible solids, however, 5 out of 163 had between 0.1 and 0.5 volume % solids. The characterization of floating solids indicated that they could be large and be smooth-surfaced aggregates. Particulates made of the aggregates could be up to 200 microns in diameter. Such solids were suspended by complex reactions of saponification, adsorption of organics, and available surface area for the adsorption of organics. The filtration studies determined that other suspended solids were simply very small particles, usually less than 2 microns in diameter, that were extremely slow to settle. These include both organic and metalloid particles.

526720 Dechloriation of Electrorefiner Chloride Salt Via Ion-Exchange Using Ultra-Stable H-Y Zeolite

Abstract:

A dechlorination process has been designed to generate durable waste forms from spent nuclear fuel electrorefiner salt waste. It is based on ion exchange of the radioactive chloride salt with dehydrated protonated ultrastable Y-type (USHY) zeolite to form metal exchanged zeolite and gaseous HCl. This process has been demonstrated using both LiCl-KCl and surrogate electrorefiner salt in a small-scale rotary tube furnace. Dechlorination was confirmed and measured using both free chloride ion analysis of samples of the ion exchanged zeolite and output of an autotitrator that continuously balances HCl generated in real time. Preliminary results showed that over 90% dechlorination was achieved in 48 hr using pelletized USHY (Si/Al=2.6) at 625^{0}C in a continuous rotary tube furnace without zeolite structure collapse. Temperatures of 650°C or higher result in rapid collapse of the zeolite structure into an amorphous phase. The rate of the ion exchange is invariant when reducing particle size below 250 micron diameter, indicating that particle mixing is likely the rate limiting process. In addition to electrorefining waste processing, this technology could be used to generate waste forms from salts used in a chloride molten salt reactor (Cl-MSR).

531021 Iodine Adsorption in Reduced Agz and Ag-Aerogel in Presence of Water and NO_X

Abstract:

Reprocessing used nuclear fuel (UNF) generates several volatile radioactive species, including ^{3}H, ^{14}C, ^{85}Kr, and ^{129}I, along with significant amount of water vapor and NO_X into the off-gas streams. The presence of water vapor and NOx impacts the performance of the adsorbents for capturing the radioactive gases, especially, the Ag-containing adsorbents for iodine adsorption. In this study, the iodine adsorption performance of reduced Ag mordenite (Ag^0Z) and Ag functionalized silica aerogel (Ag^0-aerogel) in the presence of water and NO_X were investigated experimentally and analytically. Through thin-bed adsorption experiments, uptake curves of I_2 on Ag^0Z and Ag^0-aerogel were obtained in gas streams containing I_2, H_2O, and NOx. Through chemical analyses of the iodine-loaded adsorptions including scanning electron microscopy (SEM), X-ray powder diffraction (XRD), X-ray photoelectron spectroscopy (XPS), Raman spectroscopy, and X-ray absorption fine structure (XAFS), the binding information of Ag with I_2, H_2O and NO_X during the adsorption process were obtained. Several chemical reactions occurred during the process forming silver iodide, silver nitrates and silver sulfate (in Ag^0-aerogel). Their chemical compositions were also determined. It was found that the oxidation of Ag by H_2O and NO_x occurs much faster than the iodine adsorption process due to the much higher concentration of H_2O and NO_X. As a result, the iodine adsorption performance of Ag^0Z and Ag^0-aerogel decreased significantly.

531056 Kinetic Study of Ag Mordenite and Ag Functionalized Silica Aerogel Aging in Nuclear Fuel Reprocessing Off-Gases

Abstract:

In the study of spent nuclear fuel reprocessing off-gas treatment, radioactive iodine vapor can be effectively captured by silver containing adsorbents. However, the impact of aging on adsorbents performance is one major issue during off-gas treatment operation due the presence of water vapor and NOx. To better understand the performance of the adsorbent in the off-gas treatment system, it is important to study the aging mechanism and kinetics. In this study, the impact of aging from off gas components on the kinetics of iodine capture on silver exchanged mordenite (Ag^0Z) and silver functionalized silica aerogel (Ag^0-aerogel) were investigated, two well recognized adsorbents for capture of radioactive iodine. These adsorbents were aged in dry air, humid air, 1% NO/N_2, and 2% NO_2/air, at 100 °C, 150 °C, and 200°C. Consequently, iodine loading experiments with the aged adsorbents were conducted. Results showed that humid-air, 1% NO/N_2, and 2% NO_2/air had strong impacts on the performance of Ag^0Z and Ag^0-aerogel. It was also found that the negative aging effect increased as the aging temperature increased. For example, the Ag^0Z aged in humid air showed 42% , 45% and 58% of iodine loading capacity losses at 100°C, 150°C, and 200°C, respectively. Kinetic models were developed and evaluated to simulate the aging processes in different gases. The models include reversible and irreversible n^{th} order reaction models to describe diffusion and reaction of the aging gas into the adsorbent pellets. Related parameters such as reaction rate constants and diffusivities were obtained. Work is in progress to incorporate the aging kinetic models into the iodine adsorption models for fixed beds to predict the actual adsorption process in an off-gas stream.

538319 Microflow Visualization of Tri-n-Butyl-Phosphate/Dodecane and Nitric Acid in a Centrifugal Contactor

Abstract:

Despite the common use of centrifugal contactors in nuclear solvent extraction applications, much is unknown about the underlying microflow structure. This lack of knowledge hampers the development of a modern continuum theory able to predict mixing and transport phenomena for solvent extraction processes. The ability to observe the micromixing flow and distinguish the motion and configuration of the phases involved is of great importance theoretically and practically. To that end, we report on direct, high-speed visualization of the microflow structure for the tri-n-butyl-phosphate (TBP)/dodecane/nitric acid system.
Further insight is obtained by simultaneous measurement of the electrical resistance of the phase mixture to support visual observations.
This work sheds light on a fluid microflow region at the foot of the mixing zone of the contactor resolving the microsecond/micrometer motion of the continuous and dispersed phases.
The system in this study is the most prevalent solvent extraction combination used in the nuclear reprocessing industry to extract metal ions from spent nuclear fuel from a nitric acid aqueous solution into an organic mixture of TBP and dodecane. The purpose of this work is to describe salient features of the behavior of this system, heretofore unknown, to guide the development of a predictive modeling and simulation approach.

This work also offers unique guidance and insights for continuum-based modeling of the multiphase mixture mechanics. In particular what aspects of a mixing theory may provide qualitative and/or quantitative predictions of the rich phenomena reported here. No multiphase continuum mixture theory is generic enough for all flows;
therefore, this work is fundamental in guiding the future development of a predictive theory, modeling and simulation effort at the continuum level.
In fact, the original motivation for this work was to provide experimental information for building an appropriate mixing theory for the system at hand. Finally, advances in rigorous modeling of this

system will likely provide the basis for other similar multiphase, multicomponent fluid transport used in nuclear recycling processes.

534300 Synthesis and Characterization of UiO-66-NH_2 Metal-Organic Framework Cotton Composite Textiles

Abstract:

This work details the development of a cotton fabric functionalized with UiO-66-NH_2 Metal-Organic Framework (MOF). The materials were made by seeding the growth of the MOF on the cotton by first bonding zirconium (Zr) to the surface of the fabric utilizing cyanuric chloride modified with a thiol. After seeding the fabrics with Zr, UiO-66-NH_2 was grown on the fabric using a hydrothermal method. Several different routes of attaching Zr to cyanuric chloride were examined. Scanning electron microscopy (SEM) and powder X-ray diffraction (PXRD) data are consistent with UiO-66-NH_2, and the fabrics have surface areas between 45-125 m^2/g depending on the synthetic conditions used to produce the materials. The functionalized cotton reacts with dimethyl 4-nitrophenyl phosphate (DMNP), a chemical nerve agent simulant, as monitored by UV-Vis spectroscopy. The results illustrate that MOF-fiber composites can be created using natural fibers and the resulting composites provide similar chemical warfare agent (CWA) simulant reactivity as observed on MOF-polymer composites.

537654 Role of Bed Design Characteristics on the Effective Thermal Conductivity of a Structured Adsorbent

Abstract:

The role of different design variables on the effective radial thermal conductivity of a structured adsorbent bed for potential applications in pressure swing adsorption (PSA) and temperature swing adsorption (TSA) processes is discussed. The structured adsorbent corresponds to a metal support consisting of layers of intercalated smooth and corrugated metal foils coated with 13X zeolite resulting in parallel triangular channels. The radial direction is perpendicular to the planes of the foils. The variables investigated include the thickness of the adsorbent coating, thickness of the metal, nature of contacts between smooth and corrugated foils, type of metal, presence and magnitude of an air gap between the foils, difference in alignment of the metal foil, difference in the coating methodology and effect of different void gases. The results show that the effective thermal conductivity is significantly improved if the foils are in direct contact either via imbedding or simply by point contacts. In such a situation, the thermal conductivity depends strongly on the conductivity of the metal and weakly on the conductivity of gas medium and all other design properties. In contrast, when the foils are separated either by air gaps or by sorbent coating, the effective thermal conductivity is not that different from that of a pelletized bed and depends strongly on the conductivity of the gas medium and weakly on the conductivity of the metal and all other design properties.

538050 Development of 3D-Printed Polymer-Zeolite Composite Monoliths for Gas Separation

Abstract:

Organic-inorganic materials have emerged as promising candidates for use in various gas separation processes due to the beneficial synergistic effect of both constituents. Herein, we report a novel and facile approach for fabrication of composite monoliths consisting of Torlon polymer and 13X and 5A zeolites via 3D printing technique for the removal of CO_2 from flue gas. The physical and structural properties of 3D-printed composite monoliths were systematically evaluated and compared with their pristine powders. By the incorporation of ~31 wt% zeolite particles into the polymer matrix and the addition of amine moieties into the polymer dope composite, the obtained monoliths displayed CO_2 capture capacity proportional to the zeolite loading. Most importantly, both 3D-printed Torlon-zeolite monoliths exhibited high compressive strengths of approximately 210 MPa, which was significantly higher than that of 3D-printed zeolite monoliths developed in our previous work. These results demonstrate the superiority of polymer-zeolite composite monoliths, with existing undamaged zeolite crystals, formulated by 3D printing technique as robust structures with outstanding mechanical integrity and comparable adsorption capacity for various adsorption-based gas separation processes. The printing strategy reported herein can be easily extended to other organic-inorganic composite materials with different polymers and inorganic particles with various compositional ratios.

531650 Emulsion Templated Polymers As Supports for Metal Organic Frameworks

Abstract:

Metal organic frameworks (MOFs) have been used for a variety of laboratory scale applications, including adsorption, catalysis, and gas storage. In order to become industrially relevant and to broaden their potential uses MOF powders must be made more mechanically stable and shaped into forms that are more easily handled. Emulsion templated polymers offer an ideal supporting structure, producing rigid MOF-polymer composites with permanent interconnected macroporosity. The polymer pore structure would allow the creation of large, three dimensional composite bodies while maintaining high overall permeability and access to MOF even within the bulk of the structure.

We have successfully incorporated MOF particles into two different emulsion templated polymer systems by embedding it before polymerization begins. Both surfactant stabilized and particle stabilized emulsions have been examined, as have a variety of other variables including emulsion internal phase volume, crosslinker content, MOF surface chemistry, and inclusion of solvent porogens. The effect of these variables on MOF accessibility was determined by measuring the BET surface area and the ambient temperature CO_2 adsorption capacity of the composites. Additionally, microbreakthrough measurements have been used to determine the ability of the composites to act as a barrier to toxic chemicals.

531792 UTSA-16 Growth within 3D-Printed Co-Kaolin Monolith with High Selectivity for CO_2/CH_4, CO_2/N_2, and CO_2/H_2 Separation

Abstract:

UTSA-16 Growth within 3D-Printed Co-Kaolin Monolith with High Selectivity for CO_2/CH_4, CO_2/N_2, and CO_2/H_2 Separation

Shane Lawson, Marc St. Amour, Ali A. Rownaghi, Fateme Rezaei

Department of Chemical & Biochemical Engineering, Missouri University of Science and Technology, Rolla, MO 65409-1230, United States

Abstract

3D printing technology renders direct material modification a realistic and economic prospect. In our earlier works,[1â€“3] we demonstrated the use of 3D printing for formulating various classes of solid adsorbents such as a zeolites, aminosilicates, and metal-organic frameworks (MOFs) into honeycomb monoliths as efficient gas-solid contactors for gas separation and purification processes. Using the techniques culminated previously, it has been recently shown that the combination of 3D-printing and precursor seeding may pave the way for highly selective adsorption contactors. In this work, 3D printing was utilized to dope kaolin-based monolith with UTSA-16 metal formation precursor (Co) whereupon an internal deposition was facilitated via a solvothermal synthesis approach. The cobalt weight loading in the kaolin support was varied systematically to optimize the MOF growth while retaining monolith mechanical integrity. The obtained UTSA-16 monolith with 90 wt% loading exhibited similar textural features and adsorption characteristics to its powder analogue while improving upon structural integrity. In comparison to previously developed 3D-printed UTSA-16 monoliths, the UTSA-16-kaolin monolith not only showed higher MOF loading, but also higher compression stress indicative of its robust structure. Furthermore, the 3D-printed UTSA-16-kaolin monolith displayed a comparable CO_2 adsorption capacity to the UTSA-16 powder (3.1 vs. 3.5 mmol/g at 25 °C and 1 bar) which was proportional to its loading. Selectivity values of 48, 237 and 3725 were obtained for CO_2/CH_4, CO_2/N_2, and CO_2/H_2, demonstrating good separation potential of the 3D-printed MOF monolith for various gas mixtures, as determined by both equilibrium and dynamic adsorption measurements. Overall, this work provides a novel route for the fabrication of UTSA-16 loaded monoliths which demonstrate both high MOF loading and mechanical integrity that could be readily applied to various CO_2 capture applications.

References

1. Thakkar et al. *ACS. Mater. Interfaces* **8** (41), 27753 (2016).
2. Thakkar et al. *ACS. Mater. Interfaces* **9** (8), 7489 (2017).
3. Thakkar et al. *ACS. Mater. Interfaces* **9** (41), 35908 (2017).

528187 Fibrous Carbon Molecular Sieve with 3-5 a Tunable Pores for Many Industrial Gas Separations

Abstract:

Traditional adsorbents in the form of pellet/bead used in Pressure Swing Adsorption have significant constraint to balance the needs of high mass transfer rate and low pressure drop. Structured adsorbent, in the form of fiber, film, and honeycomb are desirable, which have large external contact area, small diffusion length, and low-resistance channels for gas flow. This talk introduces a new fibrous carbon molecular sieve adsorbent made by a two-step pyrolysis of melt extruded polyvinylidene chloride copolymer (PVDC) fiber. The PVDC fiber was first pretreated in air at low temperature (< 180 C) to form conjugated double bond and crosslinking via dehydrochlorination, then pyrolyzed in N_2 at high temperatures 1000-1400 C. A high throughput kinetic adsorption tool with 48 parallel testing cells using 10 probe molecules (from CO_2, σ_K=3.3 Å to SF_6, σ_K=5.5 Å) was used to 1) characterize the effective micropore size and its changes; and 2) fast screen the adsorbents for many industrial gas separations. The new fibrous CMS adsorbents have effective micropore size in the range of 3-5 Å and show very high separation factors of 20, 26, and 88 for the industrial gas pairs of N_2/CH_4, C_2H_4/C_2H_6, and C_3H_6/C_3H_8, respectively.

529995 HETP Analysis of Structured Adsorbents for Gas Separation Processes

Abstract:

Structured adsorbents (SAs) can increase the performance of separation units when compared to traditional packed columns [1]. SAs applications span from VOC removal, hydrogen production, desiccant cooling, rapid cycle PSA and other processes where pressure drop can have a significant economic impact [2].

The aim of this work is to assess the performance of different types of SAs for gas separation processes. More specifically, the behaviour of monoliths with different channel shape, and corrugated metal foil with adsorbent coating have been investigated determining the Height Equivalent to a Theoretical Plate (HETP) and pressure drop. HETP equations for these structures have been derived with the use of the moment analysis described by Dutta and Leighton [3], and 3D numerical simulations have been produced to validate the proposed equations.

Both the moment analysis and the 3D simulations consider account an isothermal and linear system, with low concentration of the adsorbable component. In these conditions, the HETP can be written as sum of the contributions to the dispersion of the solute in the channel. Therefore, the relative weights of the single contributions can be studied.

The analysis of these structures shows that, for a gas separation process, the velocity profile inside the channel has little impact on the HETP. Furthermore, it also highlights how the main contribution to the dispersion is given by the diffusion in the solid.

In order to obtain the correct equation of the HETP Ahn and Brandani showed that for a rectangular channel of arbitrary aspect ratio a corrected wall thickness is needed [4]. The corrected wall thickness addresses the additional mass transfer resistance in the "corners" of the structures. This concept has been extended to all the structures analysed in this work and is shown to lead to the correct match between HETP and 3D simulations.

Finally, a 1-D equivalent model has been derived from the HETP correlation. The model consists of an axially dispersed plug flow with linear driving force (LDF) model. The axial dispersion is obtained from the Taylor-Aris dispersion coefficient [5], and the LDF constant is includes the corrected wall thickness. The 1-D model provides an accurate match of the 3-D breakthrough curves, leading to a quick and simple tool for the design of structures that include distributions of channel widths and wall thicknesses [6].

References

[1] Ruthven, D.M., Thaeron, C. Performance of a parallel passage adsorbent contactor. *Gas Sep. Purif.* 10, 63–73 (1996).

[2] Ruthven, D. M. Past Progress and Future Challenges in Adsorption Research. *Ind. Eng. Chem. Res.* 2127–2131 (2000).

[3] Dutta D, Leighton DT. Dispersion in large aspect ratio microchannels for open-channel liquid chromatography. *Anal Chem.* 75, 57-70 (2003).

[4] Ahn, H. & Brandani, S. Analysis of breakthrough dynamics in rectangular channels of arbitrary aspect ratio. *AIChE J.* 51, 1980-1990 (2005).

[5] Aris, R. On the dispersion of a solute in a fluid flowing through a tube. *Proc. R. Soc. Lond. A.* 235, 67-77 (1956).

[6] Ahn, H. & Brandani, S. Dynamics of Carbon Dioxide Breakthrough in a Carbon Monolith over a Wide Concentration Range, *Adsorption*, 11, 473–477 (2005).

533069 Sorption in PIM-1 from Cryogenic to Room Temperature: Experimental and Model Analysis

Abstract:

The so-called polymers of intrinsic microporosity, among the class of high free-volume glasses, show a tremendous potential as membranes for gas separations,[1] and their peculiar features provide excellent gas permeability and selectivity so that their performances are often located near the technological limit for polymeric membranes represented by the Robeson upper bound.[2] For these reasons, PIMs have been extensively studied for a variety of molecular separations,[3] ranging from CO_2 capture[4] to organic solvent nanofiltration.[5]

In this work, the sorption properties of CO_2, N_2 and Ar in glassy polymer of intrinsic microporosity PIM-1 have been thoroughly investigated in a wide temperature range, from cryogenic temperatures (77 K) up near room conditions (275 K). All isotherms show the typical behavior of gas sorption with a pronounced downward curvature on the activity axis; data at different temperatures allowed the determination of the heat of sorption for the three different penetrants inspected.

The experimental results obtained have been analyzed by the nonequilibrium thermodynamics for glassy polymers NET-GP,[6] suitable to evaluate penetrant solubility in polymer phases, treated as macroscopic, uniform and homogeneous phases out of their equilibrium conditions.[7] The model is able to provide an accurate description of all the measured data with a limited number of parameters. The model is also used to evaluate the solubility of the same gases in PIM-1 above room temperature[8] in a purely predictive fashion.

The analysis reveals that the same mechanism is driving the sorption process in the entire broad T range, from cryogenic to above room temperature. Two main regions are also identified in the cryogenic isotherm, the first is associated to an iso-volumic sorption in the excess of free volume in the low activity range, while in the second the physical dissolution is accompanied by a significant swelling of the polymer in the central portion of the isotherm. This talk will discuss both the validity of BET surface area analysis and alternative classification metrics for this group of materials.

References

(1) McKeown, N. B. *ISRN Mater. Sci.* **2012**, *2012*, 1.

(2) Robeson, L. M. *J. Membr. Sci.* **2008**, *320*, 390.

(3) Jue, M. L.; Koh, D. Y.; McCool, B. A.; Lively, R. P. *Chem. Mater.* **2017**, *29*, 9863.

(4) Pang, S. H.; Jue, M. L.; Leisen, J.; Jones, C. W.; Lively, R. P. *ACS Macro. Lett.* **2015**, *4*, 1415.

(5) Fritsch, D.; Merten, P.; Heinrich, K.; Lazar, M.; Priske M. *J. Membr. Sci.* **2012**, *401â€"402*, 222.

(6) Minelli, M.; Paul, D. R.; Sarti, G. C. *J. Membr. Sci.* **2017**, *540*, 229.

(7) Doghieri, F.; Sarti, G. C. *Macromolecules* **1996**, *29*, 7885.

(8) Li, P.; Chung, T. S.; Paul, D. R. *J. Membr. Sci.* **2014**, *450*, 380.

538575 Highly Permeable and Selective Crosslinked Polybenzoxazole (PBO) Membranes for Gas Separation

Abstract:

Gas separation using polymer membranes has rapidly grown into an area of technological importance and usefulness, because of the progresses made by synthetic chemists in designing and synthesizing new sophisticated macromolecular structures. Unfortunately, most of the polymeric membranes suffer from plasticization induced by sorption of condensable gases (e.g., CO_2) and are not stable upon exposure to chemical contaminants. Chemical crosslinking has been demonstrated as an effective pathway to improve membranesâ€™ resistance to plasticization and chemical contaminants in complex feed streams as well as physical aging. However, the much improved membrane stability resulted from crosslinking is always accompanied by significantly reduced gas permeability with little to no gain in selectivity. This is largely due to crosslinking-induced membrane densification that greatly reduces the free volume microporosity required for high gas permeability. Here we report a new series of crosslinked PBO membranes prepared via an approach involving concurrent thermal cyclodehydration and end-linking processes from telechelic oligomers with well-controlled molecular weight. This new macromolecular design of introducing bulky phenyl pendent groups at crosslink junctions upon thermal crosslinking led to disruption of chain packing counteracting the undesired densification effect of covalent crosslinking. As a result, even the most densely crosslinked membranes showed similarly high permeability compared to their linear counterparts along with expectedly improved selectivity and membrane stability. Moreover, further thermal treatment of the crosslinked membranes at 450 ^oC for 0.5 h resulted in significant increase in gas permeability while maintaining the same selectivity. For example, a thermally-treated, highly crosslinked PBO membranes (inter-crosslink chain length of 3000 g/mol) showed CO_2/CH_4 separation performance beyond the 2008 upper bound with CO_2 permeability of 563 barrer and an ideal CO_2/CH_4 selectivity of 29. The same membrane also displayed H_2/CH_4 and H_2/N_2 separation performance beyond the 2008 upper bounds and O2/N2 separation performance on the 2008 upper bound. In this talk, preparation and characterization of these new crosslinked PBO membranes will be presented. Discussions will focus on elucidating the fundamental relationship between microscopic structures and macroscopic transport properties for these innovative crosslinked polymer membranes.

527936 Polybenzimidazole-Derived Carbon Molecular Sieves with Microcavities and Ultra-Microporous Channels Achieving Superior Membrane H_2/CO_2 Separation Properties

Abstract:

Polymeric membranes are of great interest for H_2/CO_2 separation in H_2 purification and CO_2 capture, because of inherently high energy-efficiency, low cost, and easy scale-up. The key to the success of this technology is membrane materials with high H_2 permeability and H_2/CO_2 selectivity at 100 – 300 °C. However, polymers with higher permeability often exhibit lower selectivity, as indicated by Robeson's upper bound plot. Herein we report novel carbon molecular sieve (CMS) membranes derived from polybenzimidazole (PBI), a leading polymer for H_2/CO_2 separation. These membranes are prepared by pyrolysis of PBI in inert gas, forming microcavity with high gas permeability, and ultra-microporous channels with strong size-sieving ability. We systematically study the effect of the pyrolysis temperature on the physical properties and gas separation properties in the CMS. The pyrolysis leads to the formation of graphite-like structure, as confirmed by WAXD. As the pyrolysis temperature increases, the CMS density increases before decreasing, while the porosity increases consistently up to 30% for CMS with a pyrolysis temperature of 900 °C. In general, the carbonization of PBI dramatically increases H_2 permeability while retaining or slightly increasing H_2/CO_2 selectivity. For example, pure PBI exhibit H_2 permeability of 12 Barrers and H_2/CO_2 selectivity of 14 at 100 °C, while the CMS with a pyrolysis temperature of 800 °C shows an H_2 permeability of 672 Barrers and H_2/CO_2 selectivity of 18 at 100 °C. We have also demonstrated that the doping of the PBI with H_3PO_4 before the pyrolysis can dramatically increases the H_2/CO_2 selectivity. For example, the PBI doped with 0.05 wt% H_3PO_4 shows an H_2 permeability of 8.5 Barrers and H_2/CO_2 selectivity of 49, while the pyrolysis at 600 ºC increases the H_2 permeability to 100 Barrers and retaining the H_2/CO_2 selectivity at 40 at 100 ºC. The effect of temperature on gas permeability and selectivity are studied, and the fundamental gas solubility and diffusivity in these CMS are determined. This talk will also describe the mixed-gas separation properties, and the effect of water vapor on the H_2/CO_2 separation properties. These CMS with intrinsic micropores exhibit H_2/CO_2 separation properties far above the Robeson's upper bound, demonstrating their potential for practical H_2 purification and CO_2 capture.

533077 Ultra-Selective Defect-Free Interfacially Polymerized Molecular Sieve Thin-Film Composite Membranes

Abstract:

Aromatic polyamide thin-film composites made by interfacial polymerization are state-of-the-art membranes for reverse osmosis and nanofiltration. These membranes contain nano-defects that have essentially little effect on their performance for aqueous applications but reduce their selectivity for gas separation applications to very low Knudsen diffusion-based values. In this work, inexpensive high-performance gas separation membranes were fabricated by modifying the commercially successful reverse osmosis interfacial polymerization membrane production method. The effects of fabrication variables on the morphologies of the thin-film composite membranes optimized for gas separation applications will be presented. Defect-free thin-film composite membranes demonstrated unprecedented mixed-gas H_2/CO_2 selectivity of â‰ˆ 50 at 140 Â°C with H_2 permeance of 350 GPU (10^{-6} cm^3(STP cm^{-2} s^{-1} $cmHg^{-1}$), surpassing the hydrogen/carbon dioxide permeance/selectivity upper bound of all known polymer membranes by a wide margin. The combination of exceptional separation performance and low manufacturing cost makes them excellent candidates for cost-effective hydrogen purification from steam cracking and similar processes. Permeance and selectivity measurements imply an ultrathin polyamide layer with apparent thickness of â‰ˆ 10-20 nm. Potential use of the membranes in challenging reverse osmosis applications will be proposed.

522218 Gas Sorption Properties of Novel Iptycene-Based Thermally Rearranged Co-Polymers: Effect of Temperature, Mixed Gas and Pre-Treatment

Abstract:

Iptycene-containing polymers have emerged as promising materials for gas separation applications[1]. Triptycenes are three-dimensional structures formed by three aromatic rings disposed in different planes[1]. Once inserted on the polymer backbone, they provide an extra, internal free volume which improves gas permeability while preserving optimal levels of selectivity. Moreover, unlike what happens in common glassy polymers, the internal free volume provided by iptycene units is not related to the non-equilibrium conformation, but to the molecular configuration. Such configuration-based free volume is intrinsic to the polymer structure and, as such, it is noncollapsible, similar to the case of inorganic molecular sieves[1]. Conversely, free volume related to non-equilibrium conformation is subject to changes over time, due to the physical ageing. Although several iptycene-containing polymers have shown separation properties that exceed the upper bound[2], fundamental structure-property correlations for this family of polymers are scarce in the literature and, specifically, the role of sorption and diffusion coefficients to the overall permeability coefficients is still unclear.

In this study, single gas solubility isotherms for He, N_2, CH_4, CO_2 and C_2H_6 in a novel thermally rearranged co-polymer prepared from a co-polyimide precursor with controlled triptycene molar content, i.e., triptycene-dianhydride(0.25)-6FDA(0.75)-6FAP(1.0), were experimentally measured in the range 5 to 50 °C and up to 32 atm. Experimental data were analyzed in the framework of the Dual Mode Model[3]. Interestingly, the vast majority of penetrant molecules are sorbed in the Langmuir mode, which justifies the high resistance of iptycene-based materials to plasticization. Specifically, the Langmuir to Henry mode sorption ratio is greater than for other high performance materials. The dual mode parameters retrieved from the analysis of single gas sorption isotherms were used to estimate a priori the sorption behavior in mixed gas conditions[4]. Remarkably, the mixed gas solubility-selectivity is higher than ideal solubility-selectivity.

The analysis of sorption enthalpies provides additional information about polymer-penetrant interactions, as well as the plasticization resistance. Finally, to highlight the effects of thermal treatment and physical ageing, sorption experiments were run, for selected gases, on samples previously annealed at 120°C under vacuum for 1 week.

Modeling work is underway to interpret the experimental observables in the framework of the NELF model.

References

1. J.R. Weidman, R. Guo, Ind. Eng. Chem. Res. 2017, 56, 42240-4236
2. R. Guo, unpublished results
3. W.J. Koros, D.R. Paul, A.A. Rocha, J. Polym. Sci. B 1976, 14, 687-702
4. W.J. Koros, J. Polym. Sci. B 1980, 18, 981-992

534536 Synthesis and Multicomponent Permeation Evaluation of Functionalized PDMS Membranes for Enhanced NGL Recovery from Natural Gas

Abstract:

Saudi Arabia owns the 5th largest proven reserves of natural gas in the world. Most of this resource is utilized for domestic energy production. For decades, glassy membrane technology has been successfully applied to remove CO_2 from natural gas. More recently, commercially available poly(dimethylsiloxane) (PDMS) rubbery membranes, have been utilized for natural gas liquid (NGL) recovery from natural gas. NGL can be utilized as a liquid fuel for power generation and can also be used as a feedstock for chemical production. Due to their high permeability properties, PDMS membranes can at the same time reject nitrogen and concentrate heavy hydrocarbons (C_{3+}) from both wet sour gas and sweet dry gas in the membrane permeate streams, which can be then routed to a NGL recovery unit. However, the current state-of-the-art PDMS membranes exhibit low C_{3+}/methane selectivities under industrially-relevant feed streams and testing conditions. In order to achieve significant enhanced recovery of NGL from natural gas while reducing *capital and operating expenditures*, more efficient membranes with improved selectivity are desired.

This talk seeks to provide detailed investigation of separation performance of modified PDMS type composite membranes for enhanced NGL recovery under rigorous testing conditions, in order to more closely approximate those encountered in a gas plant. A new type of functionalized PDMS membrane material containing Si-C-C-Si internal network and Si-diphenyl or Si-fluorine side chains in the membrane matrix was synthesized over a porous support of polyacrylonitrile. These functionalized PDMS composite membranes demonstrate superior NGL separation performance as compared to traditional PDMS membrane at all conditions tested (e.g. multicomponent C_1-C_5 hydrocarbon mixtures, up to 850 psi feed pressure, temperature, feed gas composition, presence of aggressive BTEX contaminants). The C_{3+}/methane mixed gas selectivities were observed to be significantly enhanced by the substitution of rigid backbone and crosslinking moieties. The results of this study aim to advance the development of novel rubbery membrane materials for aggressive natural gas separations.

534489 A Scalable Method to Synthesize Zeolitic-Imidazolate Framework ZIF-8 Membranes on Polymer Hollow Fibers for Propylene/Propane Separation

Abstract:

Separation of olefin/paraffin mixtures (e.g., ethylene/ethane and propylene/propane), is one of the most challenging separations due to the similarity of their physicochemical properties. Currently, propylene/propane separation is typically achieved through energy-intensive cryogenic distillation process. Although energy-efficient membrane-based separation processes have received significant attention, however, there are no commercially available membranes for olefin/paraffin separations mainly due to the limitations of current membrane materials and lack of cost-effective and simple processing methods to fabricate membrane with large area[1-3].

A sub-class of metal-organic frameworks (MOFs), zeolitic-imidazolate frameworks (ZIFs) with zeolite topologies are crystalline nanoporous materials with well-defined aperture sizes (< 5 Å). ZIFs, which are composed of divalent metal ions (typically Zn^{2+} and Co^{2+}) and imidazole-based ligands[4], have been extensively studied especially for gas separation applications. ZIF-8 in particular is a promising membrane material for high resolution propylene/propane separation because the effective pore size of ZIF-8 (~ 4 Å) [5] lies in between the size of propylene and propane molecules. Although ZIF-8 membranes (mostly planar configurations) synthesized either by *in-situ* or seeded growth method showed excellent propylene/propane separation performances, scaling-up and commercialization of the membranes requires them to be fabricated on a cheap and scalable support (i.e., hollow fiber or tubular support). Polymeric hollow fibers have been utilized as substrates, but only a few of the resulting ZIF-8 membranes showed promising propylene/propane separation performances.[6-9]

Herein, we would like to present a simple and scalable method to synthesize well intergrown ZIF-8 membranes on polymeric hollow fibers. The method is based on substrates modification followed by solvothermal reaction in linker solution. High-quality ZIF-8 seed layers prepared from the solvothermal reaction are secondarily grown to form high quality ZIF-8 membranes. The resulting ZIF-8 membranes are tested for propylene/propane separation. In this work, the effects of different synthesis parameters on the microstructures of ZIF-8 membranes will be discussed. Finally, the gas separation performances of ZIF-8 membranes especially propylene/propane separation will be presented.

References and Notes

1 Koros, William J. and Fleming, G.K. (1993) "Membrane-based gas separation", *Journal of Membrane Science*, 83, pp l-39

2 Baker RW., (2002) "Future Directions of Membrane Gas Separation Technology" *Industrial & Engineering Chemistry Research.,* 41 (6), pp 1393-1411

3 Tanaka, K., Taguchi, A., Hao, JQ, Kita, H., and Okamoto, K., (1996) "Permeation and separation properties of polyimide membranes to olefins and paraffins", *Journal of Membrane Science*, 121, pp 197-207

4 Park KS, Ni Z, Côté AP, et al., (2006) "Exceptional chemical and thermal stability of zeolitic imidazolate frameworks", *Proceedings of the National Academy of Sciences,* 103 (27), pp 10186-10191

5 Zhang C, Lively RP, Zhang K, Johnson JR, Karvan O, Koros WJ., (2012) "Unexpected Molecular Sieving Properties of Zeolitic Imidazolate Framework-8", *The Journal of Physical Chemistry Letters*., 3 (16), pp 2130-2134.

6 Andrew J. Brown, Nicholas A. Brunelli, Kiwon Eum, Fereshteh Rashidi, J. R. Johnson, William J. Koros, Christopher W. Jones, Sankar Nair, (2014) "Interfacial microfluidic processing of metal-organic framework hollow fiber membranes", *Science*, 345, 6192, pp 72-75

7 FernandoCacho-Bailo, SilviaCatalán-Aguirre, MirenEtxeberría-Benavides, Oğuz Karvan, VictorSebastian, CarlosTéllez, JoaquínCoronas (2015) "Metal-organic framework membranes on the inner-side of a polymeric hollow fiber by microfluidic synthesis", *Journal of membrane science*, 476, pp 277-285

8 Kiwon Eum, Chen Ma, Ali Rownaghi, Christopher W. Jones, and Sankar Nair (2016) "ZIF 8 Membranes via Interfacial Microfluidic Processing in Polymeric Hollow Fibers: Efficient Propylene Separation at Elevated Pressures", *ACS Applied materials and interfaces*, 8, pp 25337 – 25342

9 Anne M. Marti, Wasala Wickramanayake, Ganpat Dahe, Ali Sekizkardes, Tracy L. Bank, David P. Hopkinson, and Surendar R. Venna (2017) "Continuous Flow Processing of ZIF 8 Membranes on Polymeric Porous Hollow Fiber Supports for CO2 Capture", *ACS Applied materials and interfaces,* 9, pp 5678-5682

534111 Purification of Exosomes Using Tangential Flow Filtration

Abstract:

Background: Increasing interest is being paid to exosomes, small membrane bound particles produced by cells, particularly due to their ability to carry proteins, RNA, and DNA. Specifically, exosomes are thought to play a large role in cancer, specifically as biomarkers and therapeutic agents [1]. However, one major difficulty in exosome research is the isolation and concentration of exosomes. The current standard method of exosome isolation is ultracentrifugation. Significant difficulties exist in ultracentrifugation of exosomes, primarily recovery. Ultracentrifugation has been shown to produce very low isolation concentrations and altered size distributions when using liposomes as a model system[2]. Additionally, ultracentrifugation is a lengthy process, requires expensive equipment, and possibly damages exosomes due to the high speeds[3]. Specifically, the high g-forces in ultracentrifugation are believed to cause the fusion of exosomes. When compared to other commercially available separation kits, ultracentrifugation has been seen to result in larger particle sizes, which is thought to be a result of exosomes fusing with other exosomes, contaminants, and proteins[4]. A high degree of variability has also been observed between different rotors. When the same conditions are run on different rotors, very different concentrations and size distributions are observed in the final concentration solution[5]. Based on these issues, a better method of isolation of exosomes is necessary. Any method should exhibit sizes consistent with the initial distribution, and should not show a drastic loss of sample. We propose the use of tangential flow filtration (TFF) which has been previously used for protein sepearations[6].

Methods: Cell supernatant is isolated from various cell lines of interest that are known to produce exosomes, HEK293, MDA 231, and CHO. After the supernatant is collected, it undergoes two benchtop centrifuge steps that are a part of the conventional ultracentrifugation process. The first, 300g for 5 minutes removes the cells, and the second 2000g for 30 minutes removes most cell debris. The supernatant after these two centrifugation steps is then loaded for filtration, with a small sample removed to allow for comparison to initial levels. Filtration is performed using a polysulfone hollow fiber filter with varying pore sizes, 10 kD, 50 kD, and 500 kD (Spectrum Labs). The supernatant is pumped through the filter, at 24 mL/min using a peristaltic pump. After passing through the filter, the retentate is returned to the initial sample container for recirculation through the system. The filtration continues until the retentate volume reaches around 3 to 5 mL remaining. The retentate of filter contains the concentrated exosomes, and the permeate media and other small components and is discarded. When collecting the retentate, the filter is flushed with 1 mL of PBS to remove exosomes that might still be on the filter. After the filtration, the initial sample, the retentate, and samples from the permeate are then tested using Nanoparticle Tracking Analysis (Nanosight) to determine the concentration and size distribution of the exosomes. We then calculate the total number of particles in both the initial and final samples to see the recovery.

Results and Discussion: Our initial trials were all run using the 10 kD filter, further experimentation is needed on the 50 and 500 kD filters. Additionally, HEK293 cells were not used in these initial runs. For the MDA 231 cells, we started with 33 mL of supernatant, which is measured to have a total particle count on the order of 10^{11} particles, with some variability between individual harvests. For most runs, the final pellet is measured to be around 6 mL, and has a similar total particle count, on the order of 10^{11} total particles. In these initial runs we are seeing approximately a 5 fold concentration of the exosomes in the sample, while reducing the volume by approximately an order of 5. Additionally, the particle size distribution is far more similar to the initial distribution than is seen in ultracentrifugation. When running the permeate samples, the concentration of particles is far lower than in the retentate, indicating very few exosomes are going through the filter. There is still a degree of variability seen in the filtration results, though far less severe than with ultracentrifugation. While most samples show a final particle count very close to the initial sample, an occasional trial shows a concentration significantly lower than the initial, or slightly higher than the initial.

The CHO samples exhibit far more variability than the MDA 231 samples. These samples started with more volume, around 45 mL, and were concentrated to around 5 mL. The final particle count is greater than the initial count consistently across the CHO samples, in many cases significantly higher. Due to the higher initial volume, we have been able to concentrate the CHO exosomes to a final concentration of around 10^{12} particles/mL. Size distribution is far more consistent to the initial sample compared to ultracentrifugation.

Conclusion: Ultracentrifugation as a method of purifying exosomes has significant drawbacks, from the cost and length of operation, to low recovery and size inconsistencies. Tangential Flow filtration has the potential to be a better method for the concentration of exosomes from cell culture supernatant. Initial runs have shown that a far greater amount of the initial exosome concentration is recovered after TFF than after ultracentrifugation, and that the size distribution after TFF is far more indicative of the initial population than an ultracentrifuged sample. There are still concerns that need to be addressed with future work. Ultracentrifugation contains an additional step that has not yet been accounted for, a 10,000g centrifugation step that is intended to remove apoptotic bodies. Further runs will be performed using larger filters, where the exosomes should pass through as the permeate, and apoptotic bodies should remain in the retentate. This will likely impact the recovery of exosomes, as some might remain in the retentate. Other concerns include shear stress on the exosomes causing damage or deforming them enough to pass through the small membrane, and the length of time required to reach the desired volume. Back pressure applied plays a role in these concerns, a higher back pressure allowed us to reach the concentrated volume faster, but it could cause more exosomes to be squeezed through the pores of the membrane leading to lower recovery. Overall, initial results indicate that there is a potential of the TFF system being a promising way to concentrate exosomes, alleviating some of the issues associated with ultracentrifugation.

[1] Guo, W., Gau, Y., et al. 2017, "Exosomes: New players in cancer", Oncol Rep. Vol 38, pp. 665-675.

[2]Lane, R., Korbie, D, et. Al. 2015, "Analysis of exosome purification methods using a model liposome system and tunable-resistive pulse sensing", Scientific Reports, Vol 5, article number 7639. doi:10.1038/srep07639

[3] Li, P. Kaslan, M. et al. 2017 "Progress in Exosome Isolation Techniques", Theranostics. Vol 7, pp. 789-804.

[4] Helwa, I. Cai, J. et al. 2017 "A Comparative Study of Serum Exosome Isolation Using Differential Ultracentrifugation and Three Commercial Reagents". PLOS ONE doi:10.1371/journal.pone.0170628

[5] Livshits, M. Khomyaova, E. et al. 2015. "Isolation of exosomes by differential centrifugation: Theoretical analysis of a commonly used protocol". Scientific Reports, Volume 5, article number 17319 doi:10.1038/srep17319

[6] van Reis, R., Leonard, L. et al. 1991. "Industrial scale harvest of proteins from mammalian cell culture by tangential flow filtration". Biotechnology and Bioengineering, Vol 38, pp 413-422.

523035 Model-Based Optimization and Design of Continuous Chromatography for Protein Capture

Abstract:

The petrochemical, chemical and food industries have practiced continuous manufacturing for many years. However, biotechnology is still late to embrace continuous processing, especially for downstream processing. Recently, multi-column periodic counter-current (PCC) chromatography has been proposed as continuous capturing technology to improve the process productivity and resin capacity utilization, as well as reduce buffer consumption and equipment footprint, which is applying for monoclonal antibody capture with Protein A affinity resin. Due to the complexity of continues chromatography, some mathematical framework should be developed to aid the optimization and process design.

To better understand and describe continuous chromatography process, a fully mechanism model was built for fitting and predicting breakthrough curves as well as optimizing the operation conditions. An integrated approach was developed to combine the Equilibrium Dispersive Model of column, General Rate Model of resin, as well as the Equilibrium Adsorption Model for protein adsorption. Firstly, breakthrough curves at different flow rates and loading concentrations are fitted to obtain the model parameters, including pore diffusion coefficient, mass transfer coefficient and so on. Then, multi-column adsorption pattern and continuous separation are evaluated and predicted through these parameters under varying operation conditions. Finally, the working window would be determined with the design strategy, considering the process productivity and capacity utilization. In addition, the performance of continuous chromatography would be compared with traditional batch process, and the advantages and limitation of continuous processing would be discussed. The model-based approach developed were tested with different continuous multi-column modes, including twin-column CaptureSMB, 3/4-column PCC from GE Healthcare and multi-column BioSMB from Pall. The interactive interface was also developed to easily input the data and visually output the results. The results indicated that model-based approach could aid the process design and enhance both the development and application of continuous chromatography.

This work was supported by the International Science & Technology Cooperation Program of China and National Natural Science Foundation of China.

530329 Sterile Filtration of Oncolytic Viruses Using Novel Nanoporous Silicon Nitride Membranes

Abstract:

Oncolytic viruses (OVs) are a growing class of bio-therapeutics that have attracted significant interest for cancer immunotherapy and are expected to be worth over $1B by 2026. In 2016 alone, there were over 40 clinical trials (phase I, II, and III) conducted, both with OVs on their own and in combination with other cancer therapies. Current Good Manufacturing Practice (GMP) standards for the manufacturing of OVs rely on sterile filtration processes using microporous polymeric membranes to remove microbial contaminants. While this is necessary to ensure the purity and safety of the formulation, sterile filtration can also lead to reduced yield and increased process costs due to entrapment of viruses within the high internal surface area and broad pore distributions of the membrane; new process technologies are required to help optimize the production of OVs.

Nanoslit silicon nitride (NSN) membranes are an emerging technology of considerable potential to address this gap. Fabricated using advanced silicon processing techniques, these ultrathin (<400 nm) membranes have precisely controlled pore size (200-500 nm) and geometry (slit or circular). It is hypothesized that the unique properties of NSN membranes will allow it to significantly reduce the titer loss during sterile filtration while also being amenable to filtration at higher fluxes and lower pressures as compared to traditional membranes. We set out to demonstrate that the novel NSN membranes would enable sterile filtration of viruses while also having titer loss less than 10% (given starting load of ~10^{10} PFU/mL) and filtration capacity greater than or equal to commercial membranes (PVDF, PES). The NPN membranes were integrated into custom-built dead-end micro-scale filtration cells and studied under constant flux filtration, while a wide range of variables were monitored and characterized for their effect on performance including membrane properties, operational parameters, and feed composition. The filtration studies were performed using an attenuated strain of the oncolytic Maraba virus, a Rhabdovirus with a unique bullet shape (170 nm x 70 nm) which is currently undergoing clinical trails and shows promise as both an immunotherapy vaccine vector and oncolytic agent. Appropriate theoretical frameworks were used to analyze the experimental results in terms of the underlying bio-physical properties of the OVs and associated impurities (e.g. DNA, host cell proteins).

537865 Docking Simulations to Predict Binding Performance of Affinity Ligands for Purification of Butyrylcholinesterase

Abstract:

An ever-increasing demand for affordable biotherapeutics manufactured under stringent regulatory guidelines is driving the development of innovative, more efficient, lower-cost downstream purification technologies. The use of an affinity separation step in downstream purification can significantly reduce the number of process steps, buffer use, production time and overall costs. Affinity resins using antibody and other protein ligands can be prohibitively expensive. Synthetic peptide ligands offer a robust, lower-cost alternative to biological affinity ligands. This work illustrates the use of computational tools in identification, development and characterization of peptide based ligands for the purification of butyrylcholinesterase - an enzyme important for its ability to degrade cocaine, hydrolyze ester-containing drugs, and as an antidote to various nerve gas and organo-phosphorous compounds. This paper describes the use of High Ambiguity Driven Molecular Docking (HADDOCK) software to identify peptides that will bind to the protein through its active site. Docking calculations were performed for *in-silico* peptide libraries of three amino acids. Using characteristics of reported butyrylcholinesterase inhibitor molecule(s) as a guideline, peptides containing aromatic residues were prioritized for creating the peptide library. Residue cysteine forms cystine through a disulfide linkage, while methionine forms methionine sulfoxide (MetO) by addition of oxygen to its sulfur atom and were not considered while designing the library. Out of a total possible 8000 peptides, a biased library of 2256 peptides was designed with peptides containing one or more aromatic residues. Based on the available description of the active site of human butyrylcholinesterase, a targeted docking of these chosen 2256 peptides to the active site of butyrylcholinesterase was performed. A three-stage approach was adopted for this docking simulation â€“ rigid, semi-flexible and finally water refined fully flexible docking. The docked structures were grouped in clusters based on root-mean-square-distance (RMSD) and scored using Firedock software to calculate free energies of binding. Procainamide, a known ligand to human butyrylcholinesterase, was also docked using the same protocol and its binding energy was used as a reference. Binding characteristics of a few chosen peptide ligands were investigated by synthesizing them on chromatographic resins via Fmoc based solid phase peptide synthesis. Experimental binding results for synthesized peptides were compared to predictions made by HADDOCK.

538173 Computational Design of Peptide Ligands for the Bioseparation of “Fab” Antibody Fragments

Abstract:

An exciting alternative therapeutic to monoclonal antibodies are antigen-binding fragments of antibodies, “Fab”, which retain the target specificity of their parental MAb but are easier and cheaper to produce, and have higher tissue penetration and quicker uptake. A key issue in Fab production is separating it from the cell cultures in which it is made during bio manufacturing. The industry standard to capture Fab from cell culture fluid is affinity chromatography using the affinity ligand Protein L which itself has a number of drawbacks including very high cost, low chemical stability, modest binding capacity, and potential immunogenicity. We describe a novel computational approach to discover new short peptides that can bind Fab specifically and selectively and then release it in eluting conditions in affinity chromatography columns thereby avoiding the need for harsh acidic conditions that denature antibodies. Two of the designed binders from the computational algorithm, were synthesized on Toyopearl Amino 650 M resin , tested for binding and purification of IgG from a cell culture harvest, and found to bind strongly and selectively .

540183 Downstream Processing in Biomanufacturing: Multimodal Chromatography, Affinity Precipitation and Integrated Bioprocessing

Abstract:

This talk will present recent developments in our lab in three areas of downstream bioprocessing; multimodal chromatography, affinity precipitation and integrated bioprocessing. Targeted experiments with different domains of mAbs with varying properties, biophysics, thermodynamics and molecular simulations will be used to shed light on the importance of protein surface cluster properties and multimodal ligand hydration and flexibility on creating selective separations. Hierarchical clustering is employed to examine preferred ligand conformations and the impact of water and dewetting phenomena in multimodal systems. NMR and molecular simulations will be used to examine the molecular basis for multimodal ligand binding to the FC region of mAbs both in solution and immobilized on SAM surfaces. In addition, pH and ligand chemistry will be employed to shift domain dominance in mAb multimodal chromatography from (Fab)2 interactions to Fc binding. Affinity precipitation using smart biopolymers for the simultaneous recovery and purification of both mAb and non-mAb biologics will then be presented. The use of ELP-Z for mAb purification as well as the development of efficient processes based on this approach will be presented with industrial feed stocks. The extension of this approach using ELP-affinity peptides as well as the application to a new class of biologics will also be discussed. Finally, results will be given on a novel approach for the rapid development of integrated downstream biomanufacturing processes for biological products from Pichia Pastoris. This approach employs linear gradient screens in concert with high-throughput uplc analytics and custom software to identify sequences of orthogonal chromatography steps likely to recover and purify the product while also being amenable to integrated processing and a continuous flowpath. To illustrate the power of this novel approach, the rapid development of 3-step purification processes for three commercial biological products will be shown.

525146 CO_2 Extraction from Simulated Air Via Temperature Swing Adsorption Using Polymer/Silica Fiber Sorbents

Abstract:

Direct CO_2 capture from atmospheric air is gaining increased attention as one the most scalable negative carbon approaches available to tackle climate change if coupled with the sequestration of CO_2 geologically. Furthermore, it can also provide CO_2 for further utilization from a globally uniform source. Supported amine based materials are effective for direct air capture (DAC) due to their high CO_2 uptakes and acceptable sorption kinetics at ambient temperature.[1] In this work, we describe the application of polymer/silica fiber sorbents functionalized with an amine rich polymer, poly(ethyleneimine) (PEI), for DAC. PEI-CA/SiO_2 fibers offer comparable CO_2 uptakes to traditional pelletized PEI/SiO_2 sorbents from ultra-dilute sources, but with lower pressure drop similar to PEI- functionalized honeycomb monoliths, which were successfully demonstrated as structured contactor systems for DAC.[2] An initial investigation to study the effect of feed flowrate, adsorption temperature, and the impact of moisture in the feed on the CO_2 breakthrough performance of a densely packed fiber sorbent module is conducted to highlight the potential application of this class of structured contactors in direct air capture. Vacuum aided temperature swing adsorption is used to estimate the CO_2 recovery and purity.

References:

(1) Sanz-PÃ©rez, E. S.; Murdock, C. R.; Didas, S. A.; Jones, C. W. Direct Capture of CO $_2$ from Ambient Air. *Chem. Rev.* **2016**, *116* (19), 11840â€"11876.

(2) Sakwa-Novak, M. A.; Yoo, C.-J.; Tan, S.; Rashidi, F.; Jones, C. W. Poly(ethylenimine)-Functionalized Monolithic Alumina Honeycomb Adsorbents for CO $_2$ Capture from Air. *ChemSusChem* **2016**, *9* (14), 1859â€"1868.

537600 Direct Air Capture of CO_2 Using Zeolites

Abstract:

Direct Air Capture (DAC) of CO_2 has been an emerging field over the past two decades. It originally started being investigated by using NaOH solutions to absorb CO_2. Today, many researchers are looking into grafting or impregnating amines onto adsorbents. However, in this field, there has yet to be a thorough investigation into using zeolites for DAC of CO_2 even though industry currently uses them in a similar process in air pre-purification.

This presentation will include an experimental analysis of CO_2 capture from ambient air using vacuum temperature swing adsorption (VTSA) as a means to capture and effectively reduce the amount of CO_2 in the atmosphere as well as to produce high purity CO_2. Commercially available zeolites have been screened using gravimetric analysis and breakthrough experiments to determine which adsorbent would be the most industrially and economically feasible one for a DAC of CO_2 plant.

532699 Silica Supported Poly(Propylene Guanidine) for CO_2 Capture in Simulated Flue Gas and Ambient Air

Abstract:

Guanidine-based compounds have been studied extensively in CO_2 capture applications in the past, due to their highly basic nature. In this work, the synthesis of poly(propylene guanidine) (PPG) is described via the polycondensation of propylene diamine and guanidine hydrochloride. It is demonstrated that the structure of synthesized PPG is composed of primarily ring structures, with the average molecular weight under 1000 Da. The synthesized polymer is impregnated into a mesoporous silica support, SBA-15, and the CO_2 capture performance of the sorbent is measured under simulated flue gas conditions and simulated ambient air conditions. It is found that SBA-15 supported PPG shows a maximum capacity of 1.1 mmol/g using 10% CO_2, and 0.61 mmol/g using 400 ppm CO_2 at 30°C with a sorbent with 50 wt% polymer loading. Meanwhile, SBA-15 supported PEI shows maximum capacity of 1.3 mmol/g and 1.1 mmol/g under similar conditions with 48 wt% polymer loading. Although its performance was lower than PEI, a benchmark polymer in the field of CO_2 capture, the first generation PPG sorbent showed good performance as a CO_2 sorbent.

535030 Development of Advanced Solid Sorbent for CO_2 Capture from Flue Gas

Abstract:

It is widely agreed that carbon capture, utilization and storage (CCUS) technologies offer a promising pathway for affordable, secure and reliable sources of clean energy. In the United States, about 30% and 21% of total US greenhouse gas (GHG) emissions come from fossil fuel-fired power plants and industrial sectors, respectively, which will continue for decades to come. Capture of CO_2 from large sources of emissions such as coal-fired power plants, is of critical importance in mitigating the GHG effect. Molecular basket sorbents (MBS) consisting of functional polymers immobilized in nano-porous materials have shown great potential as highly selective, high-capacity solid sorbent for CO_2 capture [1-3]. In this paper, we report our recent study on the influences of silica support, PEI loading amount, PEG additive, as well as the sorption temperature on the sorption performance of MBS for CO_2 capture [4]. The results suggest that the pore properties of MBS play a more important role in the CO_2 sorption capacity, rather than those of the supports alone. MBS with 3D pore structure exhibits higher CO_2 sorption capacity and amine efficiency than those with 2D-structured support. Among the sorbents studied, fumed silica (HS-5) based MBS showed the highest CO_2 sorption capacity in the temperature range of 30 to 95 °C, probably due to its unique interstitial pores formed by the aggregation of polymer-loaded SiO_2 particles. It was also found that the temperature dependence is directly related to the PEI surface coverage layers. The more PEI surface coverage layers, the higher diffusion barrier for CO_2 and the stronger temperature dependence of CO_2 capacity. 3D MBS exceeds 2D MBS at the same PEI coverage layers due to lower diffusion barrier. Adding PEG can significantly enhance the CO_2 sorption capacity and improve amine efficiency of all MBS, most likely by alleviating the diffusion barrier within PEI bulk layers through the inter-molecular interaction between PEI and PEG. Furthermore, in a practical application effort, we teamed up with the engineers from RTI International and conducted a pilot-plant scale test using a fluidized-bed adsorption-desorption system [5]. The preparation of molecular basket sorbent was successfully scaled up to pilot-plant scale in a large quantity (~150 kg per batch) through collaboration with our partner RTI under DOE support [5]. The testing data showed that that the sorbent is capable of rapid CO_2 removal and exhibits the ability to capture > 90% of CO_2 in flue gas, suggesting a potential to replace the prevalent amine-scrubbing process.

[1] Ma, X.L.; Wang, X.X.; Song, C.S. J. Am. Chem. Soc., 2009, 131, 5777-5783.

[2] Wang, X.X.; Schwartz, V.; Clark, J.C.; Ma, X.L.; Overbury, S.H.; Xu, X.C.; Song, C.S. J. Phys. Chem. C, 2009, 113, 7260-7268.

[3] Wang, X.X.; Ma, X.L.; Schwartz, V.; Clark, J.C.; Overbury, S.H.; Zhao, S.Q.; Xu, X.C.; Song, C.S. Phys. Chem. Chem. Phys., 2012, 14, 1485-1492.

[4] Zhang, L.; Wang, X.X.; Fujii, M.; Yang, L.J.; Song, C.S. J. Energy Chem. 2017, 26, 1030-1038.

[5] Nelson, T.; Kataria, A.; Soukri, M.; Famer, J.; Mobley, P.; Tanthana, J.; Wamg, D.X.; Wang, X.X., Song. C.S. Bench-scale development of an advanced solid sorbent-based CO_2 capture process for coal-fired power plants. DOE-NETL final scientific/technical project report, 2016. https://www.osti.gov/servlets/purl/1301858.

525720 Improved CO_2 Selectivity-Uptake Trade-Off Driven By Synergetic Thermodynamics, Kinetics, and Packing Effects

Abstract:

Improved CO_2 selectivity-uptake trade-off driven by synergetic thermodynamics, kinetics, and packing effects

Youssef Belmabkhout[a], Weibin Liang[a], Prashant Bhatt[a], Aleksander Shkurenko[a], Karim Adil[a], Georges Mouchaham[a], Himanshu Aggarwal[a], Arijit Mallick[a], Aqil Jamal[b], Mohamed Eddaoudi[a]

[a]Functional Materials Design, Discovery & Development (FMD3), Advanced Membranes and Porous Materials Center, Division of Physical Sciences and Engineering, King Abdullah University of Science and Technology, 4700 King Abdullah University of Science and Technology (KAUST), Thuwal 23955-6900, Kingdom of Saudi Arabia.

[b] ARAMCO, R-GC 335, Floor 3, Research and Development Center (Bldg. 2297) Dhahran, Kingdom of Saudi Arabia

In this work, we report the deliberate use of the interpenetration approach to develop a new, fluorinated MOF with the appropriate pore system (size, shape, and functionality) to enhance the efficient capture of CO_2 from flue gas at room temperature. The MOF, dptz-$CuTiF_6$, exhibits excellent volumetric and gravimetric CO_2 uptakes (65.6 $cm^3\ g^{-1}$ and 0.13 g g^{-1}) at 10% CO_2 and 298 K, which is superior to those of the reference, aqueous amine, with a significantly lower energy input for regeneration (38 kJ mol^{-1} *versus* 105 kJ mol^{-1}). In cyclic adsorption breakthrough experiments, dptz-$CuTiF_6$ achieves complete CO_2 desorption at 298 K under inert gas purging. Single-crystal X-ray diffraction studies demonstrate that the exceptional CO_2 adsorption capacity, moderate CO_2 heat of adsorption, and high CO_2/N_2 selectivity are due to the optimal packing of the CO_2 molecules within the MOF, as well as the favorable thermodynamics and kinetics from cooperative host-guest interactions.

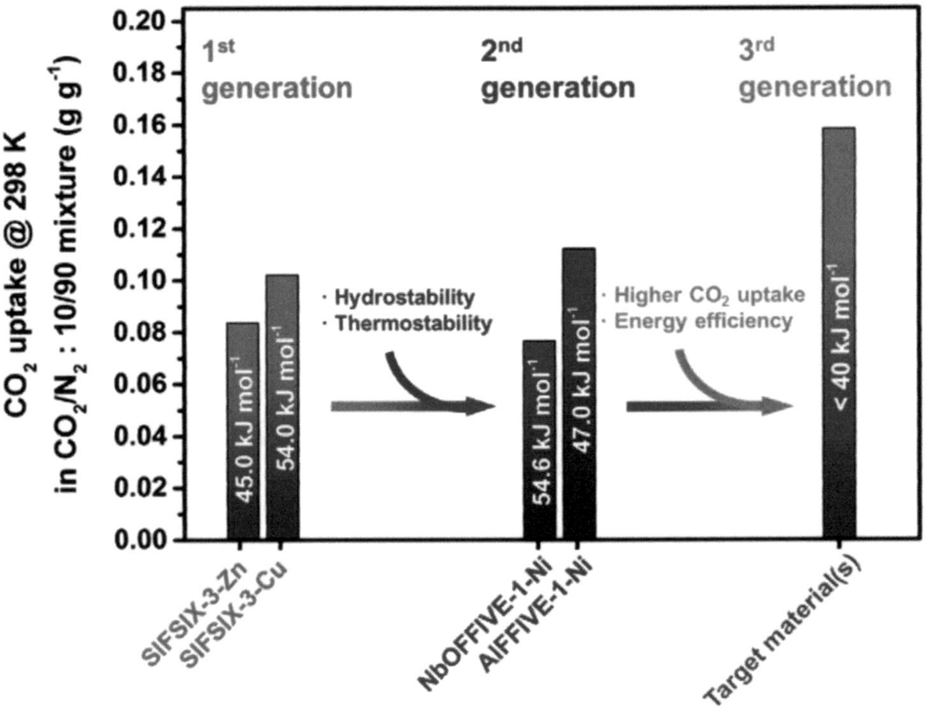

Figure 1. Evolution of fluorinated, microporous metal-organic frameworks (MOFs) as CO_2 adsorbents for flue gas treatment, showing the CO_2 adsorption capacity (g g^{-1}) at 298 K and the CO_2 isosteric heat of adsorption (Q_{st}, kJ mol^{-1}) for the first generation (**SIFSIX**-3-Zn and **SIFSIX**-3-Cu) and second generation materials (**NbOFFIVE**-1-Ni and **AlFFIVE**-1-Ni), as well as the envisioned features of the third generation materials.

527387 Impregnation of Hydrotalcite with $NaNO_3$ for Enhancement of CO_2 Sorption Uptake

Abstract:

It is well known that carbon dioxide (CO_2) emitted from fossil fuel-based energy system has caused global warming. In order to solve this environmental issue, technologies related to hydrogen production as an alternative energy carrier have been suggested. Sorption-enhanced water gas shift (SE-WGS) reaction is one of attractive methods for mass production of hydrogen. In the SE-WGS process, high-temperature CO_2 sorbent is applied and facilitates the production of high-purity hydrogen since CO_2 is a by-product of WGS reaction. Therefore, CO_2 sorbent with high capacity should be developed to improve the efficiency of SE-WGS reaction. Layered double hydroxide, also called as hydrotalcite, has been considered a promising sorbent for CO_2 capture at high-temperature range (200-500 °C). Hydrotalcite typically has fast sorption and desorption kinetics, good thermal stability, and ease of regeneration by pressure or temperature swing adsorption method. However, its low CO_2 sorption capacity should be improved for practical application. Alkali-metal impregnation using K_2CO_3 or Na_2CO_3 is a representative method to increase CO_2 sorption capacity of hydrotalcite. In this study, $NaNO_3$ was chosen for an alkali-metal precursor and impregnated into hydrotalcite through an incipient wetness method. The structural and textural properties were analyzed using X-ray diffraction and N_2 adsorption analysis, and the effect of $NaNO_3$ impregnation on CO_2 sorption behavior at high temperature (200-400 °C) was investigated using a thermogravimetric analyzer and in-situ Fourier transformed infrared spectroscopy.

533459 Preparation and Its Excellent $CO_2/CH_4/N_2$ Adsorption Selectivity of Novel Carbon Composites CPDA@A-ACs

Abstract:

Preparation and Its Excellent $CO_2/CH_4/N_2$ Adsorption Selectivity of Novel Carbon Composites CPDA@A-ACs

Wanwen Liang, Huiyu Xiao, Zhong Li*

School of Chemistry and Chemical Engineering, South China University of Technology, Guangzhou 510640, PR China

*Corresponding author e-mail address: cezhli@scut.edu.cn

Abstract

In this work, novel composites (CPDA@A-ACs) of carbonized polydopamine (CPDA) and asphalt-based activated carbons (A-ACs) were successfully synthesized. The resulting CPDA@A-ACs were characterized, and the $CO_2/CH_4/N_2$ adsorption isotherms of CPDA@A-ACs were measured separately. The isosteric heats of $CO_2/CH_4/N_2$ adsorption on the samples were estimated. The dynamic separation performances of the samples for CO_2/N_2 and CO_2/CH_4 binary mixtures were evaluated in fixed-bed breakthrough experiments. Results showed that BET area and total pore volume of CPDA@A-ACs were up to 1971 m^2/g and 0.79 cm^3/g, respectively. The N contents and the concentration of basic sites on CPDA@A-ACs are higher than those on parent A-AC due to the introduction of CPDA. CPDA@A-ACs possessed not only excellent stability, but also excellent CO_2 adsorption capacity up to 4.05 mmol/g and 6.89 mmol/g at 298 K and 273 K, respectively, at ambient pressure, much higher than the parent A-AC, most of the reported porous carbon materials and many reported metal-organic frameworks (MOFs) for CO_2 capture under identical conditions. More importantly, CPDA@A-ACs possessed excellent adsorption selectivity toward CO_2/N_2 and CO_2/CH_4 binary mixtures. Its IAST-predicted selectivity at normal pressure and temperature reached as high as 25 for CO_2/N_2 (0.15/0.85) mixture and 5 for CO_2/CH_4 (0.5/0.5) mixture separately. Fixed-bed experiments showed that the CO_2/N_2 (0.15/0.85) mixture and CO_2/CH_4 (0.5/0.5) mixture can be well separated at room temperature, suggesting great potential for industrial CO_2/N_2 and CO_2/CH_4 separation.

Keywords: Asphalt, Polydopamine, CO_2 adsorption, $CO_2/CH_4/N_2$ selectivity

539766 High Speed Production of Graphene Oxide Membranes and Their Potential Usage in Harsh Environments

Abstract:

The shape anisotropy and enhanced solubility of graphene oxide (GO) in a wide variety of solvents provides the perfect environment to showcase isotropic-to-nematic colloidal phase transitions. As a result, the fluid phase of GO starts to demonstrate properties such as higher viscosity and elasticity compared to the isotropic phase. Membranes of graphene oxide have primarily been fabricated from isotropic phase of GO by processes such as vacuum filtration; which does not attend to manufacturing speeds required by industries. Our research program endeavours to address this issue given the large possibilities in applications ranging from permeable filtration membranes, strategically constructed battery separators, strain sensors, lab-on-chip devices, and supercapacitors.

We have over the years developed quantitative polarized light imaging techniques to quantify fine structure, alignment, texture of the liquid crystalline phases and thin films of GO [1,2]; which have helped us develop processing-property correlations[3]. We have also shown that large-area GO (13 x 14 cm^2) filtration membranes can be produced in < 5 seconds using a high speed gravure printer which have demonstrated molecular sieving properties and suitability in membrane separation processes.[4] The GO membranes enable permeation, maintaining structural stability in a wide variety of polar protic, polar aprotic, and non-polar solvents, such a wide palette of solvent usage is difficult to achieve with other membranes. Flux in the GO membranes scales inversely with the dielectric constant of the solvent (i.e. an interaction parameter), and also inversely with the viscosity as per continuum predictions, enabling us to develop predictive flux correlations[5]. The GO membranes can also be cleaned with bleach for continuous re-use in separation of tannic acid from water, and means to improve the chemical oxidation resistance are being investigated. The scalability and adaptability of our membrane fabrication technique is leading to commercial adaptation and laboratory-to-market translation of products such as nanofiltration membranes.

1 Tkacz, R. *et al.* pH dependent isotropic to nematic phase transitions in graphene oxide dispersions reveal droplet liquid crystalline phases. *Chemical Communications* **50**, 6668-6671, doi:10.1039/C4CC00970C (2014).

2 Tkacz, R., Oldenbourg, R., Fulcher, A., Miansari, M. & Majumder, M. Capillary-Force-Assisted Self-Assembly (CAS) of Highly Ordered and Anisotropic Graphene-Based Thin Films. *The Journal of Physical Chemistry C* **118**, 259-267, doi:10.1021/jp4080114 (2014).

3 Sheath, P. & Majumder, M. Flux accentuation and improved rejection in graphene-based filtration membranes produced by capillary-force-assisted self-assembly. *Philosophical Transactions of the Royal Society A: Mathematical, Physical and Engineering Sciences* **374**, doi:10.1098/rsta.2015.0028 (2016).

4 Akbari, A. *et al.* Large-area graphene-based nanofiltration membranes by shear alignment of discotic nematic liquid crystals of graphene oxide. *Nature Communications* **7**, 10891, doi:10.1038/ncomms10891 https://www.nature.com/articles/ncomms10891#supplementary-information (2016).

5 Akbari, A. *et al.* Solvent Transport Behavior of Shear Aligned Graphene Oxide Membranes and Implications in Organic Solvent Nanofiltration. *ACS Applied Materials & Interfaces* **10**, 2067-2074, doi:10.1021/acsami.7b11777 (2018).

527010 New Highly CO_2-Selective Amine-Based Membranes for Carbon Capture

Abstract:

CO_2 capture from flue gas in coal- and/or natural gas-fired power plants and from <1% CO_2 concentration sources, e.g., the residual flue gas after the primary CO_2 capture system and coal-mine gas streams, requires a high CO_2/N_2 selectivity of 140 along with a very high CO_2 permeance of about 700 GPU (1 GPU = 10^{-6} cm^3 (STP)/($cm^2 \cdot s \cdot$ cmHg)) or higher in order to use a stand-alone membrane process. For the CO_2 capture, we have synthesized new amine-containing membranes, showing a high CO_2/N_2 selectivity of >140 and a high CO_2 permeance of >1000 GPU. The membrane was scaled up to 14 inches in width using continuous roll-to-roll fabrication. Aided by a material balance equation, three variables, namely the coating-knife gap setting, web speed, and coating solution concentration, were identified as the critical factors to control the membrane selective layer thickness. This resulted in the membrane with a selective layer of <200 nm. The scale-up membrane was fabricated into spiral-wound membrane modules for testing with simulated flue gas containing about 20% CO_2, 77% N_2, 3% O_2 and 1 â€“ 3 ppm. Both the scale-up membrane and the membrane modules exhibited similar results as the flat-sheet membrane tested in the lab. The membrane module has been scaled up to a size of more than 10,000 cm^2 membrane area. The module showed stable CO_2 permeance and CO_2/N_2 selectivity results for 1,700 hours. Techno-economic analysis has shown that the post-combustion capture process using the membrane is promising for meeting DOEâ€™s capture cost target set for 2025. For carbon capture from <1% CO_2 concentration sources, we have elucidated the carrier saturation phenomenon. With reducing the CO_2 concentration in the feed gas, both CO_2 permeance and CO_2/N_2 selectivity increased. These were mainly due to more available free amine carriers for CO_2 molecular transport at lower CO_2 concentration conditions.

531017 Membranes for Charge- and Aromaticity-Based Separation of Small Molecules

Abstract:

Today, there are no commercial membranes that can separate small molecules of similar size in the liquid phase based on their chemical properties. Such membranes would transform chemical manufacturing and significantly reduce energy costs associated with chemical separations. In this study, we aim to prepare synthetic polymer membranes that mimic two key features of biological pores such as ion channels and porins: Constricted pores only 1-5 nm in diameter that confine permeation, lined with functional groups lining the pore that interact with the target during passage. As an initial system, we focused on membranes capable of charge-based separation through electrostatic interactions. To build these nanostructured layers using scalable techniques, we deposited packed arrays of polymer micelles whose coronas exhibit carboxylic acid groups onto a porous support membrane. Random copolymers that combine highly hydrophobic fluorinated repeat units of trifluoroethyl methacrylate (TFEMA) with ionizable repeat units of methacrylic acid (MAA) form micelles and vesicles in methanol. When these micelles are coated onto the surface of a porous support membrane whose pores are smaller than the micelles and then immersed into water, a selective layer of micelles packed together is formed. The gaps between the micelles act as nanochannels functionalized with carboxylic acid groups. The membrane showed charge-based selectivity between organic molecules, effectively rejecting negatively charged solutes while allowing the passage of neutral solutes in two-solute separation tests. Interestingly, selectivity was enhanced when a mixture of solutes was used, likely due to competition between solutes to enter the very narrow pores.

To expand the types of separations these membranes can address, the carboxyl groups can be post-functionalized. This can enable the customization of these membranes membrane for desired separations. For instance, pi-pi interactions between aromatic groups can enable us to create membranes that preferentially interact with aromatic solutes. Such functionalization has enabled us to impart significant diffusion selectivity of ~10 between two hormones of identical size and charge, differentiated by the presence of an aromatic ring. This demonstrates the potential of using polymer self-assembly and functionality to design membranes that mimic biological pores while maintaining scalable manufacturing methods. We believe these approaches will eventually lead to novel membranes that can separate molecules of similar size but different chemical structure.

531276 Metal Organic Brush (MOB) Membranes for Organic Solvent Nanofiltration

Abstract:

Performance and characterization of a new class of synthetic membranes are presented. These membranes are based on a brush-like active layer cross-linked with copper ions. These "metal organic brush" membranes, called MOBs, are the rate limiting barrier to species transport and endow the membrane with selectivity. The semi-crystalline nature of these composite membranes are hypothesized to reduce membrane pore-size distribution and increase selectivity between molecules very close in molecular weight. Polyimide ultrafiltration support membranes were crosslinked with diamines and grafted with an ATRP initiator. A step towards "structure by design" membranes becomes possible through use of controlled radical polymerization techniques, such as ATRP. The membranes were tested for their transport properties (selectivity and permeability) of several dyes in the nanofiltration range. Several techniques, including Attenuated Total Reflectance – Fourier Transform Infrared Spectroscopy (ATR-FTIR), X-Ray diffraction (XRD), X-ray Photoelectron Spectroscopy (XPS), and Scanning Electron Microscopy (SEM) were used to characterize the membranes.

536742 Pi Electron Cloud Mediated Separation of Aromatics Using Supported Ionic Liquid (SIL) Membrane Having Antibacterial Activity

Abstract:

Supported ionic liquid (SIL) membranes using imidazolium based ionic liquids and hydrophobic PTFE base membrane were used for the π electron mediated separation of aromatic compounds including dyes in aqueous and non-aqueous medium. The FTIR of SIL membranes showed the signature of the functionalities for both base membrane and the imidazolium moiety. A significant reduction in contact angle, especially the dynamic contact angle, was observed for SIL membranes compared to the virgin membrane indicating the enhancement in surface hydrophilicity of the membrane, while considerable changes zeta potential was seen at different pH indicating the modification in surface charge. The SEM image confirmed the pore filling of base membrane with ionic liquids. The stability of the SIL membranes were studied using hexane, water and DMF as solvents. Tuning the anionic moieties, ionic liquids were utilized for aqueous and non-aqueous application through SIL membranes. For non-aqueous applications, the trend in separation efficiency was found to be divinylbenzene > styrene > toluene, which was attributed to the more π electron cloud density on extended conjugation leading to favorable interaction with ionic liquid influencing its transfer through SIL membranes. Preferential separation of Congo red (CR) and Ramazol Brilliant Blue R (RBBR) was achieved in water based application of the SIL membranes.

The ionic liquids were having affinity for the π electron cloud. This π electron cloud density for toluene is lower than styrene followed by divinyl benzene, due to the presence of additional double bonds in conjugations with the aromatic π electron cloud. Due to this extended conjugation, divinyl benzene interacted strongly with the ionic liquid sitting inside pore. This might be responsible for bringing the selectivity in the separation for the SIL membranes. The lower transport in case of allyl ionic liquid might be attributed to the presence of electron withdrawing sp^2 hybridized carbon of the ally group resulting reduction in the electron density over imidazolium ring. On the other hand, for choloro and bromo ionic liquids, butyl and hexyl groups were present in the side chain of the imidazolium ring. The (+) inductive effect of the alkyl group containing sp^3 hybridized carbon atom pushed the electron density towards imidazolium ring. More electron cloud on the imidazolium ring, it will be more favorable for the π electron interaction resulting enhancement in the separation efficiency.

Similar investigation was also carried out to understand the selectivity in separation of different dyes using SIL membranes with water insoluble ionic liquids. In this case three different dyes CR, EBT and RBBR were used. The SIL membranes were found to be more selective for CR. In case of hexyl imidazolium ionic liquid, almost 60 % CR was transferred to the receiver side within one and half hour, while for allyl imidazolium ionic liquid % transport of CR reached around 50 % after two hours. For the other dyes RBBR and EBT, the % transport through both the SIL membranes were found to be less than 5 %. Though all these dyes have aromatic moieties, their complicated structures with different functionalities and stereo-chemical arrangement might lead to different overall electron cloud density on them resulting different selectivity through SIL membranes.

The stability of the supported liquid membranes is of major concerns. Therefore, it is interesting to investigate the stability of the SIL membranes. In view of this SIL membranes were allowed to be equilibrated in presence of water, DMF and hexane and at different time interval the conductivity of the medium were measured. For SIL membranes with water soluble ionic liquid, the conductivity of the medium was found to be modified in presence of water and DMF. In presence of water, the water soluble ionic liquids came out of the membrane pore resulting changes in water conductivity. DMF being polar in nature, certainly interacts with the ionic liquid moieties in the membrane due to electrostatic interaction. As result, ionic liquid might get dissolved in DMF. This also led to changes in conductivity in DMF, while in case of hexane, upto 60 days, there is no change in the conductivity of the medium indicating the stability of the SIL membranes. Using these equilibrated membranes (30 days' equilibration), the separation of toluene, styrene and divinyl benzenes were carried out. There modification in separation efficiency were found to be within 10 – 15 %. These investigations revealed the high stability of SIL membranes with water soluble ionic liquid in hexane medium.

The anti-microbial activity of these SIL membranes were investigated using gram-positive bacteria (**staphylococcus aureus** and **pseudomonas aeruginosa)**. More the number of carbon atom on the side chain of the imidazolium ring, more antimicrobial activity was observed. For *staphylococcus aureus*, the L/D ratio for the virgin membrane was kept 52. Over the time frame of 24 hours, the virgin membrane did not show any appreciable modification in the ratio, though some fluctuation is noticed. In case of SIL membrane a drastic decrease in L/D values were observed for all the ionic liquids upto a definite time (3 hours for bromohexane and chlorobutane ionic liquids; 5 hours for allylbromide ionic liquid), followed by almost constant values for L/D upto 24 hours. This investigating clearly demonstrates the antibacterial activity of SIL membranes compared to the virgin membrane. The antibacterial activity of allylbromide SIL membrane was found to be slightly lower than that of chlorobutane or bromohexane SIL membranes. The end point was taken as 24 hours, because beyond that, there might be some appreciable modification in the L/D ratio for the original bacteria due to their life cycle.

Similar study was also carried out with another gram positive bacteria, *pseudomonas aeruginosa* starting with L/D ratio ~ 56 by monitoring the L/D ratio using luminescent spectra for 24 hours with a regular time interval. The trends were found to be similar to *staphylococcus aureus*. Though, this method is very convincing for signaturing the antibacterial activity, still it has some limitations. The bacteria having compromised membranes may be able to recover and reproduce but this method identify them as 'dead' bacteria. Similarly, some bacteria may have intact cell membrane but unable to reproduce in nutrient medium, are assigned as 'live' bacteria. Moreover, since imidazolium moiety is also able to give fluorescence, it may interfere in the determination of fluorescent intensity of the dyes indicating live and dead bacteria. Therefore, additional non fluorescent based technique should also be used to confirm the antimicrobial activity of SIL membranes.

The quantitative antibacterial analysis was carried out in terms of colony forming units per mL for bacteria. Large CFU/mL values, 2.82 E+09 and 3.5 E+09 were taken as the starting point for *staphylococcus aureus* and *pseudomonas aeruginosa*, respectively. Within 24 hours, there was almost no change in the CFU/mL values for both the bacteria for virgin membrane, while significant reduction in CFU values were observed for SIL membranes. After 5 hours of contact with the membranes the CFU/mL values become 2.82E+09, 4.4E+06, 1.10E+03 and 1.2E+04 for virgin membrane, SIL membranes with allylbromide, bromohexane and chlorobutane ionic liquids, respectively; whereas after 24 hours CFU/mL became 2.80 E+09, 9.01E+04, 2.10E+0 and 2.04E+01. The investigation clearly demonstrates, not only the antibacterial activity of SIL membranes, but also the trend of allylbromide < cholorobutane < bromohexane. This trend was found to be similar than that of earlier investigation. More the number of the carbon atom on the side chain of the imidazolium ring more pronounced antibacterial activity was observed. The quantitative antibacterial analysis was carried out in terms of colony forming units per mL for bacteria. Large CFU/mL values, 2.82 E+09 and 3.5 E+09 were taken as the starting point for *staphylococcus aureus* and *pseudomonas aeruginosa*, respectively. Within 24 hours, there was almost no change in the CFU/mL values for both the bacteria for virgin membrane, while significant reduction in CFU values were observed for SIL membranes.

After 5 hours of contact with the membranes the CFU/mL values become 2.82E+09, 4.4E+06, 1.10E+03 and 1.2E+04 for virgin membrane, SIL membranes with allylbromide, bromohexane and chlorobutane ionic liquids, respectively; whereas after 24 hours CFU/mL became 2.80 E+09, 9.01E+04, 2.10E+0 and 2.04E+01. The investigation clearly demonstrates, not only the antibacterial activity of SIL membranes, but also the trend of allylbromide < cholorobutane < bromohexane. This trend was found to be similar than that of earlier investigation.

538847 Removal and Recovery of Ammonia from Dilute Aqueous Process/Effluent Streams

Abstract:

There are many effluent/process streams containing a significant amount of ammonia. Current processes to remove and recover this ammonia involve packed tower stripping of this ammonia into air and its reabsorption in a second packed tower into say, a sulfuric acid stream to produce for example, ammonium sulfate. An improved compact membrane-based process has become commercial for some time without using packed towers. In such a process, microporous gas-filled hydrophobic hollow fiber membrane-based modules are used to strip ammonia from the aqueous solution on one side of the hollow fiber membrane and then reabsorb it into a sulfuric acid solution on the other side of the membrane. There is no current published model to describe the performance of such a process in a large module and thereby design such modules. We will describe our activities and results to achieve these goals. In order to carry out such a process, usually the feed solution pH is raised significantly by caustic addition. Alternative membrane-based processes will also be described in this presentation.

539941 Polymeric Microparticles as a New Platform for High Removal of Lipopolysaccharides

Abstract:

Lipopolysaccharides (LPS) appear to be a major drawback for the production of biomolecules that are essential for research, pharmaceuticals and industrial applications. Although there are a variety of methods for removing LPS, the question about how LPS removal can be carried out in an economical way is the most intriguing question without a satisfactory solution yet. High capacity per unit volume of adsorbents is a consideration for LPS adsorption. Here we demonstrate solid polymeric poly-e-caprolactone (PCL) microparticles (MPs) (" = 25.5 ± 10) as LPS binding sites by adsorbing these molecules on the surface of MPs. The PCL MPs removed ~82% LPS from water using only 1 milligram (mg) of MPs which is equivalent to ~2.22 × 10. endotoxin units (EU) per mg of particles. The LPS removal efficacy was relatively higher (~98%) in presence of phosphate buffered saline (PBS; 150 mM NaCl) than in water. An even more effective way was found to be the combination of MPs with a supporting layer such as a cellulose membrane that speeds up the LPS sorption process up to ~100% just by running the LPS containing water flow through it under gravity conditions. The water fluxes through the membrane were measured approximately 25 and 15 L/m2.h without and with MPs, respectively by gravity flow. A low ratio (~4.6) between the membrane thickness (~116 µm) and particle diameter (~25 µm) allowed convective solute transport that essentially enhances LPS adsorption within short residence times. The capacity per mg of MPs in the membrane was ~2.78 × 10. EU which was comparable to the adsorption capacity of MPs alone. The results indicate that PCL MPs have potential applications as LPS adsorbers in powder and membrane forms both in terms of high binding capacity as well as mass transport characteristics.

534669 Lignocellulosic Materials for Nanofiltration Membrane Synthesis

Abstract:

Lignocellulosic materials are readily available from renewable sources, and possess many interesting properties for membrane development, such as solvent resistance any hydrophilic functional groups. This work looks at two approaches for integrating cellulose and lignin materials into high performance membranes.

In the first approach, cellulose nanocomposite membranes capable of selectively separating small organic molecules were created using 1-ethyl-3-methylimidazolium acetate as a solvent. Membrane morphology has been thoroughly studied using dual beam FIB/STEM and confocal microscopy. Cellulose nanocomposite membranes have been shown to reject up to 80% methylene blue (MW 319.8 Da) while maintaining a water permeability of 10 LMH/bar. Cellulose composite membranes exhibited no swelling or deformation during permeation of isoproponal solvent, which suggests strong solvent resistance. Recovery of lignin trimers and tetramers was studied using an isopropanol solvent. High-throughput concentration of these plant derived molecules can significantly increase the value of processing lignin for chemicals.

The second approach involves functionalizing lignin sulfonate, a common byproduct of paper production, onto the surface of the DOW-Filmtec NF-270. Lignin sulfonate is baked onto the NF-270 surface for 30 minutes at 90°C. . The pure water permeability of the functionalized membrane was 14.8 LMH/bar, only slightly below the 16.3 LMH/bar measured for the commercial membrane. The LiS functionalized membrane rejected 97.1% of 1000 mg/L Na2SO4 solution compared to 98% for the unmodified NF270 membrane. After fouling with 100 mg/L BSA, 5 minutes of tangential water washing (12.2 cm/s) of LiS membranes lead to flux recovery of 100% (of the initial flux) after the first fouling cycle and 90% after the second fouling cycle. The un-functionalized commercial NF-270 membrane flux declined below 40% of the initial flux after BSA fouling and never recovered after the same water washing.

536540 Purification of Scfv Via Small Molecule Based Affinity Membrane Spin Column

Abstract:

Recently, single-chain variable fragments (scFv) have emerged as an interesting alternative to antibodies for diagnostic and therapeutic purposes. They consist of variable regions of the heavy and light chains of antibodies joined by a flexible peptide linker. Owing to its significantly smaller size (~25 kDa vs ~150kDa for IgG antibodies) and lack of post-translational modifications they can be easily expressed in bacterial systems. Additionally, lack of an Fc region makes them less immunogenic while potentially having higher tissue penetration capacity than bigger antibody fragments and retaining antigen-binding capacity and high specificity.

However, several challenges in terms of conjugation and purification limit the potential of these molecules in academia and potential commercialization. Because they lack an Fc region, common purification methods such as Protein-A/G affinity-based chromatography are not useful. Instead, scFv are typically expressed as HIS-tagged proteins which can then purified through column chromatography with Ni^{+2}-embedded resins. While this purification method yields relatively high purity products (typically >80%), it's relatively expensive and requires hours of processing for purification. Presence of other histidine-containing proteins further affects the purity of these products. Furthermore, although regeneration and re-use is possible, it's limited and requires extensive washing steps with a variety of solvents, often including highly acidic or basic solutions.

As an alternative, we've developed a novel affinity spin column capable of purifying scFv by targeting the Nucleotide Binding Site (NBS), initially identified in the variable domain of Fab and conserved in scFv. For this purpose, a small ring structured molecule with moderate affinity for the NBS was covalently conjugated to regenerated cellulose membranes, which allows scFv to be captured by the membranes. Our process allows sample purification in short times (<10 minutes) without the requirement of His-Tag or any post-purification processing while maintaining scFv activity. Multiple washing steps can be employed as required for further removal of contaminants without loss of sample and recovery can be achieved in a single elution step under mild conditions (0.5 M KCl in 20 mM TRIS pH 7.0). Here we present tests performed using several different scFv, consistently yielding >95% purity from pure or contaminated samples, including cell lysate. We study the effects of protein contaminants, amount of membranes in the column and column reusability as well as the binding activity of purified scFv. Importantly, our columns are reusable without the requirement of extensive washing or very high or low pH solutions. Additionally, compatibility of the system with resin-based separations and potential scale-up are evaluated. Overall, our system provides a simple, economic and fast tool for small-scale scFv purification without sacrificing purity or activity of the product.

535958 New ProteinÂ Î‘ and Multimodal Anion-Exchange Membranes for the Rapid Isolation and Purification of Biologics

Abstract:

Numerous campaigns are underway to develop biologics for many chronic conditions, but the process from biologics discovery to a commercial medicine is long and expensive. Thus, a key challenge in the industry is to translate biologics from research labs to FDA approved medications more efficiently and economically. Improving purification productivity by increasing the speed and capacity of capture step chromatography columns and achieving process compression by conducting polishing step purification without buffer exchange or dilution steps will play key roles in meeting this challenge.

In this study, we present performance data for Purilogicsâ€™ novel Protein A and multimodal anion-exchange membranes and compare results with advanced commercial products. We demonstrate that the Protein A membrane achieves high antibody dynamic binding capacity of >50 mg/mL at 0.6 second residence time and high impurity clearance. We further demonstrate that the multimodal anion-exchange membrane achieves high dynamic binding capacities for protein and DNA at high conductivity conditions and at high flow rates. Under conditions of 13 mS/cm conductivity in tris-base buffer, the new multimodal anion-exchange membrane offers more than twice the protein dynamic binding capacity of two leading commercial membrane products at one-half the residence time. In PBS buffer, it offers >14 times higher dynamic protein binding capacity, and >6 times higher DNA binding capacity. These new membrane columns dramatically increase isolation and purification speeds, which is needed to move biologics from discovery to market faster.

535934 Virus Filtration: Effect of Protein Fouling and Transmembrane Pressure

Abstract:

Virus filtration is extensively used in the downstream purification process for the production of protein-based biotherapeutics. Commercial virus filters can ensure robust viral clearance by removing more than 99.9% of viruses while providing nearly complete protein transmission. However the size of most therapeutic products, e.g., monoclonal antibodies, are only slightly smaller than the 20 nm parvoviruses, leading to a significant amount of protein fouling. The overall objective of this study was to obtain more fundamental insights into the fouling behavior and its effect on virus capture by direct visualization of fouled membranes using both confocal and electron microscopy.

Filtration experiments were conducted using a single-layer, highly asymmetric ViresolveÂ® Pro membrane with the skin-side facing away from the feed (as per the manufacturerâ€™s recommendation). Filters were challenged with fluorescently-labeled IgG and imaged by confocal laser scanning microscopy to determine the location of retained protein within the depth of the filter. Parallel experiments were performed by filtering a suspension of gold nanoparticles of different sizes through protein-fouled filters, with the membrane cross-section imaged using scanning electron microscopy to identify the location of individual captured nanoparticles.

IgG filtration leads to a significant decline in flux, with the extent of fouling being greatest at low transmembrane pressure. Confocal images of protein-fouled membranes show the presence of fluorescently-labeled IgG in a narrow band located immediately upstream of the filter exit, with greater fluorescent intensity observed at low transmembrane pressure (consistent with the increased flux decline). SEM images obtained after filtration of gold nanoparticles show a band of 20 nm particles located approximately 1 µm upstream of the filter exit. Protein fouling caused the location of the captured 20 nm gold particles to shift further away from the filter exit, likely due to constriction of the small pores in the Viresolve® Pro membrane. These results provide important insights into the factors controlling the performance of virus filters and the mechanisms governing protein fouling.

530861 Process Optimization of Constant Flow Series Filtration of Bioprocess Fluids

Abstract:

Normal flow microfiltration and nanofiltration of biological fluids are typically conducted under constant pressure or constant flow conditions. Often, a prefilter is installed upstream of the final, sterilizing-grade filter to increase the final filter lifetime and thereby improving filtration economics. There have been previous treatments on optimization of prefilter and final filter configurations, but primarily for the case of constant pressure operation. This work focused on process optimization for constant flow operation, which involves a different set of considerations compared to constant pressure operation.

A fouling model that accounts for both particle blocking and particle adsorption mechanisms was applied to a new class of prefilters with polyethersulfone membrane that serve to protect microfiltration and nanofiltration final filters. It was demonstrated that a fit of the constant flow filtration data at one prefilter to final filter area ratio could be used to predict the constant flow filtration performance at any other area ratio. The model can be used to determine economically optimal process designs including optimal process flow rate and optimal prefilter to final filter area ratio. A practical test methodology that can be used to rapidly and efficiently derive the model parameters that are needed for the optimization calculations will be elucidated.

536538 Understanding the Unique Sorption of Alkane-Î±,Ï‰-Diols in All-Silica Zeolites

Abstract:

Adsorption equilibria of alkane-Î±,Ï‰-diols (diols) from aqueous solution onto all-silica MFI-type zeolite (silicalite-1) with three to six carbons are obtained by simulations and experiments. After an initial slow rise, the diol isotherms (which show agreement between both approaches) exhibit steep changes in loading and eventually reach 7 to 10 molecules per unit cell (molec/uc). Thermodynamic and structural information, obtained from the simulations, allow for a mechanistic understanding of these adsorption equilibria. The step in the isotherms is caused by cooperative hydrogen-bonding that leads to a minimum in the free energy of transfer for the diol from vapor to zeolite. At $T = 323$ and 348 K, the steep change for the pentane-1,5-diol (C5) adsorption isotherm is found to be a phase transition from a low-dense phase to an ordered solid phase with a loading of 8 molec/uc. At saturation loading, nearly all C5 and C6 diols are involved in a percolating hydrogen-bonding network. With increasing temperature, the C5 isotherm becomes less steep, as the minimum in the free energy of adsorption decreases in depth while the relative population of hydrogen bonds decreases. However, the isotherm and adsorption onset do not shift significantly to higher concentration because the magnitudes of the free energies of solvation and adsorption decrease by similar and, hence, compensating amounts. The sensitivity of the calculated selectivities for C5 diol over water to the rigid MFI framework structure adopted in the simulations is investigated by randomly perturbing the framework atoms or cell parameters of experimentally determined structures. The steep change in the isotherm of C5 diol from aqueous solution is also found to occur on many other all-silica zeolites in the database available from the Structure Commission of the International Zeolite Association (http://www.iza-structure.org/databases/).

525207 A New Approach to Predict Adsorption in Metal-Organic Frameworks with Unsaturated Metal Sites

Abstract:

Recent years have seen an exponential increase in the number of computational studies aiming to predict adsorption in MOFs. A particularly promising trend is the use of large-scale computational screening of MOFs for adsorption applications, so as to identify the most promising candidates for further experimental study among a wide pool of available materials. These approaches have tremendous potential to reduce the time and money currently spent on experimental testing of MOFs, fast-tracking their translation into commercial applications. However, a crucial but often overlooked fact is that conventional molecular models for adsorption in MOFs neglect important physics of the process, and thus can lead to spectacularly inaccurate predictions. This is particularly the case for MOFs that possess coordinatively unstaturated sites (CUS), or open metal sites â€“ the specific nature of the coordination bonds formed by these sites with particular adsorbate molecules is not captured by standard force fields. The impact of neglecting these interactions can be huge, particularly considering that MOFs that contain CUS are among the most promising materials for gas storage and separation applications, precisely due to these strong and selective binding sites [1].

Our communication presents a generally applicable and transferable approach for incorporating specific coordination-type interactions between CUS and adsorbate molecules into standard molecular models, based on a combination of quantum mechanical calculations with classical Monte Carlo simulations. Crucially, our approach makes use of only a small number of QM calculations in the vicinity of the CUS, thus keeping computational expense low, while making use of conventional classical approaches to describe interactions away from the CUS [1, 2]. We demonstrate generality by applying our model to describe adsorption of both polar and non-polar adsorbates (paraffins, olefins, nitrogen and carbon monoxide) [3], and demonstrate transferability by predicting adsorption in different MOFs with the same type of metal unit (copper paddlewheel) [4]. In all cases, excellent agreement with experimental adsorption isotherms is obtained. This means that the procedure is amenable for incorporation into large-scale screening efforts, and has the potential to greatly improve the accuracy of computational screening of MOFs for adsorption applications. Overall, our new model provides detailed insight into the molecular level adsorption mechanisms on MOFs with CUS, and constitutes a useful tool to design new materials for challenging separations.

[1] Fischer, M.; Gomes, J. R. B.; Jorge, M. *Mol. Simul.* 40 (**2014**) 537-556.

[2] Fischer, M.; Gomes, J. R. B.; Froba, M.; Jorge, M. *Langmuir* 28 (**2012**) 8537-8549.

[3] Campbell, C.; Gomes, J. R. B.; Fischer, M.; Jorge, M. *submitted for publication.*

[4] Campbell, C.; Ferreiro-Rangel, C. A.; Fischer, M.; Gomes, J. R. B.; Jorge, M. *J. Phys. Chem. C* 121 (**2017**) 441â€“458.

520061 The Effect of Intrinsic Framework Flexibility on Adsorption Properties in Metal-Organic Frameworks: A Computational Exploration

Abstract:

Metal-organic frameworks (MOFs) have shown great potential for applications such as gas storage and adsorptive separation. In these applications, flexibility in these nanoporous materials plays a critical role to tune material performance. In this work, we use computational methods to explore the effect of adsorption induced framework flexibility on its adsorption properties. For the study, we selected ~300 MOFs from CoreMOFs database covering a wide range of pore diameters, surface areas and type of linkers. To obtain systematic insights into the adsorption induced flexibility on frameworkâ€™s properties, we develop a methodology combining ab-initio and molecular simulations that captures the framework dynamics at high loadings of adsorbates to predict selectivity. We have applied these methods to a range of adsorbates of industrially relevant mixtures for separations, which includes CH_4, CO_2, ethane, ethene, propane, propene, butane, ethylbenzene and xylene isomers, in selected MOFs. A detailed comparison study of adsorption properties was done with and without accounting the flexibility in ~300 MOFs for 13 adsorbates divided into 5 mixtures. We conclude that flexible simulations can be critical to correctly model adsorption in MOFs.

530254 Quasi-2D Phase Transition of Methane Adsorbed in Cylindrical Silica Mesopores

Abstract:

Recently, experimental small-angle neutron scattering (SANS) experiments identified a possible phase transition of methane adsorbed in MCM-41 and SBA-15, two cylindrical silica mesoporous adsorbents, indicated by a unexpected increased in the adsorbed fluid density at liquid-like conditions [1]. To better understand this observed phenomena, we used Monte Carlo and molecular dynamics simulations to examine the adsorption of methane in cylindrical silica mesopores of varying size and at varying temperatures and pressures. Our initial simulation results identified a roughly 10 % increase in the density of the liquidlike adsorbed phase for either an isotherm with increasing pressure or an isobar with decreasing temperature and that this densification is associated with a local maximum in the isosteric enthalpy of adsorption. Subsequent analysis of the simulated fluid, via computation of bond-orientational order parameters of specific annular layers of the adsorbed fluid, showed that the layers undergo an ordering transition from a disordered, amorphous state to one with two-dimensional hexagonal structure. Furthermore, this two-dimensional restructuring of the fluid occurs at the same thermodynamic state points as the aforementioned densification and local maximum in the isosteric enthalpy of adsorption. We thus conclude that the densification of the fluid is the result of structural reorganization, which is signaled by the maximum in the isosteric enthalpy. Owing to the qualitative similarity of the structural transitions in the simulated and experimental methane fluids, we propose this hexagonal reorganization as a plausible explanation of the densification observed in SANS measurements [2]. Lastly, we discuss why this structural transition has never been previously observed and how it may impact the transport properties of the adsorbed fluid.

[1] W-S. Chiang et al., *Langmuir*, 32:8849 (2016).
[2] D.W. Siderius, W.P. Krekelberg, W-S. Chiang, V.K. Shen, and Y. Liu, *Langmuir*, 33:14252 (2017).

533428 Molecular Model Development with Reliable Charge Distributions for Gaseous Adsorption in Nanoporous Materials

Abstract:

Nanoporous adsorbent materials possess significant potential in gas separation and storage. Advances in their development have remained challenging owing to difficulties in identifying promising ones among a myriad of possible candidates. Therefore, to effectively and efficiently facilitate the materials search, molecular simulations can play an important role. For a reliable employment of molecular simulations for predicting gaseous adsorption in nanoporous materials, an accurate description of the gas molecule of interest is critically important. To this end, we present herein a systematic and robust scheme for the model development of small gaseous molecules. In this method, all the model parameters are extensively and efficiently optimized. To ensure a reliable representation of the electrostatic potential (ESP) surrounding the molecule of interest, the number and location of pseudo-sites and the corresponding charge distribution are fitted to the ESP determined by *ab initio* density functional theory (DFT) calculations. Subsequently, the van der Waals interaction parameters are fitted to reproduce the experimental vapor-liquid equilibrium. This scheme ensures two critical features for the model development; (1) accuracy of the model by reproducing the DFT-computed ESP and (2) computational efficiency for the model development thanks to the decoupling of model parameters in optimization. While this systematic method should be applicable to general gas molecules, as a proof of concept, hydrogen sulfide (H_2S) molecular model is developed using this method, as its emission has serious impacts on the environment and humankindâ€™s health. Our results indicate that the H_2S models developed from this work can adequately present the ESP and provide accurate predictions of adsorptive properties in nanoporous materials, including all-silica zeolites and metal-organic frameworks. In addition to the adsorbed phase, we note that these models are also proven to reproduce the experimental bulk liquid properties such as radial distribution function reasonably well. We anticipate this method can largely facilitate the development of accurate molecular models, which in turn can expedite the computational discovery of optimum adsorbent materials for various applications.

526843 Computational Design of Electric Field Responsive Metal-Organic FrameworksÂ for Directional Flow

Abstract:

Stimuli responsive MOFs, which undergo physical or chemical changes under an external stimulus, such as temperature, guest adsorption, pressure, magnetic field, and light absorption, gained much attention because of their vital importance in design of new dynamic and smart materials for different applications Recent studies have shown that it is possible to use an external electric field (E-field) to induce structural changes in MOFs and these structural changes can be used for separation of gas mixtures[1-3]. In this study we propose and computationally demonstrate the concept of electric field controlled molecular gates mounted on the open-metal coordination sites in metalâ€"organic framework (MOF) materials. The MOF-molecular gate complex functions by opening and closing under the effect of an electric field. Our design involves Mg-MOF-74, a MOF with hexagonal channels with open-metal coordination sites at each corner, and a multifunctional gate molecule with permanent dipole which anchors itself on the host MOF material and responds to changes in the direction of an electric field by rotating around its backbone which acts as an axle. By carrying out density functional theory (DFT) calculations and molecular dynamics (MD) simulations we show that the MOF-molecular gate complex can be switched between two stable configurations, open and closed, by turning on and off an external electric field. We further show that the molecular gate can be controlled to block or allow the diffusion of methane molecules through the channels of the MOF like a nanoscale butterfly valve. We then extend the same concept to demonstrate how an E-field can be used to control and stabilize the rotation of functionalized ligands in IRMOFs in order to restrict the diffusion of molecules within the pores and regulate their direction of flow.

1. Tam, B.; Yazaydin, O., Design of Electric Field Controlled Molecular Gates Mounted on Metal-Organic Frameworks. *Journal of Materials Chemistry A* **2017**, *5*, 8690-8696.
2. Ghoufi, A.; Benhamed, K.; Boukli-Hacene, L.; Maurin, G., Electrically Induced Breathing of the Mil-53(Cr) Metalâ€"Organic Framework. *ACS Central Science* **2017**, *3*, 394-398.
3. Knebel, A.; Geppert, B.; Volgmann, K.; Kolokolov, D. I.; Stepanov, A. G.; Twiefel, J.; Heitjans, P.; Volkmer, D.; Caro, J., Defibrillation of Soft Porous Metal-Organic Frameworks with Electric Fields. *Science* **2017**, *358*, 347.

535423 Temperature Dependence of the Elastic Moduli of Confined Liquid Argon

Abstract:

The elastic modulus is a fundamental property of a fluid that is a measure the resistance of a fluid to being compressed due to an applied pressure. The knowledge of elastic moduli of a fluids confined in nanopores is needed for predictions of fluid flow and wave propagation in nanoporous media. The modulus of a confined fluid depends on a number of properties, such as the pressure in the pore, the size and geometry of the pore, and the surface properties of the pore walls [1-3].

Recent ultrasonic experiments have explored the temperature dependence of the elastic modulus and shown that the modulus of confined liquid argon changes linearly with temperature for a range of at least 10 degrees Kelvin [4]. Here we study the temperature dependence of the elastic modulus of confined liquid argon calculated from molecular simulations. Ultrasonic experiments typically operate under adiabatic conditions, which will provide the isentropic elastic modulus. One of the challenges is relating the experimentally determined isentropic modulus with the isothermal modulus, obtained by molecular simulations. Instead of directly simulating in isentropic conditions, we calculate the heat capacity ratio and isothermal modulus from simulations at a constant temperature, which then enables calculating the isentropic modulus.

We performed molecular simulations of argon in spherical nanopores with diameters between 3 and 8 nanometers, with temperatures between 76K and 87.3K. We operated in the grand canonical ensemble, which is typically used to simulate adsorption. Statistical mechanical relationships based on the fluctuations of the number of particles and potential energy were used determine the isothermal modulus, thermal expansion coefficient, and constant volume heat capacity. We used these values to calculate the constant pressure heat capacity using the generalized Mayerâ€™s equation. The ratio of heat capacities was used to calculate the isentropic modulus. We showed that the modulus is linear with temperature with a slope that is consistent for all simulated pore sizes and agrees well with the experimental data. We also explored how temperature affects the other thermodynamic properties for the confined fluid.

[1] G. Y. Gor, Langmuir, 30(45), 13564 (2014)

[2] G. Y. Gor, D. W. Siderius, V. K. Shen, N. Bernstein, J. Chem. Phys. 145(16), 164505 (2016)

[3] C. D. Dobrzanski, M. A. Maximov, G. Y. Gor, J. Chem. Phys. 148, 054503 (2018)

[4] K. Schappert & R. Pelster, J. Phys. Chem. C. 122, 5537 (2018)

530568 A Stochastic PBE-Model of Crystallisation Accounting for Nucleation, Growth, and Chemical Reaction

Abstract:

The stochastic nature of primary nucleation has been observed in several experimental studies, leading to the formation of various models to describe and interpret the data. Notwithstanding the progresses in understanding primary nucleation and the implications of its statistical nature, the current models cannot capture several of its features, particularly when multiple solid forms can develop in the system.

In this contribution, we have developed a Population Balance Equation model of crystallisation, which accounts for primary nucleation, secondary nucleation, and growth of the solid phase and a chemical reaction in the liquid phase. Each primary nucleation event occurs stochastically and, since growth and secondary nucleation are dependent on the primary events, the overall crystallisation process is itself stochastic.

The model structure is very flexible and allows to describe several types of systems. Here, we focus on three specific systems: first, the simplest scenario, where one single type of crystal can form; second, a system where an arbitrary number of polymorphs can form, each with its own set of kinetic and thermodynamic parameters; finally, a system with two enantiomers forming conglomerate crystals and coupled in the liquid phase by a racemisation reaction.

Using the model, we have studied the interplay of all phenomena acting in the system, their role in determining the detection times and the type of solid dominating, as well as the influence of the volume and the effect of scaling.

526768 Binding Coefficients during Condensation of Au and Mg at High Temperatures By Molecular Dynamics Simulations

Abstract:

Collisions between molecules (or atoms) take place in the early stages of particle formation (nucleation), and play an important role in environmental and industrial processes, like condensation and chemical reactions. Such collisions lead to the formation of small clusters but the impinging molecules are not always captured successfully by the cluster, especially if they are small. However, a sticking probability of unity is assumed typically, regardless of cluster size.

Here, the sticking probability during homogeneous condensation of atoms (e.g. Au and Mg) on small clusters is calculated from first principles using Molecular Dynamics (MD) simulations accounting for the detailed cluster structure and surface roughness. The effect of temperature, cluster size and orientation as well as atom velocity and collision angle on the sticking probability is investigated. Criteria for successful atom-cluster collisions are proposed based on the collision time. The sticking efficiency, defined as the ratio of successful collisions over all possible collisions, decreases with decreasing particle size.

The above method predicts accurately the sticking probability and can be used to correct the collision kernel of clusters. The revised coagulation rates can be employed in greatly simplified population balance equation models coupled with fluid dynamics to facilitate the design and operation of aerosol reactors as well as in aerosol dynamics models.

535216 Crystal Growth Kinetics of Stable and Metastable Polymorphs of Piracetam in Organic Solvents

Abstract:

Crystal growth kinetics of stable and metastable polymorphs of Piracetam in organic solvents

R. Soto[1], Å. Rasmuson*[1,2],

[1]*Synthesis and Solid State Pharmaceutical Centre (SSPC), Bernal Institute, Department of Chemical and Environmental Science, University of Limerick. Limerick V94 T9PX, Ireland.*

[2]*Department of Chemical Engineering and Technology, KTH Royal Institute of Technology, SE-100 44 Stockholm, Sweden*

**emails:* rodrigo.soto@ul.ie ; ake.rasmuson@ul.ie

The relevance of crystallization from solution as unit operation in the pharmaceutical industry is of utmost importance since the performance of crystallizers determine the required up- and downstream handling and processing operations. Polymorphism, is the ability of a substance to exist in more than one solid form and it is crucial due to its influence on the final drugs physicochemical properties, e.g. dissolution rates, bioavailability and morphology. Besides, the solvent used in the crystallization process also influences the solid form obtained and its physicochemical and morphological properties. In this sense, research on the nucleation and growth of active pharmaceutical ingredients is essential due to its immediate industrial application and its direct impact on society.

Piracetam (PCM) is a polymorphic compound used as a nootropic agent for the enhancement of memory in human beings. Its effectiveness in the treatment of ischemia, cognitive impairment, stroke, and dementia without causing sedation or stimulation is recognized widely. Up to date, five different polymorphs of PCM have been reported being only two of them stable at room temperature in it solid forms II (mestastable) and III (stable) [1]. The solubility of FII and FIII of PCM has been studied in a range of organic solvents as well as their solution mediated polymorphic transformations [2,3]. However, to the best to our knowledge there is a lack of information on the kinetics of the crystal growth of the different PCM polymorphs. The aim of the present work is to study the crystal growth of FII and FII polymorphs of PCM in organic solvents. The modeling of the experimental data to empirical equations and mechanistic ones based on the Burton-Cabrera-Frank (BCF) and Birth and Spread (B+S) theories also constitutes an important goal.

Isothermal seeded desupersaturation experiments were carried out in a 250 mL jacketed batch glass crystallizer equipped with an accurate control of temperature. The liquid concentration and the number of counts during the runs were monitored in-situ by IR and FBRM probes (both from Mettler Toledo), respectively. The solvents studied were ethanol (EtOH) and isopropanol (IPrOH). The range of temperature explored 283-308 K and the initial supersaturation ratio was always below 1.2. The seeds were prepared by cooling crystallization, the correct polymorphic forms were characterized by PXRD and SEM and eventually, they were sieved to a particle size 100-180 μm. The harvested crystals after growth experiments were analyzed by PXRD to ensure that growth was undertaken by the desired polymorph. Principal component analysis was used to treat the raw IR spectra from the experiments and to establish their relation with the liquid concentration at any instant within the crystallizer.

Experimental desupersaturation data or driving force obtained was fitted to power law empirical equations by solving the differential equation derived from the crystal growth rate definition. Thereby, calculated driving forces were computed and the involved kinetic parameters were estimated by non-linear regression using a specially designed Matlab script. An example of the fitting results is given and of the derived crystal growth rates is given in Figure 1. Using an analogous methodology, experimental desupersaturation evolutions were fitted to classical BCF and B+S theories. The goodness of the fitting was remarkably good for both, empirical and mechanistic-based growth models.

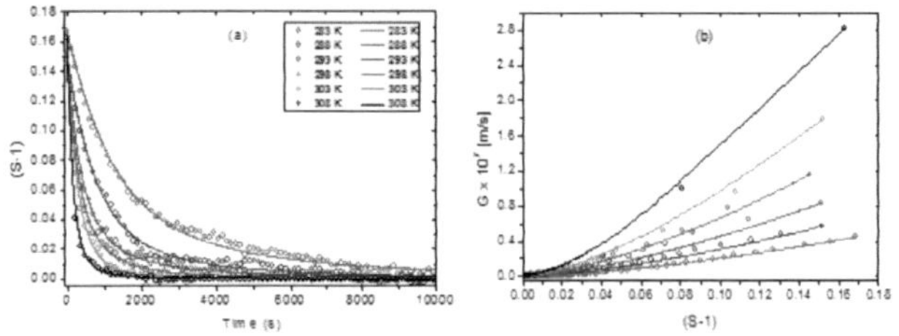

Figure 1. (a) Fitting of power-law equation to the experimental desupersaturation data in terms of (S-1) for form II of piracetam in EtOH at different temperatures. (b) Comparison of crystal growth rates calculated from the experimental driving forces and those predicted by the fitting power law equation for form II of piracetam in EtOH at different temperatures.

The estimated activation energies for the main four systems studied, polymorphs FII and FIII in EtOH and IPrOH, were in the range 55-71 kJ/mol, indicating that the growth process was controlled by the surface integration rather than by diffusion in all cases. The growth rates in terms of m/s were found to be dependent on the polymorph and the solvent studied and to increase in the order: IPrOH-FIII < IPrOH-FII < EtOH-FIII < EtOH-FII. Therefore, the metastable polymorph growth faster in both EtOH and IPrOH and the growth is also faster in EtOH than in IPrOH. The solid liquid interfacial energies estimated from B+S model ranged 1.75–2.11 mJ/m^2, indicating that there is smooth difference within the polymorphs in the solvents studied.

[1] A. Maher, C. C. Seaton, S. Hudson, D. M. Croker, Å. C. Rasmuson, and B. K. Hodnett, "Investigation of the solid-state polymorphic transformations of piracetam," *Cryst. Growth Des.*, vol. 12, no. 12, pp. 6223–6233, 2012.

[2] A. Maher, Å. C. Rasmuson, D. M. Croker, and B. K. Hodnett, "Solubility of the metastable polymorph of piracetam (Form II) in a range of solvents," *J. Chem. Eng. Data*, vol. 57, no. 12, pp. 3525–3531, 2012.

[3] A. Maher, D. M. Croker, Å. C. Rasmuson, and B. K. Hodnett, "Solution mediated polymorphic transformation: Form II to III piracetam in ethanol," *Cryst. Growth Des.*, vol. 12, no. 12, pp. 6151–6157, 2012.

536694 Templating Colloidal Crystal Growth Using Chirped Surface Relief Gratings

Abstract:

We report a method for controlling the lattice geometry of monodisperse colloidal crystals formed by confined convective self-assembly on a substrate patterned with a chirped surface relief grating. Chirped gratings were fabricated using laser interference lithography (LIL) and a curved mirror reflector to create photoresist patterns with pitch values ranging from ~500 nm to >10,000 nm spread over a planar surface. These surface nanostructures are shown to guide the formation of various lattice geometries not normally found via colloidal assembly on planar surfaces. It is shown that when the pitch of the grating is much larger than the diameter of the colloidal particles, the grating trenches serve as compartments for deposition and the particles form close-packed, linear chains. Various ordered structures are observed as the dimensions of the grating pitch decrease and approach the diameter of the particles. The grating nanostructures guide the formation of various lattice geometries due to specific particle-surface and particle-particle interactions. Observed crystal lattices include square, hexagonal and rhombic structures. The formation of these structures is explained in terms of the geometrical constraints imposed by the surface pattern and the particle diameter. These crystal lattices can be translated into large area samples when using corresponding single pitch grating substrates. The initial monolayer lattice can also serve as a template for the growth of unique, bilayer structures that include rectangular lattices, chains of particle pairs or triplets, and graphite-like structured lattices. In addition, when coated with a thin silver layer, these various lattice configurations are shown to produce optical reflection features that are precisely controlled by the underlying structure as it varies from widely spaced particle chains to close-packed lattice geometries.

534660 Mechanistic Insights into the Process of Crystallization Using a Kinetic, Multiscale Model

Abstract:

Polymorphism, an ability of a molecule to exist in more than one crystal structure is an issue usually encountered during the process of crystallization in pharmaceutical industries. To ensure maximum bioavailability, optimum dissolution rate and to comply with standards of federal institutions, it is necessary to produce a drug with uniform physical properties which depend on crystal structure. The operating conditions affect the outcome of crystallization. To predict the outcome of the crystallization, the current approaches rely on empiricism or searching for crystal structure with lowest lattice energy. Former approach, although reliable, is time consuming and uses valuable resources and the latter approach fails to properly capture the mechanism of crystallization process. To overcome this issue, we propose a multiscale model based on the kinetics of crystallization. Fundamentally, mechanism of crystallization involves attachment and detachment of molecules at different time scales. With the help of molecular dynamics simulations, the potential energy profiles of solvated and partially de-solvated molecules can be superimposed to find the energy of attachment. The information obtained from molecular dynamics simulation can then further be used to simulate the motion of the solute molecule molecule in a solvent using stochastic simulations. Proper choice of simulation technique results in obtaining most probable polymorph at certain operating conditions. The ability of this multiscale model to capture the mechanics of crystallization at different time scales will be exemplified with the help of a system containing glutamic acid molecule in aqueous environment.

533178 Insight into the Role of Piperazine in the Thermodynamics and Nucleation Kinetic of Triethylenediamine - Methyl Tertiary Butyl Ether System

Abstract:

Insight into the Role of Piperazine in the Thermodynamics and Nucleation Kinetic of Triethylenediamine - Methyl tertiary butyl ether System

Recently, additives have been taken into consideration in order to improve solubility, stability, and crystal habit. Besides the additive, a certain impurity originating from solvent or a by-product from former reaction also has an influence on the crystallization process and crystal product. In this work, we explore the role of piperazine, a major by-product from former reaction, in the nucleation behavior of triethylenediamine (referred as TEDA below) - Methyl tertiary butyl ether (referred as MTBE below) system. Different amounts of piperazine (1%, 3%, 5%), which is close to the crude material, are considered to investigate how it affects the thermodynamic and nucleation behavior of the crystallization of TEDA-MTBE system. The solubility of TEDA in MTBE in the presence of piperazine is analyzed to describe thermodynamic behavior. It is found that the solubility of TEDA in MTBE increases with increasing the amount of piperazine. Then, radial distribution function (RDF) analysis is used to uncover the interactions between piperazine, TEDA and MTBE by Materials Studio. As expected, the interaction between the solute and the solvent becomes stronger in the presence of piperazine, thus improving the solubility of TEDA in MTBE. Meanwhile, data of supercooling ¦¤T_{max}, a measure of metastable zone width (MSZWs), as a function of saturated temperature T_0, cooling rate R, and the content of piperazine are analyzed to describe the role of piperazine in the nucleation kinetics of TEDA-MTBE system. It is found that piperazine can significantly promote the nucleation rate of TEDA-MTBE system. Furthermore, based on our previous work, modified Sangwal¡¯s theory is employed to take further insight into the role of impurity in the nucleation behavior of TEDA-MTBE system. In the CNT framework, it shows that the critical Gibbs free energy is lower in the presence of piperazine, as a result it is easier to form nuclei and narrow the MSZWs. At last, the effect of piperazine on the crystal size distribution (CSD) of TEDA product is also investigated and discussed.

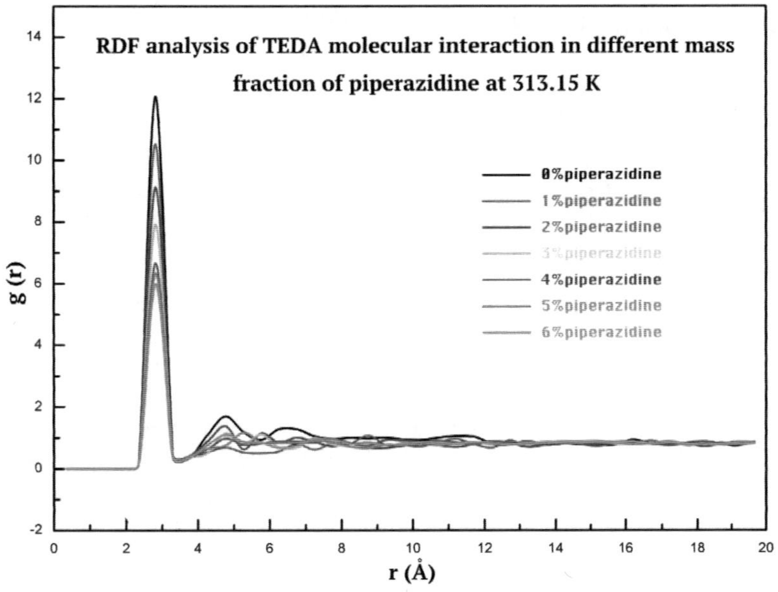

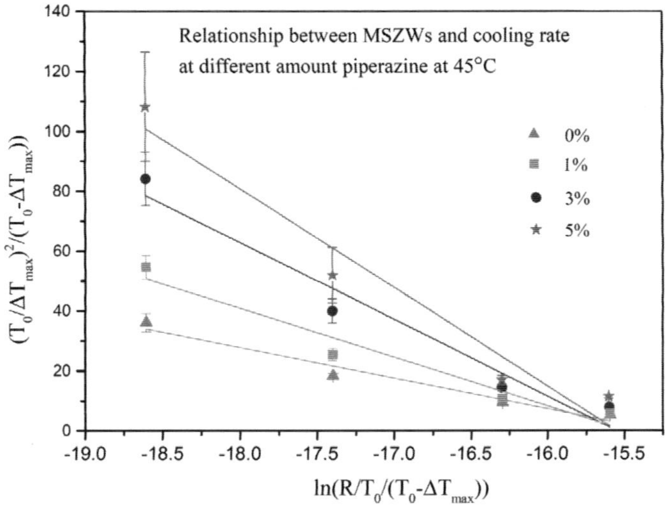

537797 Investigating the Effect of Cu and Ce Loading in Mesoporous Y Zeolite for the Adsorptive Desulfurization of 4,6-Dimethyldibenzothiohene

Abstract:

Investigating the Effect of Cu and Ce Loading in Mesoporous Y Zeolite for the Adsorptive Desulfurization of 4.6-Dimethyldibenzothiohene

Kevin X. Lee, George Tsilomelekis and Julia A. Valla

Department of Chemical & Biomolecular Engineering, University of Connecticut, 191 Auditorium Road, Unit 3222, Storrs, CT 06269-3222, USA,

Phone: +1-860-486 4602, e-mail: ioulia.valla@.uconn.edu

Sulfur emission from transportation fuels can be detrimental, if they are not controlled. Conventional hydrodesulfurization (HDS) has been the main process of sulfur removal technology ever since it was founded. However, the recent aggressive use of fossil-based fuel as energy source, particularly for the production of transportation fuels, resulted in excessive energy consumption in the HDS process. Adsorptive desulfurization using bimetal-exchanged mesoporous Y zeolites has shown to exhibit high capacity and selectivity for thiophenic compounds in liquid fuels[1]. The combination of hierarchical pore structure and ion-exchanged metal cations in the Y zeolite allows for improved mass transfer of sulfur compounds to the active sites and stronger bond formation via various adsorption modes, such as π-complexation and/or σ-bond interaction [2,3]. Our recent studies have shown that the synergy between two metal actions (e.g. Ce and Cu) plays a crucial role in high adsorption selectivity and capacity of thiophene compounds, such as thiophene, benzothiophene and dibenzothiophene [4]. In this study we demonstrate that 2%Cu10%CeSAY exhibited the highest 4,6-DMDBT capacity in the presence of aromatics, producing up to 50 mL/g of clean fuel.

The experimental results were strongly supported by characterization techniques and spectroscopic analysis. Metal-exchanged mesoporous Y zeolites were first prepared by introducing mesoporosity via a top-down surfactant-assistant technique.[5] The mesoporous zeolite was subjected to an ion-exchange procedure to replace the protons with either Cu or Ce metals. The modified zeolites were characterized using various techniques. The diffraction pattern and crystalline structure of each material were identified using an x-ray diffractometer (XRD. The crystallinity and porosity, as well as the presence of metal cations/nanoparticles were imaged using transmission electron microscopy (TEM). Surface topographical analysis and surface elemental mapping were performed by SEM with EDAX accessory. X-ray photoelectron spectroscopy (XPS) experiments were conducted to provide quantitative and oxidative information on the sample surface. The bulk metal loading in the zeolite was calculated using an inductively coupled plasma equipped with a mass spectrometer (ICP-MS). Subsequently, X-ray fluorescence (XRF) measurement was acquired to analyze the bulk concentration of zeolite elements.

Breakthrough experiments were conducted in a fixed-bed column to study the dynamic adsorption of model fuel (100 ppmw of DMDBT and 1% naphthalene in octane). The desulfurized effluent was collected periodically until saturation had been reached and analyzed using a gas-chromatograph-sulfur chemiluminescence detector (GC-SCD). Figure 1 shows the breakthrough curves of DMDBT on various metal-exchanged mesoporous Y zeolites with different metal loadings and order at which they were introduced. Figure 1(a) confirms that Ce should be introduced first, followed by Cu in mesoporous Y zeolite. Increasing the loading of Cu did not improve the desulfurization performance, as indicated in Figure 1(b). In Figure 1(c), the concentration of Ce was varied from 2 wt% to 10 wt%. All three zeolites displayed higher adsorption of 4,6-DMDBT compared to 2%Cu2%CeSAY, as about 50 mL/g of clean fuel was produced. These results suggest that in combination with the mesopores, the synergistic effects of Ce and Cu metals significantly improve the capacity and selectivity of DMDBT.

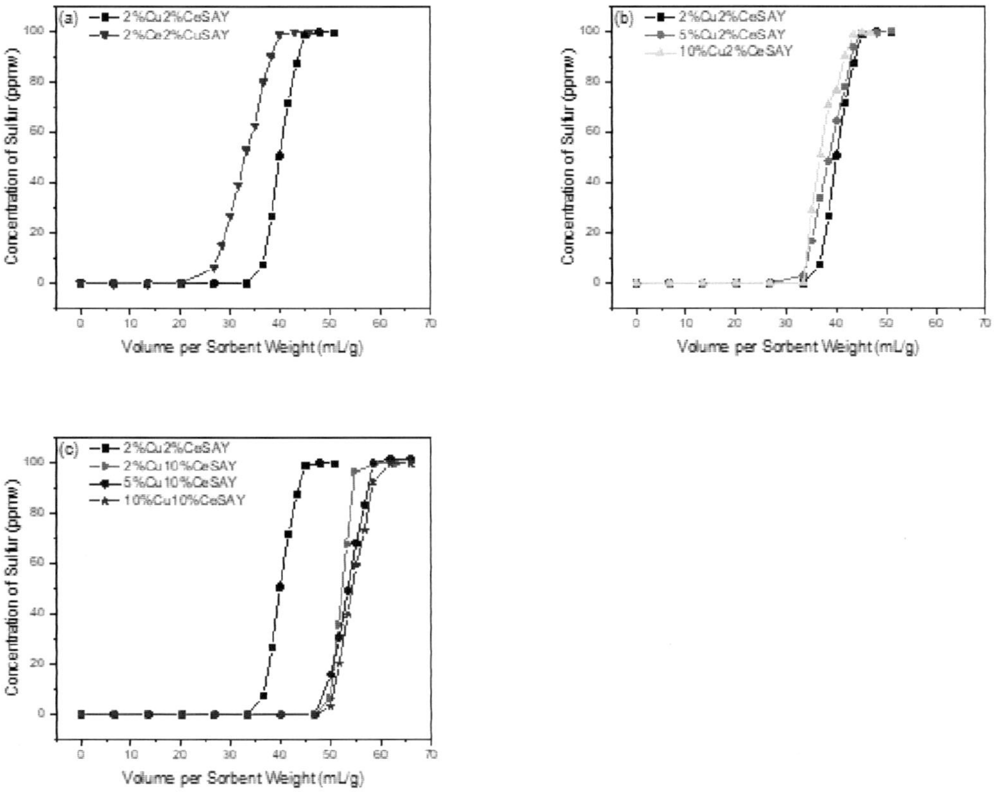

Figure 1: Breakthrough curves of 100 ppmw of sulfur in DMDBT dissolved in *n*-octane on mesoporous Y zeolites following the effects of a) metal configuration, b) Cu concentration, and c) Ce concentration.

References

1. Lee, K. X. & Valla, J. A. Investigation of bifunctional zeolites for the adsorptive desulfurization of fuels. *Appl. Catal. B Environ.* **201,** 359–369 (2017).

2. Yang, R. T., Hernández-Maldonado, A. J. & Yang, F. H. Desulfurization of transportation fuels with zeolites under ambient conditions. *Science* **301,** 79–81 (2003).

3. Velu, S., Ma, X. & Song, C. Selective Adsorption for Removing Sulfur from Jet Fuel over Zeolite-Based Adsorbents. *Ind. Eng. Chem. Res.* **42,** 5293–5304 (2003).

4. Lee, K. X., Tsilomelekis, G. & Valla, J. A. Removal of Benzothiophene and Dibenzothiophene from Hydrocarbon Fuels using CuCe Mesoporous Y Zeolites in the Presence of Aromatics. *Appl. Catal. B Environ.* (2018). In print.

5. García-Martínez, J., Johnson, M., Valla, J., Li, K. & Ying, J. Y. Mesostructured zeolite Y—high hydrothermal stability and superior FCC catalytic performance. *Catal. Sci. Technol.* **2,** 987 (2012).

524640 Role of Amine Structure on Hydrogen Sulfide Capture from Dilute Gas Streams Using Solid Adsorbents

Abstract:

The hydrogen sulfide composition in biogas ranges from 0-5% depending on the feed source to the biogas digester. To meet renewable natural gas specifications of 0-4 ppm H_2S content, technologies that remove H_2S from raw biogas feeds are needed. This contribution assessed three similar adsorbents with similar textural and physical properties, but with different amines grafted to the surface. Specifically, materials containing primary, secondary and tertiary amines at the end of a propyl surface linker grafted on a silica support are explored. H_2S adsorption isotherms and cyclic studies are presented for these materials and it is shown that secondary amines have the best amine efficiency while tertiary amines are the most stable for H_2S capture, of the materials studied. The results suggest the consideration of secondary and tertiary amines for the design of amine adsorbents suitable for H_2S removal in dilute gas streams over multiple cycles.

533635 Hydrogen Sulfide Removal from Biogas on "Molecular Basket" Sorbents

Abstract:

Biogas is a sustainable energy source which can be used as an alternative fuel. However, hydrogen sulfide (H_2S) in biogas (formed from the decomposition of organic materials) limits the application of biogas due to its corrosivity and sulfur-toxicity. Recently, "Molecular basket" sorbents (MBS) for acid gas adsorption developed in our group could achieve H_2S sorption/desorption at ambient conditions. Nevertheless, the large amount of CO_2 in biogas strongly inhibits H_2S sorption on conventional MBS. Therefore, the objective of this work is to selectively remove H_2S from biogas which contains a high-level concentration of CO_2. In this work, a series of tertiary-amine-based "Molecular Basket" sorbents were prepared and applied in H_2S removal from the gas mixture containing a high concentration of CO_2. Poly allylamine (PA), polyethylenimine (PEI), and tetramthyl hexanediamine (TMHDA) were loaded on SBA-15 to compare the H_2S sorption performance in the absence and presence of CO_2. In the presence of CO_2, the H_2S sorption dropped dramatically on PA/SBA-15 and PEI/SBA-15, whereas the selectivity of TMHDA/SBA-15 for H_2S sorption remained stable. The inlet concentration (H_2S and CO_2), the regenerability, as well as the effects of CH_4 and moisture were studied. The H_2S sorption capacity was increased with increasing inlet H_2S concentration, and the sorption on TMHDA/SBA-15 was fitted to Langmuir adsorption isotherm. However, the H_2S sorption capacity was slightly decreased with the increase of inlet CO_2 concentration. The regenerability was good under mild regeneration conditions. In addition, other components in biogas, such as CH_4 and moisture, showed little effect on the H_2S sorption performance over MBS. This work demonstrated that tertiary-amine-based MBS is selective for the H_2S removal from biogas.

531284 Preparation and Characterization of Novel Clinoptilolites: Treatment of Natural Gas Type Mixtures Using Inorganic Membranes

Abstract:

A significant quantity of natural gas reserves in the United States do not meet the pipeline specification of having less than 4% N_2 content and thus require further refinement. This proves to be tough for smaller producers due to the economies of scale necessary for conventional separation techniques such as cryogenic distillation. The integration of adsorption technology within these upgrading processes may achieve a more economically efficient separation. To that end, the characterization of adsorbents is important to optimise adsorption selectivity for this separation.

Clinoptilolite is a natural zeolite that may be modified through cation exchange to induce favourable adsorption characteristics such as high CH_4/N_2 separation properties. This adsorbent material is a promising candidate for the application of natural gas upgrading.

In this work, experimental equilibrium adsorption isotherms and uptake measurements of CO_2, CH_4 and N_2 gases are presented for various clinoptilolites. Additionally, the effects of competitive adsorption of binary CH_4-N_2 mixtures on the studied adsorbents is shown. The applicability of this material in novel inorganic membranes is discussed, and a novel membrane preparation technique is explored using clinoptilolite particles sintered within a glass binder. The prepared membrane surface was characterized using SEM and XRD analyses and the membranes were tested for the permeation properties of several constituent gases present in natural gas type mixtures.

534784 Copper Modified Activated Carbon for Hydrogen Purification

Abstract:

Copper modified activated carbon for hydrogen purification

Frederico Relvas[1]*, Carlos M. Silva[2], Roger D. Whitley[3], Adélio Mendes[1]**

[1]LEPABE - Laboratory for Process Engineering, Environmental, Biotechnology and Energy, Faculdade de Engenharia, Universidade do Porto, Rua Dr. Roberto Frias, 4200-465 Porto, Portugal

[2]CICECO – Aveiro Institute of Materials, Department of Chemistry, University of Aveiro, Campus Universitário de Santiago 3810-193 Aveiro, Portugal

[3]Air Products & Chemicals, Inc., 7201 Hamilton Boulevard, Allentown, Pennsylvania 18195, United States

(*) Presenting author: frelvas@fe.up.pt (**) Corresponding author: mendes@fe.up.pt

Keywords: Adsorption; Kinetics; Hydrogen purification

<>1. Introduction

Several studies have been proposing the use of π-complexation adsorbents for different applications, including paraffin/olefin separations and hydrogen purification. In a recent work, the authors modified a commercial activated carbon with $CuCl_2{\cdot}2H_2O$, which showed an enhanced capacity for carbon monoxide adsorption. The adsorbent was tested in a lab-scale pressure swing adsorption unit and proved to be very effective for improving the PSA performance for producing fuel cell grade hydrogen from reformate.[1] However, using such adsorbents, the double contribution of the adsorption on the activated carbon surface and π-complexation bonds formation turns the process difficult to simulate using a single kinetic model.

In the current work, the contributions of both the activated carbon surface and the copper adsorption sites are assessed. A dual kinetic model was proposed based on the linear driving force (LDF) approximation, attributing a fraction of the total capacity to the activated carbon and the remaining fraction to the π-complexation bonds formed with copper salt, each presenting distinct adsorption kinetics. The model was used to describe experimental fractional uptake and breakthrough curves.

<>2. Experimental <>2.1 Adsorbent preparation and characterization

Several samples of commercial activated carbon Kuraray 2GA-H2 (AC) were modified by wet impregnation of $CuCl_2 \cdot 2H_2O$ using different copper loadings. After drying, the samples were activated under hydrogen atmosphere for reducing the copper state to Cu^+, to which improved adsorption capacities are attributed. Adsorption equilibrium isotherms of H_2, CO_2, CH_4, N_2 and CO were measured afterwards by volumetric method. The adsorbents were subsequently exposed to oxygen and the adsorption isotherms were measured again. The objective of this step was to oxidize the copper to Cu^{2+}, which does not form π-complexation bonds, thus isolating the contribution of the pristine AC.

The initial adsorption isotherms (before oxygen exposure) were fitted by dual-site Langmuir model according to eq. (1),

$$q_i = \frac{q_{\max,1} b_1 P_i}{1 + q_{\max,1} b_1 P_i} + \frac{q_{\max,2} b_2 P_i}{1 + q_{\max,2} b_2 P_i}$$

where the first set of sites was attributed to the contribution of the activated carbon surface, which corresponds to the equilibrium assessed after oxygen exposure, and the second set of sites was endorsed to the presence of the copper salt.

<>2.2 Kinetic model

The analytical solution proposed by Kočiřík *et al.*[2] for describing simultaneous mass and heat transfer under variable pressure conditions was used to fit the experimental uptake curves by nonlinear optimization. The model was adapted for considering two parallel processes according to eq. (2), thus obtaining the effective LDF mass transfer coefficients $k_{LDF,AC}$ and $k_{LDF,Cu-AC}$,

$$F = \frac{\bar{q}^*_{AC}(0_+) - \bar{q}_{AC}(t)}{\bar{q}^*_{AC}(0_+) - \bar{q}_{AC}(\infty)} \times f_{AC} + \frac{\bar{q}^*_{AC}(0_+) - \bar{q}_{AC}(t)}{\bar{q}^*_{AC}(0_+) - \bar{q}_{AC}(\infty)} \times f_{Cu-AC}$$

Where F is the fractional uptake, k is the effective LDF mass transfer coefficient, $\bar{q}(t)$ is the average adsorbate concentration in the solid at instant t, $\bar{q}^*(0_+)$ and $\bar{q}^*(\infty)$ are the are the solid concentrations in equilibrium with the initial gas (t_{0+}) and with feed (t_∞), respectively, and f is the contribution factor of each mechanism. The subscripts AC and Cu-AC denote the activated carbon and copper contributions, respectively.

<>3. Results and conclusions

Figure 1 presents the adsorption capacity of the studied gases on the pristine AC and the modified adsorbent Cu-AC-2 (2 mol kg^{-1} of copper loading). A substantial increase in the carbon monoxide adsorption was observed for the modified adsorbent, while the opposite effect was observed for the other gases. As reported elsewhere,[1] such effect is even more evident with copper loading increase. The improvement is attributed to π-complexation bonding of the Cu^+ with CO, which does not happen for the other gases under study. This fact is confirmed by the adsorption experiments after O_2 exposure: when contacting oxygen, the copper is oxidized to its valence state Cu^{2+}, which cannot form chemical bonds with CO and thus decreases the CO adsorption capacity. Since the other gases cannot form π-complexation bonds, their adsorption capacities remain approximately the same before and after O_2 exposure.

The same conclusions can be drawn from the fractional uptake curves and breakthrough curves, Figures 2 and 3 respectively. While the adsorption of CO_2, CH_4, N_2 and H_2 can be described using a single-kinetic model, carbon monoxide exhibits a bi-modal behavior, suggesting the presence of two different mechanisms. The dual-kinetic model suggested in this work proved to efficiently describe the adsorption dynamics inside the adsorbent material.

The development of a simulator based on stable adaptive multiresolution methods and comprising this kinetic approach is ongoing to simulate the breakthrough curves and further PSA applications.

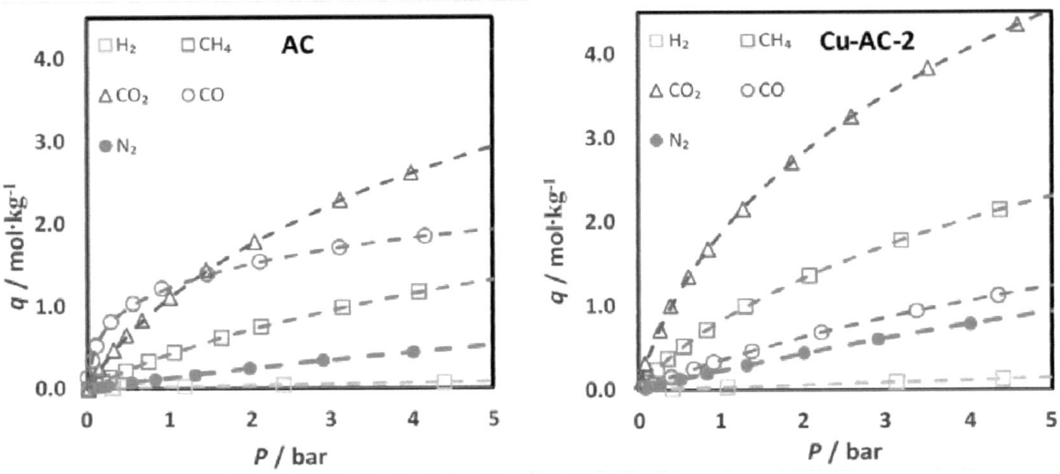

Figure 1 - Experimental adsorption isotherms in commercial (AC) and modified (Cu-AC) adsorbent at 40 °C for H_2, CO_2, CH_4, N_2 and CO. Dashed-lines denotes the dual-site Langmuir model.

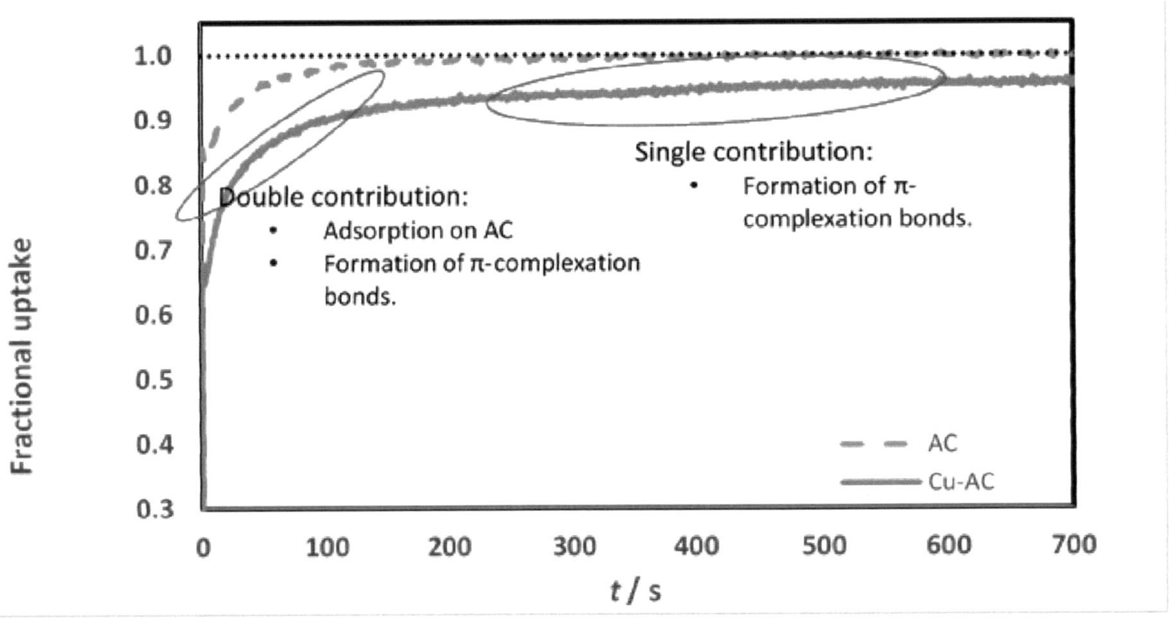

Figure 2 - Fractional uptake curves of CO on the pristine AC and Cu-AC-2.

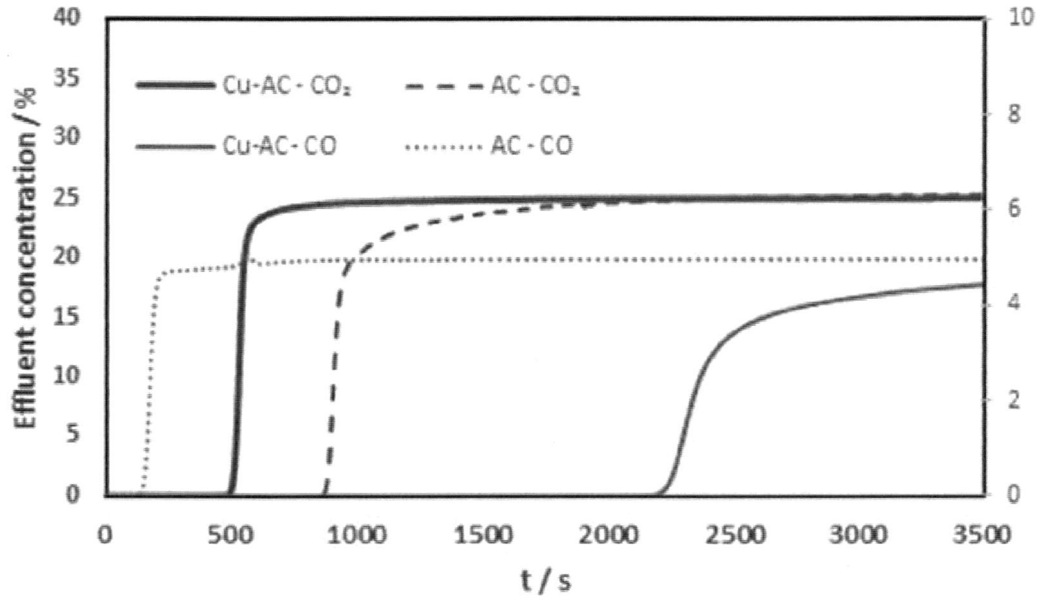

Figure 3 - Experimental monocomponent adsorption breakthrough of CO (95 % He; 5 % CO) and CO_2 (75 % He; 25 % CO_2). Dashed-lines – Pristine AC; Solid lines: Cu-AC-2. Primary axis: CO_2; Secondary axis: CO.

<>Acknowledgements

This work was partially supported by projects POCI-01-0145-FEDER-006939 (Laboratory for Process Engineering, Environment, Biotechnology and Energy – UID/EQU/00511/2013) funded by the European Regional Development Fund (ERDF), through COMPETE2020 - Programa Operacional Competitividade e Internacionalização (POCI) and by national funds, through FCT - Fundacao para a Ciencia e a Tecnologia. F. Relvas would also like to thank to PhD grant NORTE-08-5369-FSE-000028 supported by North Portugal Regional Operational Programme (NORTE 2020), under the Portugal 2020 Partnership Agreement and the European Social Fund (ESF). Authors also thank project CICECO-Aveiro Institute of Materials (Ref. FCT UID/CTM/50011/2013), financed by national funds through the FCT/MEC and when applicable co-financed by FEDER under the PT2020 Partnership Agreement.

<>References

1. Relvas, F.; Whitley, R. D.; Silva, C. M.; Mendes, A., Single-stage pressure swing adsorption for producing fuel cell grade hydrogen. *Industrial & Engineering Chemistry Research* **2018**.

2. Kočiřík, M.; Struve, P.; Bülow, M., Analytical solution of simultaneous mass and heat transfer in zeolite crystals under constant-volume/variable-pressure conditions. *Journal of the Chemical Society, Faraday Transactions 1: Physical Chemistry in Condensed Phases* **1984,** *80* (8), 2167-2174.

535947 Carbon Dioxide Adsorption at Elevated Temperatures for Vehicle Exhaust Gas Treatment

Abstract:

To forestall the effects of climate change, due to an excess of carbon dioxide in the atmosphere, scaled carbon capture and storage (CCS) operations are needed to separate CO_2 from the exhaust gases generated by fossil fuel combustion at electric power plants and from the internal combustion engines of automobiles. Many different classes of porous materials have been considered for use as carbon capture adsorbents. Many published studies of these materials focus on their CO_2 adsorption capabilities at near-ambient conditions; e.g., at 25 °C and 1 atm [1]. However, higher discharge temperatures, in the range from 40 to 75°C, are relevant for CO_2 removal from combustion stack gases and vehicle exhaust streams. Fewer studies have been reported [2-4] for CO_2 uptake on adsorbents at these elevated temperatures.

To assess the performance of prospective carbon capture adsorbents at the higher temperatures characteristic of vehicle exhaust trains, CO_2 uptake measurements were performed on eight commercially available porous adsorbents at temperatures of 25, 40, 55, and 70 °C using a Micromeretics ASAP 2050 extended adsorption gas analyzer. The materials tested included two activated carbons, two zeolite molecular sieves, two metal-organic framework adsorbents (MOFs), one zeolitic imidazolate framework (ZIF), and one elastic-layered MOF (ELM). The CO_2 adsorption capacities of the two zeolites were additionally evaluated at temperatures of 85 and 100 °C. A decrease of 23% to 27% in the CO_2 adsorption capacity was observed at 1 atm pressure for each 15 °C stepwise increase in the measured isotherm. From application of the Clausius-Clapeyron equation to the adsorption isotherms, isosteric heats of adsorption are obtained for each of the porous solids that are in good agreement with adsorption enthalpies reported for these materials at lower temperatures. Among the materials considered, the reduction in CO_2 adsorption capacity with increasing temperature is most pronounced for zeolites 5A and 13X, which correspondingly have the largest heats of adsorption for carbon dioxide.

1. Keskin, S., van Heest, T. M., & Sholl, D. S. (2010). Can Metal–Organic Framework Materials Play a Useful Role in Large Scale Carbon Dioxide Separations. *Chemistry and Sustainability, 3*, 879-891.

2. Yazaydın, A. O., Snurr, R. Q., Park, T. H., Koh, K., Liu, J., LeVan, M. D., & Low, J. J. (2009). Screening of metal– organic frameworks for carbon dioxide capture from flue gas using a combined experimental and modeling approach. *Journal of the American Chemical Society*, *131*, 18198-18199.

3. Singh, V. K., & Kumar, E. A. (2016). Comparative studies on CO_2 adsorption kinetics by solid adsorbents. *Energy Procedia*, *90*, 316-325.

4. Yong, Z., Mata, V. G., & Rodrigues, A. E. (2001). Adsorption of carbon dioxide on chemically modified high surface area carbon-based adsorbents at high temperature. *Adsorption*, *7*, 41-50.

524978 Adsorption of Rare Earth Elements in Phosphorous Functionalized Nanoporous Carbon

Abstract:

Due to a worldwide shortage of supply of rare earth elements and their critical applications in energy and defense sectors, researchers are currently looking for a sustained supply of those elements. Neodymium (Nd) and Dysprosium (Dy) are two of those highly valued rare earth elements and one of the possible ways to maintain a constant supply is to recover them from waste permanent magnets. Selective adsorption in a suitable nanoporous material is considered to be a benign and inexpensive way to recover those precious elements. In this research, phosphorus functionalized nanoporous carbon was synthesized from biomass with the BET surface area of 837 m^2/g and 0.9 atom% phosphorus. The adsorption of neodymium and dysprosium were in the range of 335â€"344 mg/g upto their initial concentration of 500 ppm in water and such adsorption amount is higher than that of majority of the adsorbents reported in literature. The adsorption capacity of iron (Fe(III)) under the similar conditions was one order of magnitude lower and it suggests a possible separation of these rare earth elements from iron present in the waste magnets. X-ray photoelectron spectroscopy (XPS) studies on both Nd and Dy-adsorbed carbons revealed a small shift in the P-$2p_{3/2}$ energy level of phosphorus towards a higher energy level hereby suggesting a possible formation of metallic phosphates. Additionally, XPS also indicated a small amount of carbonates and oxides of rare earth elements that might have been formed on the surface of adsorbents. The overall results suggest that this carbon can be used as a potential sorbent for enrichment and separation of Nd and Dy.

533874 Large-Area Single Layer Graphene Membrane By Crack-Free Transfer for Gas Mixture Separation

Abstract:

Nanoporous single-layer suspended graphene films have been predicted to be the ultimate membranes, capable of providing orders of magnitude higher permeance that that from the state-of-the-art membranes.[1-3] This is mainly because of the ultrashort diffusional path of the molecules across the two-dimensional graphene film. However, the development of such membranes is impeded by two major bottlenecks; (i) incorporation of molecular-size nanopores in otherwise impermeable graphene, (ii) crack- and tear-free transfer of graphene to a porous substrate. Here we report novel approaches that overcome these bottlenecks, leading to formation of 1 mm^2-sized single-layer graphene membranes (40000-fold larger than the state-of-the-art single-layer graphene membranes).[4]

The membrane was stable during multiple cycles of heating and cooling, and at least up to moderate transmembrane pressure differences (7 bar). Attractive gas sieving performance (H_2 permeance up to 4.1 x 10^{-7} mol m^{-2} s^{-1} Pa^{-1} with H_2/CH_4 selectivities up to 25) were achieved from graphene film with a minuscule porosity of 0.025%. Use of gas mixture feed did not reduce either the H_2 permeance or the H_2/CH_4 separation selectivity. Finally, we report post-synthetic performance tuning of the graphene membrane by ozone-functionalization, improving H_2 permeance (up to 300%) as well as H_2/CH_4 selectivities (up to 150%). Overall, the methods developed here can be instrumental to realize large-scale deployment of the nanoporous two-dimensional membrane.

References

1. Jiang, D.-E., Cooper, V. R. & Dai, S. Porous graphene as the ultimate membrane for gas separation. ***Nano Lett.*** 9, (2009).
2. Koenig, S. P., Wang, L., Pellegrino, J. & Bunch, J. S. Selective molecular sieving through porous graphene. ***Nat. Nanotechnol.*** 7, 728â€"732 (2012).
3. Celebi, K. *et al.* Ultimate permeation across atomically thin porous graphene. ***Science*** 344, 289â€"92 (2014).
4. Agrawal, K. *et al.,* Single-Layer Graphene Membranes by Crack- Free Transfer for Gas Mixture Separation. ***Nat. Commun.*** 2018

533983 Rapid Synthesis of Mixed Linker Zeolitic-Imidazolate Frameworks (ZIFs) Membranes with Tunable Gas Separations

Abstract:

Metal organic frameworks (MOFs) are emerging organic-inorganic crystalline microporous solids for gas separation application owing to their tunable pore structure and functionality with diverse metal nodes and organic linkers.[1] Zeolitic-imidazolate frameworks (ZIFs),[2] a subclass of MOFs, in particular have attracted many attentions due to their chemical/thermal stabilities, their ultra-microporosities, and high surface area when compared to other MOFs material.[2] ZIFs possess zeolite-like topologies in which the tetrahedral Si or Al and the bridging O in zeolite structure are replaced by transition metals (such as Zn,[2] Co,[3] and Cd[4]) and imidazolate-derived ligands. A common drawback for gas separation is the limited availability of aperture pore size and functionalities of the porous materials. Furthermore, the common approach to fabricate ZIF polycrystalline membranes are lengthy requiring up to weeks of synthesis.[5-11] Studies have shown that through mixing metals[12,13] and linkers,[14,15] one can continuously tune the MOFs/ZIFs (termed hybrid ZIFs) properties to match with the characteristic of specific gas mixture. However, no study have shown the tuning effect of the hybrid ZIFs grown as polycrystalline membrane.

Here, we plan to present a rapid microwave-assisted in situ synthesis of well-intergrown mixed linker ZIF-7-8 membranes in under â^¼90 s.[16] ZIF-7-8 consists of Zn^{2+} metal nodes bridged by a mixture of benzimidazolate (bIm, ZIF-7 linker) and 2-methylimidazolate (mIm, ZIF-8 linker) linkers.[15] To the best of our knowledge, this is the fastest synthesis of any polycrystalline MOF membranes reported up to now. Furthermore, the gas separation performances (separation factor and permeance) of hybrid ZIF-7-8 membranes were systematically tuned by varying the bIm-to-mIm ratio incorporated into the framework. This is attributed to the fact that an increase in bulky benzimidazolate linker incorporation reduces the effective aperture size of the hybrid framework, consequently improving the separation factor at the expense of reduction in permeance. The unprecedentedly rapid synthesis of ZIF-7-8 membranes with tunable molecular sieving properties is an important step forward for their practical applications in the separation of gas mixtures of interest.

References:

1. Li, J.-R.; Kuppler, R. J.; Zhou, H.-C. Selective gas adsorption and separation in metal-organic frameworks. *Chem. Soc. Rev.* **2009**, *38*, 1477-1504.
2. Park, K. S.; Ni, Z.; Cote, A. P.; Choi, J. Y.; Huang, R.; Uribe-Romo, F. J.; Chae, H. K.; O'Keeffe, M.; Yaghi, O. M. Exceptional chemical and thermal stability of zeolitic imidazolate frameworks. *Proc. Natl. Acad. Sci. U. S. A.* **2006**, *103*, 10186-10191.
3. Banerjee, R.; Phan, A.; Wang, B.; Knobler, C.; Furukawa, H.; O'Keeffe, M.; Yaghi, O. M. High-Throughput Synthesis of Zeolitic Imidazolate Frameworks and Application to CO2 Capture. *Science* **2008**, *319*, 939-943.
4. Karagiaridi, O.; Bury, W.; Sarjeant, A. A.; Stern, C. L.; Farha, O. K.; Hupp, J. T. Synthesis and characterization of isostructural cadmium zeolitic imidazolate frameworks via solvent-assisted linker exchange. *Chem. Sci.* **2012**, *3*, 3256-3260.
5. Pan, Y.; Lai, Z. Sharp separation of C2/C3 hydrocarbon mixtures by zeolitic imidazolate framework-8 (ZIF-8) membranes synthesized in aqueous solutions. *Chem. Comm.* **2011**, *47*, 10275-10277.
6. Bux, H.; Chmelik, C.; Krishna, R.; Caro, J. Ethene/ethane separation by the MOF membrane ZIF-8: Molecular correlation of permeation, adsorption, diffusion. *Journal of Membrane Science* **2011**, *369*, 284-289.
7. Bux, H.; Feldhoff, A.; Cravillon, J.; Wiebcke, M.; Li, Y.-S.; Caro, J. Oriented Zeolitic Imidazolate Framework-8 Membrane with Sharp H2/C3H8 Molecular Sieve Separation. *Chem. Mater.* **2011**, *23*, 2262-2269.
8. Bux, H.; Liang, F.; Li, Y.; Cravillon, J.; Wiebcke, M.; Caro, J. Zeolitic Imidazolate Framework Membrane with Molecular Sieving Properties by Microwave-Assisted Solvothermal Synthesis. *J. Am. Chem. Soc.* **2009**, *131*, 16000-16001.
9. Kwon, H. T.; Jeong, H.-K. Highly propylene-selective supported zeolite-imidazolate framework (ZIF-8) membranes synthesized by rapid microwave-assisted seeding and secondary growth. *Chem. Comm.* **2013**, *49*, 3854-3856.
10. Kwon, H. T.; Jeong, H.-K. In Situ Synthesis of Thin Zeoliticâ€"Imidazolate Framework ZIF-8 Membranes Exhibiting Exceptionally High Propylene/Propane Separation. *J. Am. Chem. Soc.* **2013**, *135*, 10763-10768.
11. Kwon, H. T.; Jeong, H.-K.; Lee, A. S.; An, H. S.; Lee, J. S. Heteroepitaxially Grown Zeolitic Imidazolate Framework Membranes with Unprecedented Propylene/Propane Separation Performances. *J. Am. Chem. Soc.* **2015**, *137*, 12304-12311.
12. Wang, C.; Yang, F.; Sheng, L.; Yu, J.; Yao, k.; Zhang, L.; Pan, Y. Zinc-Substituted ZIF-67 Nanocrystals and Polycrystalline Membranes for Propylene/Propane Separation. *Chem. Comm.* **2016**, *52*, 12578-12581.
13. Kaur, G.; Rai, R. K.; Tyagi, D.; Yao, X.; Li, P.-Z.; Yang, X.-C.; Zhao, Y.; Xu, Q.; Singh, S. K. Room-temperature synthesis of bimetallic Co-Zn based zeolitic imidazolate frameworks in water for enhanced CO2 and H2 uptakes. *J. Mater. Chem. A* **2016**, *4*, 14932-14938.
14. Eum, K.; Jayachandrababu, K. C.; Rashidi, F.; Zhang, K.; Leisen, J.; Graham, S.; Lively, R. P.; Chance, R. R.; Sholl, D. S.; Jones, C. W.; Nair, S. Highly Tunable Molecular Sieving and Adsorption Properties of Mixed-Linker Zeolitic Imidazolate Frameworks. *J. Am. Chem. Soc.* **2015**, *137*, 4191-4197.
15. Thompson, J. A.; Blad, C. R.; Brunelli, N. A.; Lydon, M. E.; Lively, R. P.; Jones, C. W.; Nair, S. Hybrid Zeolitic Imidazolate Frameworks: Controlling Framework Porosity and Functionality by Mixed-Linker Synthesis. *Chem. Mater.* **2012**, *24*, 1930-1936.
16. Hillman, F.; Brito, J.; Jeong, H.-K. Rapid One-Pot Microwave Synthesis of Mixed-Linker Hybrid Zeolitic-Imidazolate Framework Membranes for Tunable Gas Separations. *ACS Applied Materials & Interfaces* **2018**, *10*, 5586-5593.

537303 Suppression and Reversal of Physical Aging in Carbon Molecular Sieve Hollow Fiber Membranes

Abstract:

Carbon molecular sieve (CMS) materials can provide high performing membranes for many challenging gas pair separations. By pyrolyzing polymer precursors under an inert atmosphere, *ultramicroporous* sheets are formed. These sheets stack imperfectly, thereby producing *micropores* between the sheets that control diffusion jump lengths and sorption. The constituent sheets transition to a closer packing state to reach a more thermodynamically stable state. The closer packing leads to lower permeability and is referred to as physical aging. Our group has discovered that physical aging in CMS can be suppressed by storage under moderate (100 psig) carbon dioxide. After a physical aging process occurs in a simple amorphous polymer glass, rejuvenation can be achieved via temperature cycle or sorption cycle processes. CMS materials are complex glasses due to their sheet like nanostructures, and it has been thought previously that CMS rejuvenation would be impossible due to irreversible reduction in *micropore* volumes. In this presentation, we will provide data and analysis to explain the CMS permeability loss and subsequent rejuvenation in terms of rearrangement within the *ultramicropore* distribution.

538720 Vapor Phase Processing of Zeolitic Imidazolate Framework Membranes

Abstract:

Zeolitic imidazolate framework (ZIF) membranes possess unique properties for the separation of important industrial gas mixtures, such as propylene/propane. High quality ZIF membranes are typically fabricated with solution-based methods involving the use of at least one solvent. These methods are difficult to scale up due to the uncontrolled homogeneous nucleation/growth of ZIFs and the possible dissolution of membrane materials during membrane synthesis. Here, we report a vapor phase membrane fabrication approach that combines atomic layer deposition (ALD) and ligand vapor treatment. In this method, ZnO is deposited on porous ceramic support by ALD and then converted to ZIF-8 by exposure to 2-methylimidazole vapor. The membranes exhibit excellent and stable performance for propylene/propane separation. Under an equimolar propylene/propane mixture feed and ~7 atm feed pressure, the membranes show a combination of high separation factor (>45) and highest propylene flux reported (>10^{-2} mol m^{-2} s^{-1}). Without the use of any solvents, the vapor phase processing method has environmental, cost, and scalability advantages.

539051 ALPO-18 Membrane for Gas Separations

Abstract:

CO_2 separation property of AEI-type zeolite, a kind of small pore zeolites having 8-membered ring (0.38 × 0.38 nm) was investigated. $AlPO_4$-18 membrane was prepared on a tubular a-alumina support by a secondary growth method using seed crystals. $AlPO_4$-18 seed crystals were supported on a surface of support by a dip coating method. Seeded support was soaked in synthesis gel ($1Al_2O_3$: $1P_2O_5$: 1.8DIPEA : $100H_2O^1$) and a hydrothermal treatment was carried out at 453 K for 4 h. Gas permeation tests were performed for the mixture of CO_2 and CH_4. As a result, the AlPO4-18 membrane exhibited high CO_2 permselectivity. CO_2 permeance of 5×10^{-7} mol m^{-2} s^{-1} Pa^{-1} with separation factor for CO_2/CH_4 of 39 at 313 K. Kinetic diameters of CO_2 and CH_4 are 0.33 and 0.38 nm, respectively, suggesting that CO_2/CH_4 separation by $AlPO_4$-18 membrane is due to a molecular sieving effect.ã€€In addition, $AlPO_4$-18 membrane separated hydrogen from the mixture of hydrogen and propane. We will discuss on the possibilities of applying such a type of inorganic membrane to apply to separations of gas mixture containing small molecules like hydroge, CO_2 and light hydrocarbons.

530483 Two-Stage Varying-Temperature Synthesis of High-Performance SAPO-34 Membranes for H_2/N_2 Separation at High Temperature

Abstract:

Hydrogen has attracted worldwide attention in last decades for its combined advantages of high energy transferring efficiency and zero pollution emission. Especially, the demand of high-purity H_2 for PEM fuel cells requires new techniques for H_2 production with higher separation efficiency and lower cost. Membrane separation can be used for hydrogen purification with low energy consumption and ease of operation. Among those membrane-based separation techniques, the separation of gases in harsh conditions with zeolite membranes attracted tremendous interest due to its well-defined pore size, chemical, thermal and mechanical stability. In this work, high-performance SAPO-34 membranes on ceramic hollow fibers were successfully prepared by a two-stage varying temperature synthesis approach. The influence of synthesis parameters, such as precursor composition, synthesis temperature and duration, and seeding, on the microstructure evolution and separation performance of SAPO-34 membranes were investigated systematically. Finally, SAPO-34 membranes with high H_2 permeance of 4.0×10^{-7} $mol \cdot m^{-2} \cdot s^{-1} \cdot Pa^{-1}$ and H_2/N_2 selectivity of 15 were achieved at 400 °C and pressure drop of 1.1 MPa. Long-term stability testing results indicated that the obtained membranes were stable at 400 °C for over 12 h.

531946 Molecular Layer Deposition Modified SAPO-34 Membranes on Ceramic Hollow Fibers for Separation of H_2/N_2 Mixtures at High Temperature and High Pressure

Abstract:

Zeolite membranes have drawn research interest in gas and liquid separations due to their well-defined pores and channel structure and high chemical, thermal and mechanical stability. In this study, SAPO-34 membranes were successfully synthesized on the outside of ceramic hollow fibers by secondary growth method. The influence of synthesis parameters, such as precursor composition, seed particle size, morphology, concentration and stabilizer concentration in the seed suspension, and seeding conditions, and of pore size of support, on the separation performance of SAPO-34 membranes was studied in detail. Results indicated that high quality SAPO-34 membranes with thickness of < 5 μm could be achieved by using high aspect ratio seeds. The as-synthesized membranes exhibited a high H_2 permeance of 2.9×10^{-7} mol/(m^2·s·Pa) and H_2/N_2 separation selectivity of 19 at 400 °C and pressure drop of 100 psig. To further increase the selectivity, a microporous Al_2O_3 coating was deposited on SAPO-34 membrane using molecular layer deposition (MLD). The microporous Al_2O_3/SAPO-34 membrane showed higher selectivity compared SAPO-34 membrane. The thickness of the MLD layer is being optimized for optimal permeance and selectivity of the membranes, which will be discussed in detail in the presentation. These high-quality membranes may be used for high purity H_2 production from decomposition of NH_3.

532356 Enhanced Gas Separation Performance of Mixed-Linker Zeolitic Imidazolate Framework ZIF Membranes Via Post Synthetic Ligand Exchange

Abstract:

Zeolitic-imidazolate frameworks (ZIFs), a subclass of MOFs, consisting of transition metals (Zn, Co, etc.) and imidazole-derived ligands, have drawn attention mainly due to their ultra-micropores (< 5 Å) and relatively high thermal/chemical stabilities.[1] A prototypical ZIF, ZIF-8, consisting of Zn and 2-methylimidazole with sodalite (SOD) zeolitic topology, is one of the most widely studied ZIFs mainly due to its robust synthesis and potential for gas separations due to its well-defined pore size. ZIF-7, consisting of Zn and benzimidazole with SOD, exhibits smaller pore size (ca. 3.2 Å)[2] than ZIF-8 (ca. 4.0 Å)[3], therefore promising for helium or hydrogen separation.[4]

Properties of MOFs including gas transport properties are affected by structures of MOFs. To obtain MOFs with new properties, one can obtain via 1) direct (*de novo*) synthesis of new MOFs and 2) post-synthetic modifications (PSMs) of existing MOFs. PSMs include post-synthetic ligand and metal exchange[5]. Compared to the complex and sensitive nature of MOF formation process via *de novo* synthesis,[6] PSMs are a simple and facile approach to impart new properties since daughter MOFs with new properties can be obtained without compromising the structures of mother MOFs. Recent studies have shown that PSMs can modify pore size and surface properties, thereby separation performances.[7-9]

In this talk, we would like to present improved gas separation performances of ZIF-8 membranes via post-synthetic ligand exchange by systematically reducing the effective aperture size of ZIF-8. Mother ZIF-8 membranes prepared by a secondary growth were subjected to post-synthetic linker exchange with ZIF-7 linkers, i.e., benzimidazole. The effects of the PSM on the gas separation performance and water stability of the modified membranes will be discussed.

References

1 Park KS, Ni Z, Côté AP, et al., (2006) "Exceptional chemical and thermal stability of zeolitic imidazolate frameworks", *Proceedings of the National Academy of Sciences,* 103 (27), pp 10186-10191

2 Thompson, J. A., Blad, C. R., Brunelli, N. A., Lydon, M. E., Lively, R. P., Jones, C. W., Nair, S., "Hybrid Zeolitic Imidazolate Frameworks: Controlling Framework Porosity and Functionality by Mixed-Linker Synthesis", *Chemistry of Materials*, 2012, 24 (10), pp 1930–1936

3 Zhang C, Lively RP, Zhang K, Johnson JR, Karvan O, Koros WJ., (2012) "Unexpected Molecular Sieving Properties of Zeolitic Imidazolate Framework-8", *The Journal of Physical Chemistry Letters*., 3 (16), pp 2130-2134.

4 Yan Shuo Li., YS., Liang, FY., Bux, H., Feldhoff, A., Yang, WS., Caro, J., "Molecular Sieve Membrane: Supported Metal–Organic Framework with High Hydrogen Selectivity", *Angewandte Chemie International Edition*., 2010, 122, pp 558-561

5 Tanabe, KK. and Cohen, SM., (2011), "Postsynthetic modification of metal-organic frameworks-a progress report", *Chemical Society Reviews,* 40 (2), pp 498-519.

6 Kim, M., Cahill, JF., Fei, H., Prather, KA., and Cohen, SM., (2012) "Postsynthetic Ligand and Cation Exchange in Robust Metal–Organic Frameworks", *Journal of the American Chemical Society,* 134 (43), pp 18082-18088

7 Lee, M. J., Kwon, H. T., Jeong, HK., "High Flux Zeolitic Imidazolate Framework Membranes for Propylene/Propane Separation by Postsynthetic Linker Exchange", *Angewandte Chemie International Edition*, 2018, 57, pp 156-161

8 Karagiaridi, O., Lalonde, M. B., Bury, W., Sarjeant, A. A., Farha, O. M., Hupp, J. T., "Opening ZIF-8: A Catalytically Active Zeolitic Imidazolate Framework of Sodalite Topology with Unsubstituted Linkers", *Journal of the American Chemical Society*, 2012, 134 (45), pp 18790-18796

9 Zhang, H., James, J., Zhao, M., Yao, Y., Zhang, Y., Zhang, B., Lin, Y.S., "Improving hydrostability of ZIF-8 membranes via surface ligand exchange", *Journal of Membrane Science*, 2017, 532, pp 1-8

523176 Relevance of "Cross-over" Pores in CMS Membrane Transport Properties

Abstract:

So-called "cross-over pores" are believed to exist at imperfect joints between adjacent micropores in carbon molecular sieve (CMS) membranes, thereby providing unusually large ultramicropores between neighboring micropores. Ideally, CMS membranes can be described by a bimodal pore-size distribution curve comprising two different classes of pores - micropores (7 Å < $d_{micropore}$ < 20 Å) and ultramicropores ($d_{ultramicropore}$ < 7 Å). While micropores are envisioned to regulate the sorption characteristics and diffusion jump lengths for penetrant gas molecules, the molecular sieving characteristics are attributed to the ultramicroporous slits in plates between micropores. Cross-over pores need not be undesirable and may provide additional tuning tools to engineer transport of penetrant molecules, especially for bigger hydrocarbon gas-pairs. Such pores do not contribute significantly to the sorption capacities of the CMS membranes, but double the diffusion jump lengths (λ) of the molecules for the idealized case of equal-sized micropores. In the direction of transport, crossover pores act "in-parallel" with standard ultramicropores within the micropore walls. The crossover or "defect" pores can be tightened by optimizing processing parameters during CMS formation post-formation tuning strategies. Beyond size-based sieving abilities, these pores appear to present opportunities for subtle entropic discriminations as well, by restricting the degrees of freedom for large, complex gas molecules.

524948 Mixed Matrix Membranes Formed from Branched HKUST-1 for Improved Plasticization Resistance and Transport Performance

Abstract:

Mixed Matrix Membranes Formed from Branched HKUST-1 for Improved Plasticization Resistance and Transport Performance

Won Seok Chi[1], Benjamin Sundell,[2] Ke Zhang,[2] Daniel Harrigan,[2] Steven C. Hayden,[2] Zachary P. Smith[1,*]

1. MIT, Department of Chemical Engineering, Cambridge, MA, USA
2. Aramco Services Company, Aramco Research Company, Cambridge, MA, USA

Polymer membranes are promising systems for addressing energy-efficiency and the environmental concerns related to gas separations. However, polymer membranes have property sets limited by the Robeson upper bound, which describes the tradeoff relationship between permeability and selectivity. Additionally, glassy polymers, which are often used as gas separation membranes, exhibit plasticization behavior for polarizable gases (e.g., CO_2) at high gas feed pressure, resulting in decreased selectivity when tested under realistic conditions. Mixed-matrix membranes (MMMs), which incorporate inorganic fillers in polymer matrixes, could potentially address these current limitations for polymer membranes. As an effective inorganic filler, we have synthesized a metal-organic framework (MOF) containing coordinatively unsaturated metal sites (i.e., open metal sites), which can increase gas transport properties because of their high surface area, proper pore dimensions, and Lewis acidic metal sites. The MOF, known as HKUST-1, was transformed from its typical octahedral, micron-sized structure into a branched nanoparticle structure by introducing a modulator to control the nucleation and growth mechanism. The branched architecture of HKUST-1 nanoparticles shows an interconnected, high aspect ratio structure that is a few hundreds of nanometers in length but with branch widths of only a few tens of nanometers. When dispersed in a 6FDA-DAM polymer matrix, 3D reconstructed images from FIB-SEM tomography demonstrate that well-dispersed, high loadings of 30 wt% MOF lead to a partially percolated MOF network. In contrast, bulk HKUST-1 particles do not have high aspect ratios and settle in the polymer solution, leading to undesirably phase-separated membranes into MOF-rich and MOF-lean domains. The 30 wt% branched HKUST-1 nanoparticles in 6FDA-DAM significantly increase H_2, CH_4, N_2, O_2, and CO_2 gas transport properties without a large loss in selectivity for gas pairs of interest (i.e., CO_2/CH_4 and CO_2/N_2). Of particular interest, the MMMs with branched HKUST-1 nanoparticles demonstrate exceptional plasticization resistance compared to the pure polymer for high-pressure CO_2 permeation measurements. This study describes the benefits of using branched MOF nanoparticle structures for gas separation membranes compared to conventional, nearly spherical bulk MOFs. Additionally, extensive results are presented on the design, synthesis, and characterization of these MOFs and their corresponding MMMs.

530351 Engineering the Nanochannels in Reduced Graphene Oxide Membranes for Dye Desalination

Abstract:

Membranes that are highly permeable to salts (such as NaCl and Na_2SO_4) and reject organic dyes are of interest for dye desalination in textile industries. Commercial nanofiltration (NF) and ultrafiltration (UF) membranes cannot efficiently desalinate the dye solutions because NF membranes reject both divalent salts and dyes and UF membranes have a low rejection to dye molecules. Graphene oxide (GO) based membranes with tunable nanochannels formed by stacking nanosheets in parallel present a new opportunity for this application. Current research is often focused on reducing the size of nano-channels to enhance the salt rejections, which cannot be adopted for dye desalination. Herein we develop a facile approach in preparing solvated reduced GO (S-rGO) membranes with the increased nanochannel size, which lies between the size of the salt ions and dye molecules (such as Direct Red 80). Specifically, these membranes were prepared by filtration of rGO solutions through microporous Nylon support. The membranes were always kept in the wet and swollen state to control the channel sizes for the dye desalination. The S-rGO membranes can reject about 99% of Direct Red 80 with almost no rejection for NaCl and Na_2SO_4. More importantly, the nanochannels in S-rGO membranes can also be controlled by adjusting the salt concentration in the dye solutions. For example, as the Na_2SO_4 content increases to 60 g/L, the rejection of Direct Red 80 in the S-rGO membranes increases to 99.9%, while the rejection of Na_2SO_4 remains negligible. The reduction in the channel size is also validated by the decrease in the molecular weight cutoff (MWCO) of the S-rGO membranes in the salt solution. These membranes exhibit water permeance as high as 80 $L/m^2 \cdot h \cdot bar$, which is about eight times that of NF membranes and two times that of UF membranes (MWCO: 2000), suggesting their promise for practical applications of dye desalination.

532227 Tuning Water Nanofiltration Performance of Few-Layered, Reduced Graphene Oxide Membranes By Oxygen Plasma

Abstract:

Ultrathin, graphene oxide (GO) membranes have shown great potential for water nanofiltration application. However, due to the difficulties in controlling the intrinsic oxygen-containing functional groups on GO and the swelling of GO interlayers in aqueous solution, it is highly challenging to tune GO membranes' nanofiltration performance by changing their hydrophilicity and interlayer nanochannel size. In this study, oxygen plasma was reported as an effective technique to tune water nanofiltration performance of few-layered, reduced graphene oxide (rGO) membranes, by improving rGO membranes' hydrophilicity and/or adjusting effective permeation channel/pore sizes. The 5-nm thick GO membranes were fabricated through a layer-by-layer deposition method and subsequently reduced at 220 °C under vacuum. No detectable water permeance (< 1.0 L/(m^2·h·bar)) under 5 bar pressure drop suggests the high quality of the rGO membranes and/or possible existence of very small hydrophobic nanochannels/pores. Exposure to oxygen plasma for different times (0~30s) gradually improved the hydrophilicity of the rGO membranes, with the water contact angle decreased from 74 to 31°. This is consistent with x-ray photoelectron microscopy (XPS) results showing gradually increased oxygen-containing functional groups with oxygen plasma time. By tuning membrane properties upon different oxygen plasma exposure times, the 5-nm thick rGO membranes showed the optimum nanofiltration performance with water permeance of 44 L/(m^2·h·bar) and 98% rejection for methylene blue. Oxygen plasma treatment, therefore, may provide a viable way for tuning few-layered, graphene-based membranes for highly efficient water nanofiltration application.

536236 Graphene Oxide Nanoplatelets Embedded Polyamide Thin Films for Water Desalination

Abstract:

Polyamide thin film membranes are used in water desalination applications. Incorporation of graphene oxide nanoplatelets in the polyamide layer can alter the surface characteristics, enhance the mechanical properties, and can improve the permeability and selectivity of the membranes. Here, we present a modular fabrication method to incorporate nanomaterials in polyamide thin films. In this study, a layer-by-layer (LbL) synthesis technique has been used for embedding graphene oxide nanoplatelets (GONPs) in the polyamide thin films. We have synthesized alternating layers of GONPs and polyamide, and GONPs on top of the polyamide layer. The LbL technique allowed us control the thickness of these films, also the locations of GONPs within the films. Incorporation of GONPs resulted in an increase of surface hydrophilicity, as captured by the change of contact angle. The presence of GONPs in the synthesized membranes has been confirmed by SEM and TEM. Water flux and salt rejection properties of synthesized membranes have been investigated by using a dead-end cell. The GONP embedded membranes have been found to enhance the salt rejection while keeping water flux similar compared to the pristine membranes also fabricated through the LbL synthesis technique.

532613 Adhesion and Scaling Mechanism of Gypsum on the Commercial Microfiltration Membranes

Abstract:

In view of the scaling problem in the membrane water treatment process, the adhesion mechanism of gypsum crystals on different commercial microfiltration membranes in aqueous solution was firstly investigated using experimental methods and molecular dynamics simulations. The membrane chemistry is decisive to its adhesion potential to gypsum crystals. Compared to the hydrophobic membranes, hydrophilic membranes have higher tendency to scaling by the gypsum crystals. The number of polar groups in the membrane molecules has higher effect on the adhesion energy than the variety of polar groups. The anti-adhering properties of the membranes increase with decreasing adhesion energy. Physical structure of membranes was also crucial to its adhesion tendency under the condition that the adhesion energy of membranes was close to each other. The larger pore that exceeds a proper range leads to the decline of adhesion mass. Subsequently, the membranes potentially used for membrane distillation were predicted in terms of their anti-scaling potential using the same simulations. This article provides an insight for the mechanism of gypsum scaling, and gives the prediction on the anti-scaling performance of the potential polymer membranes in membrane distillation.

536380 Permselective Transport of Organic Molecules in Perfluorosulfonic Acid Polymer Membranes for Personal Protective Equipment

Abstract:

Personal protective equipment (PPE) typically employ air-permeable textiles that incorporate activated carbon or other microporous adsorbents and coated by a liquid-repellent outer shell. Alternately, full encapsulation by an air-impermeable material provides the greatest level of protection per OSHA Level A standard. Either system is heavy and bulky resulting in an experience of heat stress even under mild exertion. Permselective membranes and coatings are believed to be the logical next advance by eliminating activated carbon while improving breathability. Prior research with perfluorosulfonic acid (PSA) polymers and membranes such as Nafion attempted to exploit the heterophase morphology of these materials to achieve toxic agent-water permselectivity. In particular, water is known to diffuse through the hydrophilic subphase formed by the sulfonate side chains while, ideally, organic agents would be trapped or impeded by the hydrophobic subphase formed by the backbone. Such permselectivity would allow sweat from the wearer to escape while the transport of toxic agents would be impeded. Instead, water is found to significantly promote agent transport. Molecular dynamics (MD) simulations and indirect experimental evidence suggest that this is due to the "Janus" structure of toxic agents such as dimethyl methyl phosphonate (DMMP) and di-isopropyl methyl phosphonate (DIMP) which contain a hydrophilic "head" and a hydrophobic "tail". This structure is hypothesized to permit the agents to persist at the interphase between water and the hydrophobic subphase and transport through the membrane. In this presentation, we provide diffusivity data for a variety of organic molecules with a "Janus" structure (e.g., ketones, phosphonates, phenols and organic acids) that exhibit Nafion permeability ranging from un-impeded transport in the case of ketones to near-immobilization in the case of organic acids. Such behavior demonstrates that, despite years of research and significant understanding of how PSA polymer morphology impacts the transport of ions and low-molecular weight solvents, substantial uncertainty remains concerning the permeation behavior of higher molecular weight organic molecules. Our efforts to develop mechanistic understanding in this area are discussed.

536885 Phosphate-Functionalized Membranes for the Selective Sequestration of Uranium from Seawater

Abstract:

Phosphate-functionalized membranes for the selective sequestration of uranium from seawater

Priyanka Suresh, Christine E. Duval

Department of Chemical and Biomolecular Engineering, Case Western Reserve University, Cleveland OH

Increasing global energy demand and diminishing fossil-fuel resources have led to the exploration of alternative energy sources. In spite of research advances in renewable energy sources such as wind, tidal and solar, nuclear energy has captured extensive attention due to its economic viability and absence of CO_2 emissions. For a sustainable development of nuclear energy, the availability and accessibility of uranium, the most prevalent nuclear fuel, is vital. Since terrestrial ores of uranium are limited, there is an interest to sequester some of the 4.5 billion tons of uranium that exists in seawater. Use of traditional adsorbents to recover U(VI) from seawater is a challenge because of uranium's low concentration (~3.3ppb) and the presence of numerous competing ions. Amidoxime-based (PAO) sorbents have been extensively studied for this purpose due to their fast kinetics, high capacity and facile synthesis; however, they lack uranium selectivity. In fact, when placed in seawater, PAO sorbents bind greater quantities of magnesium, calcium and vanadium than uranium. One strategy to improve sorbent capacity is to develop sorbents with high uranium selectivity. Among the various functional groups, phosphate groups have attracted increased attention due to their stability, irradiation resistance, and high complexation stability.

In this work, we present phosphate-functionalized ultrafiltration membranes that selectively adsorb uranium from seawater. Membranes were prepared by grafting ethylene glycol methacrylate phosphate (EGMP) from the surface of polyethersulfone (PES) membranes by UV-initiated graft polymerization. The surface morphology of the membranes was studied by scanning electron microscopy before and after grafting. The presence of phosphate functional groups on the membrane was supported by Fourier transform infrared spectroscopy (FTIR). The degree of grafting (DoG) and the number of acidic sites were quantified by acid-base titrations. Grafting conditions were optimized by evaluating the effect of multiple solvents on the DoG. Grafted membranes were characterized in terms of water uptake as well as uranium capacity and selectivity in seawater conditions. Membrane stability and reuse is demonstrated through uranium desorption cycles. This work demonstrates the importance of employing selective chemistries for the trace-level concentration of analytes in the environment.

533246 Highly Water Selective Mixed Matrix Membranes with UiO-66-Type MOFs in 6FDA-HAB/DABA Polyimide for Alcohol Dehydration Via Pervaporation

Abstract:

This study demonstrated the development of a highly water selective mixed matrix membranes (MMMs) consisting of porous UiO-66-NH_2 nanoparticles and 6FDA-HAB/DABA polyimide for isopropanol dehydration. The water in permeate can be concentrated to almost 100%. Results suggest incorporation of UiO-66-NH_2 nanoparticles not only can enhance separation factor but also greatly improve the normalized flux if the particle loading is within the range of 0-20wt%. The UiO-66-NH_2 based MMMs also exhibited stable isopropanol dehydration performance in uninterrupted 200-hour tests due to the good compatibility between UiO-66-NH_2 particles and the 6FDA-HAB/DABA polyimide resulting from the intermolecular bonding among the amine and imide groups. Besides, in order to investigate the effects of functional groups on UiO-66 particles on alcohol dehydration, MMMs containing the other two UiO-66-type MOF nanoparticles (UiO-66 and UiO-66-F_4) were also fabricated and compared with UiO-66-NH_2 based MMMs for alcohol dehydration. UiO-66-NH_2 based MMMs have the best separation performance among these three MMMs because the hydrophilic amino-functional groups can facilitate water transport. The excellent alcohol dehydration performance of UiO-66-NH_2 based MMMs opens up their great potential for alcohol dehydration via pervaporation.

530850 Solution-Phase Adsorption of Furan and Carboxylic Acid in Hierarchical Zeolites:Insights from Molecular Simulation

Abstract:

Oleo-furans[1] are a novel class of bio-renewable surfactants with superior hard-water stability compared to petroleum-derived surfactants. During their synthesis, sugar-derived furans react with triglyceride-derived fatty acids via acid-catalyzed acylation. Their synthesis is promoted by hierarchical zeolites, such as self-pillared pentasil (SPP) zeolites[2], which combine microporous and mesoporous functionalities to perform selective catalysis and accommodate diffusion of large molecules.

To improve fundamental understanding and enable improvements in catalyst and process design, the co-adsorption of furan, hexanoic acid, and various solvents from solution into SPP was investigated using Monte Carlo simulations in the Gibbs Ensemble. Solution composition and reaction conditions significantly affect adsorption selectivity, as well as the distribution of species in the meso- and micropores of SPP. Silanols on the surface of SPP strongly hydrogen bond with hexanoic acid, promoting selective adsorption of acid from the solution phase onto the mesopore walls, while furan and hexane solvent are more evenly distributed throughout the mesopore and micropore volumes. The influence of several process variables on the adsorption selectivity and spatial distribution in the zeolite was also investigated, including solution composition, the chain length and degree of branching of the solvent, and adsorption in both subcritical and supercritical regimes.

References

[1] D.S. Park, K.E., Joseph, M. Koehle, C. Krumm, L. Ren, J.N. Damen, M.H. Shete, H.S. Lee, X. Zuo, B. Lee, W. Fan, D.G. Vlachos, R.F. Lobo, M. Tsapatsis, and P.J. Dauenhauer. "Tunable Ole-Furan Surfactants by Acylation of Renewable Furans," ACS Central Science, **2016**, 2, 820-824.

[2] X. Zhang, D. Liu, D. Xu, S. Asahina, K.A., Cychosz, K.V. Agrawal, Y.A. Wahedi, A. Bhan, S. A. Hashimi, O. Terasaki, M. Thommes, M. Tsapatsis, "Synthesis of Self-Pillared Zeolite Nanosheets by Repetitive Branching," Science, **2012**, 335, 1684-1687.

524649 First-Principles-Derived Force Fields for CH_4 Adsorption and Diffusion in Siliceous Zeolites

Abstract:

Most previous studies on development of force fields for molecules in porous materials focus on prediction of adsorption properties. However, accurately reproducing adsorption data is not sufficient to guarantee the accuracy of other properties such as diffusivities. We demonstrate an approach to develop force fields based on periodic first-principles calculations that can accurately predict both adsorption and diffusion properties in crystalline nanoporous materials using CH_4 in siliceous zeolites as an example. First, multiple dispersion-corrected density functional theory (DFT) methods were tested for describing CH_4 in siliceous chabazite, and two most important configurations that are relevant to CH_4 adsorption and diffusion were investigated. By comparing with the results from high level random phase approximation (RPA) results, DFT/CC (coupled cluster) was found to be the optimum method for force field development. The DFT/CC-derived force fields that accurately predict both adsorption and diffusion of CH_4 in several commonly studied siliceous zeolites are then developed.

535440 Investigating High Pressure Methane Storage Using Lennard-Jones Crystals

Abstract:

Decades of research have yet to yield porous adsorbents that meet the US Department of Energy's methane storage targets. To investigate this lingering challenge, we simulated high pressure methane adsorption in large libraries of randomly generated porous crystals, or "pseudomaterials." These pseudomaterials are periodic configurations of Lennard-Jones (LJ) spheres which are configured at random. Methane adsorption was compared for all materials against a range of other properties: average ε (potential well depth) and σ (van der Waals radius) value, number density, helium void fraction, and volumetric surface area. The results reveal structure-property relationships that resemble those observed previously for MOFs and other porous materials. We find that our computational methodology can be useful for discovering structure-property relationships related to gas adsorption without requiring detailed structural data of real (or even realistic) materials.

523273 Formaldehyde Adsorption Performance of Selected Metal-Organic Frameworks from High-Throughput Computational Screening

Abstract:

Volatile organic compounds (VOCs) have caused severe air pollutions due to their harmfulness to human health. Among the most popular VOCs, formaldehyde (HCHO), is very allergenic and carcinogenic even at very low concentrations. As an emerging class of porous crystalline materials, Recently, metalâ€"organic frameworks (MOFs), a class of crystalline nanoporous materials composed of inorganic metal nodes and organic ligands, have attracted increasing research interest due to their high specific surface areas and pore volumes, well-defined porosity, and tunable pore properties, which has been utilized for gas storage, separation, and catalysis. So, the utility of MOFs for formaldehyde removal is an intersting and meaning research topic. With the rapidly increasing number of reported metal-organic frameworks (MOFs), high-throughput computational screening (HTCS) strategy based on grand canonical Monte Carlo (GCMC) simulation is proposed to quickly explore the top-performing MOFs with high adsorption capability towards formaldehyde. In this work, the computation-ready experimental (CoRE)-MOF database consisting of 2932 MOF structures carrying density derived electrostatic and chemical (DDEC) charges obtained from density function (DFT) theory calculations, were employed in high-throughput GCMC simulations for formaldehyde adsorption from the air. The structure-property relationship from HTCS revealed that the MOF candidates with high formaldehyde uptakes exhibited small pore sizes, relatively high selectivity and moderate heat of adsorption (Qst). Afterwards, the top MOFs with both high uptake and selectivity towards formaldehyde were chosen for further experimental evaluation. Three selected MOFs Y-BTC, ZnCar and Ni-BIC were successfully synthesized and characterized by Powder X-ray diffraction (PXRD) and BET surface area analysis. In order to validate our HTCS strategy, the representative Cu-BTC and activated carbon (AC) were also adopted as controls. The formaldehyde adsorption test was performed in a sealed container with the formaldehyde concentration of 100 mg/m
24 h adsorption, the formaldehyde uptakes of the adsorbents were obtained according to the concentration changes prior to and after formaldehyde exposure by UV-vis spectrometer. It was found that the adsorption capacities of Y-BTC, ZnCar and Ni-BIC are 0.38, 0.25 and 0.11 mol/kg, respectively, which were remarkably higher than Cu-BTC (0.08 mol/kg) and AC (0.06 mol/kg). The recyclability of the best performer Y-BTC was also verified. These findings open up the possibility of employing HTCS strategy for highly efficient exploration of MOF adsorbents for formaldehyde removal.

533611 Modeling and Simulation of Multicomponent Adsorption Columns

Abstract:

Adsorptive separation of a multicomponent gaseous mixture in a suitable nanoporous material is a non-hazardous, inexpensive, sustainable, and environmentally benign process. The present work focuses on developing a comprehensive theoretical approach to separate methane and carbon dioxide from other gases such as nitrogen using packed-beds filled with nanoporous adsorbent. Previous work by this group considered binary mixtures, and isothermal as well as isobaric conditions. Here, emphasis is given in multicomponent mixtures and temperature and pressure effects. The adsorption of gaseous species onto the surface of a porous solid is simulated using kinetic Monte Carlo methods on a square lattice. Species from the gas phase deposit onto the sites of the surface thus forming a monolayer of adsorbed species. The parameters required in the rate expressions of the different molecular micro-processes are estimated from experimental pure-component adsorption data. Adsorptive separation of the components of a multicomponent gaseous mixture is achieved in fixed-beds filled with porous adsorbent. As the gas stream flows through the column, the species deposit onto the porous solid at different degrees. The objective of the modeling is the calculation of the concentrations at the column outlet in terms of time (breakthrough analysis). Modeling of the column comprises conservation statements for the species (mass balances) that contain the amounts adsorbed. As already emphasized, the present work considers non-isothermal conditions and pressure losses. The resulting conservation equations can be solved numerically using a finite difference/element technique in which the equations are discretized in space. The system of equations is advanced in time by calculating the amount adsorbed from kinetic Monte Carlo simulations using parameters that correspond to the condition (concentrations, temperature, and pressure) of the gas phase at each node. The model will be tested and used in the sweetening (separation of sour gases) and inert rejection (separation of nitrogen) of natural gas.

534598 Integrated Kinetic Monte-Carlo Method to Find Face Specific Growth Rates

Abstract:

The morphology or shape of crystalline materials is a critical determinant of their physical properties such as bulk density, stability, wettability, flowability, and dissolution rates. For example, when plate- or needle-like morphologies appear in pharmaceutical crystals, granulation and compression are difficult when making tablets. The ability to use theoretical models to predict how the crystallization conditions will affect the crystal morphologies is imperative for developing control strategies to prevent the formation of undesired morphologies. Knowledge of face-specific growth rates of crystals is key for predicting dynamic and steady-state crystal morphologies. Current models rely on empiricism and calculation of interfacial free energies which are often difficult. In this talk, I will describe the multi-scale model for determining face specific growth rates using a Kinetic Monte-Carlo method. The kinetic, multi-scale model is used to capture the most fundamental mechanism that occur during crystallization â€“ the association and dissociation of molecules â€“ and determines the most probabilistic realization of the surface. Our method considers probabilities of events such as the nucleation of 2D islands, formation of screw dislocations, and the spreading of the layers. The velocity of the edges is based on the kinetic approach which calculates the energy barrier for integration of molecules into the crystal structure. The approach described in this talk will be exemplified on a glutamic acid crystal.

536424 High Aspect Ratio Metal Organic Framework Synthesis Using Microfluidic Devices

Abstract:

Metal organic frameworks (MOFs) are a promising class of material for use in applications ranging from separations, storage, drug delivery, optical devices and more. The diversity of synthesis techniques for MOFs with controllable shape and size makes the material interesting to explore. Recently, addition of modulators and structure directing agents has gained momentum as a technique for producing high aspect ratio crystals such as nanorods, nanotubes and nanosheets. Here, a microfluidic device was used to control the synthesis environment of metal organic frameworks. We discuss the creation of different microfluidics devices to control the crystallization of MOFs, and the effect device characteristics have on subsequent MOF crystallization. Channels with variable depth and width can create crystals with different morphologies. The crystallinity of the resulting crystals is characterized. The use of microfluidic synthesis for multiple types of MOFs is discussed, as well as their potential use in future applications.

533970 Towards Modelling Facet-Specific Impurity Incorporation Rates Using Molecular Dynamics Simulations – a Case Study on the Urea/Biuret/Water System

Abstract:

Crystallization is one of the most widely used separation technologies for the purification of fine chemicals, agrochemicals, and pharmaceuticals. In the pharmaceutical industry, as demands on product purity can be extremely high (up to 99.9%)[1], it is of vital importance to minimize impurity incorporation during the crystallization process. To this end, this study is aimed towards a better understanding of the molecular mechanisms underpinning impurity incorporation into organic crystals.

Crystals grown from solution expose multiple facets, where each facet is a cessation of the crystal structure. The unique arrangement of host molecules and hence functional groups on a developing crystal face leads to varying affinities for the solute and impurity molecules present in solution. This, in turn, results in facet-specific growth rates and inherently anisotropic impurity distribution in the crystal lattice.

This work has quantitatively investigated impurity incorporation into an organic crystal using the classical example of urea crystal growth from aqueous solution, with biuret (a reaction by-product) present as an impurity. Molecular dynamics (MD) simulations were performed in GROMACS to determine incorporation of biuret on the (001) and (110) facets respectively. Keeping the initial supersaturation constant among the different cases, we have varied the concentration of the biuret impurity present in the solution. The computational method developed previously by Salvalaglio *et al.*[2] to quantify the 'crystallinity' of urea molecules (the 'crystallinity' parameter was calculated by taking into consideration the C=O bond angles between a urea/biuret molecule incorporated from solution and that of neighboring urea molecules in the crystal lattice) was adapted to allow quantifying the number of biuret molecules incorporated during the growth of different crystal facets.

It was previously shown experimentally and from simulation work[2,3] that the growth rate of the {001} facet is strongly affected by the presence of biuret in the solution, while the growth rate of the {110} facet is largely unaffected. However, an analysis of the number of biuret molecules incorporated during crystal growth is to date absent from the literature.

References:

[1] "FDA Guidance for Industry Drug Substance Chemistry, Manufacturing, and Controls Information," Structure no. Cvm, **2004**.

[2] M. Salvalaglio, T. Vetter, F. Giberti, M. Mazzotti, and M. Parrinello, "Uncovering Molecular Details of Urea Crystal Growth in the Presence of Additives," *J. Am. Chem. Soc.*, 134 (41), 17221–17233, **2012**.

[3] M. K. Singh, V. S. Tiwari. "Uncovering the Mode of Action of Solvent and Additive Controlled Crystallization of Urea Crystal: A Molecular-Scale Study", *Cryst. Growth Des.* 15(7), 3220–3234. **2015**.

536041 Designing Inhibitors of Mineral Scale: A New Platform Based on Cooperative Microfluidic and Computational Assays

Abstract:

Mineral scale occurs in processes ranging from water treatment and purification to oil and gas production systems, posing significant challenges to the upstream petroleum industry. Designing effective biodegradable chemical treatments to reduce scale formation requires understanding the molecular-scale interactions of inhibitors during nucleation and growth of scale. Here we apply new approaches to design and test environmentally friendly inhibitors of scale formation using a combination of surface science techniques, microfluidic technology, and molecular modeling. Crystallization can be inhibited in highly supersaturated solutions using dilute quantities of macromolecular modifiers. How the presence of flow, which alters the dynamics of modifier-crystal interactions, affects inhibitor efficacy is not fully understood. Our studies focus on barium sulfate (barite), a common scaling mineral. In this presentation we will describe a microfluidic platform that is used to perform multiple bulk crystallization assays simultaneously and capture crystal growth and inhibition in real time in the presence of biomacromolecules. Using *in-situ* scanning probe microscopy, we probe the interfacial behavior between the biomacromolecules and different surfaces of barite and elucidate the role of these putative inhibitors during the growth phase of mineralization. Concurrently, we use molecular modeling of inhibitor adsorption of barite surfaces to probe the interactions that govern inhibitor specificity and efficacy. Collectively, these studies generate new understanding of the mechanism(s) of crystal growth modifiers and apply this knowledge to design improved generations of scale inhibitors.

526095 Secondary Nucleation and Growth Kinetics of Aluminum Hydroxide Crystallization from Potassium Aluminate Solution Using FBRM

Abstract:

Aluminate hydroxide crystallization is a significant procedure in extraction and recovery of aluminum from aluminum-containing ore with alkaline method. The alumina-rich potassium aluminate liquor is separated from the leaching solution[1,2], then the alumina is crystallized (precipitated) as hydrate by seeding with aluminate hydroxide. Secondary nucleation and growth are main phenomena in industrial crystallization of aluminum hydroxide.

There are many different experimental techniques that have been used to measure the secondary nucleation and growth kinetics. As a promising apparatus of process analytical technologies(PAT), focused beam reflectance measurement (FBRM) can directly monitor the chord length distribution (CLD) at a very short time interval, such as 2s, so more detailed information can be recorded during crystallization without sampling[3]. A method is developed to calculate the low-order moments from the CLD directly without converting the CLD to PSD[4]. Therefore, secondary nucleation and growth rates were measured by FBRM directly.

In this work, moments of crystal population were estimated from the measured CLD, generated by the FBRM[4], and the supersaturation was calculated from the measured solution concentration by ICP-AES. The moments were uesed to approximate the secondary nucleation and growth rates. Meanwile, the secondary nucleation and growth rates were also generally represented by a power-law function of the main driving force for crystallization process, supersaturation. Therefore, kinetics parameters for secondary nucleation and crystal growth were calculated based on experimental data at different supersaturations and temperatures. Also, the kinetic model was verified in further experiments by comparison of experimental data and predicted data.

Another method was perfomed to estimated gibbsite secondary nucleation and crystal growth using a semiempirical power-law model by BET and ICP-AES[5]. Instead of measuring total particle surface area by BET, moments of crystal population by FBRM were uesed to calculate total particle surface area in this work. And the kinetics calculated by FBRM showed a same trend in comparision with the results by BET.

References:

[1] Mengjie Luo,Chenglin Liu,Jin Xue,Ping Li,Jianguo Yu.Leaching kinetics and mechanism of alunite from alunite tailings in highly concentrated KOH solution[J]. Hydrometallurgy, 2017, 174: 10-20.

[2]Mengjie Luo,Chenglin Liu,Youfa Jiang,Jin Xue,Ping Li,Jianguo Yu.Green recovery of potassium and aluminum elements from alunite tailings using gradient leaching process[J]. Journal of Cleaner Production,2017,168:1080-1090.

[3]Yuzhu Sun, Xingfu Song, Jin Wang,Yan Luo,Jianguo Yu. Determination of seeded supersolubility of lithium carbonate using FBRM. J. Cryst. Growth. 294â€"300, 311(2010).

[4]Milana Trifkovic,Mehdi Sheikhzadeh,Sohrab Rohani.Multivariable real-time optimal control of a cooling and antisolvent semibatch crystallization process[J]. AIChE Journal, 2009, 55(10): 2591-2602.

[5]Jun Li,Jonas Addai-Mensah, Alagu Thilagam, Andrea R.Gerson.Growth Mechanisms and Kinetics of Gibbsite Crystallization: Experimental and Quantum Chemical Study[J]. Crystal Growth & Design, 2012 Jun, 12(6): 3096-3103.

528006 Crystallization Behaviors of Lithium Carbonate in Strong Alkaline Solution: Solubility, Nucleation Mechanism, and Particle Size Control

Abstract:

Crystallization behaviors of lithium carbonate in strong alkaline solution: solubility, nucleation mechanism, and particle size control

Youfa Jiang, Chenglin Liu, Jin Xue, Mengjie Luo, Ping Li, Jianguo Yu**

State Key laboratory of Chemical Engineering, National Engineering Research Center for Integrated Utilization of Salt Lake Resources, East China University of Science and Technology,

Meilong Road 130, Shanghai, China, 200273.

jiangyoufa2012@126.com, liuchenglin@ecust.edu.cn, jgyu@ecust.edu.cn

Abstract:

Lithium carbonate is extensively applied in high energy density batteries,[1] pharmaceuticals[2] and ceramics industry.[3] In particular, with the rapid development of lithium-ion battery, the crystallization of lithium carbonate has drawing increasing attention under various situations, such as the Li_2CO_3-NaCl-Na_2SO_4-H_2O system,[4] reaction between LiCl and CO_2 by use of solvent extraction process,[5] and the gas-liquid reactive crystallization of CO_2 and LiOH.[6]

In practice, two natural lithium source are typically utilized for production of Li_2CO_3: salt brine and lithium-containing minerals,[7] among which the later accounts for ~70% of total lithium production in China.[8] The industrial process for extraction of lithium from minerals includes acid digestion, alkaline digestion and ionic exchange method.[9] And the soluble lithium in strong alkaline solutions is obtained during alkaline digestion.[10] The objective of this work is to investigate the reactive crystallization process of Li_2CO_3 in strong alkaline solution.

In this study, the solubility of lithium carbonate in alkaline solutions of different concentrations over the temperature range from 303.15 K to 363.15 K is firstly investigated. In addition, the Pitzer model is adopted to accurately estimate the solubility of Li_2CO_3 in NaOH solutions, which provides a fundamental basis for recovery of lithium from the alkaline leaching liquor.

The induction period for lithium carbonate primary nucleation, as a function of caustic concentration, supersaturation, temperature and string rate, is subsequently measured by use of focused beam reflectance measurement (FBRM). The primary nucleation includes homogeneous nucleation and heterogeneous nucleation, which can be verified according to the correlation between induction period and supersaturation. The determination of induction period not only helps obtain important properties such as interfacial tension, and critical nucleus size, but also contributes to manipulate the crystallization process to produce products of high quality.

Furthermore, the operating parameters such as caustic concentration, temperature, the feed rate of Na_2CO_3 solution and the concentration of LiOH are subsequently studied to investigate the influence on the lithium recovery and the quality of product. And the mechanism of crystallization of lithium carbonate in strong alkaline solutions is proposed to highlight the effect of alkali on the morphology and crystal size distribution (CSD) of lithium carbonate.

Keywords: lithium carbonate, strong alkaline solution, solubility, nucleation, particle size control

Literature Cited

1. Sun Y, Song X, Wang J, Luo Y, Yu J. Unseeded Supersolubility of Lithium Carbonate: Experimental Measurement and Simulation with Mathematical Models. *J. Cryst. Growth.* 2009;311(23¨C24):4714-4719.

2. Yu W, Sun X, Li Y, Li M, Wang K, Li H. Population Pharmacokinetics of Lithium Carbonate in Young Male Healthy Chinese Volunteers. *Pharmacopsychiatry*. 2016;49(05):177-181.

3. Han P, Jiang X, Zhang L, et al. Effect of Li_2CO_3 flux on the preparation temperature, particle micro morphology and light absorption performance of samarium borate by solid state method. *J. Mater. Sci.: Mater. Electron.* 2014;26(2):1-5.

4. Wang J, Wu X, Zhang S. Development of a thermodynamic model for the Li_2CO_3-NaCl-Na_2SO_4-H_2O system and its application. *J. Chem. Thermodyn.* 2018;123:62-73.

5. Zhou Z, Liang F, Qin W, Fei W. Coupled Reaction and Solvent Extraction Process to Form Li2CO3: Mechanism and Product Characterization. *AIChE J.* 2014;60(1):282-288.

6. Sun Y-Z, Song X-F, Jin M-M, Jin W, Yu J-G. Gas¨CLiquid Reactive Crystallization of Lithium Carbonate by a Falling Film Column. *Ind Eng Chem Res.* 2013;52(49):17598-17606.

7. Swain B. Recovery and recycling of lithium: A review. *Sep. Purif. Technol.* 2017;172(Supplement C):388-403.

8. Jaskula BW. U.S. GEOLOGICAL SURVEY MINERALS YEARBOOK¡ª2015. 2015.

9. Rosales GD, Ruiz MDC, Rodriguez MH. Novel process for the extraction of lithium from ¦Â-spodumene by leaching with HF. *Hydrometallurgy*. 2014;s 147¨C148:1-6.

10. Averill WA, Olson DL. A review of extractive processes for lithium from ores and brines. Energy. 1978;3(3):305-313.

527110 Electrospun Nanoparticle/Nanofiber Composites (ENNCs) As New Core-Shell Adsorbent Materials

Abstract:

Core-shell structures have potentials to increase selectivity and/or attenuate contaminants uptake by forming a protective shell on the external surfaces of a porous adsorbent particle. These materials function by a kinetically selective mechanism, where the shell provides a barrier that slows the uptake rate of an undesirable component through either a solution-diffusion or molecular sieving mechanism. One critical factor to consider is the shell thickness, which controls the diffusion barrier. However, currently used models do not address the shell thickness as a parameter at all. It is essential for predictive kinetic equations to be developed for the field of core-shell sorbent design to progress. In this work, we solve the diffusion equations for core-shell sorbents using finite-element methods and compare to a derived analytical model to describe the uptake equations for core-shell sorbents. We also use CO2/N2 separation as a good example to demonstrate the potential of electrospun nanoparticle/nanofiber composites (ENNCs) as new core-shell adsorbent materials.

537651 TheÂ Structure-Property Relationships of Metal-Organic Frameworks for Ethylene/Ethane Separation

Abstract:

Oxidative coupling of methane (OCM) is a promising route to convert natural gas (CH_4) to value-added ethylene. However, the existing OCM catalysts will typically produce a large amount of by-products like ethane that is difficult to separate from ethylene due to their similar physical properties. In industry, these two gases are separated via high energy-consuming cryogenic high-pressure distillation. Therefore, there is significant motivation to develop a simple and effective separation method to afford low energy consumption and high ethylene purity. Metal organic frameworks (MOF) are promising candidates for use in this application due to their high surface areas, large pore volumes, high adsorption capacities and selectivity for specific adsorbates. DOMF and its isostructural family were synthesized to determine the structure-property relationship of metal-organic frameworks for ethylene/ethane separation due to the ability of their framework ligands to possess a myriad of functional groups. The results show that both DMOF and DMOF-TMBDC demonstrate preferential selective adsorption of ethane over ethylene. Since the pore size of DMOF-TMBDC is closer to the kinetic diameter of ethylene and ethane than DMOF, it can reach saturation faster with higher IAST predicted selectivity, indicating the methyl groups facilite the interactions between gas molecules via increasing van der Waals forces. Furthermore, DMOF-TMBDC exhibits much higher ethane uptake in the low-pressure region than DMOF, which is helpful for the ethylene production from OCM reaction where the partial pressure of ethylene and ethane is low. In addition, both heats of adsorption of ethylene and ethane are lower than 40 kJ/mol, excluding chemisorption interactions in the framework. Low heats of adsorption suggest adsorbed gas can be regenerated easily without consuming large amount of energy, which is desirable for efficient purification of ethylene.

515280 Enhanced $SF_{6Â}$ adsorption Kinetics By Hierarchically Structured Nanoporous Materials

Abstract:

Research on greenhouse gas capture have been extensively conducted in view of the rising concerns of global warming phenomena. Nonetheless, sulfur hexafluoride (SF_6), which is commonly incorporated in metal-working industries and electronic sectors possess much higher global warming potential (GWP) and atmospheric lifetime than carbon dioxide (CO_2), a common greenhouse gas. Thus, recovery of SF_6 is necessary to prevent strong release of this gas to the atmosphere. Conventionally, liquefaction process which had been utilized industrially to remove SF_6 are considerably energy intensive due to low SF_6 content in the feed (typically 10%, with N_2 as the balance). Therefore, adsorptive-based separation which uses zeolites and metal-organic frameworks displays its promising behavior in overcoming the limitation. However, detailed studies and analyses are required so as to assure its practicability in industrial process.

In this study, HKUST-1 which possess coordinatively unsaturated open metal sites was conducted for its potential application in SF_6/N_2 separation. It was observed that the creation of hierarchically-structured HKUST-1 nanocrystal demonstrates improved SF_6 adsorption, SF_6/N_2 selectivity as well as faster SF_6 adsorption kinetics as compared to conventional HKUST-1. The creation of mesoporosity as well as high accessible surface area allows strong facilitation and transport of SF_6 to the active sites, leading to such enhancement. The analysis of idealized vacuum swing adsorption further validates its capability of hierarchically-structured HKUST-1 in this process.

Besides, using similar strategy, a series of hierarchically-structed zeolite MFI was also studied for its potential utility in SF_6/N_2 separation. In this study, enhanced SF_6/N_2 selectivity at slight elevated temperature (40 °C) was observed as compared to bulk MFI. Most importantly, a clear enhancement in SF_6 adsorption kinetics was seen, owing to the presence of mesoporosity that allows strong facilitation of SF_6 to the available active sites. Further evaluation through the SF_6/N_2 breakthrough analysis have verified that a sharper breakthrough curves can be seen on hierarchically-structured MFI, indicating that the diffusion of SF_6 in such structure can be facilitated easily under dynamic condition.

References:

1. C. Y. Chuah, K. Goh, T. H. Bae, Hierarchically Structured HKUST-1 nanocrystals for enhanced SF_6 capture and recovery, J. Phys. Chem. C, 121 (12), 6748-6755

2. C. Y. Chuah, S. Yu, K. Na, T. H. Bae, Enhanced SF_6 recovery by hierarchically structured MFI zeolite, J. Ind. Eng. Chem., *in press*

533691 Development of Small-Pore Rho Zeolites for Gas Separation Applications

Abstract:

Gas separations, such as the kinetic separation of oxygen from either nitrogen or argon by adsorption, requires adsorbent structures with very specific pore dimensions. The Lennard-Jones 6-12 kinetic diameter of Ar (3.4Å) is smaller than that of O_2 (3.46Å). Unlike Ar, O_2 is not spherical and has a minimum molecular dimension of 2.8Å, and this 0.6Å difference in diameters between O_2 and Ar can be exploited for kinetic separation of these gases. For example, a small-pore zeolite with an effective diameter of 2.8Å would allow O_2 diffusion and restrict Ar. With such an adsorbent, an Ar pressure swing adsorption (PSA) process could be developed for the production of crude argon in an efficient and safe manner. In this paper, we will discuss the development of novel RHO zeolites with a Si/Al ratio of 3.9, which are useful as kinetically selective adsorbents for oxygen and/or nitrogen.

The RHO zeolites useful for Ar PSA, which will be the focus of this work, can contain a variety of extra-framework cations. The composition of these zeolites is tunable through ion-exchange, and we will discuss how the choice of extra-framework cation(s) causes structural change in the 8-ring windows that act as the selective pores. This structural change is the key component to controlling the selectivity of the zeolite toward gas molecules of interest. These RHO zeolites have superior O_2 uptake rate, capacity, and selectivity relative to carbon molecular sieves at sub-ambient to ambient temperatures.

523030 Molecular Blends of Methylated-Poly(ethylenimine) and Amorphous Porous Organic Cages for SO_2Â Adsorption

Abstract:

Porous organic cages (POCs) are emerging porous materials that exhibit intriguing properties in the areas of self-assembly, host-guest interaction, and solution processability. In this work, we explore the applicability of POCs as molecular porous supports for polymeric amines. We find that primary amines in poly(ethylenimine) (PEI) can undergo metathesis with the imine bonds present in POCs, resulting in non-porous products. This problem can be overcome by transforming the primary amines in PEI to tertiary amines via methylation. The methylated PEI (mPEI) forms homogeneous composites with amorphous scrambled porous organic cages (ASPOCs) without undesired reactions or phase separation. The microscopic structure of the composites is studied using molecular dynamics simulations. These composite materials are evaluated as adsorbents for low concentration SO_2 (200 ppm) adsorption and show good thermal and cyclic stability.

533796 Acetic Acid Extraction Using Amine Grafted OMS: Effects Due to Degree of Amine Methylation

Abstract:

Cellulosic biomass represents a potential feedstock for the production of sustainable fuels, but currently requires a large energy input to be depolymerized into sugars suitable for enzymatic digestion. Organic acid side products, with the majority being acetic acid, produced during depolymerization of cellulose are a large energy sink for the process because they must be selectively removed in order to raise the pH to a range that enzymes can efficiently produce alcohols. Liquid amine sorbents have been traditionally used for this due to their inherent basicity, but suffer from a large energy penalty during separation. Additionally, the nature of the extraction is not fully understood with many competing molecular forces at play. By grafting amine functional groups on high-surface area supports, the separation could be significantly less energy intensive while maintaining high uptake and be used as a model substrate to provide a better mechanistic understanding of the extraction process.

In this work, amine grafted ordered mesoporous silica(OMS) was used to extract acetic acid and the acid uptake of primary, secondary, tertiary, and quaternary methylated amines were compared. SBA-15 and MCM-41 were used as the OMS supports due to their pore uniformity and well-characterized ordered structure. Amines were grafted on the OMS surface by post-synthetic grafting of their respective (3-amino)propyltrimethoxysilane with the exception of the quaternary amine, which was synthesized by first grafting (3-iodo)propyltrimethoxysilane on the surface, then reacting with trimethylamine in a dry environment. OMS structure was confirmed by x-ray diffraction and nitrogen adsorption; then thermo-gravimetric analysis was used to confirm loading of amine groups.

The amine grafted OMS materials were tested for uptake at various concentrations that are relevant to bio-fuels, low concentrations are found in acid hydrolysis and high in pyrolysis cellulose depolymerization, and an uptake isotherm was obtained. Results showed that primary and quaternary amines had the highest uptake with uptake values for both equal to 0.21 mmol acetic acid/g SBA-15. For secondary and tertiary loaded amines, the uptake values were 0.185 and 0.18 mmol/g, respectively. These values were recorded for a 0.5%, by volume, acetic acid in water solution and 0.5 mmol of amine loaded/g SBA-15 for all samples. Uptake as a function of the mesopore size, ligand loading, and pH will also be described. Additionally, uptake was shown to significantly decrease for all amines with increasing salt concentration and a significant decrease in uptake was observed with increasing pH. Competing uptake of varying alkyl chain length organic acids was also tested and it was found that the most hydrophobic acid was preferentially bond to the grafted amine groups.

Ongoing work, which will also be described, are findings from pulsed-field gradient NMR. These experiments will provide a complimentary microscopic picture through which to interpret the macroscopic uptake measurements. From this the self-diffusion coefficient of acetic acid was calculated and binding properties were investigated when amine grafted OMS was introduced to the system. Further, by using diffusion ordered spectroscopy these same characteristics were probed in the presence of multiple organic acids.

533615 Low Temperature Synthesis of Magnetic Carbonaceous Materials Coated with Silica for the Effective Adsorption of Methylene Blue for Aqueous Solution

Abstract:

Low temperature synthesis of magnetic carbonaceous materials coated with silica for the effective adsorption of methylene blue for aqueous solution

Reshma Babu, Hemant Mittal and Saeed M Alhassan*

Department of Chemical Engineering, Khalifa University of Science and Technology, PO Box 2533, Abu Dhabi, United Arab Emirates

*E-mail corresponding author: salhassan@pi.ac.ae

This work reports the synthesis of silica coated magnetic carbonaceous adsorbents (MCA-SiO_2) using low temperature hydrothermal carbonization technique and the feasibility of its capacity to adsorb methylene blue (MB) dye from aqueous solution. Initially, the carbonaceous precursor (CP) was prepared from corn starch under saline conditions at 453 K using hydrothermal method followed by the magnetization of the synthesized CP again using hydrothermal carbonization at 453 K. Finally, MCA was coated with silica and characterized using different techniques such as SEM/EDX, TEM, XRD and FTIR. The effect of adsorption parameters on MB dye removal such as MCA-SiO_2 dose, solution pH and dye concentration were also investigated. The adsorption behaviour of MCA-SiO_2 was best defined using Langmuir adsorption isotherm model and pseudo-second-order kinetics rate equation. The mechanism of the dye diffusion was governed by the combination of intraparticle diffusion as well as the film diffusion mechanisms. The versatility of MCA-SiO_2 was also checked for the adsorption of a number of cationic and anionic dyes and it showed much better performance for the cationic dyes as compared to anionic dyes. Furthermore, the reusability and regeneration properties of MCA-SiO_2 were examined and found to be successful for consecutive five cycles of adsorption–desorption.

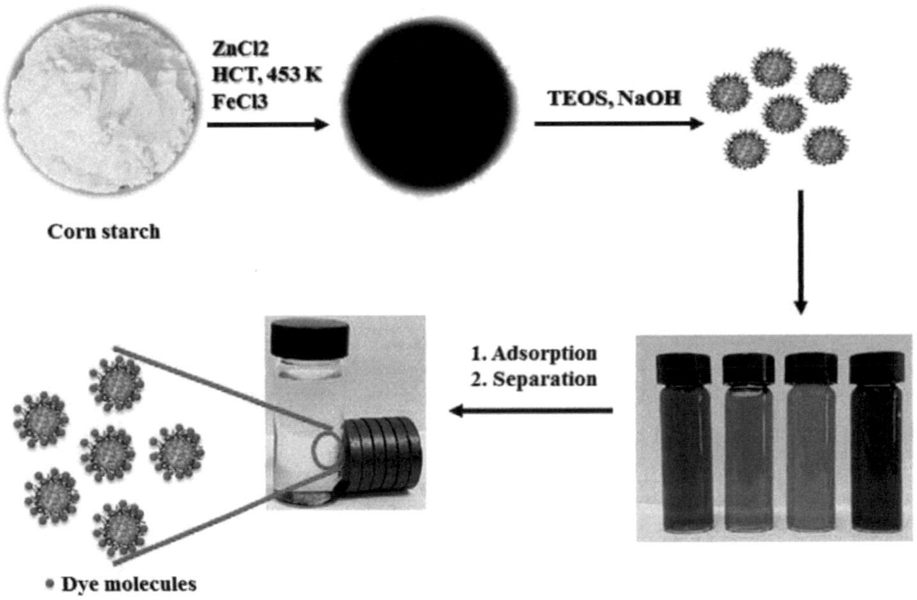

536197 Modeling of the Flow Dynamics through Incompressible Porous Media in Solid-Liquid Filtration

Abstract:

Solid-liquid filtration is a long-standing engineering practice and has been widely used in the chemical, process and mineral industries. Current models are semi-empirical in nature, thus they require significant experimental and/or computational resources in order to determine the empirical quantities. In contrast, in this work we provide a model to predict the dynamic behavior for both the liquid and solid phase of a filtration process without the requirement of empirical parameters. Instead, our model relies solely on the to-be-captured particle size distribution of contaminants as well as the pore size distribution of the filtration media. Our new algorithm is capable of describing filtration based on both â€œâ€™sterică€ capture of contaminants as well as capture within dead-end pores in the material. Here we we show the performance of this model in modeling beds comprised of high void fraction materials (diatomaceous earth) that is used for the removal of multi-modal mixtures of contaminant. By formally accounting for the complex pore size distribution, we predict flow dynamics that are much closer to our experimental results than the predictions of the traditional Kozeny-Carmen (KC) model and show that this approach is viable for both statically formed and evolving (dynamic) beds. In an effort to understand the relationship between flow dynamics and pore size distribution more fully, we built a dynamic filter cake model that continuously modifies the pore size distribution as contaminants (polydispere spheres) are deposited. In short, the predicted flow dynamics of this new model match the dynamic experimental results remarkably well, setting the stage for a priori prediction of filtration times, flow-rates, particle capture, and pressure requirements from simple measurements of the size distribution of both the filter media pores as well as the contaminant particles/droplets.

525196 Two-Drop Model of Depth Coalescing Filter Performance

Abstract:

Depth coalescing filters are used to enlarge droplets in liquid aerosols and in liquid-liquid emulsions. The enlarged drops are subsequently separated from the continuous phase by interfacial mechanisms, gravity, or inertial forces. Mathematical models of coalescing filters are useful to predict performance and to guide filter design. Models using CFD are successful at predicting performance at the cost of significant computational effort. For many practical applications, simpler less computationally intensive models are desired.

In this work, a â€œtwo-dropâ€ model is described. The drops are of two sizes. The small drop, carried by the continuous phase, is captured on fibers of the filter. Once captured, the small drops are immobilized. The small drops coalesce to form the second, larger drops. The larger drops move through the filter due to fluid drag force. The model aims to account for significant mechanisms including capture, coalescence, drag, and capillary forces.

A unique feature of the model is the mathematical description of the capillary forces as a gradient of a capillary potential function. This capillary potential function enables prediction of the effects of the capillary forces near the surfaces of the filter medium that can cause liquid accumulation at the surfaces.

536342 Cake Filtration of Catalyst Materials

Abstract:

Cake filtration is widely used in catalyst production. This traditional process is still largely empirical. There is often a lack of knowledge about the required pressure drop or the resultant flow rate or the actual rate of filter cake formation, especially for fine particles. Our approach is to develop a laboratory scale test that can provide insight into how a particular material of interest would behave in a manufacturing scale filtration process. This study will examine the dependence of the Darcy permeability coefficient on the operating parameters during filtration. An additional aim is to map the filter cake microstructure and the variation in particle size distribution and bulk density at different axial locations along the filter cake.

In this work, we have developed tools and protocols for filtration experiments using a benchtop Nutsche filter device, a model material of 5 microns diameter nonporous glass beads and porous catalyst materials. The applied pressure is varied and we aim to relate filtration behavior to particle physical properties, such as bulk density, air permeability, size distribution and zeta potential. Particle size distribution and bulk density within filter cake layers were analyzed. Furthermore, additional recycling and washing steps of the filtration process were included and the displacement of fine particles through the layers of the cake was obtained as well as changes in the bulk density profile.

535856 The Role of Pore Structure and Chemistry on Particle Deposition during Membrane Filtration

Abstract:

Many relevant industrial fluids (e.g. cell culture broths, paints, milk, etc.) can be characterized as colloidal suspensions (i.e. nanometers to micrometers) and are of great interest for microfiltration. This pressure-driven membrane process is one of the oldest, yet, remains not fully understood. A major limitation is membrane fouling. Particle deposition *onto* the microporous membrane, resulting in pore constriction and/or blockage, and eventually in cake formation, has been largely investigated and mitigation strategies have been proposed.

In this study, we focus on particle deposition *into* a microporous membrane, with particular attention for the role of pore structure and chemistry. The membranes used were 0.2 µm and 5 µm pore size poly(ether sulfone) (PES), since PES is widely used due to ease of casting and mechanical stability. Silica nanoparticles, 40 nm and 1 µm, respectively, were selected as model colloidal suspensions.

In vitro studies focused on 2D and 3D characterization of pore distributions, by atomic force microscopy (AFM), scanning electron microscopy (SEM), and focused ion beam SEM (FIB-SEM). In order to understand the forces in play, (i) zeta potential for both membranes was measured using 1-10-100 mM electrolyte solutions in a pH range 4-10; and (ii) AFM in force mode was used to estimate the adhesion force between the membranes and a silica microsphere attached to an AFM probe. Finally filtration experiments were performed using dilute colloidal suspensions (i.e. <1% v/v), and array tomography, combined with SEM imaging, was applied to visualize the distribution of particles into the fouled membranes.

Taking advantage of the experimental characterization, *in silico* modeling using COMSOL Multiphysics software was conducted in parallel to build a better understanding of particle interactions with the membrane walls in the pores. Prototype models will combine an accurate pore geometry representation coupled with experimental information on particle-wall forces to obtain particle trajectories through the structure and deposition rates onto the pore wall. The information will be used to narrow down the key features pointing toward optimal membrane performance.

515518 Anion Exchange Membranes: Towards Extreme Stability and High Conductivity

Abstract:

As an attractive energy conversion technology, fuel cells generate electric power from fuels with high energy density and slight chemical emissions. Anion exchange membrane fuel cells have been developed as a desirable alternative to proton exchange membrane fuel cells because materials cost-effectiveness and comparable performance. As a pivotal module of AEMFCs, anion-exchange membranes (AEMs) serve as both the fuel barrier and hydroxide carrier. It is required to make membranes with properties of superb hydroxide conductivity, good dimensional and mechanical stability, and prominent alkaline resistance. Here we present that how we can achieve extreme chemical/mechanical stability and high ionic conductivity. The goals has been achieved by two different approaches: 1. introducing free volume promoting moiety to induce effective phase separation and 2. delocalizing charges over the polymer backbone networks so minimizing charge barriers and increasing stability. Hydroxide ion conductivity of membranes were measured to be over 0.1 S/cm for working temperature ranges (60-80 °C). We have achieved those high performance AEMs from the functionalized triptycene copolymers. Advantages of using triptycene containing polymers includes 1) high stability to chemicals and heat for long term usability, 2) nanoscale phase separation which would facilitate nanoscale transport, 3) materials functionalization on-demand, and 4) readily controllable degree of functionalization. Extensive studies from X-ray scattering, TGA, DSC corroborated the above findings.

529637 Anion Exchange Membranes with Responsive Properties

Abstract:

Stimuli responsive membranes have been developed as sensors,[1,2] actuators,[3] and drug delivery systems[4,5]. In particular, polymers with reversible or "switchable" responses to changes in their environment are sought for advanced applications where membranes change their transport properties depending on conditions. With these capabilities, new polymer membraness and polymeric surfaces can increase the functionality afforded by current polymer membranes.

Various stimuli responsive membranes have been developed, especially those which respond to pH,[6] temperature,[7] and light.[8] In most cases, porous membranes are modified by either "grafting from" or "grafting to" techniques, by incorporating the functional or responsive moiety to the surface of the pre-formed membrane. Consequently, the responsiveness of the film is usually due to changes in the size of the pores that arise from the swelling or shrinkage of the grafts. For example, quadri-stimuli responsive membranes were developed by surface initiated atom transfer radical polymerization of commercial Nylon-6 porous membranes.[9] The responsive gates were designed as poly(N-isopropylacrylamide)-*block*-poly(methacrylic acid) (PNM) block copolymers. Because poly(N-isopropylacrylamide) presents a lower critical solution temperature (LCST) of 32ºC, it undergoes a reversible phase transition from swollen to shrinkable above or below its LCST. In the case of poly(methacrylic acid), the same occurs upon protonation/deprotonation. In consequence, the effective pore size and permeability change reversible at different pH and temperatures. pH responsive nanofiltration membranes for sugar separations have also been developed by grafting poly(acrylic acid) onto commercial NF membranes by UV initiated free radical polymerization.[10] In this case, increased hydrogen bonding interactions of the poly(acrylic acid) with glucose compared to sucrose upon deprotonation are responsible for the different permeation rates of the two types of sugar at different pH levels.

Dense, ion containing membranes are also interesting candidates to present changes in their transport properties in response to changes in their environment. For example, an actuator was developed by plating both sides of a perfluorocarboxylic acid film with gold as electrode layers.[11] The actuator bent in a 90º angle when a pulse voltage of 2.0 V between the gold electrodes was applied. It was proposed that the bending motion is driven by differences in water content in the membrane carried by the ions. Additionally, potentiometric humidity sensors have been prepared from anion exchange membranes and polypyrrole composite membranes.[3] The strip prepared from the composite membrane generates an electrical potential dependent on the relative humidity and the anion exchanged.

Redox responsive membranes have also been proposed as candidates for drug and gene delivery.[12]

Glutathione is the most abundant reducing agent in most cells, thus, polymeric membrane with thiol/redox responsive properties have been developed for drug delivery that specifically release drugs upon entry to the cells. A redox sensitive microporous membrane was prepared by grafting viologen-containing side chains to poly(vinyldene fluoride) porous membranes.[13] Viologen moieties are chemically stable and reversibly convert from a dicationic to a radical cationic state when immersed in a reducing agent. These porous membranes presented higher permeation rates of 4-styrenesulfonic acid sodium salt in the radical cationic state compared to the dicationic state. It was hypothesized that increased solubility of viologen in water in the dicationic state may cause the side chains to extend into the pores, reducing the free volume.

In this work, we present the development of anion exchange membranes that contain a viologen moiety which provides reversible changes in resistance and permselectivity at the reduced and oxidized state. Previously, anion exchange membrane with viologen moieties have been developed by Sata[14,15], but transport properties such as ionic resistance and permselectivity haven't been investigated. In our study, membranes were prepared by light induced free radical polymerization of poly(ethylene glycol diacrylate) (PEGDA, 30 wt %), diurethane dimethacrylate (DUDMA, 50 wt %), and vinyl benzyl chloride (VBC, 15 wt %) as a precursor for functionalization. Dipentaerythritol penta-/hexa-acrylate was used as crosslinker (5 wt %). To this base resin, 1 wt % of phenylbis(2, 4, 6-trimethylbenzoyl) phosphine oxide and 1 wt % of 1-hydroxycyclohexyl phenyl ketone were added as initiators. The resin was spread on a casting glass plate and light was projected to initiate the polymerization. Once the membrane was obtained, it was reacted with a 5 wt % solution of 4,4'-bipyridine in ethanol at 60ºC for 24 h. Later, in order to methylate unreacted amines, the membrane was immersed in a 2 M solution of iodomethane in ethanol at 60ºC for 24 h. The final membrane was washed with ethanol and stored in NaCl. The ionic resistance of the membranes was measured using a DC 4-point method and the permselectivity was measured in a similar cell arrangement by measuring the potential drop across the membrane. For each sample, both ionic resistance and permselectivity were measured in the oxidized (dicationic) and reduced (radical cationic) states. For all the samples, it was observed that the ionic resistance increased from the dicationic to the radical cationic state. As the membrane becomes less positively charged when changing from the oxidized to the reduced state, it is expected that the ionic resistance of the membrane increases, as observed in the results. Additionally, a decrease in permselectivity was observed when switching from dicationic to radical cationic state. Usually, in ion exchange membranes, permselectivity increases as ionic resistance increases due to a lower number of ions being transported through the membrane. However, in this case, the membrane also experiences changes in charge when redox cycling. Permselectivity is reduced as a consequence of a decrease in its charge density, which reduced the ion repulsion of co-ions, as indicated by the Donnan exclusion principle. A cycling experiment was also performed in which the membrane was reduced and re-oxidized 10 times, to observe the degree at which the stimuli responsive property is reversible. Overall, this works presents an anion exchange membrane with varying ionic resistance and permselectivity dependent on its redox state.

(1) Qiu, Y.; Park, K. Environment-Sensitive Hydrogels for Drug Delivery. *Adv. Drug Deliv. Rev.* **2001**, *53*, 321–339.

(2) Napoli, A.; Valentini, M.; Tirelli, N.; Müller, M.; Hubbell, J. A. Oxidation-Responsive Polymeric Vesicles. *Nat. Mater.* **2004**, *3* (3), 183–189.

(3) Sata, T. Possibilty for Pontentiometric Humidity Sensor of Composite Membranes Prepared from Anion-Exchange Membranes and Conducting Polymer. *Sensors Actuators A ...* **1995**, *23*, 63–69.

(4) Langer, R. Drug Delivery and Targeting. *Nature* **1998**, *392* (6679), 5–10.

(5) Cerritelli, S.; Velluto, D.; Hubbell, J. A. PEG-SS-PPS: Reduction-Sensitive Disulfide Block Copolymer Vesicles for Intracellular Drug Delivery.

(6) Han, Z.; Cheng, C.; Zhang, L.; Luo, C.; Nie, C.; Deng, J.; Xiang, T.; Zhao, C.; Zhao, C. Toward Robust pH-Responsive and Anti-Fouling Composite Membranes via One-Pot in-Situ Cross-Linked Copolymerization. *DES* **2014**, *349*, 80–93.

(7) Peng, T.; Cheng, Y. Temperature-Responsive Permeability of Porous PNIPAAm-G-PE Membranes. *J. Appl. Polym. Sci.* **1998**, *70* (11), 2133–2142.

(8) He, D.; Susanto, H.; Ulbricht, M. Photo-Irradiation for Preparation, Modification and Stimulation of Polymeric Membranes. *Prog. Polym. Sci.* **2009**, *34* (1), 62–98.

(9) Chen, Y. C.; Xie, R.; Chu, L. Y. Stimuli-Responsive Gating Membranes Responding to Temperature, pH, Salt Concentration and Anion Species. *J. Memb. Sci.* **2013**, *442*, 206–215.

(10) Himstedt, H. H.; Du, H.; Marshall, K. M.; Wickramasinghe, S. R.; Qian, X. PH Responsive Nanofiltration Membranes for Sugar Separations. *Ind. Eng. Chem. Res.* **2013**, *52* (26), 9259–9269.

(11) Sewa, S.; Onishi, K.; Asaka, K.; Fujiwara, N.; Oguro, K. Polymer Actuator Driven by Ion Current at Low Voltage, Applied to Catheter System. *Sensors Actuators, B Chem.* **1995**, *23* (63), 148–153.

(12) Bulmus, V.; Woodward, M.; Lin, L.; Murthy, N.; Stayton, P.; Hoffman, A. A New pH-Responsive and Glutathione-Reactive, Endosomal Membrane-Disruptive Polymeric Carrier for Intracellular Delivery of Biomolecular Drugs.

(13) Liu, X.; Neoh, K.; Kang, E. Redox-Sensitive Microporous Membranes Prepared from Poly (Vinylidene Fluoride) Grafted with Viologen-Containing Polymer Side Chains. *Macromolecules* **2003**, No. Scheme 1, 8361–8367.

(14) Sata, T.; Matsuo, Y.; Yamaguchi, T.; Matsusaki, K. Preparation and Transport Properties of Anion-Exchange Membranes Containing Viologen Moieties as Anion-Exchange Groups in the Presence or Absence of Photoirradiation. *J. Chem. Soc. Faraday Trans.* **1997**, *93* (15), 2553–2560.

(15) Sata, T. Anion Exchange Membrane with Viologen Moiety as Anion Exchange Groups and Generation of Photo-Induced Electrical Potential from the Membrane. *J. Memb. Sci.* **1996**, *118* (1), 121–126.

535434 Nano- and Mesoscale Transport and Mechanics in Ionomers

Abstract:

The quintessential fuel-cell membrane is a perfluorinated sulfonic-acid (PFSA) ionomer. These copolymers phase separates into bicontinuous morphology with a hydrophobic phase that provides structural integrity and a water-filled hydrophilic phase that forms a network of domains, which facilitates solvent and ion transport. The mesoscale properties of the hydrophilic transport network (e.g. domain size, distribution, and connectivity) determine the effective macroscale transport properties of these membranes. Similarly, the mesoscale topology of the hydrophobic network determines the mechanical properties. These two networks communicate through hydrophilic domain swelling and propagation of stress to neighboring domains through the membrane. Consequently, the relationship between properties of these two network and macroscopic transport determines fuel-cell membrane efficacy. In this work, we instantiate this relationship by simulating solvent and ion transport along with local stress distribution throughout the hydrophilic and hydrophobic phases.

In this work, resistor and spring networks coarse-grains the interconnected hydrophilic and hydrophobic domains, respectively, as segments connecting nodes. Conservation of species flux and stress at each node in the hydrophilic and hydrophobic networks dictates electrochemical potential and strain in segments, respectively. A mean-field nanoscale model of the domains parametrizes resistances of the segments as a function of size (thickness and length) and water content. The average and distribution of domain sizes and connectivity are consistent with experiments. Accordingly, there is a distribution of segment resistances and modulus in the networks. Congruent with Onsager-Stefan-Maxwell concentrated solution theory, we treat chemical and electrostatic gradients as driving water and ion transport and their corresponding coupling (electroosmotic effect). We validate the predicted modulus, diffusion and electroosmotic coefficients, and conductivity against experiments.

We find macroscopic transport and mechanics behavior emerges out of the mesoscale. For example, how nanoscale resistance depends on domain size dictates which pathways a species will take across the network. Accordingly, different modes of transport utilize different network pathways and have different effective tortuosities. Moreover, water gradients at the macroscale induce mesoscale electrostatic potential gradients (and visa-versa). These results show how design of emergent behavior improves membrane performance.

526710 Elucidating the Effects of Pattern Geometry on Ion Transport through Charge Patterned Membranes

Abstract:

Charge mosaic membranes consist of cationic and anionic domains that traverse the membrane thickness. Due to their mosaic structure, these membranes are capable of transporting ionic solutes more rapidly than neutral particles of comparable or smaller sizes. To date, few charge mosaic membranes have been fabricated successfully and as a result, efforts to prepare such a membrane with high permselectivity are lagging. We recently developed an inkjet printing process that enables the ready fabrication of charge mosaic membranes with controlled patterns of varying surface chemistries. This unique capability was utilized to develop a fundamental understanding of the transport phenomena and membrane-solute interactions that result in the novel properties of charge mosaic membranes. In particular, this talk will focus on the effects of charge patterning on membrane throughput.

Parent nanofiltration membranes based on a poly(acrylonitrile-*co*-[oligo(ethylene glycol) methyl ether methacrylate]-*co*-[3-azido-2-hydroxypropylmethacrylate]) [P(AN-OEGMA-AHPMA)] copolymer were prepared. These parent membranes possess pore walls lined by reactive azido moieties that made them amenable to post-synthetic modification via a printing device. Charge patterning was accomplished by deposition of alkynyl-terminated reactants on the surface of azido-functionalized parent membranes to form equal areas of cationic and anionic domains through the copper(I)-catalyzed alkyne-azide cycloaddition (CuAAC) reaction mechanism. By varying the pattern geometry (e.g., stripes, cubes, hexagons) and feature size between 300 μm - 900 μm, a series of charge patterned membranes were generated. Each of the membrane was distinct from others in respect of its interfacial packing density, which was defined as the total length of the border between oppositely charged domains in a unit area. By executing single salt rejection experiments using these charge patterned membranes, we deomonstrated that the chemically patterned membranes were able to facilitate transport of salt more effectively than the single charge-functionalized membranes. Specifically, the rejection values of symmetric salts (e.g., KCl and $MgSO_4$) decreased monotonically with increasing interfacial packing density. Notably, a value of -90% rejection for potassium chloride was observed when using a patterned membrane with the highest interfacial packing density. This result revealed the role of electrostatic interactions near the interfacial region between the oppositely charged domains in governing the coupled ion transport and will enable further development of charge mosaic membranes that can be deployed in the many established and emerging technologies where the selective transport of ionic solutes is of critical importance.

532397 Ion Diffusion Coefficients in Ion-Exchange Membranes: Significance of Counter-Ion Condensation

Abstract:

Despite the long history of literature on ion exchange membranes and their industrial importance, fundamental understanding of ion transport in such materials remains incomplete. Experimental techniques for characterizing ion sorption and transport in charged membranes have been firmly established, but a simple, unifying theoretical framework for interpreting, and ultimately predicting, the experimental results is missing. Such understanding would facilitate rational design of charged membranes with transport properties specifically tailored for a given application. Recently, we have proposed a theoretical framework to describe ion partitioning and diffusion in charged membranes using concepts from Manningâ€™s counter-ion condensation theory for polyelectrolyte solutions. The framework accurately predicted equilibrium ion partitioning and salt permeability coefficients of homogeneous ion exchange membranes for various single electrolytes over a broad salt concentration range. This presentation will focus primarily on the impact of counter-ion condensation on ion diffusion in charged membranes. Counter-ion and co-ion diffusion coefficients in a series of commercial and synthesized ion exchange membranes were extracted from experimental ion sorption, salt permeability, and ionic conductivity results. For the membranes in which counter-ion condensation occurred, counter-ion diffusion coefficients were systematically greater than co-ion diffusion coefficients even after accounting for differences in ion mobilities due to differences in ionic hydrated radii. This observation was rationalized within the framework of the proposed model by assuming that condensed counter-ions migrate along the polymer backbone and contribute to the total measured counter-ion mobility. Diffusion coefficients of the condensed counter-ions were approximately two times greater than those of the uncondensed counter-ions. Interestingly, for membranes in which counter-ion condensation did not occur, the ratio of counter-ion to co-ion diffusion coefficients in the membrane was similar to that in aqueous solutions, in accordance with predictions from the model.

531549 Novel Charge Mosaic Membranes for Desalination

Abstract:

Membrane-based desalination has been extensively developed during the last 50 years as a promising solution for the growing concern of clean water scarcity in many parts of the world. In the same period, ion exchange membranes (IEM) have evolved from laboratory stage to industrial products with remarkable technical and commercial impacts. The evolution of ion exchange membranes has enabled their use in a variety of separation processes including electrodialysis for desalination. In contrast IEMs which possess either positive or negative fixed charges, a charge mosaic (CM) membrane consists of arrays of anion and cation exchange elements with fixed and mobile ions of both charges. Adjacent ionic channels of opposing charge provide a continuous pathway from one side of the membrane to the other. Negative osmosis and salt permeability occur in CM membranes much greater than non-electrolyte permeability.

This work describes a new formation technique to develop CM membranes, which offers the potential to increase the density of contact regions and thereby is expected to greatly enhance performance. The proposed membranes are prepared by adding Poly(sodium 4-styrenesulfonate) (PSS) and Poly(allylamine hydrochloride) (PAH) as two countercharge polyelectrolyte solutions to an ultrafiltration support. Diffusion of polymers inside the pores of the support layer provides the opportunity to build continuous channels, capable of flowing negative and positive ions through their respective pathways. The new method also seeks to maximize the already-achieved salt enrichment through CM membranes by operating under Gibbs_Donnan equilibrium condition. Therefore, one ionic polymer is added to the feed solution. Mathematical models are developed for quantification purposes of salt and polymer transport. The salt enrichment of outlet stream is calculated based on the developed models.

The challenges of producing these structures are discussed, and membrane performance results are reported illustrating the potential of the new membranes for desalination.

534688 Influencing Transport Properties in Polymerized Ionic Liquids through Ion Chemistry

Abstract:

We characterize diffusion trends in polymerized ionic liquids using atomistic molecular dynamics simulations of imidazolium cations with eight counterion species. We report diffusivity as it relates to glass-transition-normalized temperature. Results suggest that ion mobility is decoupled from mechanical strength for these polymers compared with their pure analogs. We examined the structural coordination and dynamic timescales of ion association, relating such measures to the mobility of counterion species. The effects of size and symmetry on ion mobility and structure/dynamics decoupling, along with the mechanisms responsible for such effects, are revealed in this report.

529376 From Batch to Continuous Reactive Crystallization: A Case Study on Beta-Lactam Antibiotics

Abstract:

Using data from experiments on batch crystallization to design a continuous system is recognized to be a challenge. Additional complications exist when crystallization is coupled with a chemical reaction and reaction and crystallization kinetics are linked through the generation of a supersaturated environment. Should this coupling of kinetics lead to a Class I behavior (i.e. one in which there is significant supersaturation in the product stream), then the product yield and crystal quality, especially as measured by crystal size distribution, become linked [1].

In the present work, penicillin G acylase (PGA) catalyzes the formation of the antibiotic but also catalyzes the degradation of the product into a slightly soluble by-product. In batch, the PGA-catalyzed reactive crystallization process improved selectivity and productivity by 50% and 25%, respectively [2]. Adapting the batch process for continuous operation can improve productivity further while ensuring better quality consistency, as measured by purity and size distribution. However, PGA must be retained within the crystallizer for the process to be economically feasible, complicating product removal. As a stepping-stone that circumvents this issue, a fed-batch process has been used to tune models accounting for nucleation and growth kinetics, reaction kinetics, and catalyst deactivation kinetics.

Achieving high yields while controlling crystal size distribution requires independent control of the residence times of the solutes (reactants) and solids. Issues encountered include washout, where nucleation or reaction or both are insufficient to maintain a stable slurry; enzyme loss, where catalytic activity diminishes due to either PGA escaping the crystallizer or permanently unfolding; and fines-rich size distributions, where crystal residence time is too short for desired growth. Control over the enzyme reaction rate can be used to reach the desired set point within a feasible space defined by the batch operation. The reaction rate of PGA can be tuned by adjusting temperature, pH, and reactant concentration, however all three of these have been found to impact the antibiotic solubility or crystallization kinetics, further complicating the development of a continuous process. A model of an MSMPR with independent crystal and liquor residence times has been used to optimize the continuous process, and an experimental continuous apparatus has been demonstrated to validate the model while operating under more constrained conditions (imposed by issues of scale and equipment). Compared to the previously published batch process, the *in silico* productivity of the continuous process increases by approximately 2-fold at the sacrifice of approximately 10% of the conversion. The *in silico* average crystal size is small, on the order of tens of microns, comparable to that obtained in batch. The *in vitro* experimental results confirm a decrease in conversion compared to batch; however better purity, consistency, and productivity offset this loss. Experimental crystal sizes were also small, however it may be possible to alleviate this issue with fines destruction.

References

[1] Tavare, N. S. (2013) *Industrial crystallization: process simulation analysis and design*, Springer Science & Business Media.

[2] McDonald, M. A., Bommarius, A. S., and Rousseau, R. W. (2017), *Chem. Eng. Sci. 165*, 81-88.

530097 A New Slurry Reactive Crystallization to Improve Process Robustness and Scalability

Abstract:

Several challenges were faced during the development of development of a drug candidate isolated as a salt. These challenges included high solvent levels entrapped in isolated solids, disproportionation and agglomeration. A root-cause analysis pointed to dimethyl sulfoxide (DMSO), a co-solvent utilized in the crystallization as the main cause for solvent entrapment and agglomeration. DMSO was introduced in the procedure because it enhances drastically the solubility of the drug allowing for a clarifying filtration to be performed at the final step. However, because of its low volatility, DMSO is removed at a slower pace during drying causing partial solubilization of crystal surfaces during drying followed by recrystallization. It is believed that this behavior causes solvent entrapment and agglomeration during drying. In order to address these issues, a study was conducted to identify an alternative co-solvent to DMSO. In this presentation, the development of two types of crystallizations, namely an anti-solvent crystallization and slurry-slurry reactive crystallization to address the issues mentioned above is described. In particular, a new slurry-slurry reactive crystallization, based on an approach recently reported (Derdour et al[â€¡]) was found to be superior in terms of improving process robustness and scalability. That procedure performed successfully when executed at industrial scale.

â€¡ Derdour L., Reckamp J.M. and Pink C., Development of a reactive slurry salt crystallization to improve solid properties and process performance and scalability, Chem Eng Res Des, 121, pp 207-2018, 2017

532745 Crystallization Process Development of Large Size TMP Crystals

Abstract:

Abstract

2,2,6,6-Tetramethyl-4-piperidinol (TMP) crystal is a very important intermediate in organic synthesis. To obtain an improved crystallization yield and desirable crystal-size characteristics, several commonly used organic solvents were investigated to determine the solubility of TMP using a gravimetric method. An ethanol solvent and a cooling crystallization method were chosen to obtain an improved theoretical yield. The cooling crystallization process of TMP in the ethanol solution was further optimized to crystallize larger crystals with a narrow crystal-size distribution using an orthogonal array design. Comprehensive analysis showed that water content in ethanol and seed loading are respectively the main influence factors of the average diameter and the diameter span. Final TMP crystals with an average diameter of 1110 Î¼m and a diameter span of 0.715 were obtained under laboratory conditions using the optimized process.

534105 Polymorph Selection in Continuous Crystallizers in the Presence of Wet Milling

Abstract:

The design of controlled processes that ensure the consistent production of a desired polymorph is an important goal in the context of pharmaceutical manufacturing.[1] The most stable crystalline form can be obtained from a batch crystallization process either through seeding or by allowing enough time for metastable forms to convert to the stable form. On the other hand, producing metastable forms in a batch process relies on delicate process control strategies (temperature, solute concentration and batch time control). As shown in the work by Farmer et al.[4] which polymorph is obtained from a continuous crystallizer at steady state can be controlled by selecting appropriate combinations of residence time, operating temperature and inlet concentration. For cases where the crystallization process is dominated by nucleation and growth, they have shown that the polymorphic outcome in such a continuous process can be accurately predicted once two modified Damköhler numbers are known. In essence, these dimensionless groups represent ratios of the characteristic time for nucleation and growth, the process time (residence time of the crystallizer) and the feed concentration.

In this contribution, we are expanding on this previous works and consider the case where wet milling is used to tune the particle size distribution and potentially alter the polymorphic outcome. We thus present a mathematical model of continuous crystallization with wet milling for polymorphic systems. The model relies on one-dimensional population balance equations including breakage terms. Making the process model dimensionless, we identified a surprisingly small number of additional dimensionless groups that describe the process outcome in terms of polymorphic form when breakage is present. The resulting model is solved using a finite volume scheme (FVS)[5]. Operating regions where the metastable/stable polymorph can reliably be obtained can again be identified. In comparison to the work by Farmer et al.[4], we show that the operating region where the metastable/stable polymorph is obtained can be enlarged by tuning key dimensionless numbers in the breakage kinetics. We combine this type of analysis with an investigation of the productivity of such systems and report regions of attainable particle sizes[4,5]

We show that introduction of a wet suspension mill therefore represents an attractive way to obtain the desired polymorph even if the underlying kinetics of nucleation and growth are disadvantageous. Apart from this finding, we show that introducing breakage increases process productivity and allows to tune particle size more extensively.

[1] Hermanto, M. W.; Chiu, M. S.; Woo, X. Y.; Braatz, R. D. Robust optimal control of polymorphic transformation in batch crystallization. *AIChE J.* 2007,53,2643â€"2650.
[2] Farmer, T. C.; Carpenter, C. L.; Doherty, M. F. Polymorph selection by continuous crystallization. *AIChE J.*2016, 62, 3505â"3514.
[3] Kumar, R.; Kumar, J. Numerical simulation and convergence analysis of a finite volume scheme for solving general breakage population balance equations. *Appl Math Comput.* 2013, 219, 5140-5151.
[4] Vetter, T.; Burcham, C.L.; Doherty, M.F. Regions of attainable particle sizes in continuous and batch crystallization processes, Chem. Eng. Sci., 2014, 106, 167-180.
[5] Vetter, T.; Burcham, C.L; Doherty, M.F. Designing robust crystallization processes in the presence of parameter uncertainty using attainable regions, Ind. Eng. Chem. Res., 2015, 54, 10350-10363.

536947 Polymorph Dynamics of 2-Aminobenzoic Acid System in Continuous Oscillatory Baffled Crystallizer

Abstract:

The polymorph dynamics of 2-Aminobenzoic acid is investigated in a continuous oscillatory baffled crystallizer (COBC). The COBC is a tubular crystallizer containing periodically spaced orifice baffles with oscillatory motion superimposed on the net flow.[1] A Nitech DN15 COBC consisting of eight segments with each segment and elbow jacketed for temperature control is used for the studies. Preliminary experiments show the feasibility of generating the two polymorphs in the COBC via continuous cooling crystallization. However, in order to obtain higher yield combined cooling and antisolvent crystallization (CCAC) is used. In pharmaceutical industry traditionally an antisolvent is added to the solution to increase the yield of the product. However, the presence of the antisolvent can influence the polymorphic outcome.[2] Design of experiments are done to investigate the effects of the parameters – temperature, solvent composition and residence time. The poster will show additional detail studies such as the effect of mixing parameters – piston amplitude and frequency of oscillation on the polymorphic outcome and yield.

[1]Caldeira, R., Ni, X. 2009 Evaluation and Establishment of a Cleaning Protocol for the Production of Vanisal Sodium and Aspirin using a Continuous Oscillatory Baffled Reactor. Org. Process Res. Dev. 13(6), 1080–1087.

[2]Lindenberg, C., Krättli, M., Cornel, J., Mazzotti, M., & Brozio, J. 2008. Design and Optimization of a Combined Cooling/Antisolvent Crystallization Process. Cryst. Growth Des. 9(2), 1124–1136.

535157 In-Situ Optical Imaging and X-Ray Diffraction Techniques to Probe Organic Molecule Thin Film Crystallization

Abstract:

Crystallization of organic materials as thin films is an area of interest in multiple fields, such as in the pharmaceutical industry, organic electronics and for the multiple applications afforded by metal organic frameworks (MOFs). However, many traditional techniques of studying crystal growth and morphology becomes difficult to use for thin film studies, due to the nanometer to micron scale thickness of the films. To overcome this problem, optical and X-ray diffraction techniques can be utilized to probe crystallization of thin films in organic materials. In the crystallization of organic molecules and MOFs, in-situ grazing incidence X-ray diffraction can provide valuable insight in the mechanism of crystal nucleation. In-situ optical studies can then provide complementary information about the crystal growth process in the micron â€“ centimeter scale. Together, in-situ optical and X-ray diffraction technique can provide a clear picture of crystallization spanning multiple orders of magnitude in both temporal and spatial dimensions. This talk will focus on the use of in-situ X-ray diffraction and optical imaging to study crystallization from solution of organic molecules and MOFs. The future of in-situ techniques using novel instrumentation will also be discussed.

523533 Imaging Crystallization Using Deep Learning to Quantitatively Track the Polymorphic Transformation of Carbamazepine

Abstract:

Solvent mediated polymorphic transformation (SMPT) is a crucial phenomenon for a specific polymorph crystalsâ€™ production and stability control in pharmaceutical crystallization processes. In this study, a completely new method based on computer vision was developed using deep learning to quantitatively track crystal individuals. For the model compound carbamazepine (CBZ), two polymorphic forms with different morphologies were identified using deep learning image segmentation. The geometry models of two crystal forms were built to calculate the volume of each crystal in the real-time tracking process. The spatial valid vision volume of a high-speed camera probe was calculated and then verified by glass bead. Furthermore, the mass of two polymorphic crystals in suspension were fitted by f_1 and f_2 through comparing real-time tracking and off-line measurement. At last, the real-time quantitative tracking of polymorphic transformation of CBZ was realized through deep learning-based imaging analysis. The transformation process including dissolution of form II, nucleation and growth of form III, crystal size distribution (CSD) etc. was analyzed and identified by off-line measurement. Our work in this study attempts to bridge the gap between the advanced image tracking technology available today and the specific needs of solution crystallization to track, count and measure the crystal individuals. It is a great opportunity to proceed intelligent manufacturing through visualizing traditional crystallization process with the help of artificial intelligence (AI).

Acknowledgement

The authors acknowledge the financial support provided by the Natural Science and Engineering Research Council (NSERC) of Canada, National Natural Science Foundation of China (NNSFC 21576187, NNSFC 21776203, NNSFC 21621004, NNSFC 81361140344), and the China Scholarship Council.

537186 Applying the Wave Theory to Fixed-Bed Dynamics of Metal-Organicframeworks Exhibiting Stepped Adsorption Isotherms: Water/Ethanolseparation on ZIF-8

Abstract:

Zeolitic Imidazolate Framework-8 (ZIF-8) is a promising material for the energy-efficient recovery of bioalcohols from aqueous mixtures. Its adsorption isotherms have inflection points and breakthrough curves obtained in dynamic column separation experiments exhibit stepped profiles. With the aim to optimize this adsorptive alcohol recovery process using ZIF-8, the correlation between the shape of the adsorption isotherms and the breakthrough curves was further explored in this work by means of liquid phase batch measurements and breakthrough experiments, with diluted ethanol/water mixtures. The shape of the concentration profiles, in adsorption mode as well in desorption mode, have been carefully analyzed making use of simple theoretical tools, including an axial dispersion model and a wave theory based approach. It was shown that, as a direct expression of the adsorption isotherm and the feed concentration, the concentration profiles were composed of different "shock waves" and "dispersive waves" propagating at different velocities along the column length. The description of the breakthrough and desorption curves in terms of propagation velocities enabled to gain insights into the fundamental role of the adsorption isotherm into the characteristics and the formation of the concentration profiles. The concept of "fractional used bed capacity" (FUBC) was introduced to illustrate the effect of the adsorption isotherm shape on the dynamic capacity at breakthrough point. According to the shape of the adsorption isotherm, efficient desorption could be achieved with the ZIF-8 material. This work provides, with limited computational effort, new perspectives for the further development and optimization of adsorptive separation processes with MOF materials, such as the recovery of bio-alcohols from aqueous mixtures with the ZIF-8 adsorbent.

J. Cousin-Saint-Remi, J.F.M. Denayer, Chemical Engineering Journal 324 (2017) 313–323.

537049 Microscopic Diffusion of Mixed and Pure Gases in Mixed Linker ZIF-7-8

Abstract:

Zeolitic imidazolate frameworks (ZIFs), a subset of metal organic frameworks (MOFs), are promising materials for applications in separations and catalysis for their high porosity and the ability to be easily modified using various linkers and metal ions. Recently, there have been many successful attempts to mix different linkers and incorporate them into the same framework. These mixed linker ZIFs have the potential to give tunable material parameters distinct from their parent materials thereby offering a possibility to optimize molecular sieving properties. This work focuses on ZIF-7-8, i.e. a mixed linker ZIF containing ZIF-8 (methylimidazolate, mIm) and ZIF-7 (benzylimidazolate, bIm) linkers. Since the effective pore aperture size accounting for linker flexibility is much smaller in ZIF-7 (~3 Å) than ZIF-8 (~4 Å), lower gas diffusivity values and higher diffusion selectivity are expected in ZIF-7-8 in comparison to ZIF-8.

^{13}Carbon, ^{1}H and ^{129}Xe pulsed field gradient (PFG) NMR with high magnetic fields (14 T and 17.6 T) and high field gradients (up to 30 T/m) was used to measure intra-ZIF self-diffusivities (D) of the following mixed and/or pure gases: C_2H_6, C_2H_4, CO_2, CH_4 and Xe. For each studied sorbate the characteristic (mean) self-diffusivities in ZIF-7-8 were found to be smaller than the corresponding self-diffusivities in ZIF-8. Unexpectedly, the PFG NMR signal attenuation curves measured for gas diffusion inside ZIF-7-8 crystals showed deviation from the monoexponential behavior corresponding to normal diffusion with a single diffusivity. This deviation was attributed to an existence of a difference between intracrystalline gas diffusivities in different ZIF-7-8 crystals. Such difference was confirmed by complementary measurements of ethane transport diffusion performed using infrared microscopy (IRM) for different single ZIF-7-8 crystals. Comparison of the IR bands associated with ZIF-8 and ZIF-7 in ZIF-7-8 crystals allowed assigning the observed diffusivity difference to a difference in the fraction of mIm and bIm in different ZIF-7-8 crystals.

The diffusion selectivity (i.e. a ratio of the characteristic self-diffusivities of sorbates x and y in the same material, D_x/D_y) for CO_2/CH_4 mixture was found to be larger in ZIF-7-8 than in ZIF-8. At the same time, the diffusion selectivity for C_2H_4/C_2H_6 was found to be similar in both ZIF-7-8 and ZIF-8. These diffusion selectivity data in combination with the observed dependence of the self-diffusivities in ZIF-7-8 on the sorbate size and loading reveal a prominent role played by the framework flexibility in diffusion inside ZIF-7-8 crystals.

531384 Process Analysis of Amine-Functionalized Adsorbents for CO_2 Capture Applications

Abstract:

Amine-functionalized adsorbents such as MIL-101(Cr) impregnated with poly (ethyleneimine) (PEI) and Mg_2(dobpdc) functionalized with N,Nâ€™-dimethyl ethylene diamine (MMEN) have shown high capacity and stability for CO_2 capture under different concentrations. The stepped CO_2 adsorption isotherm in mmen-Mg_2(dobpdc) also allows for a high working capacity in a temperature swing process. However, nearly all analyses of such systems have focused on equilibrium adsorption assessments, whereas a robust process must take into account equilibrium capacity, kinetics, and the shape of the isotherm. In this work, CO_2 adsorption in these materials was studied using packed bed and thermogravimetric systems. The effects of flowrate and temperature on the breakthrough profiles were studied. Mass transfer during CO_2 adsorption was modeled using a linear driving force model and the performance of the system was evaluated using the productivity of the bed (mmol/g/h) for different concentrations and flowrates. Overall, this talk will address many aspects that can enable system design considering the adsorbent combined with specific CO_2 application conditions.

530083 Transport of Carbon Dioxide in a Carbon Molecular Sieve

Abstract:

Transport in carbon molecular sieves (CMSs) is of practical importance in air separation for the production of N_2, but the literature includes models that typically include a combination of surface resistances and micropore diffusion.

We present a careful analysis of volumetric experiments and show that adsorption kinetics of CO_2 is due to a distribution of time constants, ie it cannot be described by a single linear driving force (LDF) constant or a combination of LDF and micropore diffusion. This should not be a surprise, given the way in which CMSs are produced, which is likely to produce a distribution of micropores that are then narrowed at the neck by repeated coke deposition.

Taking this into account we present a simple model that introduces a distribution of time constants (inverse of the LDF constant), which can be represented using a log-normal distribution with Î¼ = 0. This results in an integral model that has only 2 parameters, the characteristic time constant and the standard deviation of the distribution. The linear model for the dynamic response of a volumetric system that includes the flow resistance through the valve can be solved analytically in the Laplace domain, but direct inversion is not possible via the method of residues given the continuous distribution of time constants. The analytical solution is therefore inverted numerically to obtain the time domain solution. The comparison with the kinetic experiments shows that the transport in the CMS can be described with very high accuracy using a surface barrier model with a simple distribution of time constants characterized by a single additional parameter.

532211 Gas Sorption, Kinetics of Sorption and Characterization of the Gas Permeation in the Pores of Microporous Metal Organic Frameworks (MOFs)

Abstract:

The separation and conversion of CO_2 from industrial processes are an important focus for research aimed at mitigating the effects of greenhouse gases on the environment. To this end, it is important to develop experimental techniques for the characterization and study of the transport in porous crystalline materials that present potential in CO_2 removal at stationary sources. This study reports on the sorption equilibrium and kinetics of CO_2, N_2 and CH_4 on a Zn(II)-based metal organic framework, and the fabrication and characterization of the gas transport of supported MOF membranes. Sorption equilibrium and kinetics were studied to evaluate its efficacy for CO_2 capture under post-combustion capture conditions. Adsorption and desorption equilibria were measured gravimetrically for temperatures between 308 K and 338 K and pressures up to 5 bar. The Langmuir adsorption model was used to fit sorption data and the Ideal Adsorbed Solution Theory (IAST) was used to calculate selectivity from single-component isotherms suggesting that separation can be enhanced by a decrease in temperature. The MOF exhibited preferential CO_2 adsorption based on the high enthalpy of adsorption and adsorption selectivities of CO_2 over N_2 and CH_4. Kinetics of adsorption and desorption of CO_2 (308Kâ€"338 K, pressures up to 1 bar) were fitted to the linear-driving force (LDF) kinetic model, showing a relatively fast adsorption and a low activation energy for adsorption and desorption. Diffusion inside the pores was found to be the rate-limiting step based on fits to the LDF model and the micropore diffusion model. Desorption kinetics studies at 1 bar indicated that CO_2 has greater average residence times at all temperatures and lower values of activation energy for desorption than N_2 and CH_4. This suggests the selective adsorption and capture of CO_2 on (1) will be favored. Zn(II)-based membranes were used to develop an approach to study membrane quality and determine the transport mechanisms through the pores of the crystalline membrane. MOF membranes were consistently synthesized via a solvothermal method with structural defects sealed by a low-permeability polymer coating. This polymer coating does not modify the separation properties of the MOF, but acts to seal pin-holes and cracks in the membrane, allowing for the measurement of permeation in materials that do not form uniform, defect-free films. The Zn(II)-based MOF membrane permeation of CO_2, CH_4, N_2 and H_2 was proportional to the inverse square root of the molecular weight of the permeant gases, indicating that diffusion occurs via Knudsen diffusion. Membrane quality was studied by permselectivity measurements of CO_2, CH_4, N_2 and H_2 as a function of temperature showing that the MOF membranes are crack-free. A new approach, based on studying gas permeation through a polymer coated sparse MOF membrane, was used to corroborate that transport occurs through the pores of the MOF, rather than through pinholes or defects in the structure.

529984 Fast and Highly Selective CO_2 Adsorbent: Li^+/ZSM-25 Zeolite

Abstract:

Fast and highly selective CO_2 adsorbent: Li^+/ZSM-25 zeolite

Jianhua Zhao,[1] Ke Xie,[1] Gang (Kevin) Li,[1] Ranjeet Singh,[1] Gongkui Xiao,[2] Qinfen Gu,[3] Penny Xiao[1] and Paul A. Webley[1,4]

[1]Department of Chemical Engineering, The University of Melbourne, VIC 3010, Australia

[2]Fluid Science & Resources Division, School of Mechanical & Chemical Engineering, The University of Western Australia, WA 6009, Australia

[3]Australian Synchrotron, 800 Blackburn Rd, Clayton, VIC 3168, Australia

[4]The Peter Cook Centre for Carbon Capture and Storage, The University of Melbourne, VIC 3010, Australia

Abstract

CH_4 is one of the most important fossil fuels because it is efficient and clean. The major source of CH_4 is landfill gas, natural gas and biogas. However, those methane-rich gases may contain 15-50 vol.% CO_2 which reduces the grade of the fuel and also corrodes the pipeline. Therefore, the capture and removal of CO_2 from CO_2/CH_4 mixtures is attracting world-wide interest. The adsorption method is a promising approach due to its low operating and capital costs.

Compared with various adsorbents such as activated carbons and metal-organic frameworks, zeolites are more favourable for CO_2/CH_4 separation due to their high adsorption capacity, tuneable pore structure and the molecule affinities, high stability and ease of mass production.

The structure of Na/TEA-ZSM-25 (TEA refers to tetraethylammonium bromide), a RHO family zeolite, was resolved recently by Guo et al. (Nature, 2015, 74-78). Na-ZSM-25 has been recently prepared by removing the TEA template. Na-ZSM-25 shows promising CO_2 capture properties such as high CO_2 working capacity and excellent CO_2/CH_4 ideal selectivity for pressure swing adsorption. However, the previous reported single-gas adsorption performance of Na-ZSM-25 is unable to provide reliable information on CO_2/CH_4 separation behaviour in practical scenarios where the adsorption is performed in a mixed gas. Furthermore, Na-ZSM-25 suffers from low adsorption kinetics that limits its industrial applications. In this presentation, Li^+/ZSM-25 zeolites (LZZ) is developed to enhance the CO_2 adsorption rate, and the adsorption performance of LZZ is then studied using both single-(isotherm) and binary-(breakthrough) gas measurements. The results indicate the CO_2 adsorption rate of the LZZ is 4.5 times (Table 1) that of Na-ZSM-25 with a simultaneous increase in the CO_2 adsorption capacity by 6.1% at 303K and 9.5 bar. Remarkably, the very high CO_2/CH_4 adsorptive selectivity is successfully inherited from Na-ZSM-25 zeolite in both single- and binary-gas adsorption (Fig. 1 and Table 2). In addition, in-situ XRD reveals CO_2-induced framework expansion (Fig. 2), which explains the origin of the high working capacity and type-II like isotherms of the ZSM-25 based zeolites. This study suggests promising application of LZZ in natural gas purification and biogas upgrading.

Table 1 Adsorption kinetics of CO_2 on Na-ZSM-25, Li(0.063)-ZSM-25, Li(0.10)-ZSM-25 and Li(0.13)-ZSM-25 at different pressures.

Temperature	Pressure	D_c/r^2 ¡Á 10^4 (s^{-1})			
		Na-ZSM-25	Li(0.063)-ZSM-25	Li(0.10)-ZSM-25	Li(0.13)-ZSM-25
303K	7.23kPa	0.98	2.67	2.87	4.4
303K	12.07kPa	2.73	4.09	4.47	6.53

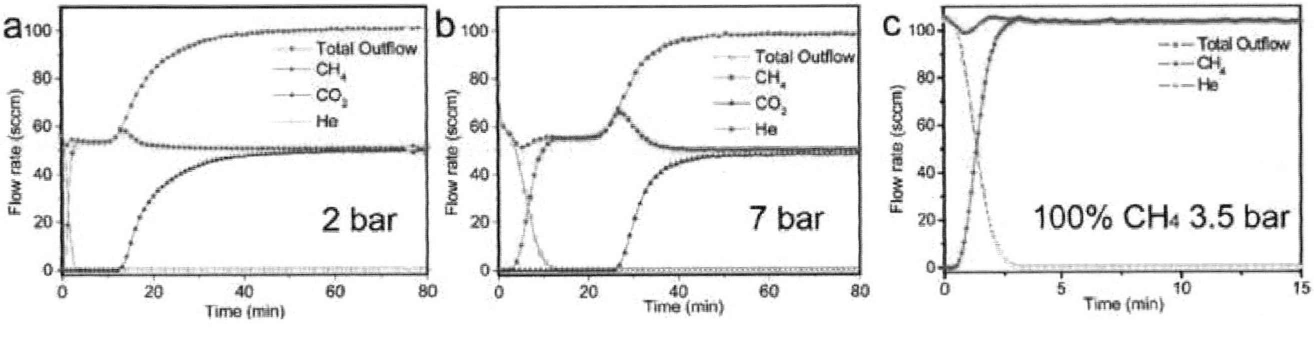

Fig. 1 Breakthrough profiles of Li(0.13)-ZSM-25 recorded in 50%CO_2/50% CH_4v/v mixed gas (a, b) at total pressures of 2 bar (a) and 7 bar (b) and pure CH_4 at 3.5 bar (c). Total gas flow rate: 100sccm; temperature: 303 K.

Table 2 Adsorbed amounts and selectivity derived from breakthrough curves of 50% CO_2/50% CH_4v/v mixed gas at 303 K and different pressures on Li(0.13)-ZSM-25.

Total Pressure (bar)	Total CO_2 uptake (mmol/g)[a]	Initial CH_4 uptake (mmol/g)[b]	Overall CH_4 uptake (mmol/g)	CO_2/CH_4 Selectivity
2	1.94	0.19	0.029	66.9
7	3.16	0.61	0.25	12.6

[a] This CO_2 uptake is slightly lower than that measured in isotherm probably due to the difference in instrument and the incomplete activation of a small portion of the adsorbent.

[b] The uptake before the breakthrough of CO_2.

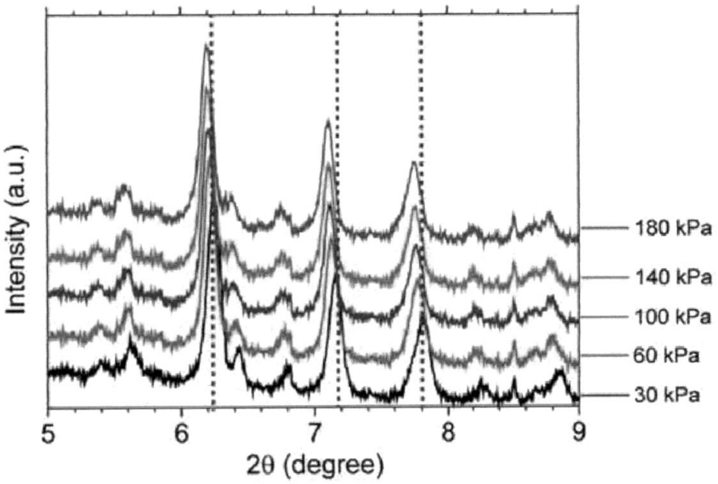

Fig. 2 In-situ synchrotron PXRD of Na-ZSM-25 during CO_2 adsorption after activation from 0.30 bar to 1.80 bar at 303K.

527158 Rapid CO_2 Capture from Ambient Air By Electrospun Particle/Fiber Composite Adsorbents

Abstract:

Direct air capture (DAC), the process of removing carbon dioxide directly from ambient air, has the potential to become a powerful tool for combating climate change. To use these sorbents for direct air capture, they must be embedded in a macroscopic scaffold that effectively exposes them to the air. Ideally, the scaffold should disperse the sorbent well, provide minimal impedance to the kinetic uptake, and not severely restrict the natural movement of the bulk air. In this talk, we would like to present our recent progress in the study of using electrospun particle/fiber composite adsorbents to provide a rapid CO2 capture performance in moisture swing adsorption. Electrospun fibers can provide fiber diameters as low as 20 nm, allowing for fibers to be readily designed to be smaller than most sorbents. Also, fiber matrices may provide enhanced heat removal, and high inter-fiber porosity allows easy fluid transport through the scaffold. The ion-exchange resin embedded in porous electrospun fibers fabricated and characterized in our recent work shows the highest carbon dioxide uptake per gram-hour found in literature as far as we are aware.

535006 Molecular Simulation Insights on Xe/Kr Separation in a Set of Nanoporous Crystalline Membranes

Abstract:

Separation of xenon and krypton is highly relevant to several applications such as spent nuclear fuel processing. Molecular simulation has been extensively used to understand the Kr/Xe separation performance of nanoporous materials for adsorption-based technologies but less frequently for membrane-based technologies. Motivated by recent experimental reports on krypton-selective membranes, herein, we present grand canonical Monte Carlo and biased molecular dynamics simulations (using adaptive biasing force) to elucidate the nature of adsorption- and diffusion-based Kr/Xe separation mechanisms in a set of nanoporous materials: SAPO-34, ZIF-8, UiO-66, and IRMOF-1. Xenon is found to preferentially adsorb on all materials, but diffusion selectivity for krypton is found to dominate the overall membrane separation selectivity. To increase adsorption selectivity for krypton, large pore cages are found to be desirable. To increase diffusion selectivity for krypton, stiff pore windows with a diameter smaller than xenon (but larger than krypton) are found to be desirable. No perfect molecular sieving was found, but the relatively rigid SAPO-34 was more effective at excluding xenon than the more flexible ZIF-8. Indeed, during xenon “window crossing,” the SAPO-34 window opened to only 3.8 Å, while the ZIF-8 window opened to 4.1 Å, resulting in a lower free energy “diffusion” barrier for xenon in ZIF-8. Therefore, an ideal membrane material for Kr/Xe separation should be rigid and have large pore cages and small pore windows. Temperature was found to have opposite effects on adsorption and diffusion selectivity, but because of the dominance of diffusion selectivity, our simulations indicate that it is preferable to operate membranes for Kr/Xe separation at lower temperatures than at higher ones.

1. *ACS Appl. Mater. Interfaces*, 2018, 10 (1), pp 582–592

523521 The Effect of the Pore Entrance on Particle Motion in Slit Pores: Implications for Ultrathin Membranes

Abstract:

Membrane rejection models generally neglect the effect of the pore entrance on intrapore particle transport. However, entrance effects are expected to be particularly important with ultrathin membranes, where membrane thickness is typically comparable to pore size. In this work, a 2D model was developed to simulate particle motion for spherical particles moving at small *Re* and infinite *Pe* from the reservoir outside the pore into a slit pore. Using a finite element method, particles were tracked as they accelerated across the pore entrance until they reached a steady velocity in the pore. The axial position in the pore where particle motion becomes steady is defined as the particle entrance length, PEL. PELs were found to be comparable to the fluid entrance length, larger than the pore size and larger than the thickness typical of many ultrathin membranes. Results also show that, in the absence of particle diffusion, hydrodynamic particle-membrane interactions at the pore mouth result in particle â€˜funnelingâ€™ in the pore, yielding cross pore particle concentration profiles focused at the pore centerline. The implications of these phenomena on predicted rejections from ultrathin membranes have also been examined.

526833 Non-Equilibrium Molecular Dynamics Simulations of Composite MOF/Polymer Membranes

Abstract:

We recently developed a new non-equilibrium molecular dynamics simulation method [1] in order to simulate the permeation of pure fluids and separation of mixtures through membranes, namely "Concentration Gradient Driven Molecular Dynamics" (CGD-MD). This new method works by employing bias forces to fix the concentration of fluids (pure or mixture) at the inlet and outlet of a membrane in order to create and maintain a concentration gradient and drive the diffusion of molecules through the membrane. This is aimed at mimicking membrane separation experiments; for instance, high concentration/pressure at the feed side and vacuum at the permeate side. CGD-MD addresses two main shortcomings of previous non-equilibrium MD methods used for simulating membrane separation processes at the molecular scale. First, it avoids the feed depletion issue and allows running steady state and continuous simulations for unrestricted simulation times, and second, it maintains the feed composition at a target value without the need of any complex Monte Carlo-MD coupling. In this study we extend our method to include gas transport and separation in composite MOF/Polymer membranes. Accurate structural models of Polymer of Intrinsic Microporosity(PIM)-1 and ZIF-8 layers have been assembled and the separation of CH_4/H_2 mixture has been modeled by using the CGD-MD method.

[1] Ozcan, A., Perego, C., Salvalaglio, M., Parrinello, M., **Yazaydin, O.***; "Concentration gradient driven molecular dynamics: a new method for simulations of membrane permeation and separation", *Chemical Science,* **2017**, 8, 3858â €"3865.

538421 Module-Scale Simulation of Hollow Fiber Vacuum Membrane Distillation Using Openfoam

Abstract:

Membrane distillation is a thermal process using a hydrophobic membrane as a physical barrier. MD processes are classified based on the phase condition in the distillate side. Vacuum membrane distillation (VMD) often provides higher distillate flux in comparison to direct contact membrane distillation (DCMD) and other types of MD processes. A VMD module consists of liquid feed, low-pressure gas, and porous membrane regions. In each phase, at least two transport phenomena are coupled. Momentum and heat transfer in the liquid feed side can be predicted using Navier-Stokes and Fourier equations, respectively. Through the membrane pores, migration of vapor molecules and heat transfer through the void and solid membrane regions can be explained using convective and diffusive heat transfer equations. In the low-quality vacuum phase of the distillate, a set of thermodynamic governing equations is necessary to adequately describe the momentum, heat, and mass transfer of the rare gas phase.

Due to the complexity of the multiple transport phenomena and their continuity at the two interfaces at the feed-side and distillate-side of the membrane, the performance modeling of VMD has been limited to a sample module system of one or several fibers. In reality, the number of hollow fibers in a cylindrical vessel ranges from (at least) a few hundred to thousands. In this case, the distillate flux is sensitively influenced by the configuration of hollow fibers, aligned in parallel. Even distribution of feed flow to each hollow fiber is a challenging task, and therefore the performance of the VMD process is highly dependent on geometrical configurations of packed hollow fibers and flow patterns. In this vein, the current paper investigates the VMD module of real size, having hollow fibers of an order of 100, and simulate the coupled transport of momentum, heat, and mass in a seamlessly combined manner. An open-source CFD package, OpenFOAM is used mainly to solve for the thermal fluid and gas, and specific solver for MD is developed to investigate the coupled transport phenomena without theoretical approximations.

528786 Understanding Polyether Sulfone Membrane Formation Via Nonsolvent Induced Phase Separation By Dissipative Particle Dynamics (DPD) Simulations

Abstract:

Nonsolvent induced phase separation (NIPS), in which a polymer collapses from solution into a thin film upon exposure to a nonsolvent, is the most widely adopted method to fabricate porous polymeric membranes. The link between the NIPS process variables and final membrane pore structure remains surprisingly empirical and relies mainly on continuum thermodynamics and transport models. In this work, we take a molecular approach using dissipative particle dynamics (DPD) simulation to investigate the early stages of membrane formation via the NIPS process. The model parameters were chosen to represent the polyether sulfone/methyl-2-pyrrolidinone/water (PES/NMP/water) system, which has been applied commonly in industry. Millions of particles were used to create an initial model of the sharp interface between the nonsolvent (water) and the polymer solution (PES/NMP), and the dynamics were followed for approximately hundreds of microseconds. The effects of polymer concentration and polymer molecular weight were investigated. The simulations showed clear evidence of microstructure formation, with the growth of polymer-rich and polymer-lean (fluid) domains, due to the mass transfer between the solvent and nonsolvent. An overall compaction of the polymer in the direction normal to the interface as well as a porous sub-layer were also observed. Compared with the polymer chain length, the polymer concentration had a stronger influence on the morphology; the surface layer became significantly denser, with smaller fluid domains, as the polymer concentration increased. Our findings are consistent with several observations and hypotheses about the interfacial processes that occur during a NIPS process and their effects on membrane pore structure.

534447 Computationally Investigating Zeolite Nanosheets As Pervaporation Membranes for Ethanol Extraction and the Role of Membrane Surfaces

Abstract:

Computationally Investigating Zeolite Nanosheets as Pervaporation Membranes for Ethanol Extraction and the Role of Membrane Surfaces

Changlong Zou[1] and Li-Chiang Lin[1]*

[1]William G. Lowrie Department of Chemical and Biomolecular Engineering, The Ohio State University, Columbus, OH 43210, USA

Discovering more energy-efficient and cost-effective approaches for anhydrous ethanol extraction from dilute biomass fermentation broth is of utmost importance due to the increasing demand for a green alternative to the fossil fuels. For this, pervaporation, driven by the vapor pressure difference, using siliceous zeolite membranes may hold a great promise owing to their hydrophobic nature. As reported in the literature, zeolite membranes can offer outstanding separation factors (i.e., selective toward ethanol over water). However, their corresponding membrane flux is still relatively slow. Zeolite nanosheets, emerging ultrathin-film materials that are synthesized with a thickness of a few nanometers, may overcome this limitation because of their intrinsically short diffusion paths. To date, the potential of zeolite nanosheets as pervaporation membranes for ethanol extraction remains unknown. In this study, by employing molecular dynamics techniques, we studied the separation performance of zeolite nanosheets for ethanol extraction. The total flux of zeolite nanosheets was found to be orders of magnitudes higher than those zeolite membranes reported in the literature, while a high separation factor can be still achieved. Further, we have also explored the role of nanosheet surfaces at an atomic level, and the hydrophilicity of the membrane surface was identified to impact the overall separation factor significantly. Our results suggest that strategically functionalizing the surface of nanosheets and/or selecting structural topologies to maximize the ethanol concentration at the entrance region of pore/channels are the key toward the design of highly selective zeolite nanosheets membranes. Moreover, the multilayer effects of nanosheets on the separation performance were also studied. Although the existence of multiple layers was found, as expected, to decrease the overall flux, multilayer structures can accordingly increase the selectivity of ethanol and therefore offer an additional design variable to improve the separation factor. Overall, this work has demonstrated the enormous potential of zeolite nanosheets as pervaporation membranes. We believe that the insights obtained here can help guide the future design and discovery of the zeolite nanosheet membranes for ethanol/water pervaporation separation and potentially for other organics separation applications.

534904 A New Design Strategy for Membrane Gas Separation: Engineered Anisotropic Mass Diffusion

Abstract:

The design of membrane materials for gas separations has primarily relied on tuning the membrane chemical properties with the aim of tailoring the permeability and selectivity for the compounds of interest. From a physical point of view, manipulating the permeability and selectivity allows controlling the flux magnitude of the species in a prescribed way. Although this approach has been extensively used to design membrane and separation systems, the possibility of controlling flux direction to perform separations has remained largely unexplored. In this work, we show how metamaterial theory and effective medium approaches can be used to design novel anisotropic membranes that can be exploited for membrane-based gas separations. The proposed membranes consist of multilayer systems made of isotropic materials in which rationally designed anisotropic diffusion is controlled by the structural arrangement of the constituent materials rather than their chemical composition. Our novel anisotropic membranes operate by directing molecules of different species toward different places in the permeate side. Separation occurs as a result of the rerouting of mass flux in addition to differences in solubility and diffusivity. In terms of efficiency, we show that the selectivity of anisotropic membranes can be increased by orders of magnitude in comparison with typical isotropic materials while retaining similar permeabilities. Our work paves the way for the development of a new approach to design membrane materials and separation processes by taking advantage of an often overlooked intrinsic property of mass flux, its directionality.

530500 Amine-Functionalized Graphene Oxide Hollow Fiber Membranes for CO_2 Capture

Abstract:

Membrane technology for CO_2 capture from flue gas exhibits the superior separation performance and attractive levelized cost of energy to achieve the post combustion capture goal set by the US Department of Energy (DOE), 90% capture of CO_2 with purity >95% with less than 35% increase in energy cost. Novel membrane materials with potential to greatly improve membrane performance and, such as 2-dimensional graphene and graphene oxide (GO), were extensively studied in fabricating ultrathin and chemical-resistant membranes. In this work, we developed a novel deposition method to prepare ultrathin GO-based hollow fiber membranes. Single layered graphene oxide (GO) sheets were uniformly deposited on the inner surface of polymeric hollow fibers with high quality. Furthermore, ethylenediamine (EDA) was selected as CO_2-philic agent and incorporated into the GO interlayer nanochannels by chemical grafting. To characterize GO and GO-EDA hollow fiber membranes, field emission scanning electron microscopy (FESEM), X-ray photoelectron spectroscopy (XPS), attenuated total reflectance-Fourier transform infrared spectroscopy (ATR-FTIR) and X-ray diffraction analysis (XRD) were performed. Mixed gas (CO_2 in 15 vol.% and N_2 in 85 vol.%) separation was conducted for both membranes under wet condition. Results show superior CO_2 separation performance with CO_2 permeance of > 600 GPU and CO_2/N_2 selectivity of >500. This work, therefore, may provide a facile and scalable fabrication method of GO-based hollow fiber membranes for efficient CO_2 capture from flue gas.

532828 High CO_2 Separation Performance of Hydroxide-Ceramic Dual-Phase Membrane

Abstract:

High CO_2 separation performance of hydroxide-ceramic dual-phase membrane

Azadeh Amiri[1], Li Sze Lai[1], Maira R. Cerón[2], Matthew Merrill[3], Patrick G. Campbell[2], Sangil Kim[1†]

[1]Department of Chemical Engineering, University of Illinois at Chicago, Chicago, IL 60607, USA

[2]Lawrence Livermore National Laboratory, Livermore, CA 94550, USA

[3]Luna Innovations Inc., Charlottesville, VA 22903, USA

[†]- sikim@uic.edu

Dual-phase membrane based on molten electrolyte has been proposed as a solution to efficient CO_2 capture from flue gas because of their high CO_2 selectivity/permeance at high temperature (> 600 °C). There is, however, no existing CO_2 separation membrane technology which can operate in the 300 to 600 °C temperature range of incipient flue gas (in between the combustion turbine and the primary heat exchanger). It has been recently reported that CO_2 absorption by alkali hydroxides (e.g., KOH) is reversible in the presence of H_2O due to a phase change near 250 °C. This reaction can be used as a basis for developing highly CO_2 selective dual-phase membranes.

To validate this concept, we have developed a novel dual phase membrane consisting of molten hydroxide phase and a ceramic solid oxide phase. We investigated the gas separation performance of molten hydroxide separation phase loaded in the Y_2O_3-ZrO_2 (YSZ) membrane supports. Single and binary gas permeation results showed that hydroxides loaded YSZ membranes have high CO_2/N_2 selectivity (>1000) and CO_2 permeance is an order of magnitude higher than the best values of other carbonate-based dual-phase membranes. Besides, the membrane is stable for more than 80 hours and the results are repeatable by switching testing condition between dry and wet sweep condition. The hydroxide-based dual-phase membrane can operate temperature ranges of 500 ^{oC} to 650 °C which is close to the temperature range of the flue gas.

533082 Atomistic Investigation of CO_2-Induced Plasticization in PIM-1 Polymer

Abstract:

Membrane plasticization has been a major concern over years in the industrial application of gas separation membranes. At elevated gas pressures, condensable gases such as CO_2, induce swelling of polymeric membranes resulting reduced selectivity hence loss of separation performance. Prediction of the plasticization pressure requires elaborate high-pressure experiments. While molecular simulation is a powerful tool to predict materials performances, modelling gas separation performances of polymeric structures is proven to be challenging due to the complex and flexible structures. In this work, CO_2-induced plasticization of promising PIM-1 membranes were investigated using a complete molecular simulation approach.

A detailed study on modeling of PIM-1 was accomplished for accurate simulations of the material to be performed. Quantum-level charge calculations were integrated wih all-atom force fields for PIM-1 model. Molecular Dynamics (MD) and Monte-Carlo (MC) simulations were integrated in order to mimic PIM-1 chain behaviors with and without CO2 presence and also CO_2 sorption on the membrane. The methodology was validated in previous studies [1], where plasticization pressure of several polyimides were predicted. CO_2 sorption simulations up to 40 bar were carried out using integrated MD-MC simulation and validated using experimental data.

Plasticizing effect of CO_2 was then investigated using specific fractional free volume (FFV) analysis that evaluates interconnected CO_2-accessible free volume. Experimental physical properties of PIM-1 such as density, FFV and glass transition temperature, were successfully reproduced via simulations. Accurate modeling of rigid PIM-1 chains were accomplished and structural analyses on plasticization were performed.

This work was supported by the Scientific and Technological Research Council of Turkey - TUBITAK (Grant# 217M630)) and the Scientific Projects Unit of Istanbul Technical University (Grant# MDK-2018-41400).

[1] Balcik M., Ahunbay M.G., *Prediction of CO2-induced plasticization pressure in*

polyimides via atomistic simulations, J. Memb. Sci. 547 (2017) 146-155.

534691 Novel Facilitated Transport Membrane for Post-Combustion Carbon Capture: From Membrane Synthesis to Process Design

Abstract:

Novel Facilitated Transport Membrane for Post-combustion Carbon Capture: From Membrane Synthesis to Process Design

Yang Han and W.S. Winston Ho

William G. Lowrie Department of Chemical and Biomolecular Engineering, The Ohio State University

Large-scale application of membrane in post-combustion carbon capture has been limited by the trade-off between CO_2 permeance and CO_2/N_2 selectivity of most polymeric membrane materials. In order to overcome this limitation, research efforts on facilitated transport membrane (FTM) have been devised with the objectives of (1) developing carriers with high CO_2 loading capacity and reactive diffusivity, and (2) designing membrane processes that can capitalize on the outstanding selectivities shown by FTM. In this presentation, novel FTM was synthesized in a composite membrane configuration with an 170 nm selective layer coated on a polyethersulfone nanoporous substrate. In the selective layer, polyvinylamine with amine-sites covalently bound to the polymer backbone was used as fixed-site carrier and an amino acid salt, synthesized by deprotonating sarcosine with 2-(1-piperazinyl)ethylamine, was blended as mobile carrier. The membrane demonstrated a CO_2 permeance of 1500 GPU (1 GPU = 10^{-6} cm^3 (STP)$\cdot cm^{-2}\cdot s^{-1}\cdot cmHg^{-1}$) and a CO_2/N_2 selectivity >140 at 67 °C with 1 atm feed and permeate pressures. Two membrane processes were designed for the FTM to decarbonizing coal derived flue gas with 90% CO_2 recovery. In the first membrane process, i.e., air sweep process, the flue gas flowed to a vacuum membrane stage to produce a >95% pure CO_2 permeate; the remaining CO_2 was removed by an air-sweep membrane stage. The CO_2-laden sweep air was fed to a power plant boiler as combustion air. In the second membrane process, i.e., retentate recycle process, the flue gas was passed to two membrane stages in an enriching cascade. A portion of the CO_2-depleted retentate of the first membrane stage was recycled as its own sweep gas, while a vacuum was pulled for the second membrane stage. Both membrane processes yielded a capture cost <40/tonne CO_2 in the 2007 dollars, which met the target set by the Department of Energy (DOE) for 2025. A common element in these two processes was a vacuum membrane stage with a permeate pressure as low as 0.3 atm. Under this testing condition, however, the elastic selective layer of the synthesized FTM sunk into the nanoporous substrate, which resulted in a drastically reduced CO_2 permeance. To address this issue, multi-walled carbon nanotubes (MWNTs) wrapped by a copolymer poly(vinylpyrrolidone-*co*-vinyl acetate) were dispersed in the selective layer as reinforcement fillers. The gas permeation measurements showed that the incorporation of MWNTs strengthened the polymer matrix and the selective layer penetration was refrained by a 3 wt.% MWNTs loading. The presence of MWNTs also mitigated the polymer compaction if the feed gas was compressed. The reinforced membrane demonstrated stable separation performance for a feed pressure up to 4 atm with a vacuum of 0.3 atm pull on the permeate side.

538405 Fabrication and Optimization of SAPO-34 Membranes for Efficient Separation of CO_2/CH_4 Gas Mixture

Abstract:

Zeolites are crystalline materials with angstrom scale pores and channels. Zeolite powders have been widely used as shape-selective catalysts, ion exchangers and adsorbents. Zeolite membranes, as emerging membrane materials in recent years, can offer a great potential for a variety of applications such as gas and liquid separation, membrane catalysis, chemical sensor, corrosion-resistant coating and micro-electronic device. The ways to manipulate the membrane microstructure, including grain shape and intergrowth, thickness, crystal orientation, and even defect minimization, are quite important for its performance improvement. Oriented zeolite membranes usually exhibit superior performances compared with randomly oriented counterparts. For example, the appropriate orientation of crystals can effectively enhance mass transport within the membranes, which is closely related to separation and catalysis. Despite great efforts made over the past decade, challenges still exist to fabricate high-quality oriented zeolite membranes.

In the present work, a thin seed layer was prepared the α-Al_2O_3 substrate by spin-coating by using ca. 400 nm-sized SAPO-34 crystals, based on which preferentially oriented and thin SAPO-34 membrane were synthesized under microwave irradiation. The fabrication process was optimized by changing a variety of parameters, including heating method, water content, aging time, crystallization temperature, and even the mixing order in preparing for the synthesis solution.

Without doubt, the preferentially oriented SAPO-34 membranes could be synthesized using the seeded growth method under microwave irradiation. Under the optimized condition in this study, the synthesized SAPO-34 membrane was only ca. 1.5 μm thick with a preferential (*h0l*)-orientation. This membrane exhibited a very satisfactory CO_2/CH_4 separation performance, through which a CO_2 permeance of 1.8×10^{-6} mol/(m^2•s•Pa) with the selectivity of 89 was achieved for the 50/50 CO_2/CH_4 gas mixture under 138 kPa feed pressure and 303 K. The results we obtained were among the best ones compared with the counterparts in the literature. It seemed that both the preferential orientation and thickness of the membrane played an important role for its separation performance.

This work was supported by the National Natural Science Foundation of China (21476171).

533096 Ionic Liquids Based Composite Membranes for CO_2 Separation

Abstract:

A large number of CO_2 emissions have caused global warming, bringing some serious disasters like ice melting, sea level rising, floods, et al. It is a worldwide concern and urgent to reduce CO_2 emissions. Among available measures, CO_2 capture and storage (CCS) is recognized as one of feasible methods, in which efficient CO_2 capture and separation is critical. Additionally, the removal of CO_2 is an important part in the gas purification of syngas, natural gas and biogas. Therefore, the efficient capture and separation of CO_2 is greatly significant for sustainable development of ecosystem, economy and society. Ionic liquids (ILs) as a novel solvent possess negligible vapor pressure, good stability and structural design, making them widely applied in gas separation (CO_2, SO_2, NH_3 etc.), especially CO_2 separation[1]. Due to these unique properties, ILs have been used in membrane systems including supporting liquid membranes, membrane contactors and composite membranes to improve the stability and gas performance[2]. Three-component composite membranes containing ILs, nano-materials and polymers, will be a promising approach to simultaneously obtain optimal gas separation performance and excellent stability of membranes.

In this work, a series of phosphonium based ILs ([P_{66614}][NTf_2], [P_{66614}][DCA], [P_{66614}][Triz] were synthesized and blended with polyether copolyamide (Pebax) to prepare IL/Pebax composite membranes for CO_2 separation. The results showed that the [P_{66614}][NTf_2]/Pebax composite membrane has the best performance, with the CO_2 permeability up to 244.3 Barrer, when the IL content reached 40 wt%. With the increase of IL content, the CO_2 permeability of [P_{66614}][NTf_2]/Pebax composite membrane increased. In order to further improve the separation performance and strengthen the mechanical property of the membranes, ZIF-8 porous particles were added to form three-component composite membranes and the single gas permeability and CO_2 selectivity was systematically investigated. The results showed that the gas permeability of the membrane increases as the content of [P_{66614}][NTf_2] increases. When the content of [P_{66614}][NTf_2] reaches 30 wt% and the content of ZIF-8 is 10 wt%, the CO_2 permeability of the membrane is up to 386.8 Barrer while the CO_2 selectivity keeping constant. The various characterization including FTIR, DSC and SEM revealed that the addition of IL increases the free volume of the membrane and reduces the agglomeration of ZIF-8 particles, which promotes gas diffusion and hence improves the CO_2 permeability of the membranes.

References:

1. Zeng, S. J.; Zhang, X. P.; Bai, L.; Zhang, X. C.; Wang, H.; Wang, J. J.; Bao, D.; Li, M. D.; Liu, X. Y.; Zhang, S. J., Ionic-Liquid-Based CO_2 Capture Systems: Structure, Interaction and Process. Chemical Reviews, 2017, *117*: 9625-9673.

2. Dai, Z.; Noble, R. D.; Gin, D. L.; Zhang, X.; Deng, L., Combination of ionic liquids with membrane technology: A new approach for CO_2 separation. Journal of Membrane Science, 2016, *497*: 1-20.

Corresponding authorâ€™s e-mail address: *xpzhang@ipe.ac.cn*

533181 Automated Process Design and Techno-Economic Assessment of Membrane-Based CO_2 Capture Systems

Abstract:

Considerable efforts from academic and industrial communities have been made to develop cost-effective and energy-efficient CO_2 capture processes. A wide range of separation techniques, for example, amine solvents, dry sorbent and membranes, are available, and Techno-Economic Assessment (TEA) is necessary to evaluate technical feasibility and compare economics in a systematic and fair manner. In this presentation, Techno-Economic Assessment model will be addressed, with which economic evaluation of membrane-based CO_2 capture processes is made and then comparison is made with other capture technologies, for example, amine-based capture process, for gaining conceptual insights in the development required for membrane process in the context of CO_2 capture. Based on TEA model presented, CAPEX and OPEX of membrane-based CO_2 capture process is obtained, subject to economic parameters and process design constraints. CO_2 capture cost and avoidance cost are also estimated. TEA also accommodates economic benefits gained from the application of process integration between CO_2 emitting plant and CO_2 capture process, while process optimization technique is fully used for minimizing overall cost at the capture process level. Sensitivity analysis is made to understand techno-economic impact of process design and their performance on capture cost. The membrane-based CO_2 separation process is simulated and optimized with authorsâ€™ in-house process design framework developed in MATLAB$^{Â®}$ environment, which can effectively identify the cost-effective configuration of multi-stage membrane systems, together with full consideration of CO_2 liquefaction. Case study will be given to demonstrate the benefits of TEA for critical and systematic evaluation of different capture technologies in a holistic manner.

Acknowledgement

This research was supported by the Korea Carbon Capture & Sequestration R&D Center (KCRC) through the National Research Foundation of Korea (NRF) funded by the Ministry of Science, and ICT (NRF-2014M1A8A1049338)

538696 Enhancement of Adsorbed Natural Gas Storage in Defect-Engineered MOFs during Long-Term Cycling

Abstract:

Vehicular adsorbed natural gas (ANG) technology is urged by the demand to balance the increasing development of vehicle industry and the environmental sustainability. Metal-organic frameworks (MOFs) have been considered as one of the most promising materials for ANG applications. However, MOF-based ANG is hampered by two challenges, i.e. the impurities from natural gas not only accumulate in MOFs, but also cause structural degradation of the materials (see **Figure 1a**) during long-term cycling, leading to significant reduction of ANG capacity (deliverable energy reduces up to 50% after 200 cycles). To overcome these challenges, combining molecular simulation, process modeling and experimental procedures, (1) we designed a class of Zr-based dual-ligand MOFs (see **Figure 1b**) by well-tuning the distribution (short-range order) of labile ligand-B in the structures, which were attacked and split preferentially to create stable and defect-controllable (defect-engineered) MOFs during ANG cycling, increases the MOF porosity and working capacity (deliverable capacity increase about 20% compared to single-ligand MOFs, as denoted in **Figure 1c**), (2) developed a mathematical ANG model to quantitatively optimize the regeneration strategy for the materials, which recovers the storage capacity (up to 99% after 673 K/2 h/vacuum treatment) and lengthens the working life time period compared to the single-ligand MOFs.

This work focuses on (1) investigating the mechanism (transition states) of metal -ligand connection breaking by the impurity attack using DFT simulations, and select a serial of appropriate ligand pairs to construct dual-ligand MOFs; (2) optimizing the ligand composition and distribution in the dual-ligand MOFs to control the defects and maintain structural stability; (3) developing a ANG process model to predict the performance of the materials, and optimize the regeneration period, temperatures and pressures to resume their ANG performance. The results of this work provide a novel strategy to design MOFs in overcoming the drawback of ANG capacity degeneration, facilitating the popularization of vehicular ANG technology in practical applications.

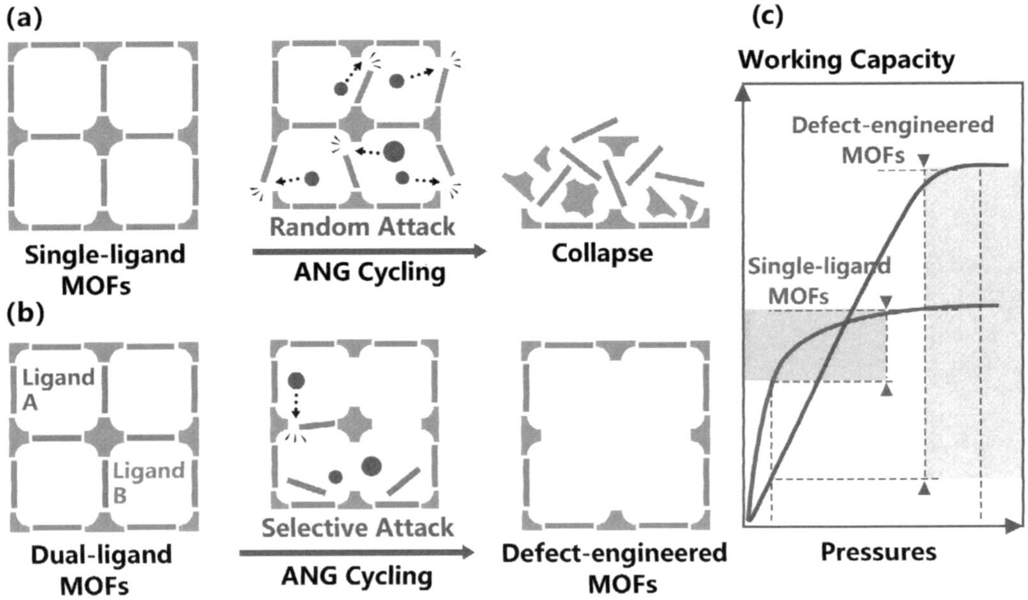

Figure 1. (a) Impurities randomly attack single-ligand MOFs and cause structural collapse, (b) impurities selectively attack the labile ligand (ligand B) in the dual-ligand MOFs and create controllable defect materials (Defect-engineered MOFs), (c) natural gas working capacity of defect-engineered MOFs and single-ligand MOFs.

527518 Investigation of Missing-Cluster Defects in UiO-66 and Ferrocene Deposition on Defective Sites for O_2/N_2 Separation

Abstract:

Due to the high porosity and function diversity of metal-organic frameworks (MOFs), MOFs are recognized as the promising candidates for many adsorptive separation processes, including CO_2/N_2, CO_2/CH_4, and O_2/N_2 separations. Among reported MOFs, UiO-66 is such a special case due to its extraordinary stability. However, the framework inertness limits its potential for many applications. Recently, the investigation of the defects in UiO-66, which significantly increase the activity of framework, have made UiO-66 become much more attractive than ever before. In this presentation, we will discuss the formation of missing-cluster (MC) defects is highly related to the partially-deprotonated ligands. A series of syntheses controlling the defect formation are provided to support the discussion. The number of MC defects are sensitive to the addition of the deprotonation reagent, synthesis temperature, and reactant concentration. Pore size distribution derived from the N_2 adsorption isotherm at 77K allows accurate and convenient characterization of the defects in UiO-66. The existence of defects in UiO-66 framework can cause significant deviations in its pore size distribution from the results derived from the crystal structure file. The generated extra porosity allows the deposition of large functional molecules in the defective framework to manipulate the adsorption properties. Compared to the defect-free UiO-66, which is a N_2-selective adsorbent, a O_2-selective adsorbent is demonstrated by introducing ferrocene into the defective framework of UiO-66.

537302 Chemical Potential Difference between the Large and Narrow Pore Forms of MIL-53 (Al)

Abstract:

Flexible adsorbents are interesting class of materials. Upon adsorption of guest molecules, some of these materials exhibit phase transition. The isotherms usually exhibit a step and phase transition occurs over a range of pressures. It is important to correctly quantify the chemical potential difference (Î”Î¼$_{S}$) between the solid in the two phases to be able to predict and/or model binary gas adsorption (using IAST for example).

In this work we present a consistent method to quantify Î”Î¼$_{S}$ as a function of temperature. As an example, MIL-53 (Al) is chosen as a solid. It will be shown that Î”Î¼$_{S}$ is independent of the gas used in the measurement, as is to be expected.

Moreover, the adsorption grand potential in the transition region can not be obtained using the thermodynamic framework for rigid solids. Some insights into the hysteresis and width of the transition region will also be presented.

525424 Unveiling Metal Organic Framework for Handling Indoor and Gas Containing Humidity

Abstract:

Unveiling Metal Organic Framework for handling indoor and gas containing humidity

Youssef Belmabkhout, Prashant M. Bhatt, Amandine Cadiau, Karim Adil Rasha Abdulhalim, Sk Md Towsif Abtab, Dalal Alezi and Mohamed Eddaoudi

Functional Materials Design, Discovery & Development (FMD3), Advanced Membranes and Porous Materials Center, Division of Physical Sciences and Engineering, King Abdullah University of Science and Technology, 4700 King Abdullah University of Science and Technology (KAUST), Thuwal 23955-6900, Kingdom of Saudi Arabia.

Highly stable and porous water vapor adsorbents are of primary importance in many water adsorption related applications such as moisture control, adsorption desalination and adsorption heating/cooling pumps. However, various difficulties have been experienced in synthesizing porous material combining high porosity and exceptional stability and adequate structural properties relationships.[1,2,3] In this paper, I will demonstrate the last development in Molecular Building Block (MBB) approach to assemble novel extremely highly porous and stable Metal Organic Frameworks (MOFs) with water uptake ranging from 0.5 to 2 g/g (Figure1). In particular, the structural properties with relation to the shape of water adsorption isotherms and the mechanism of adsorption-desorption involved, will be debated.

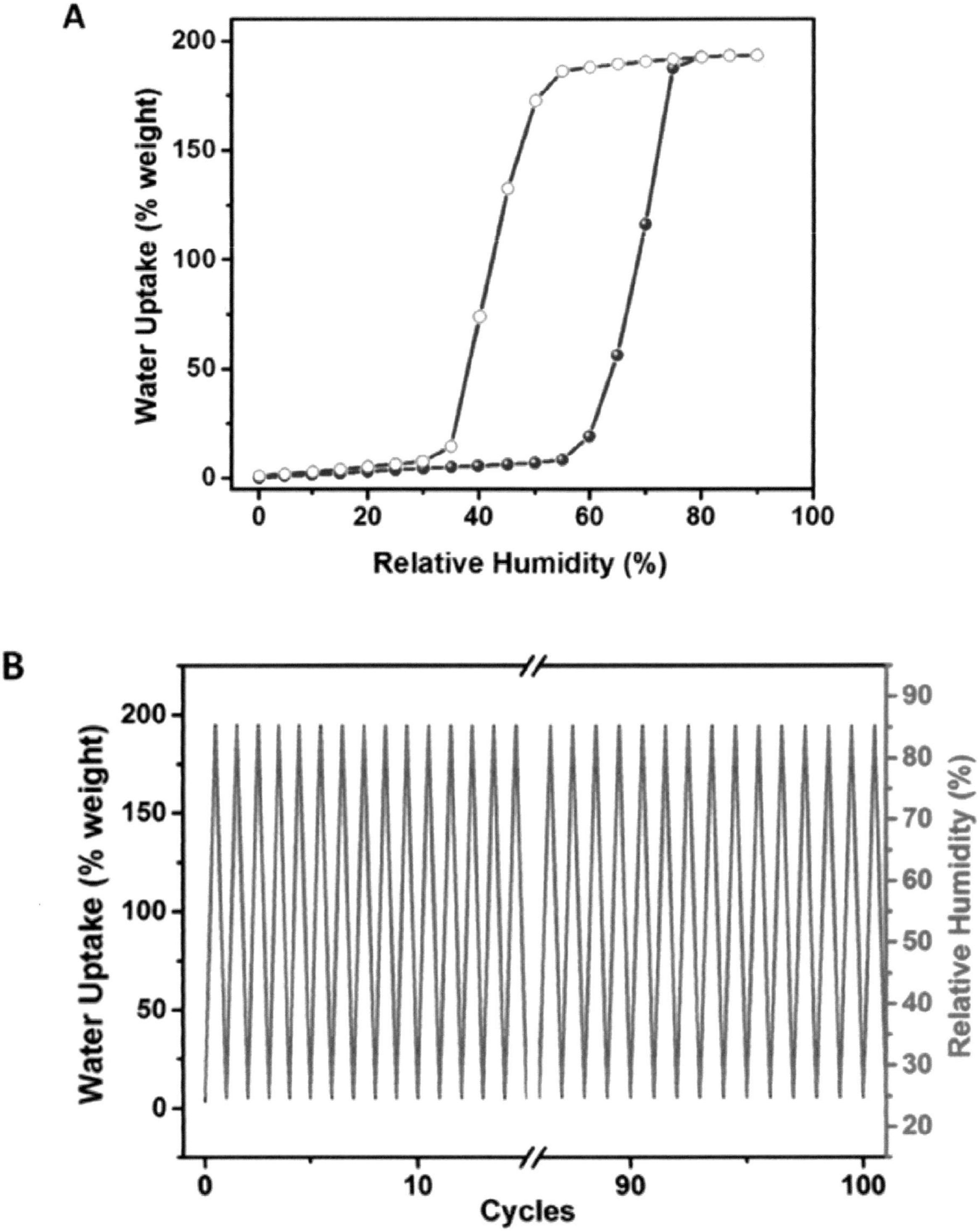

Figure 1. Water adsorption study of Cr-soc-MOF-1[2]. (A) Water adsorption (solid spheres) and desorption (empty circles) isotherms at 298K for activated Cr-soc-MOF-1. (B) 100 cycles of water uptake profile vs relative humidity of the Cr-soc-MOF-1 at 298 K.

References

[1] Rasha G AbdulHalim et al. A fine-tuned metal–organic framework for autonomous indoor moisture control. *Journal of the American Chemical Society* 139 (31), 10715-10722.

[2] Sk Md Towsif Abtab et al. Reticular chemistry in action: a hydrolytically stable MOF capturing twice its weight in adsorbed water. *Chem* 4 (1), 94-105

[3] Amandine Cadiau et al. Hydrolytically stable fluorinated metal-organic frameworks for energy-efficient dehydration. *Science* 356 (6339), 731-735.

538958 CO/N_2 Separation Using MOFs: Investigation of the Role of Metal Type and Metal Density

Abstract:

Carbon monoxide (CO)/Nitrogen (N_2) separation is a particularly challenging separation, yet one with great industrial relevance for the synthesis of high-value chemicals. While an expensive cryogenic step can be used to perform such separation, it remains ineffective in purifying CO from syngas streams with high concentrations of N_2.

Taking advantage of the lower energy requirement of adsorption processes, we have explored the use of metal-organic frameworks (MOFs) as adsorbents for this challenging separation. Due to the multitude of possible MOF structure, we used a multi-scale approach â€“ combining molecular simulation and process system modelling â€“ to identify suitable MOFs for the targeted separation. We will present our work on a particular sub-set of MOFs, *i.e.* MOF-74 family. The effect of different types of metals as part of the MOF structure or impregnated on their surface was evaluated. The materials were characterised by analytical, spectroscopic and imaging tools and their performance in terms of capacity and selectivity was evaluated using dynamic and equilibrium tests. High recovery and purity of the CO stream were achieved, though variations were observed depending on the type of metals. The trends enabled to identify the adsorption mechanisms and guide the design of new MOF structures.

533388 Towards Systematic Assessment of Porous Adsorbents for Post-Combustion CO_2 Capture Via Multiscale Simulation Strategies

Abstract:

The discovery of new classes of porous adsorbents such as metal-organic frameworks (MOFs), Zeolitic Imidazolate Frameworks (ZIFs), porous organic cages (POCs), covalent-organic frameworks (COFs), and porous polymer networks (PPNs) has opened access to a very large number of porous structures with a wide range of functionalities, which can be potentially exploited in different separation applications. Experimental evaluation of all these materials for specific applications is not feasible, and as a result, this prompted the development of high throughput computational screening methods.

So far, these screening methods have been predominantly based on certain adsorbent metrics, such as pore volume and surface area, and equilibrium and dynamic properties, such as adsorption isotherms, selectivity and diffusivity, obtained from molecular simulations. It is now becoming apparent that a more realistic picture of the performance of porous materials in a PSA or VSA process should be obtained from the actual process simulation.

In our study, we reflect on the development of the multiscale strategies that combine molecular simulations and pressure swing adsorption modelling and optimization to predict performance of the materials on the process scale. Several studies that employ these strategies have already emerged, ranking MOFs and other materials for post-combustion carbon capture. Here we focus on the **challenges** associated with the interface between molecular and process levels of description and demonstrate that the emerging picture is quite complex. In particular, we will discuss (a) the effect of the protocol for fitting experimental adsorption data with analytical adsorption models (e.g. dual-site Langmuir model), (b) influence of the pellet porosity and (c) influence of the pellet size on the process performance and material raking. Another aspect of the multiscale strategies we intend to explore is the accuracy of the molecular force fields, particularly in reproducing nitrogen isotherms, and how this affects predictions for the performance of the material in a process and the resulting ranking.

As a case study, we consider a well-known 4-step vacuum swing adsorption (VSA) cycle with light product pressurization (LPP), and Zeolite 13X as adsorbent in application to carbon dioxide removal from a typical flue gas stream (15% CO_2, 85%N_2, 1 atm).

In the last part of the presentation we attempt to evaluate performance of four widely studied materials (Zeolite 13X, Silicalite-1, MOF74-Ni and Cu-BTC) and compare process performance predictions based on molecular simulation predictions and experimental data. Our results demonstrate that simulation-based ranking of materials qualitatively agrees with that based on the experimental data, however the actual results can be sensitive to the variation of parameters mentioned above.

535953 Treatment Approaches for Produced Water Re-Use and Surface Discharge

Abstract:

Treatment Approaches for Produced Water Re-use and Surface Discharge

Madison Wenzlick, Alexandra Hakala, Nicholas Siefert

Nearly all methods for extracting oil and gas resources produce a high salinity brine byproduct [1]. Half of the high salinity brines produced in the U.S. are currently managed by injection into salt water disposal wells, which is a practice that has been linked to increased seismicity and has prompted new regulatory policies. One option for reducing the amount of reinjected brine is to remove fresh water from the brine, which allows for development of a tailor-made brine of a desired density or TDS [2]. These near-saturation NaCl brines could be sold as a feedstock to the chemical manufacturing industry. For example, saturated NaCl brines are called "10-lb brines" and are a tradeable commodity in parts of the U.S, particularly in Texas and Louisiana. Dewatering produced brines from oil & gas wells creates both a concentrated, lower-volume brine and fresh water.

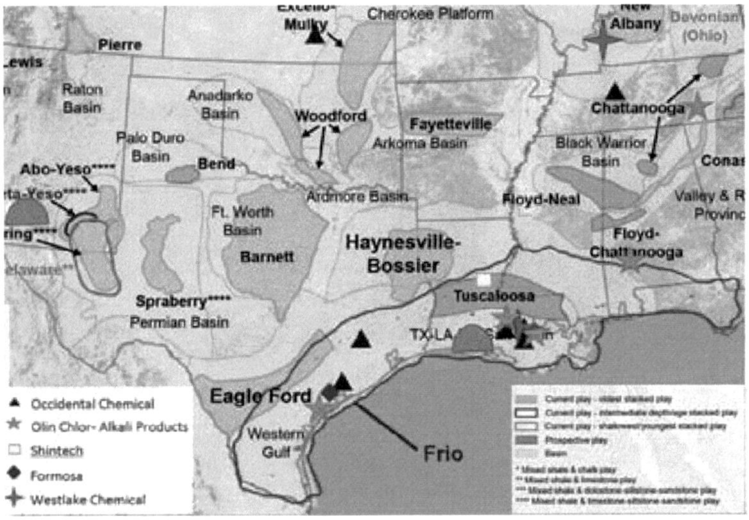

Figure 1: Relative locations of formations and basins, chemical plants, as well as locations of known salt dome collapses.

Base map from eia.gov

The main commercially available technology for concentrating brines is mechanical vapor recompression (MVR). However, typically the energy usage and cost of this technology are high and there is still limited data available on applying MVR to oil and gas produced brines.

In this presentation, the following topics will be covered. First, we will discuss the region (TX & LA) and shale formations chosen for this analysis. Second, we will present a geochemical analysis of the brines produced in the formations in this region. The analysis will include geographic considerations of brine concentration, determination of necessary pretreatment and chemical use, transportation, and potential benefits from reuse. Next, we will present a baseline techno-economic analysis of concentrating these brines into 10-lb brines while generating fresh water for surface discharge. The baseline case is a process using commercially-available treatment technologies, such as MVR.

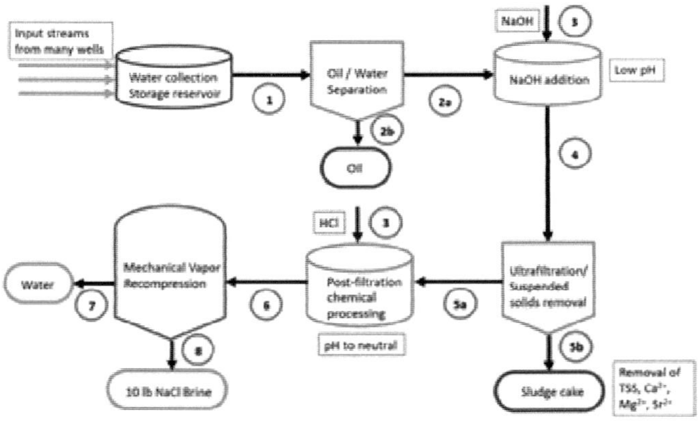

Figure 2: The baseline process flow diagram for the centralized water treatment process

Next, we present results from a sensitivity analysis around a number of the assumptions in the model, such as the cost of pretreatment chemicals, electricity costs, and brine disposal fees. Finally, we present an advanced case in which we model advanced membrane processes for both the pretreatment and brine concentrating steps of the process. Preliminary results from the baseline case suggest that, for those brines with low concentrations of divalent cations (Ca^{+2}, Mg^{+2}, Sr^{+2}), there is a positive rate of return on investment for dewatering oil and gas produced brines in TX & LA into 10-lb brine and fresh water.

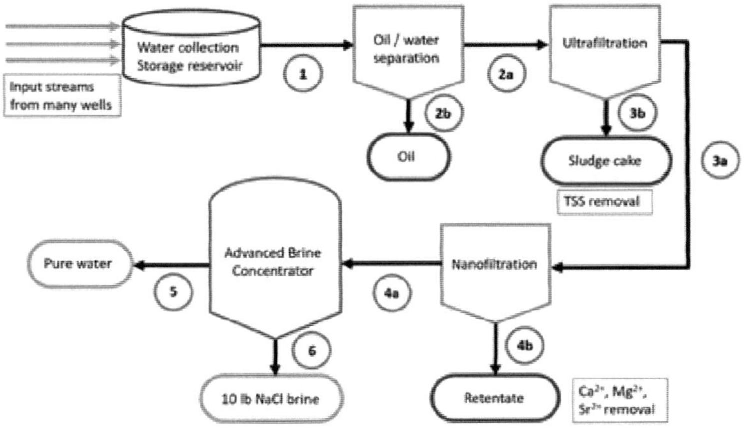

Figure 3: Advanced case process flow diagram

References

[1] J. Veil, "U.S. Produced Water Volumes and Management Practices in 2012," 2015.

[2] J. T. Arena, J. C. Jain, C. L. Lopano, J. A. Hakala, T. V. Bartholomew, M. S. Mauter and N. S. Siefert, "Management and dewatering of brines extracted from geologic carbon storage sites," *International Journal of Greenhouse Gas Control,* no. 63, pp. 194-214, 2017.

529250 Advanced Supercritical Water-Based Process Concepts for Treatment and Beneficial Reuse of Brine Generated By Oil/Gas Production

Abstract:

Both U.S. energy and economic security rely upon continued development of U.S. unconventional shale plays, which require access to suitably clean water resources. Produced water (brine) is the nation's largest industrial waste stream, with approximately 14 billion barrels generated annually by the U.S. oil and gas sector. The recent surge and continued development of unconventional U.S. shale plays will likely result in greater volumes of produced water. In addition, an additional brine waste stream will be created by future sequestration of CO_2 emissions generated by fossil-based power plants into saline aquifers. Utilizing this brine in development of unconventional shale resources represents a beneficial reuse of this waste stream. This approach can not only lower stress on local watersheds but can also address public/local government concerns regarding long-term ground water contamination potential and seismic activity associated with Class II salt water disposal wells (SWDs).

However, direct reuse of brines in shale development activities currently is limited due to constituents found in this waste stream. Specifically, the high levels of dissolved solids found in the brine may cause scaling within the shale or within production casings, reducing well productivity. Current brine treatment technologies including membrane- and thermal-based processes are ineffective in treating brine containing concentrations of dissolved solids greater than 80,000 ppm due to fouling or cost/sizing, respectively.

To address these limitations, Ohio University (OHIO) with funding from the U.S. Department of Energy's National Energy Technology Laboratory (NETL) Crosscutting Research Program Project DE-FE0026315, has been developing an advanced supercritical water (SCW)-based process for treatment and beneficial reuse of brine waste streams. This SCW process offers an advantageous media for brine treatment, as its lower fluid density and decreased hydrogen bond strength provides a means to simultaneously remove dissolved solids and hydrocarbons.

Previous SCW-based brine treatment systems have been plagued by internal scaling, resulting in inefficient heat transfer, plugging, and process downtime. To address this issue OHIO has been developing both externally- and internally-heated SCW reactor design concepts, which utilize advantageous fluid dynamics and electrically driven Joule-heating mechanisms, respectively. OHIO's new SCW reactor designs offer the potential to provide a field deployable brine treatment process and a resultant product which may be reused in shale development or other beneficial reuse applications, thereby supporting goals of NETL's Strategic Center for Oil and Natural Gas.

To evaluate process potential, OHIO has been conducting both experimental investigations using prototype SCW reactors and process simulations/techno-economic assessments using Aspen Plus^{TM}. Removal of naturally occurring radioactive materials (NORM) using commercial ion exchange resins and natural adsorbent has been investigated. Further, removal of dissolved solids from simulated and field-derived produced water has been investigated at pressures ranging from 23-28 MPa, respectively, demonstrating the ability to recover greater than 99 percent of salts from brines containing dissolved solid concentrations greater than 240,000 mg·L^{-1}. Techno-economic summaries have identified important brine treatment cost sensitivity parameters and heat recovery schemes to lower brine treatment costs. This presentation will review experimental and techno-economic study results from both U.S. DOE supported projects.

516908 Permian & Anadarko Basin Produced Water Recycling: Keys to Success

Abstract:

Demand for recycling of produced water in the Permian & Anadarko Basin are expanding. Although 2017 hasnâ€™t necessarily been a drought year, some areas remain short on available water. Moreover, the considerations that producers must account for is vastly expanded over prior years. Issues such as Induced Seismicity, limited disposal zones for Class II wells, water availability, costs for fresh water, environmental concerns, and more are all expanding interest in produced water recycling! However, there are a multitude of options available that can be used for recycling and many companies have spent years vast amounts of money trying to test or evaluate various technologies. This paper discusses a very basic and successful approach to water recycling in some key development areas of the mid-continent and associated keys to success. For instance, Class II Disposal Wells can play a critical role in making a recycling project successful.

533288 Model Predictive Control of 4-Zone Simulated Moving Bed Chromatography for the Separation of Bicalutamide Enantiomers: Experimental Validation

Abstract:

Simulated moving bed (SMB) technology has been applied in petrochemical, pharmaceutical, and fine chemical industries to isolate products in high purity and high yield since it was introduced in 1960s [1]. In recent decades, different control concepts were applied to SMB processes to optimize the operating conditions dynamically [2-4]. Since the SMB process is operated periodically, i.e. the process outputs do not reach a steady-state, it is difficult to observe the process states in detail. Therefore, the control of SMB process is still challenging.

To estimate the SMB process states, a simple cell model, a flux limiter, and competitive Langmuir/Bi-Langmuir isotherms were used. The cell model is equivalent to the explicit finite difference method, and the nonlinear isotherm models can be solved Newton's method. The computation time can be remarkably reduced by parallel computation algorithms. From the estimated concentration profiles in the SMB process, the actual and estimated process outputs, e.g. the product stream concentrations, can be compared to decide the model parameters, i.e. isotherm parameters and dispersion coefficients. A model predictive control concept was applied to search the optimized future operating conditions to obtain the desired target purity and yield.

Bicalutamide is a racemic drug for prostate cancer (CASODEX, AstraZeneca). Since (R)-bicalutamide is more efficacious than (S)-bicalutamide, a four-zone SMB process with chiral stationary phase (Chiralpak IA, 20μm) was used to purify the eutomer, (R)-bicalutamide [5]. In this work, an analytical HPLC unit was used to analyze the observed stream concentrations. To design the four-zone SMB process, basic information obtained from the batch preliminary experiments was provided. The four-zone SMB process performed in a pilot plant scale (4 columns, 2.5 cm I.D., 15 cm length) was successfully controlled to obtain the target (extract) and waste (raffinate) components up to 99% purity (Figure 1).

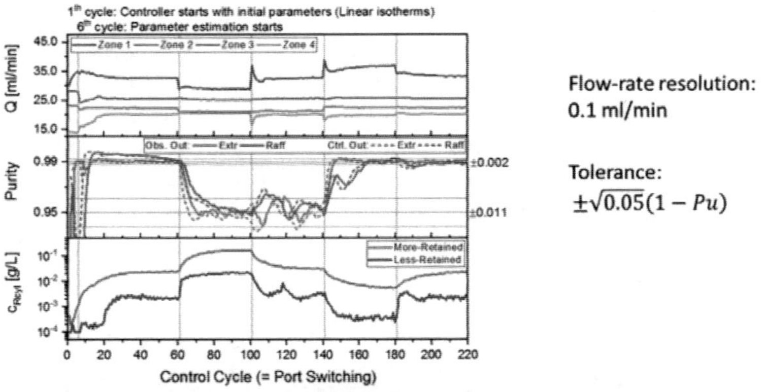

Set-point	1-60	61-100	101-140	141-180	181-220
Pu_{Extr}/Pu_{Raff}	0.99/0.99	0.95/0.95	0.95/0.95	0.99/0.99	0.99/0.99
R_{S1}/R_{S4}	0.5/0.1	0.5/0.1	0.1/0.02	0.1/0.02	0.5/0.1

Figure 1. Controlled operation histories (feed concentration = 14 g/L, Pu_{Extr}: desired purity of extract product, Pu_{Raff}: desired purity of raffinate product, R_{S1}: raffinate stream impurity concentration ratio in recycle stream, R_{S4}: extract stream impurity concentration ratio in recycle stream).

[1] Broughton, D.B., Neuzil, R. W., Pharis, J. M., Brearley, C. S. (1970). The parex process for recovering paraxylene. Chem. Eng. Prog., 66, 70.

[2] Natarajan, S., Lee, J.H. (2000). Repetitive model predictive control applied to a simulated moving bed chromatography system. Comput. Chem. Eng., 24, 1127.

[3] Schramm, H., Grüner, S., Kienle, A. (2003). Optimal operation of simulated moving bed chromatographic processes by means of simple feedback control. J. Chromatogr. A, 1006, 3.

[4] Grossmann, C., Langel, C, Mazzotti, M., Morari, M., Morbidelli, M. (2010). Experimental implementation of automatic 'cycle to cycle' control to a nonlinear chiral simulated moving bed separation. J. Chromatogr. A., 1217, 2013.

[5] Kaemmerer, H., Horvath, Z., Lee, J. W., Kaspereit, M., Arnell, R., Hedberg, M., Herschend, B., Jones, M. J., Larson, K., Lorenz, H., Seidel-Morgensten, A. (2012) Separation of Racemic Bicalutamide by an Optimized Combination of Continuous Chromatography and Selective Crystallization. Org. Process Res. Dev., 16, 331.

533844 Separation of Anhydrosugars and Phenolic Species in a Fast Pyrolysis Aqueous Product Stream Using Resin Adsorbents and Simulated Moving Bed Technology

Abstract:

Fast pyrolysis is a promising technology for the thermo-chemical conversion of lignocellulosic biomass into liquid fuels and chemicals. Bio-oil is the liquid product of fast pyrolysis and is a complex mixture of water, anhydrosugars, organic oxygenates including carboxylic acids, alcohols, aldehydes, esters, ethers, furans, ketones, and phenolic monomers, dimers, and oligomers with a variety of functional groups. The chemical complexity of bio-oil necessitates several separation and upgrading strategies be employed before it can be used as a fuel or specific groups of chemicals can be isolated for other applications.

Fractional condensation of the vapors and aerosol products exiting the pyrolyzer makes use of successive condenser stages at progressively lower temperatures and is a practical first step at separating the bio-oil components. A distinction is made between heavy components that condense in the early stages above 120 °C and the remaining light components that condense in the latter, colder stages. The heavy fraction, or â€œheavy endsâ€, contains a few percent water and consists of low to high molecular weight phenolic species and anhydrosugars. A water extraction is performed on the heavy ends yielding an organic phase containing the larger water insoluble phenolic species and an aqueous phase containing phenolic monomers to their extent of solubility in water and nearly all of the sugars present. The aqueous phase contains approximately 8 wt% phenolics and 10 to 25 wt% anhydrosugars, depending on the biomass substrate utilized and pyrolysis reactor conditions. One intended application is to hydrolyze the anhydrosugars to glucose and perform fermentation using traditional microorganisms; however, fermentation experiments indicate that the presence of phenolic monomers at concentrations above 0.1 wt% completely inhibit the microorganisms.

This work explores the use of liquid phase adsorbents for the separation of phenolic species and other small organics from sugars in the aqueous phase. Adsorption equilibrium and kinetic parameters were determined for a variety of commercial polymeric resin adsorbents. This information was used to model continuous adsorption processes and assist in the design of a simulated moving bed (SMB) for this application. Single column experiments demonstrate that effective separation of sugars and phenolics is feasible and the target of producing an aqueous sugar product stream with less than 0.1 wt% phenolics is achievable. Construction of a lab-scale SMB is underway, and we intend to report on system parameters such as the purity and recovery of sugar and phenolics in the separated product streams, solvent consumption, and the overall efficiency of the separation.

534295 Constant-Pattern Design of Displacement Chromatography

Abstract:

Displacement chromatography has been used in a variety of applications, especially those related to the biopharmaceutical industry. Compared to traditional elution chromatography, displacement chromatography can deliver products at higher concentrations, higher purity, and higher yield, which makes it an attractive option in many separation systems. However, the design of displacement chromatography systems is a complex problem that requires the specification of many design parameters. While previous work has been done to determine the column length, and loading required to form an isotachic train in ideal systems, it is still unknown when the mixed band regions between components reach a constant pattern. Additionally, the design and optimization of displacement systems is done using trial and error, requiring many simulations or experiments.

In this study, a general method was developed to design displacement systems to obtain high purity products with high yield and high productivity. The number of design parameters were greatly reduced through dimensionless groups analysis. Then, rate model simulations were used to examine the effects of the dimensionless groups to determine the conditions which are required to reach the constant pattern state. An expression for yield as a function of dimensionless variables was derived as well. The yield and productivity are greatly affected by the selectivity of the components, overall mass transfer coefficient, feed composition, loading fraction, volume of feed loaded, and the displacer concentration. Given a target yield, displacer concentration, feed volume and composition, as well as column properties, this method can be used to determine the necessary column length and operating flow rate to reach the target yield. Because the constant-pattern design method utilizes dimensionless groups, the results are easily scalable while maintaining high purity, high yield, and high productivity.

519148 ZIF-8 As an Efficient Adsorbent for Ethane/Ethylene Separation By Gas Phase Simulated Moving Bed

Abstract:

Growth of global population, along with rising living standards lead to increasing demands for synthetic solid and liquid chemicals such as plastics and lubes with designed properties. Ethylene is the key building block in chemical industry as it is capable to catalytically co-oligomerize to form controlled polymeric hydrocarbon chains, and thus can be used in polymer production with designed physical properties, and also as a starting material/intermediate for the synthesis of other high demand chemicals. Annual production of ethylene achieves 150 Mt and expected to grow with rates of 4 to 6% per year making this olefin one of the most important chemical product in terms of volume, number of derivatives and sales value.

There are mainly two categories of feedstocks for ethylene production. One includes those derived from crude oil, and another includes those derived from natural gas. The raw materials from crude oil are mainly paraffins (alkanes), aromatics and naphtha, being the latter a mixture rich in paraffinic, naphthenic and aromatic compounds, with a boiling point between 20 and 200 ºC, besides having between 4 and 12 carbons in the carbonic chain. The composition of natural gas is basically methane with other higher alkanes, such as ethane, propane and butane. The composition of the final products is intrinsically related with the feedstock selected to produce ethylene, with implications on the costs of the olefin purification. The feedstock used depends on availability, price, and what other products are intended to obtain.

Thermal cracking is the main industrial method for producing ethylene at reaction temperatures around 850-900 °C. Light hydrocarbon feeds such as ethane, LPG or light naphtha is dehydrogenated to give complex mixture containing product alkenes such as ethylene, propylene and butadiene mixed with hydrogen, unreacted alkanes such as ethane and propane and other by-products. Heavier hydrocarbon feeds (full series and heavy naphtha as well as other refinery products) give some of these, but also give products rich in aromatic hydrocarbons and hydrocarbons suitable for inclusion in gasoline or fuel oil. The cracked product mixture are processed with successive distillation steps to initially pre-separate the ethane and ethylene mixture from other lighter and heavier molecules and finally separate ethylene from ethane with different olefin proportions, which can range from 50 to 80%.

Ethylene purity above 99.9% is required for polymerization processes due to presence of catalysts. The significant fraction of ethylene production cost is associated with separation steps primarily due to energy intensive cryogenic liquefaction of gaseous molecules, as well as similar relative volatility of the components. Separation processes consume more energy than thermal cracking of feeds and, in this way, determining the end price of many essential commodities in our daily lives.

The rising demand of ethylene combined with the large economic and energy production costs, make it crucial to search for more efficient and profitable separation processes. In this field, the separation of high value products based on adsorption at near ambient temperature appears as a strong substitute, to the currently practiced cryogenic distillation process in use, in particular cyclic adsorption processes, such as simulated moving bed (SMB) and pressure swing adsorption (PSA). Generally, it is possible to find an adsorbent for which the adsorption separation factor is much greater than the relative volatility, so in such cases an adsorption separation process may be economically more feasible.

More recently, a breakthrough was attained with the emerging of new MOFs with a paraffin selective peculiar characteristic, and some results in this type of materials are already available in the literature. ZIF-7, ZIF-8, ZIF-9 and ZIF-11 were the first examples of a microporous adsorbent exhibiting selective adsorption of paraffins over olefins offering the advantageous separation process configuration. These MOFs also display an interesting performance due to its structural flexibility.

Based on this new concept of paraffin selective adsorbents, we tested ZIF-8 granulates purchased from Materials Center of the Technische Universität Dresden, Germany, to separate ethylene from ethane, at high purities from a mixture having a composition very close to that obtained at the outlet of the industrial ethylene/ethane splitters.

The ZIF-8 material was characterized preliminarily by SEM/EDS, X-ray powder pattern diffraction, mercury intrusion porosimetry, and sorption of nitrogen at 77 K and carbon dioxide 273 K. Particle average diameter is about 1.1 to 1.4 mm and the crystal radius of 460 nm, observed by SEM analyses. The mercury intrusion porosimetry and the helium picnometry provided the apparent (339 kg/m^3), solid (935 kg/m^3) and skeleton (1328 kg/m^3) density values. The textural properties, such as Langmuir surface area (1653 m^2/g), micropore area (1635 m^2/g), external surface area (17.9 m^2/g) and micropore volume (0.58 cm^3/g) were determined from the nitrogen and carbon dioxide adsorption equilibrium isotherms.

The adsorption equilibrium isotherms of the ethane, ethylene, propane and carbon dioxide at 303, 333 and 373 K on the ZIF-8 granulates were determined. The isotherms were obtained by a gravimetric method using a magnetic suspension balance (Rubotherm). The sample is placed in a basket within the microbalance measurement cell and regenerated, at 423 K and under vacuum, to remove any chemical species that have previously been adsorbed. Each point of the isotherm was obtained by increasing successively the system pressure. The presence of hysteresis was evaluated by measuring the adsorption and desorption isotherm branches. The Langmuir model was selected in order to describe the observed equilibrium data.

Two SMB cycles for ethane/ethylene separation with a four zone configuration, in open loop, were experimentally performed in the bench scale SMB unit with eight columns. Each column in stainless steel were individually packed with approximately 4.7g of ZIF-8. Activation of the adsorbent was performed by feeding 0.500 SLPM of nitrogen for 16 h at a temperature of 423 K. After the activation, the temperature of the oven was set to the desired value of 323 K. In each experiment, the desorbent gas cylinder (propane or carbon dioxide) was set to 600 kPa, to impose a higher pressure at the beginning of the Zone I, and the back pressure controller was configured to operate at 500 kPa and placed at the end of the Zone IV to keep the gas flow in only one direction.

The equilibrium data was assessed in order to allow the desorbent selection. It was concluded that propane would behave as a strong desorbent, since it presents an adsorption capacity higher than both components of the mixture, on the full ranges of the studied pressure and temperature. Its saturation capacity presents the value of 4.56 mol/kg, an affinity constant of 4.7×10^{-8} kPa^{-1}, and a heat of adsorption of 35.1 KJ/mol. For carbon dioxide (weak desorbent), a lower adsorption capacity was observed when compared with the C_2 components, possessing a saturation capacity of 9.72 mol/kg, an affinity constant of 8.0×10^{-8} kPa^{-1} and an adsorption heat of 23.3 kJ/mol. The desorbent strength influences drastically the SMB performance therefore the importance of obtaining a suitable desorbent to ensure a separation as efficient as possible.

These experiments had as main focus the study of the influence of the desorbent strength on the ethane/ethylene separation using ZIF-8 sample, under similar operating conditions. In both cycles a 2-3-2-1 columns distribution and a switching time of 50 s were adopted. Propane was used as a desorbent for separation of 0.40/0.60 ethane/ethylene (run

1) feed mixture and a total feed flow rate of 0.127 SLPM, while carbon dioxide was used to separate 0.41/0.59 ethane/ethylene (run 2) feed mixture and a total feed flow rate of 0.081 SLPM. The desorbent flow rate in the run 1 was 0.780 SLPM, while in the run 2 was imposed a desorbent flow rate of 1.176 SLPM. The pressure range was very similar for the two experiments, since for run 1 the range was 495 to 520 kPa and for run 2 it was 500 to 540 kPa.

In the experiment, with carbon dioxide as desorbent, ethylene was obtained with 99.6% purity and 94.2% recovery, while in the experiment using propane as desorbent only 77.1% of ethylene was recovered with a purity of 82.7%. With the cycle using carbon dioxide desorbent it was possible to obtain better performances, but, above all, these experimental data allowed the validation of the mathematical model.

A process modelling software was used to simulate and to validate the gas-phase SMB model against the experimental data, and to compute the separation regions that lay within the purity and recovery restrictions, a purity of raffinate (ethylene) higher than 99.9% and recovery of ethylene of at least 98%, corresponding to an extract purity higher than 97.2%. The obtained regions are drastically different. While for the case in which propane is used as desorbent, the obtained separation region is non-existing; for the case considering carbon dioxide the separation, obeying to the desired specifications, is possible within a significant separation region.

525388 Aromatics/Alkanes Separation: Simulated Moving Bed Process Model Development By a Concurrent Approach and Its Validation in a Mini-Plant

Abstract:

SMB process design for multicomponent systems is often challenged by the difficulty of establishing a reliable adsorption model, especially when the adsorption equilibrium is governed by multi-component nonlinear isotherms. In this talk, we discuss a "concurrent" iterative approach to simultaneously determine the adsorption model while optimizing SMB process design, thereby greatly reducing the experimental effort for process development. Here we specifically focus on a concurrent approach for separation of multicomponent aromatics/alkanes mixtures that approximate hydrocarbon streams often encountered in petroleum refining. These multicomponent streams display strong competition for adsorption in mesoporous or microporous adsorbents. Using an approach that combines adsorption measurements and model fitting and SMB model predictions that allow targeted selection of compositions for adsorption measurements, we demonstrate convergence of the concurrent method with only a small number of multicomponent adsorption measurements. Then, we demonstrate a second iterative methodology to refine the SMB model by estimating the non-thermodynamic (i.e., mass transfer) and other (dead volume) system parameters, using only a small number of runs on a 16-column SMB "mini-plant" designed and constructed in-house at the Georgia Institute of Technology. The final process design, validated in the SMB mini-plant, maximizes the throughput while maintaining the desired aromatic purity and recovery.

533058 Atomistic Investigation of Polymer-MOF Interfaces in Mixed Matrix Membranes

Abstract:

Membrane processes are promising alternatives to conventional amine absorption process for CO_2 separation from gas mixtures. While polymers have been preferred materials due to their processability, their separation performance needs to be enchanced in order to make them viable for industrial applications. This has motivated development of Mixed Matrix Membranes (MMM's) that combines polymer processibility with high separation performance of fillers. Metal Organic Frameworks (MOF's) are one of the best candidates as filler materials with their excellent gas adsorption properties.

While separation performances of individual components are main concern in MMM's, it has been proven that compatibility of materials used in MMM's are as important. Incompatibility between composite materials can result in interfacial voids leading to deviation from ideal separation performance. An interface with molecular scale or sub-molecular scale extra free volume between segments can also occur, and this produces a small decrease in selectivity below that of the pure polymer while still demonstrating an increase in permeability.

Another type of nonideality is matrix rigidification, which results in reduced free volume near the sieve surface. Rigidified matrix surrounding the sieves have lower permeability than that of bulk polymer, resulting in decreased membrane permeability.

In this study, in order to contribute to further understanding of MMM's in molecular level, a full atomistic study of Matrimid 5218/ZIF-8 MMM was studied. CO_2 adsorption up to 30 bar was performed to observe the effect of CO2 induced swelling on the interface of polymer and filler and to understand the behavior of polymer-MOF interactions with increasing CO_2 concentration.

Quantum-level charge calculations were incorporated with all-atom force fields in order to model polymer and MOF behavior accurately. Subsequent Monte Carlo and Molecular Dynamics Simulations were performed using LAMMPS software. GCMC simulations combined with NPT and NVT ensemble MD simulations were used for simulating sorption behavior of the polymers. Polymers swell with increasing CO_2 concentration and this swelling increases the sorption capacity. To account this swelling, NPT and NVT MD runs after each GCMC run was added for simulation process. Sorption simulations were done at 35 °C in line with most experimental data in literature. A CO_2 specific Free Volume approach was used in order to estimate plasticization behavior of MMM. Interfacial Dynamics of both unplasticized and plasticized structure was investigated for further understanding of compatibility issues in MMM's.

537221 Microscopic Gas Diffusion inside ZIF-11 Crystals Dispersed in Different Polymers to Form Mixed Matrix Membranes

Abstract:

Mixed matrix membranes (MMMs) consisting of zeolitic imidazolate framework (ZIF) filler particles embedded in a polymer matrix have shown great potential for gas separations. These membranes were introduced to combine the gas-separation properties of the filler particle with the ease and scalability of fabrication of pure polymeric films. ZIFs represent a subgroup of metal organic frameworks (MOFs), which exhibit great chemical and thermal stability and functionality.

^{1}H and ^{13}C pulsed field gradient (PFG) NMR was used to study diffusion of ethane and ethylene inside a loosely packed ZIF-11 bed and various ZIF-11-based MMMs at equivalent loading pressures. Diffusion measurements were performed by utilizing a high magnetic field (17.6 T) to obtain sufficiently high signal-to-noise ratios for gas diffusion data and high magnetic field gradients (up to 25 T/m) to enable measurements of diffusivities for the length scales of displacements smaller than the mean size of the ZIF-11 crystals. It was observed that depending on the polymer type, the confinement of the ZIF-11 crystals in the polymer can lead to a reduction in the intra-ZIF diffusivities of the studied gases. This effect was investigated as a function of temperature as well as gas loading inside MMMs. It was tentatively explained by a reduction in the framework flexibility of ZIF-11 crystals due to the crystal confinement in a polymer.

532877 Tailoring Separation Properties of Mixed-Matrix Membranes Via Combined Use of Two- and Three-Dimensional Fillers

Abstract:

Mixed-matrix membranes containing both two-dimensional (ns-CuBDC) and three-dimensional (ZIF-8) metal-organic frameworks were fabricated to investigate their potential capabilities in CO_2/CH_4 separation. The mixed-matrix membrane containing ns-CuBDC alone was capable of improving CO_2/CH_4 selectivity whereas the mixed-matrix membrane containing ZIF-8 alone ensured improved CO_2 permeability. However, by combining both fillers in a mixed-matrix membrane aimed at tailoring the CO_2/CH_4 separation properties, both CO_2 permeability and CO_2/CH_4selectivity were successfully improved by 16.6% and 30.5%, respectively, which indicates a highly desirable method of performance enhancement. Analysis of solubility-diffusivity in the mixed-matrix membranes revealed that the use of both fillers could improve both solubility selectivity and diffusivity selectivity. Our overall results imply that the separation properties of gas separation membranes can be readily adjusted to meet the requirements of real-world applications by the combined use of two fillers with different geometries.

523950 Room-Temperature Synthesis of Functionalized Two-Dimensional Metal-Organic Frameworks for Enhanced CO_2/CH_4 Separation

Abstract:

Removal of CO_2 from raw natural gas is an essential process prior to the actual utilization of natural gas. Mixed-matrix membranes (MMMs) have showed advantages in CO_2/CH_4 separation, but great challenges exist in controlling the morphology and functionality of filler materials, which plays a key role in gas permeation and compatibility between filler materials and polymer matrix. Metal-organic frameworks (MOFs), due to their versatile synthetic procedures, exhibit tunable structures and chemical functionalities, and thus are good candidate for preparing MMMs. Herein, we demonstrated a one-step room-temperature synthesis of 2D MOFs based on Cu (II) paddle-wheel units with mixed ligands, where the intrinsically anisotropic building blocks led to the anisotropic growth of 2D MOF nanoparticles. This synthetic method was also successfully extended to preparing functionalized 2D MOFs by selecting ligands with different functional groups, with the thickness of the thinnest particles around 3 nm. To illustrate the advantage of using 2D MOFs in MMMs, the synthesized 2D MOFs were incorporated into polysulfone, a commercial polymer for CO_2/CH_4 separation, and the resultant MMMs exhibited improved CO_2/CH_4 separation performance with CO_2/CH_4 ideal selectivity increased by approximately 400% at 5 wt% loading of amino-functionalized 2D MOFs. Because the functionality of 2D MOFs can be readily tuned based on the selection of ligands, the presented strategy to prepare MMMs can be further extended to other energy and environment-related separations.

525403 Adsorption-Enhanced, Plasticization Resistant Composite Membranes Using Metal–Organic Framework Nanocrystals

Abstract:

Here, we have developed a method for improving the gas separation performance for membranes in a broadly applicable and synergistic way. Specifically, we address two fundamental problems with conducting gas separations in neat polymer materials, namely selectivity/permeability tradeoffs and membrane plasticization. These two phenomena limit the application of membrane materials in numerous processes, and one severe case is the separation of ethylene from ethane. As the condensability and kinetic diameter of these gases are similar, there is no good chemical handle to conduct a separation. Introducing adsorption-based selectivity by compositing nanocrystals of the metal–organic frameworks M_2(dobdc) (M = Mg, Mn, Co, Ni; dobdc^{4-} = 2,5-dioxido-1,4-benzenedicarboxylate) with the polyimide 6FDA-DAM (6FDA = 4,4¢-(hexaflouroisopropylidene)diphthalic anhydride; DAM = 2,4,6-trimethyl-1,3-phenylenediamine), a unique chemical handle is used to improve the permeability of ethylene and the selectivity for ethylene over ethane. Additionally, we found that the metal–organic frameworks physically crosslink the polymer, which addresses plasticization, the second major issue with polymers for this gas separation.

We then expanded on this work by addressing another important gas separation that is limited by plasticization - natural gas purification - by again applying the crosslinking effects of the metal–organic framework nanocrystals. In this case, nanocrystals of the metal–organic framework Ni_2(dobdc) were incorporated into six different polyimides that can be used for this gas separation. This resulted in significantly improved plasticization response in the majority of materials, as measured by CO_2 permeation hysteresis and binary gas permeation measurements. Notably, high selectivity is retained under pressures of CO_2 in excess of 20 bar. These pressures represent some of the most aggressive feed environments that might be encountered in the field. Together, these works introduce a novel handle to improve permselectivity in addition to a new crosslinking mechanism to improve membrane stability.

527619 Synthesis of 2D MOF Nanosheets with Large Aspect Ratio through Control over Hydrogen Bond Formation for Membrane Separation

Abstract:

2D Metal-organic frameworks (MOFs) have tremendous application prospect in membrane gas separation. MOF nanosheets with high aspect ratio have potential to overcome the compatibility issue with polymer in the mixed-matrix membranes (MMMs) and to provide a much better permeability compared with other filler morphologies. To fabricate the high-performance MMMs, the large-scale synthesis of ultrathin 2D MOF nanosheets is a critical step, which can facilitate their property investigation and industrial applications. In this presentation, we will discuss the role of hydrogen bond in controlling the dimensions of synthesized MOF nanosheets. We will present a bi-phase synthesis method, which is capable of producing mBDC (m=Zn and Cu) 2D MOFs at large scales. The tightly coupled MOF structure is effectively prohibited and this synthesis route can easily produce high quality nanosheets in either crystal powder or depositing onto various substrates for separation applications.

537768 Polymorphic Selection of Biominerals By Anionic Polyelectrolytes

Abstract:

In most biomineralization processes many soluble biopolymers exist in minute quantities; these control the nucleation, growth, orientation, and polymorph selection, leading to the formation of minerals with diverse biological functions and distinct properties but also to pathological mineralization, e.g., bone disease, dental calculus, and renal stone formation. In this work, we study the effects of supersaturation, temperature, and additives on the nucleation and growth of calcium oxalate (the major constituent of kidney stones) in aqueous solutions. We have demonstrated that certain anionic polyelectrolytes can be very effective toward kidney stone inhibition by reducing the growth rate, altering the crystal morphology, and favoring the formation of the metastable dihydrate (COD) over the thermodynamically stable monohydrate (COM) which is mainly responsible for kidney, gall, and bladder stones. We compare our results to citrate, a naturally occurring inhibitor in the urinary system and currently the standard pharmacological treatment for stones. Our findings offer a better understanding of the calcium oxalate crystallization kinetics and provide information on the viability of synthetic polymers as potential calcium oxalate stone therapies.

538910 Coupling Experimental Results with Molecular Dynamics Simulations to Describe Polymorphism Obtained Using Solution Shearing

Abstract:

During pharmaceutical drug development, it is important to carry out an extensive search to identify all drug forms. Stabilizing metastable polymorphs is necessary to ensure that the correct drug form is selected for development to balance form stability and drug bioavailability. However, searching for rare forms is a nontrivial process. Here, we show that a combination of fluid dynamics and surface chemistry can be used to overcome some of the challenges associated with searching this extensive landscape of polymorphism. In this study, multiple forms of glycine and acetaminophen were isolated, including metastable forms that are difficult to obtain using conventional methods. The mechanisms through which multiple polymorphs are obtained are described herein. Molecular dynamics simulations and in-situ grazing incidence wide angle X-ray diffraction lend insight to the phase transitions that occur during the crystallization process for fundamental understanding of solution shearing.

533104 Investigation of the Polymorphy Phenomenon of the Carotene Lutein

Abstract:

Carotene is a special group of hydrocarbon pigments characterized with π-π conjugated double bonds in the molecular backbone. This group of compounds is supposed to have great potential in pharmaceutical, healthcare and optical materials because of its excellent properties in antioxidative, light-harvesting, photo-protective, and conductive. Carotene is unstable, due to the conjugated double bonds intend to abstract electron by photo-oxidation. Several methods such as encapsulation have been employed to enhance the stability of carotene. However, seldom studies concerned crystal engineering of carotene to improve its stability has been reported. Nonetheless, H and J aggregation phenomenon of carotene moleculars have been investigated with specific solvents. Recently, β-carotene were reported to have the ability to bind decanuclear homo-and heterometal chain through μ_{10}-briding π-coordination. These findings suggested that the π-π conjugated bond in carotene molecular might have strong plasticity in molecular self-assembling process. As a member of the carotene family, lutein also posseses two hydroxy group (which is easy to occurs polymorph due to the H-bonds between molecular chains) in the molecular structure besides the conjugated double bonds.Thus, cooling crystallization, diffusion zone process and vapor diffusion crystallization methods with selected solvents were tested for lutein, and crystal powders of lutein with needle like, platelet and pyramids morphylogy were obtained. By XRD, TG and DSC analyzing, three crystalline polymorphic forms of lutein were identified. The three crystalline forms of lutein exhibited distinct characters in morphylogy and stability, and the correlation between the stability and the crystal form will be investigated further.

A Kinetic Study of Crystallization Process of Imatinib Mesylate with Polymorphic Transformation Phenomenon

Mengxing Lin, Yuanyi Wu, Sohrab Rohani *
The University of Western Ontario, Department of Chemical and Biochemical Engineering, London, Ontario N6A 5B9, Canada
* Corresponding author: srohani@uwo.ca

Abstract

For pharmaceutical crystallization design and control, the kinetics of nucleation and growth of crystals are significant parameters, especially for the system exhibiting polymorphic transformation. In this study, imatinib mesylate, whose solubility, nucleation and growth kinetics are lacking in the literature, was thoroughly explored in the aspects of characterization, solubility and polymorphic transformation, and evaluation of nucleation and growth rate. Two forms of imatinib mesylate, α and β, were characterized by X-ray powder diffraction, scanning electron microscopy and differential scanning calorimetry. The solubility measurement of the two forms was performed with ultraviolet–visible spectroscopy in three solvents from 278.15 to 333.15 K at atmospheric pressure. The results indicated the β-form is more stable in methanol,1-propanol, and 2-propanol under the experimental condition. The solubility order is methanol >1-propanol >2-propanol. The solvent-mediated polymorphic transformation (SMPT) of imatinib mesylate from α-form to β-form was studied with the in-situ Raman spectroscopy and conductivity meter. It is found that the SMPT process of imatinib mesylate from α-form to β-form in methanol is controlled by nucleation and growth of β-form. Finally, the nucleation and grow rate of β-form of imatinib mesylate were estimated by minimizing the difference between the calculated and experimental solution concentration with MATLAB optimization function.

Keywords: A1. Crystal morphology, A1. Solubility, A2. Growth from solutions, B2. Organic compounds

1. Introduction

Depending on the arrangement of molecules, a solid can be a crystal, a quasicrystal or an amorphous. A phenomenon called polymorphism, exists widely in the crystalline solids, especially in organic compounds[1]. Even though polymorphs have the same molecular formula, they are different in molecular conformations and/or packing arrangements in the solid state, resulting in the distinct physical and chemical properties, such as morphology, melting point, flow properties, mechanical properties, solubility, and Gibbs free energy[2]. In terms of active pharmaceutical ingredients (API), polymorphism would affect the processability, bioavailability and stability of the API[3]. Therefore, the identification and control of polymorphs of a given API are essential for the pharmaceutical industry.

Comparing with the stable polymorph, the metastable polymorph has higher solubility and Gibbs free energy. The divergence in Gibbs energy drives polymorphism to transit from the metastable form to the stable state. Generally, there are two ways to realize the transition: solid-state polymorphic transformation (SST) and solvent-mediated polymorphic transformation (SMPT)[4]. The SMPT process consists of three steps: dissolution of the metastable form, nucleation of stable form, and growth of stable form. The slowest step of the SMPT process limits the whole transition rate. Based on the relative rate of dissolution, nucleation, and growth, O'Mahony et al. rationalized the SMPT process into four possible scenarios: dissolution-controlled, nucleation-dissolution-controlled, growth-controlled, nucleation-growth-controlled [5].

In order to model and control a crystallization process, the kinetic parameters, namely nucleation rate, growth rate, dissolution rate, agglomeration, and breakage have to be determined first through experiments. For the nucleation and growth rate, there are three experimental methods commonly used to determine the nucleation and growth rate. The first one relies on the crystal size distribution of the final product obtained from a Mixed-Suspension, Mixed-Product Removal (MSMPR) crystallizer. The slope and intercept of the "particle number density vs. crystal size" curve provide the information necessary to extract the nucleation and growth rates [2]. The second method which is somewhat approximate, is based on the count and chord length distribution measured by the Focused Beam Reflectance Measurement (FBRM) device [6]. The change in the count of fine crystals presents the nucleation rate, while the change in average chord length refers to the growth rate. The last method uses a rigorous parameter estimation algorithm and works by finding a set of proper constants in the kinetic equations to minimize the difference between the experimental data (measured solute concentration and/or crystal size distribution) and modeling results [7].

The compound studied in this work is imatinib mesylate. Imatinib mesylate (Figure 1), chemically 4-[(4-methylpiperazin-1-yl) methyl]-N-[4-methyl-3-[(4-pyridin-3-ylpyrimidin-2-yl) amino]-phenyl] benzamide methanesulfonate, was developed by Novartis under the trade name "Gleevec/Glivec®" in the late1990s. It was designed as a protein tyrosine kinase inhibitor and shows a remarkable therapeutic effect in Philadelphia chromosome positive chronic myelogenous leukemia (CML) and metastatic malignant gastro intestinal stromal tumors (GISTs) [8]. Many polymorphs of crystalline imatinib mesylate have been observed by precipitating from a solution that consists of an organic solvent, imatinib base and methanesulfonic acid. The different solvents lead to different polymorphs of the final product[9]–[16].

Among the dozens of polymorphs of imatinib mesylate, the α-form and the β-form proposed in WO99/03854 [9] are the most frequent and commercialized. The cubic-shaped β-form has the lower hygroscopic ability, higher flow properties, and higher stability at temperatures below 140° C than the needle-shaped α-form. Therefore, the β-form shows better performance than α-form in terms of processability, manufacturing, pharmaceutical preparation, and storage [17].

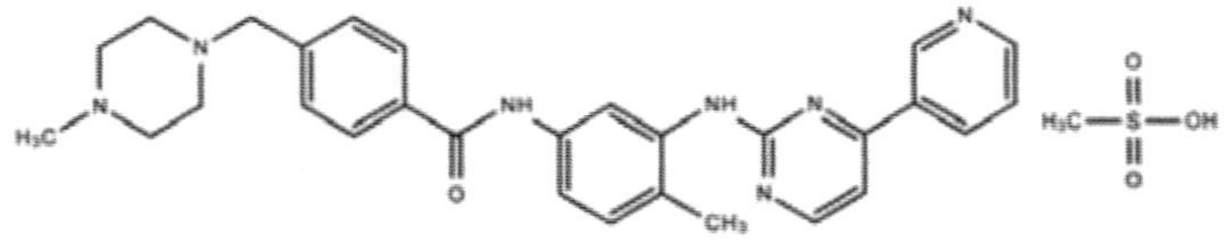

Figure 1. Schematic of imatinib mesylate

Grillo et al. revealed the crystal structure of α-form and β-form imatinib mesylate and found these two polymorphs have significantly different molecular conformations. By grinding them, the amorphous imatinib mesylate was produced that converted to a crystalline phase by either thermal treatment or aging [18]. Veverka et al. modified the morphology of α-form imatinib mesylate with various organic solvent mixtures by precipitation [19]. Srivastava et al. presented a comprehensive study of the vibrational spectroscopies of β-form with the FT-Raman and FT-IR spectra [20]. The polymorphic composition of two polymorphs mixture was determined quantitatively with PXRD, ATR-FTIR, and DSC in the literature [21]. The results show that PXRD technique is more accurate than ATR-FTIR and DSC.

Despite the mentioned studies, the investigation of the imatinib mesylate properties, particularly those related to the crystallization process, was still lacking. The solubility data has not been published up to date. In this study, we measured the solubility of two forms of imatinib mesylate in three solvents (methanol, 1-propanol, and 2-propanol) from 278.15 to 333.15 K at atmospheric pressure to study their relative stability. Subsequently, the solvent-mediated polymorphic transformation (SMPT) from metastable α-form to stable β-form was studied in-situ with Raman spectroscopy and conductivity meter. Finally, the nucleation and grow rate of β-form imatinib mesylate were estimated by minimizing the difference between the calculated and experimental solute concentration with MATLAB optimization function.

2. Material and Experiments

2.1 Material

The α-form imatinib mesylate was supplied by Apotex Pharmachem Inc. (Brantford, ON, Canada) and used without any further processing. The β-form was obtained by the solvent-mediated polymorph transformation from α-form in methanol. The solvents (methanol, ethanol, isopropanol, n-propanol, 1-butanol, acetone, acetonitrile, and tetrahydrofuran) with HPLC grade were purchased from Sigma-Aldrich (Milwaukee, WI,) and no further purification was performed.

2.2 Characterization

2.2.1 Powder X-Ray Diffraction (PXRD)

The polymorphs of solid-state α-form and β-form imatinib mesylate were identified with powder X-ray diffraction (Rigaku, Miniflex, Tokyo, Japan) with the Cu-Kα source ((λ for Kα = 1.54059 Å) at room temperature. The conditions of voltage (30kV), current (15mA), scan angle (2° to 50°) and scan speed (2°/min) were kept the same for all samples.

2.2.2 Differential Scanning Calorimetry (DSC)

Thermal properties of crystals, such as melting point and enthalpy, were measured by differential scanning colorimeter ((DSC, Mettler Toledo, Chicago, United States). The samples (5-10 mg) were placed in a 40-μl aluminum crucible with a pierced lid for gas escaping and then heated from 25°C to 250°C with 10°C/min heating rate under nitrogen purge at a flow rate of 40 ml/min.

2.2.3 Optical microscopy and Scanning Electron Microscopy (SEM)

The optical microscopy (Axioskop 40, Carl Zeiss Light Microscopy, Germany) with a magnification of 500 times was used as an off-line method to give a rough judgement of the polymorphic composition of solid-state imatinib mesylate in suspension. Scanning electron microscope (LEO 1450XB , Zeiss, Oberkochen, Germany) was employed for detailed morphologic information. The samples have been coated with coated with 5 nm of osmium metal in Osmium Plasma Coater (OPC80T, Filgen Inc. Japan) before SEM testing.

2.2.4 Raman spectroscopy

A Raman RXN Analyzer spectroscopy (Kaiser Optical Systems, Inc., Ann Arbor, MI, USA) equipped with a diode laser (784.8-nm) and a fiber optic probe was applied to monitor the polymorphic composition of solid-phase during the SMPT process from α-form to β-form. The characteristic peaks of α-form and β-form of imatinib mesylate are at 1664 cm^{-1} and 1656 cm^{-1}, and the single point baseline is at 1640 cm^{-1}.

2.3 Solubility measurement

To study the stability of two forms, we used Ultraviolet-visible spectroscopy (UV-Vis, Cary Bio 100 spectrophotometer, Varian, Mississauga, ON) to measure the solubility of imatinib mesylate in methanol, 1-propanol, and 2-propanol. The calibration curve of "Absorbance-Concentration" should be determined with a series of known concentration solutions at room temperature at first. The characteristic UV-Vis absorption band at 271-nm and 275-nm was applied for quantification in methanol and propanol, respectively. Then, an excessive sample was added to the selected solvent in a 5-ml glass vial. The vials were immersed in a constant temperature bath with an accuracy of 0.1 °C and agitated for several hours at a specific temperature to reach equilibrium. After 30 mins standing, the supernatant was taken by a syringe and filtered with a 0.45-μm membrane syringe filter (VWR, Mississauga, ON), followed by proper dilution and concentration measurement with UV-Vis. The excess solids in the vials were filtered, dried, and checked with PXRD.

As the α-form imatinib mesylate transformed to β-form in methanol during the solubility measurement, the solubility of α-form imatinib mesylate was determined during the transition process (see section 2.4 for more details).

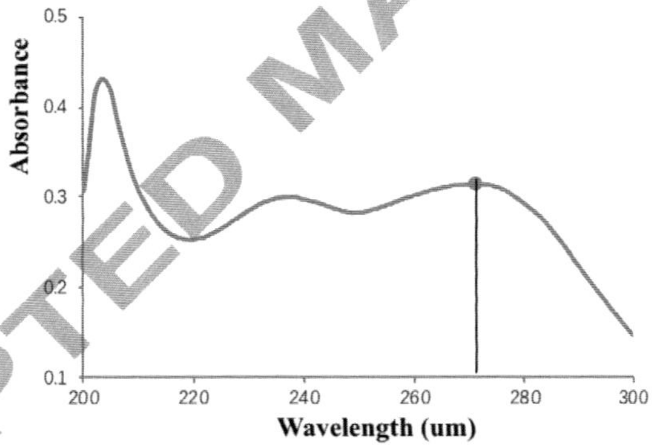

Figure 2. UV-Vis absorption spectra of imatinib mesylate in methanol with a characteristic peak at 271-nm

2.4 SMPT experiments

Eight solvents (methanol, ethanol, isopropanol, n-propanol, 1-butanol, acetone, acetonitrile, and tetrahydrofuran) were screened for mediating the polymorphism transformation. An excessive amount of α-form imatinib mesylate was added to the solvents and agitated in a thermostatic shaker for 2 weeks. The start time of transition was determined by the morphology of the crystals remaining in the solution with the microscope. After filtration and drying, the crystal polymorph was determined by PXRD.

Subsequently, the SMPT process from α-form to β-form imatinib mesylate was investigated in methanol at different temperatures. A carefully excess weighed amount of α-form was added to 35-ml methanol in a 50-mL double-jacketed crystallizer. The crystallizer temperature was kept constant by a water bath circulator (FP-50, Julabo LABORTECHNIK GMBH, Germany). The solute concentration and solid-phase polymorphic composition were monitored by a conductivity meter (Pinnacle Series: M541P) and Raman spectroscopy. The start and end time of the transition process was determined through the morphology of the undissolved crystals with the optical microscope. The solubility of α-form was detected during the SMPT process when the transition did not happen.

3. Theory

3.1 Solubility models

In this work, the solubility of two forms of imatinib mesylate in the selected solvents was fitted with the modified Apelblat equation, as shown in Eq. (1)

$$lnx = A + \frac{B}{T} + ClnT \quad (1)$$

where x is the molar fraction of solute in the solution, T is the absolute temperature in Kelvin, and A, B, and C are dimensionless parameters. The value of three parameters were estimated with MATLAB curve fitting toolbox.

The solubility, in molar fraction, can be calculated using Eq. (2):

$$x = \frac{m/M}{m/M + m_s/M_S} \quad (2)$$

where m and m_s are the mass of solute and solvent, M and M_S are the molecular mass of solute and solvent, respectively.

Solubility reflects the Gibbs energy and stability. The more stable a polymorph is, the more stable a polymorph is, the lower the solubility it is at a given temperature. The difference of Gibbs free energy between two polymorphs relates to the solubility by Eq. (3)

$$\Delta G = RTln\left(\frac{a_1}{a_2}\right) \approx RTln\left(\frac{x_1}{x_2}\right) \quad (3)$$

where ΔG is the difference of Gibbs free energy (J/mol), and a is the solute activity in the solution[22].

3.2 Transition process

The SMPT process consists of three steps: the dissolution of metastable form, nucleation of stable form and growth of stable form. The empirical equations for secondary nucleation and growth rate are expressed as Eq.(4)-(5) [23]:

$$B_{sn} = \frac{dn_0}{dt} = k_{sn}{M_T}^j S^b \quad (4)$$

$$G = \frac{dL}{dt} = k_g S^g \quad (5)$$

where B_{sn} is secondary nucleation rate [#/(m^3· s)], G is growth rate (m/s), S is relative supersaturation, M_T is suspension density (kg solute /kg solvent). k_{sn}, k_g, j, b and g are parameters, representing the nucleation rate coefficient, growth rate coefficient, nucleation order on suspension density, nucleation order on supersaturation, and growth order on supersaturation.

$$S = \frac{C - C^*}{C^*} \quad (6)$$

where C is the solution concentration, while the $C^*$is the equilibrium concentration, namely solubility. In this work, their units are kg solute /kg solvent.

3.3 Parameter estimation

The population balance equation (PBE) coupled with solute mass balance provides a mathematical framework of the crystallization processes. The PBE with the Lagrangian approach[23] describes the change in crystal population density with respect to time and space, as shown in Eq. (7).

$$\frac{\partial n}{\partial t} + \nabla.(v_e n) + \nabla.(v_i n) = B - D \tag{7}$$

where n is population density, v_e is external velocity vector, v_i is internal velocity vector, B is crystal birth rate, and D is crystal death rate. Assuming that a crystallizer has a uniform distribution of suspended solids, a size-independent growth rate, and negligible breakage and agglomeration, Eq. (7) can be simplified to a 1-D PBE as following when $L \neq 0$:

$$\frac{\partial n(t,L)}{\partial t} + G(t)\frac{\partial n(t,L)}{\partial L} = 0 \tag{8}$$

The boundary condition at $L = 0$ is

$$n(t,0) = \frac{B(t)}{G(t)}|_{L=0} \tag{9}$$

Hu et al.[24] proposed a methodology by converting Eq. (7) from a partial differential equation (PDE).to a set of ordinary differential equations (ODE). The method was applied in this study to solve Eq. (7) in MATLAB software. The single objective to minimize the sum of squares of the difference between the calculated and experimental solute concentration of all independent experiments is defined:

$$min\ \ F = \sum_{k=1}^{k}\sum_{i=1}^{N}\left(C_{i,k}^{exp} - C_{i,k}^{cal}\right)^2 \tag{10}$$

where N is the sample number and k is the experiment number. $C_{i,k}^{exp}$and $C_{i,k}^{cal}$ experimental and calculated solution concentration, respectively. In this work, the last step of SMPT process at different temperature was used to estimate the parameters of nucleation and growth rate of β-form imatinib mesylate. The single objective function was solved with the MATLAB optimization function, *fmincon*, based on the sequential quadratic programming (SQP) method.

The solute mass balance is

$$\frac{dC(t)}{dt} = -3k_v\rho_c\int_0^\infty G(t,L)\,n(t,L)L^2 dL \tag{11}$$

where k_v is the crystal volume shape factor (k_v=1 for cube), and ρ_c is crystal density.

The average relative deviation (ARD) was introduced to evaluate the accuracy of estimation,

$$ARD = \frac{1}{N}\sum_{i=1}^{N}\left|\frac{C_{i,k}^{cal} - C_{i,k}^{exp}}{C_{i,k}^{exp}}\right| \quad (12)$$

4. Results and discussion

4.1 SEM, PXRD and DSC

α-form imatinib mesylate supplied by Apotex Pharmachem Inc, and β-form transformed from α-form in methanol at 293K were used for SEM, powder XRD, and DSC test.

Figure 3 shows the SEM images of the α-form and β-form imatinib mesylate. Obviously, the α-form is needle-shaped while β-form is cubic-shaped. According to the morphology, we can distinguish the crystal polymorph easily.

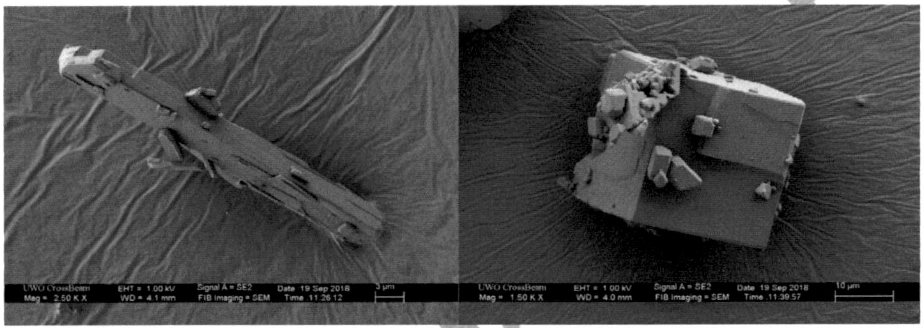

Figure 3. SEM images of imatinib mesylate. Left: α-form (3-μm), Right: β-form (10-μm)

Figure 4 displays the powder XRD patterns of α-form and β-form imatinib mesylate. The results are consistent with the α-form and β-form single-crystal XRD patterns, which were collected at 293K by Grillo et al.[18] and deposited in the Cambridge Crystallographic Data Center with publication numbers CCDC 821869 and 821871. α-form shows strong peaks at scattering angle (2θ) of 4.8°, 10.4°, 14.8°, 19.0°, 24.8° and 28.5°, whereas β-form has characteristic peaks at 9.7°, 13.9°, 18.2°, 20.0° 21.1° and 30.8°. The distinct differences on the XRPD spectrum can be used for polymorph identification of imatinib mesylate.

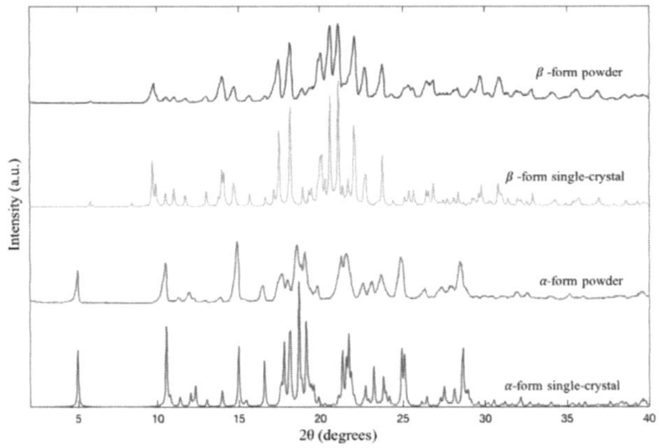

Figure 4. The XRD patterns of α-form and β-form imatinib mesylate

For both of α-form and β-form imatinib mesylate, there is only one endothermic peak observed in the DSC curves (Figure 5). The peaks are related to the melting event, which was confirmed with hot-stage microscopy and thermogravimetric analysis. The onset (start point) of the peak corresponds to the melting point, and the heat of fusion is equal to the integral of heat flow over time divided by sample mass. According to the heat of fusion rule [25], α-form and β-form imatinib mesylate are enantiotropic, as the α-form imatinib mesylate has higher melting point and lower heat of fusion than β-form.

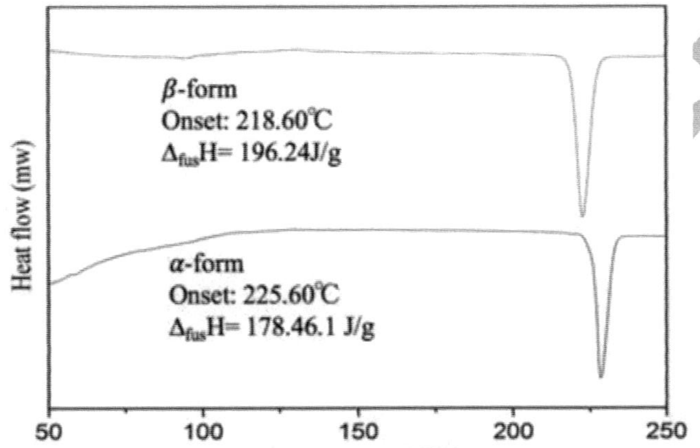

Figure 5. The DSC curves at a heating rate of 10 K/min of α-form and β-form imatinib mesylate

4.2 Solubility

The PXRD results of the residual solids demonstrated that no transformation or conversion occurred during the solubility experiments, except for the α-form in methanol. Figure 6 shows the solubility of α-form and β-form imatinib mesylate in methanol, 1-proponal and 2-proponal. As shown in Figure 6, the α-form has higher solubility than β-form in methanol,1-proponal and 2-proponal. The solubility order for both forms is methanol>1-propanol >2-propanol. The fitting parameters are listed in Table 1 and the results indicate that the modified Apelblat equation matches the solubility well with all R^2 greater than 0.99.

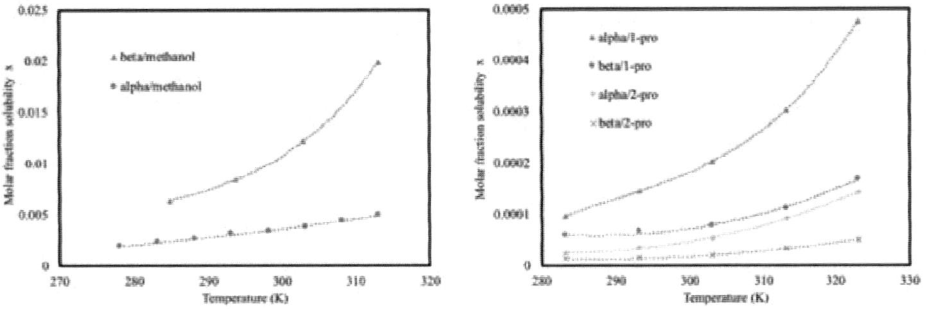

Figure 6. The molar fraction solubility of α-form and β-form imatinib mesylate in methanol, 1-proponal and 2-proponal. Dashed line: calculated curve with modified Apelblat equation

Table 1. Parameters of the modified Apelblat equation for imatinib mesylate in three solvents

		A	B	C	R^2
Methanol	α-form	-685.8	2.73E+04	103.5	1
	β-form	144.2	-8.61E+03	-21.24	0.9938
1-proponal	α-form	-279.2	9116	42.11	0.9983
	beta	-762.2	3.18E+04	113.4	0.9972
2-proponal	α-form	-455.2	1.66E+04	68.38	0.9982
	β-form	-933.4	3.89E+04	139	0.9949

As concluded above, α-form and β-form imatinib mesylate are enantiotropic. Figure 7 presents that the difference of Gibbs free energy between α-form and β-form imatinib mesylate increases with temperature. So, the transition temperature should be lower than 283.15K, which means that the β-form is always more stable than α-form above 283.15K.

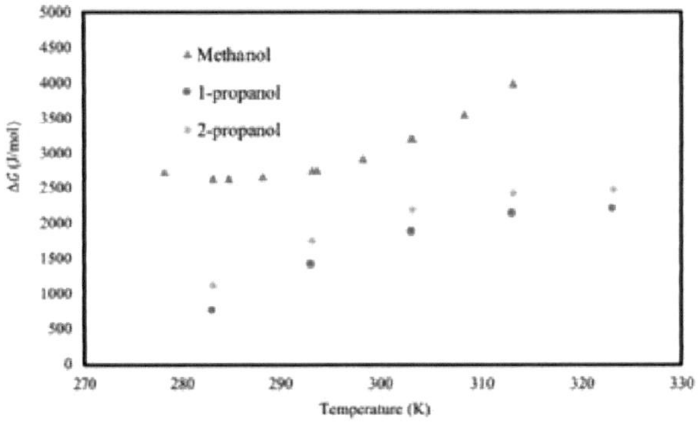

Figure 7. The difference of Gibbs free energy between α-form and β-form imatinib mesylate in methanol, 1-proponal and 2-proponal

4.3 Solvent-mediated polymorphic transformation (SMPT)

Through the solvent screening, we found that the SMPT from α-form to β-form imatinib mesylate could occur in methanol, ethanol, isopropanol, n-propanol, and 1-butanol. The transition time increased with the solvent in the order of methanol, ethanol, 1-propanol, 2-propanol and 1-butanol. Thus, methanol was chosen as the solvent to investigate the SMPT process at different temperatures by in-situ Raman spectroscopy and conductivity meter.

Due to the apparent fluctuations in the baseline, the relative heights of the Raman spectrum were used for quantitative analysis to reduce the effect produced by disturbance variables [26]. Here, the peaks at 1664 cm^{-1} and 1656 cm^{-1} were selected as the characteristic peaks of α-form and β-form imatinib mesylate, and the single point baseline was at 1640 cm^{-1}. Generally, the conductivity of a binary conducting solution is a function of temperature and solute concentration. For an isothermal process, conductivity increase in direct proportion to concentration, so the conductivity meter can be used for real-time measurement of the concentration.

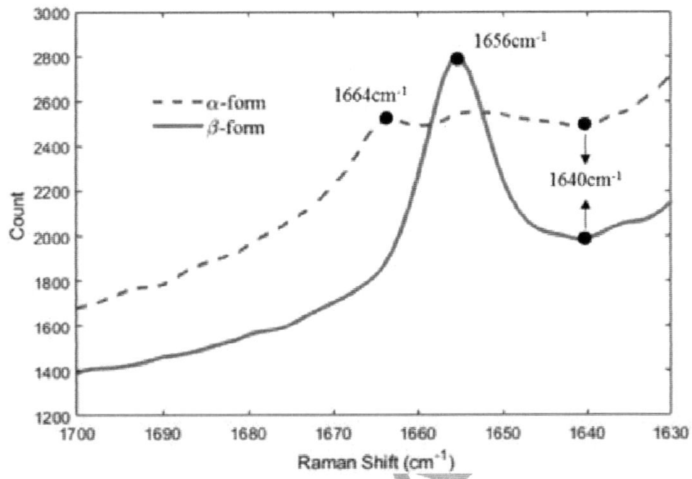

Figure 8. Raman spectra of α-form and β-form imatinib mesylate

Figure 9 displays how the concentration and solid composition change during the SMPT process in methanol at 293.15K. The process can be divided into four steps. At the beginning, the concentration increased rapidly as α-form dissolved, and the undissolved solid was pure α-form. The second step involves the nucleation of β-form. Meanwhile, the conductivity and the Raman intensity for both forms kept constant. The dissolution of α-form and growth of β-form happened during the third step, resulting in a decreased Raman intensity of α-form and an increased intensity of β-form. In the last step, the Raman intensity of α-form did not change, illustrating that all α-form crystals had been dissolved and the solid-phase were pure β-form. The nucleation and growth of β-form led to the drop of conductivity and a gradual increase of Raman intensity of β-form.

During the SMPT process, namely the third step, the concentration was close to the solubility of α-form, which indicated dissolution rate of α-form is much faster than growth rate of β-form. Therefore, it can be concluded that the SMPT from α-form to β-form imatinib mesylate is nucleation-growth controlled of β-form.

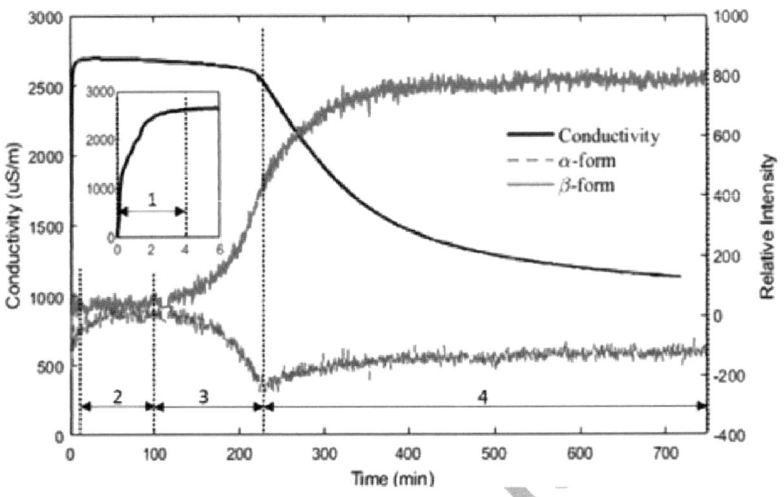

Figure 9. Solvent-mediated polymorphic transformation from α-form to β-form imatinib mesylate at 293k

4.4 Kinetic parameter estimation

As mentioned in section 4.3, all α-form crystals had dissolved and only β-form crystals existed in the solid-phase in the last step (step 4) of the solvent mediated polymorphic transformation. Therefore, the last step of the SMPT process can be regarded as a seeded isothermal batch crystallization. During the process, the solute concentration was measured by the conductivity meter and the suspension density was calculated by solute mass balance. Hence, the nucleation and growth can be estimated through minimizing the difference between the calculated and experimental solute concentration measured in all independent experiments. The estimated parameters are listed in Table 2.

The nucleation order with respect to supersaturation (b) is much higher than the nucleation order with respect to suspension density (j), indicating that supersaturation affects the secondary nucleation rate more significantly than slurry density does. $b > g$ shows that the nucleation rate has a stronger relationship with supersaturation than the growth rate, hence the mean crystal size will decrease with supersaturation. These estimated kinetic parameters would be very useful for optimizing and controlling the cooling crystallization process of β-form imatinib mesylate.

Table 2 Kinetic parameter estimated from optimization

Kinetic parameter	Estimated value	Units
k_b	9.22×10^{11}	$[(\#/m^3 \cdot s)/(kg/kgsolvent)^j]$
j	0.754	$[-]$
b	3.12	$[-]$
k_g	9.12×10^{-11}	$[m/s]$
g	1.23	$[-]$

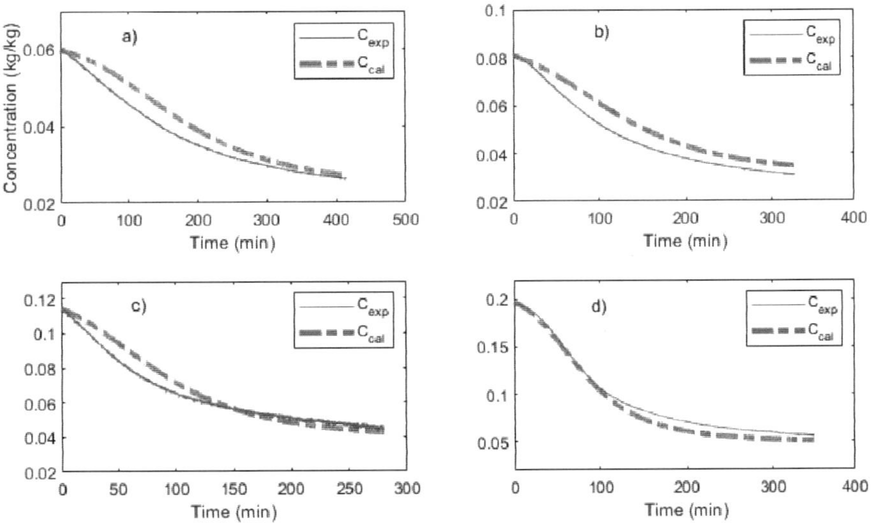

Figure 10. Calculated and experimental solution concentration a)284.75 K, b)293.65 K, c)302.95 K and d) 313.15K

The calculated and experimental solute concentration at different temperatures are plotted in Figure 10. The ARD of the four experiments are 7.6%, 12.19%, 6.44 %, and 8.62%, respectively, which indicates the empirical equations (Eq. 4, 5, 7) used in this work are suitable for β-form imatinib mesylate. The PBE combined with agglomeration and breakage would give better estimation performance with considerable increase in computation time.

5. Conclusions

In this work, the α-form and β-form imatinib mesylate, were characterized by powder X-ray diffraction, scanning electron microscopy and differential scanning calorimetry firstly. Then, we measured their solubility in methanol,1-propanol, and 2-propanol with UV-Vis. The modified Apelblat equation was used to fit the solubility data. According to the DSC and solubility results, we found that α-form and β-form imatinib mesylate are enantiotropically related and the transition temperature is lower than 283.15K. Above 283.15K, α-form always has a higher solubility and lower stability than β-form in methanol, 1-proponal and 2-proponal. The solubility order in selected solvents is methanol>1-propanol>2-propanol. After solvent screening, the solvent-mediated polymorphic transformation (SMPT) from α-form to β-form imatinib mesylate was investigated in methanol. The solute concentration was monitored by a conductivity meter, while the solid phase was monitored by Raman spectroscopy. As the concentration was close to the solubility of metastable form during the SMPT process, it can be concluded that SMPT process of imatinib mesylate from α-form to β-form in methanol is controlled by nucleation and growth of β-form. Finally, the nucleation and grow rate of

β-form of imatinib mesylate were estimated by minimizing the difference between the calculated and experimental solution concentration with MATLAB optimization function.

6. Nomenclature

a	solute activity	-
b	nucleation order on supersaturation	-
B	crystal birth rate	$\#/m^3 \cdot s$
B_s	secondary nucleation rate	$\#/m^3 \cdot s$
C	solution concentration	kg solute /kg solvent
C^*	equilibrium concentration--solubility	kg solute /kg solvent
C^{cal}	calculated solution concentration	kg solute /kg solvent
C^{exp}	experimental solution concentration	kg solute /kg solvent
D	crystal death rate	$\#/m^3 \cdot s$
F	objective function	
g	growth order on supersaturation	-
G	crystal growth rate	m/s
ΔG	difference of Gibbs free energy	kJ/mol
j	nucleation order on suspension density	-
k	number of experiments included in the objective function F	-
k_g	growth rate coefficient	m/s
k_{sn}	secondary nucleation rate coefficient	$(\#/m^3 \cdot s)/(\text{kg/kg solvent})^j$
k_v	the crystal volume shape factor	
L	characteristic size of crystal	m
m	mass of solute	kg
m_s	mass of solvent	kg
M	molecular mass of solute	kg/mol
M_s	molecular mass of solvent	kg/mol
M_t	suspension density	kg solids /kg solvent
n	population density	$\#/m^3 \cdot m$
N	number of samples in each experiment included in the objective function F	-
R	universal gas constant	kJ/mol·K
S	relative supersaturation	-
t	time	s
x	molar fraction of solute	-
ρ_c	crystal density	kg/m^3

7. Reference

[1] G. P. Stahly, "Diversity in Single- and Multiple-Component Crystals. The Search for and Prevalence of Polymorphs and Cocrystals," *Cryst. Growth Des.*, vol. 7, no. 6, pp. 1007–1026, Jun. 2007.

[2] A. Myerson, *Handbook of Industrial Crystallization*. Butterworth-Heinemann, 2002.

[3] A. Y. Lee, D. Erdemir, and A. S. Myerson, "Crystal Polymorphism in Chemical Process Development," *Annu. Rev. Chem. Biomol. Eng.*, vol. 2, no. 1, pp. 259–280, 2011.

[4] J. W. Mullin, *Crystallization*. Elsevier, 2001.

[5] M. A. O'Mahony, A. Maher, D. M. Croker, Å. C. Rasmuson, and B. K. Hodnett, "Examining Solution and Solid State Composition for the Solution-Mediated Polymorphic Transformation of Carbamazepine and Piracetam," *Cryst. Growth Des.*, vol. 12, no. 4, pp. 1925–1932, Apr. 2012.

[6] M. Trifkovic, S. Rohani, and M. Sheikhzadeh, "Kinetics Estimation and Polymorphic Transformation Modeling of Buspirone Hydrochloride," *J. Cryst. Process Technol.*, vol. 02, no. 02, pp. 31–43, 2012.

[7] G. Morris, G. Power, S. Ferguson, M. Barrett, G. Hou, and B. Glennon, "Estimation of Nucleation and Growth Kinetics of Benzoic Acid by Population Balance Modeling of a Continuous Cooling Mixed Suspension, Mixed Product Removal Crystallizer," *Org. Process Res. Dev.*, vol. 19, no. 12, pp. 1891–1902, Dec. 2015.

[8] B. M. H. Al-Hadiya, A. H. H. Bakheit, and A. A. Abd-Elgalil, "Chapter Six - Imatinib Mesylate," in *Profiles of Drug Substances, Excipients and Related Methodology*, vol. 39, H. G. Brittain, Ed. Academic Press, 2014, pp. 265–297.

[9] J. Zimmermann, B. Sutter, and H. Buerger, "Crystal Modification of a N-Phenyl-2-Pyrimidineamine Derivative, Processes for Its Manufacture and Its Use," WO9903854 (A1), 28-Jan-1999.

[10] R. B. Parthasaradhi, R. K. Rathnakar, R. R. Raji, R. D. Muralidhara, C. R. K. Subash, and H. D. Limited, *Novel polymorphs of imatinib mesylate*. 2003.

[11] K. Amala, R. T. Srinivasa, K. S. B. R. Adibhatla, S. Rachakonda, C. N. Venkaiah, and K. Podili, "Novel polymorphic form of imatinib mesylate and a process for its preparation," WO2005077933A1, 25-Aug-2005.

[12] S. L. Pathi, R. Puppala, R. N. Kankan, D. R. Rao, C. Limited, and C. Wain, *Stable crystal form of imatinib mesylate and process for the preparation thereof*. 2005.

[13] H. Patel, R. Jani, and R. Thennati, "Imatinib Mesylate Crystal Form and Process for Preparation Thereof," WO/2006/048890, 12-May-2006.

[14] A. K. Kompella, K. S. B. R. Adibhatla, K. Podili, and C. N. Venkaiah, "Polymorphic forms of imatinib mesylate," WO2006054314A1, 26-May-2006.

[15] Mutz M., "Delta and Epsilon Crystal Forms of Imatinib Mesylate," WO/2007/023182, 02-Mar-2007.

[16] M. Mutz, "F,g,h,i and K Crystal Forms of Imatinib Mesylate," WO/2007/059963, 01-Jun-2007.

[17] J. Zimmermann, B. Sutter, and H. M. Bürger, "Crystal modification of a N-phenyl-2-pyrimidineamine derivative, processes for its manufacture and its use," US6894051B1, 17-May-2005.

[18] D. Grillo, G. Polla, and D. Vega, "Conformational Polymorphism on Imatinib Mesylate: Grinding Effects," *J. Pharm. Sci.*, vol. 101, no. 2, pp. 541–551, Feb. 2012.

[19] M. Veverka, P. Šimon, J. Lokaj, and E. Veverková, "Crystal habit modifications of imatinib mesylate under various precipitation conditions," *Monatshefte Für Chem. - Chem. Mon.*, vol. 143, no. 1, pp. 65–71, Jan. 2012.

[20] A. Srivastava, B. D. Joshi, P. Tandon, A. P. Ayala, A. K. Bansal, and D. Grillo, "Study of polymorphism in imatinib mesylate: A quantum chemical approach using electronic and vibrational spectra," *Spectrochim. Acta. A. Mol. Biomol. Spectrosc.*, vol. 103, pp. 325–332, Feb. 2013.

[21] E. Bellur Atici and B. Karlığa, "Quantitative determination of two polymorphic forms of imatinib mesylate in a drug substance and tablet formulation by X-ray powder diffraction, differential scanning calorimetry and attenuated total reflectance Fourier transform infrared spectroscopy," *J. Pharm. Biomed. Anal.*, vol. 114, pp. 330–340, Oct. 2015.

[22] A. Grunenberg, J.-O. Henck, and H. W. Siesler, "Theoretical derivation and practical application of energy/temperature diagrams as an instrument in preformulation studies of polymorphic drug substances," *Int. J. Pharm.*, vol. 129, no. 1, pp. 147–158, Mar. 1996.

[23] A. Ranodolph, *Theory of Particulate Processes: Analysis and Techniques of Continuous Crystallization*. Elsevier, 2012.

[24] Q. Hu, S. Rohani, and A. Jutan, "New numerical method for solving the dynamic population balance equations," *AIChE J.*, vol. 51, no. 11, pp. 3000–3006, Nov. 2005.

[25] A. Burger and R. Ramberger, "On the polymorphism of pharmaceuticals and other molecular crystals. I," *Microchim. Acta*, vol. 72, no. 3–4, pp. 259–271, May 1979.

[26] A. Caillet, F. Puel, and G. Fevotte, "In-line monitoring of partial and overall solid concentration during solvent-mediated phase transition using Raman spectroscopy," *Int. J. Pharm.*, vol. 307, no. 2, pp. 201–208, Jan. 2006.

8. Acknowledgment

Thanks to Apotex Pharmachem Inc. (Brantford, ON, Canada) for supplying the raw imatinib mesylate, Dr. Tim Goldhawk from Western University Nanfabrication Facility for help with SEM imaging, and Chenxi Li for her assistance in the solubility measurement. The authors acknowledge the financial support provided by the Natural Science and Engineering Research Council (NSERC) of Canada and the China Scholarship Council.

Highlights

- Two forms of imatinib mesylate, α and β, were thoroughly characterized.
- Solubility of the two forms was measured in methanol, 1-propanol, and 2-propanol.
- Developed kinetics of polymorphic transformation from α-form to β-form in methanol.
- The nucleation and grow rates of β-form of imatinib mesylate were estimated.

532399 Polymorphism of D-Mannitol: Selective Nucleation and Crystal Growth Mechanism

Abstract:

Polymorphism of D-mannitol: Selective Nucleation and Crystal Growth Mechanism

Weiyi Su, Chunli Li, Honghai Wang, Jing Fang

School of Chemical Engineering, Hebei University of Technology, Tianjin 300130, PR China

Polymorphism has gained more and more attention in the field of pharmaceuticals[1]. D-mannitol is a natural hexahydric alditol which has been widely used as a nephropathy treatment medicine and also an excipient in the formulation of various tablets and granulated powders[3]. It has been reported that there are three anhydrous polymorphs[4-6] and a hemihydrate[7] of mannitol. However the nomenclature of these polymorphs is often diverse in different literatures[8, 9] in despite of similar unit cell parameters. Therefore a review of mannitol polymorphism is initially presented along with the crystal structure parameters. Then the influence of supersaturation on the polymorphic nucleation of mannitol is investigated in aqueous solutions and ethanol-water mixtures. Moreover, a method reported by Kuldipkumar[10] was extended to determine the crystal growth of metastable ¦Ä mannitol, which was found to follow the 2D nucleation-mediated mechanism.

1. Polymorphs of D-mannitol

It has been confirmed that there are only three pure anhydrous polymorphs of mannitol even though numerous names have been used[6, 8]. Hereby, for clarity and to avoid any confusion, the unite cell parameters of mannitol polymorphs in various literatures are summarized in Table 1. It is clear in Table 1 that the first six ¦Â references have a similar structure, which is named ¦Â in our work. While focusing on the polymorphs from the 7th to 9th in Table 1, the parameters were nearly the same disregarding what they were called in the original literature. So this form is referred to as ¦Á mannitol. Finally, the last three items are significantly distinct from the others, and they are referred to as the ¦Ä form of mannitol in this paper.

Table 1. The review of unit cell parameters and nomenclature of D-mannitol anhydrous polymorphs in literatures

Mark	Literature	Space group	a (Å)	b (Å)	c (Å)	β (°)	V ($Å^3$)	Z	D_x ($kg \cdot m^{-3}$)
β	Marwick, 1931[11]	$P2_12_12_1$	8.65	16.90	5.56	--	203*	4	1.488
β	Rye, 1952¤[12]	$P2_12_12_1$	8.65	16.90	5.56	--	203*	4	1.488
β	Berman, 1968[13]	$P2_12_12_1$	8.672	16.875	5.560	--	--	4	1.487
β	Walter-Levy, 1968[4]	$P2_12_12_1$	8.672	16.89	5.549	--	--	4	1.489
β	Kaminsky, 1997[14]	$P2_12_12_1$	8.694(7)	16.902(8)	5.549 (6)	--	815	4	1.490(2)
β	R. Fronczek, 2003[5]	$P2_12_12_1$	5.5381 (10)	8.580(2)	16.795 (5)	--	789.0 (3)	4	1.516
α	Walter-Levy, 1968	$P2_12_12_1$	8.839	18.778	4.896	--	--	4	1.470
K	Kim, 1968[15]	$P2_12_12_1$	8.942	18.798	4.893	--	--	4	1.471
α	R. Fronczek, 2003	$P2_12_12_1$	4.8653 (10)	8.873(2)	18.739 (3)	--	809.0 (3)	4	1.496
δ	Walter-Levy, 1968	$P2_1$ or $P2_1/m$	5.095	18.254	4.919	118.36	--	2	1.501
δ	R. Fronczek, 2003	$P2_1$	4.899(2)	18.268(6)	5.043 (2)	118.39(2)	397.0 (3)	2	1.524
δ	Botez, 2003[16]	$P2_1$	5.08941(5)	18.2504 (2)	4.91702(5)	118.303	402.103	2	1.377

* volume per molecule, and others are volume per unit cell.

¡è the author also mentioned the ¦Â form they obtained was in agreement with Marwick¡¯s result.

2. Selective Nucleation of D-mannitol Polymorphs

The influence of the supersaturation on mannitol polymorphic nucleation is investigated in cooling crystallization, and the results are displayed in Figure 1 with a comparison to the solubility curves. As illustrated in Figure 1, it is quite interesting to find that the nucleation sequence is ¦Á, ¦Ä, ¦Â as the decrease of concentration while the solubility order is always ¦Ä > ¦Á > ¦Â along with temperature.

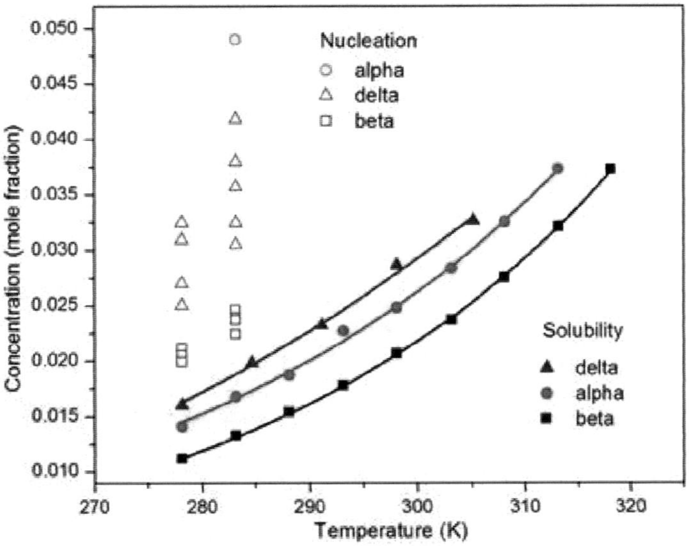

Figure 1. Polymorphic nucleation results of mannitol (open symbols, labeled as ¡°Nucleation¡±) at different initial concentrations and temperatures shown in aqueous solution with the solubility curves of the three polymorphs (solid symbols, labeled as ¡°Solubility¡±).

To clarify the nucleation zone for different polymorphs, the nucleation parameters (*DG**, *r**, and *n**) at various initial concentrations at 10 ¡ãC are obtained as shown in Table 2 based on the measured induction time and supersaturation data. It is apparent that when the initial concentration is higher than 0.0305, the critical excess free energy forming the ¦Ä form nuclei is lower than that of the ¦Â form. It indicates that the kinetic properties play a more important role than the thermodynamic properties. While the concentration is lower than 0.0246, the supersaturation is the dominant factor, that is why ¦Â mannitol is easier to nucleate in this region.

Table 2. Nucleation properties of two mannitol polymorphs at 10 ¡ãC, the subscripts ¦Ä and ¦Â refer to the polymorph ¦Ä and ¦Â forms respectively while *s* stands for the supersaturation

initial mole concentration	s_δ	s_β	ΔG^*_δ (10^3 J/mol)	ΔG^*_β (10^3 J/mol)	r^*_δ (10^{-10} m)	r^*_β (10^{-10} m)	n^*_δ	n^*_β
0.0418	2.22	3.16	*1.37*	3.12	*1.60*	2.05	*2*	3
0.0380	2.02	2.87	*1.76*	3.82	*1.81*	2.23	*3*	4
0.0357	1.90	2.70	*2.12*	4.32	*1.99*	2.37	*3*	4
0.0325	1.73	2.46	*2.92*	5.27	*2.33*	2.62	*5*	5
0.0305	1.62	2.31	*3.73*	6.09	*2.67*	2.82	*7*	7
0.0246	1.31	1.86	12.07	*11.02*	4.74	*3.79*	39	*16*
0.0239	1.27	1.81	15.31	*12.17*	5.34	*3.98*	55	*18*
0.0237	1.26	1.79	16.44	*12.52*	5.54	*4.04*	61	*19*
0.0234	1.24	1.77	18.43	*13.09*	5.86	*4.13*	73	*20*

The values highlighted in italics in Table 2 show lower critical excess free energy values, smaller radius and molecular number of the critical nucleus for either the ¦Ä or ¦Â form at certain concentrations, leading to the selective nucleation of this polymorph at those conditions. In this way, controlling to get the desired polymorphs can be fulfilled.

A reverse anti-solvent crystallization of adding saturated mannitol solution into cold ethanol is also performed. The nucleation results at -10¡æ are shown in Figure 2. It is similar to Figure 1 that high concentration favors the metastable ¦Á form and the least stable ¦Ä mannitol can only be obtained at relatively low concentration.

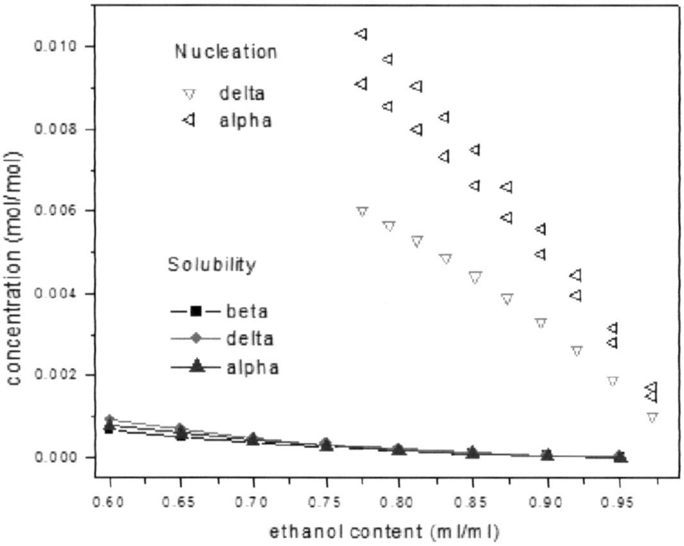

Figure 2. Polymorphic nucleation results of mannitol (open symbols, labeled as ¡°Nucleation¡±) at different concentrations in ethanol-water mixture shown with the solubility curves of the three polymorphs (solid symbols, labeled as ¡°Solubility¡±).

3. Crystal growth

After nucleation, the nuclei should grow to be crystals by consuming supersaturation. But in bulk aqueous solution, the solvent-mediated polymorphic transformation is quite obvious for mannitol metastable forms, which makes it even more difficult to clarify the growth mechanism. In order to enlarge the acquaintance to the growth of ¦Ä mannitol, the induction time are used to provide the growth mechanism following a reported method by Kuldiplumar et al[10]. In this method, a function F(s) is defined for normal, spiral, and volume diffusion-controlled growth mechanisms:

$$F(s) \equiv \ln\left\{s^{\frac{1}{n}}[f(s)]^{\frac{n-1}{n}} t_{ind}\right\} = \ln A_p + \frac{B}{n\, ln^2 s} \quad (1)$$

Similarly for 2D nucleation-mediated growth, the function is defined as:

$$F(s) \equiv \ln\left[(s-1)^{\frac{2(n-1)}{3n}} s^{\frac{n+2}{3n}} t_{ind}\right] = \ln A_u + (n-1)\frac{B_2D}{3n\ \ln s} + \frac{B}{n ln^2 s} \quad (2)$$

According to equation 1 and 2, the calculated F(s) against $1/(\ln^2 s)$ for the normal, spiral, and volume diffusion-controlled growth mechanism, as well as F(s) against 1/(lns) following the 2D nucleation-mediated growth mechanism are shown in Figure 3.

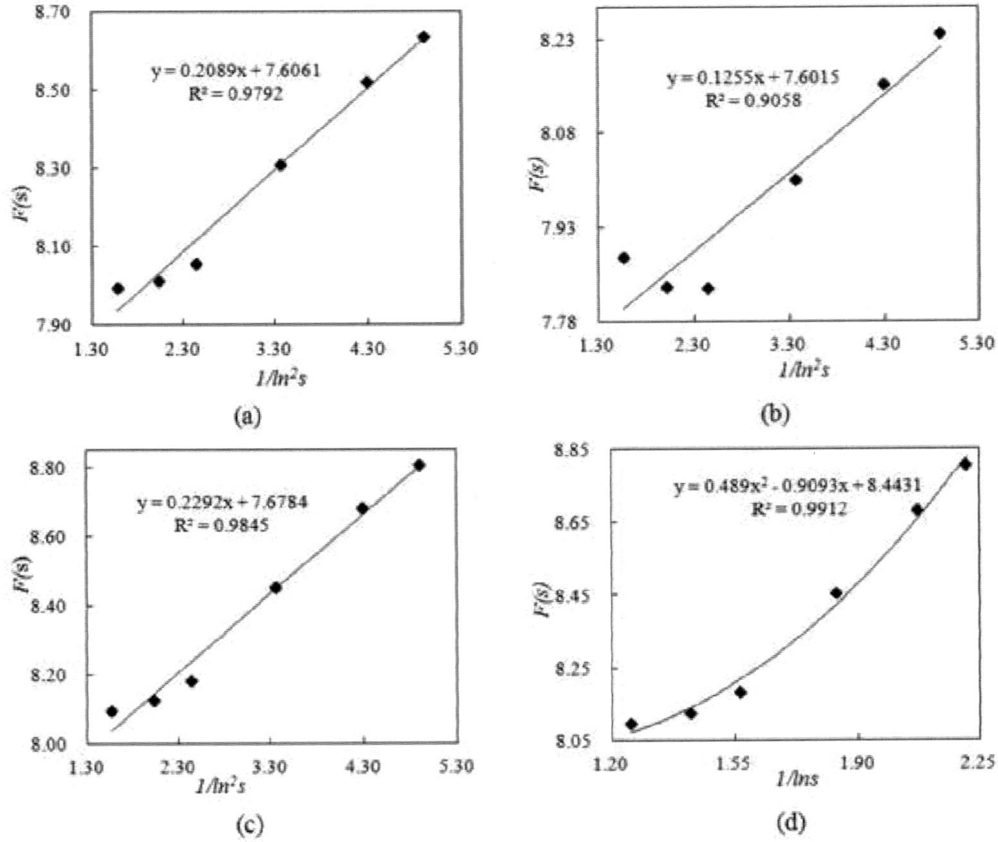

Figure 3. Plots of *F(s)* versus $1/ln^2 s$ for ¦Ä form of mannitol under the normal growth mechanism (a), the spiral growth mechanism (b), the volume diffusion controlled growth mechanism(c), and *F(s)* versus *1/lns* under the 2D nucleation-mediated growth mechanism (d).

It is clear in Figure 3 that the largest R^2 of 0.9912 appears when the data are fitted to the 2D nucleation-mediated mechanism. Therefore it is reasonable to believe that the growth mechanism of the metastable ¦Ä form of mannitol in aqueous solution should be 2D nucleation-mediated.

Reference

[1] J. Bernstein, Polymorphism in molecular crystals, Oxford University Press, USA, 2002.

[2] S. Datta, D.J.W. Grant, Nat. Rev. Drug Discov. 3 (2004) 42-57.

[3] B. O'Sullivan, The application of in situ analysis to crystallization process development, University College Dublin, Dublin, 2005, p. 177.

[4] L. Walter-Levy, C. R. Acad. Sc. Paris, Ser. C 267 (1968) 1779.

[5] F.R. Fronczek, et al., Sect. C: Cryst. Struct. Commun. 59 (2003) o567-o570.

[6] A. Burger, etal., J. Pharm. Sci. 89 (2000) 457-468.

[7] C. Nunes, et al., J. Pharm. Sci. 93 (2004) 2800-2809.

[8] J. Cornel, et al., Industrial and Engineering Chemistry Research 49 (2010) 5854-5862.

[9] W Su, et al., J. Raman Spectrosc. 46 (2015) 1150-1156.

[10] A. Kuldipkumar, et al., Cryst. Growth Des. 7 (2007) 234-242.

[11] T.C. Marwick, Proc. R. Soc. London, Ser. A 131 (1931) 621-633.

[12] A. Rye, H. Sorum, Acta Chem. Scand 6 (1952) 1128-1129.

[13] H.M. Berman, et al., Acta Crystallogr., Sect. B: Struct. Sci B24 (1968) 442-449.

[14] W. Kaminsky, Z. Kristallogr. 212 (1997) 283-296.

[15] H.S. Kim, et al., Acta Crystallogr., Sect. B: Struct. Sci B24 (1968) 1449-1455.

[16] C.E. Botez, et al., Powder Diffr. 18 (2003) 214-218.

525689 Can the Solvation Manner Leads to a Nucleation Diversity for Polymorphic System? the Case of D-Mannitol

Abstract:

Can the solvation manner leads to a nucleation diversity for polymorphic system? The case of D-mannitol

There has been much interest in the role of solution chemistry plays in determining the polymorphic nucleation process, especially for the field requires a precise control and even design of the final crystalline product, such as pharmaceutical and fine chemical industries. In this study we focused on a quite challenging subject: D-mannitol (D-man). An interesting solvent-dependent nucleation behavior was found by us: in pure methanol the alpha form can be easily harvested regardless of the supersaturation and temperature, in water the alpha can also be sporadically observed but the solution mainly crystallized beta form, whereas for mixtures of water and methanol, interestingly, the pure delta form was exclusively yielded instead of either alpha or beta. We wondered the reason for this solvent dependent crystallization outcomes, especially the unexpected appearance of delta form, however, the nearly identical molecular conformations and extensive hydrogen bonds throughout the crystal structures making D-man a system difficult to resolve [1], as shown in Fig. 1.

Two points are mainly concerned in this study : 1. exploring is there any structural feature could differentiate three anhydrous polymorphs of D-man properly; 2. based on such difference, whether a ¡°link¡± could be detected between the nature of intermolecular interactions present in solution and the crystallization outcomes, especially the unexpected appearance of delta form.

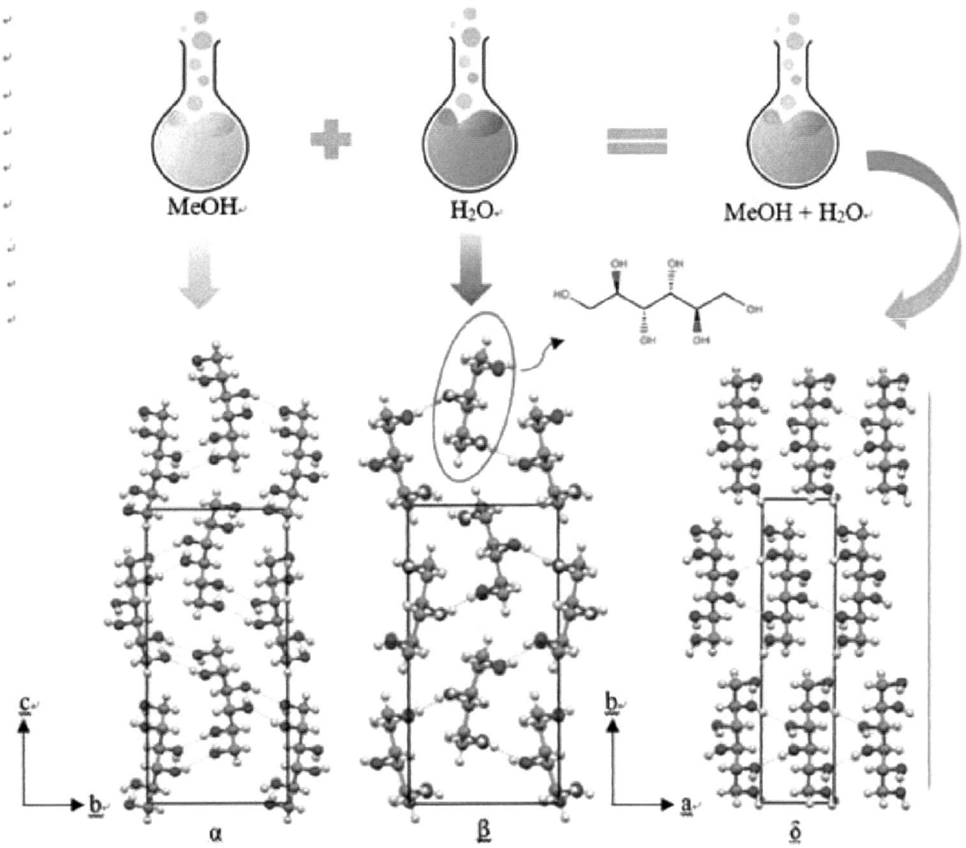

Fig. 1 Molecular structure and unit cell packings of three anhydrous D-man polymorphs (the dimers are overlapped in this view)

Though a Hirshfield Surface analysis [2], we found two important characteristics can be used to differentiate three polymorphs, namely the dimerization effect and terminal packing intensity, as illustrated in Fig.2 and TABLE 1.

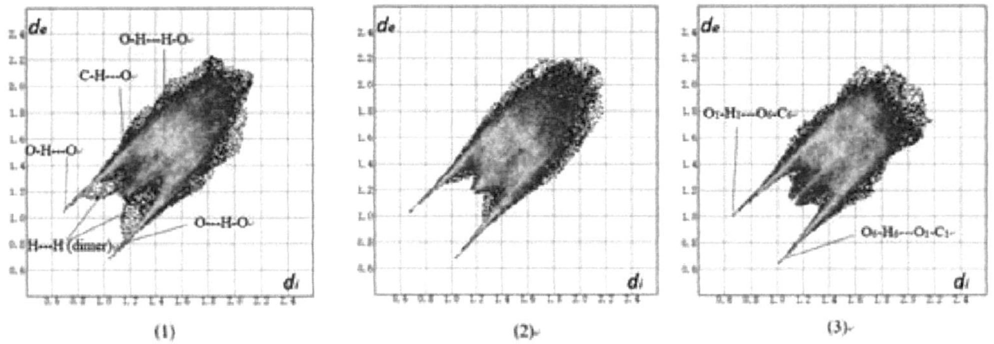

Fig. 2 2-D Finger plots for three D-man anhydrous polymorphs: α (1), β (2) and δ (3).

TABLE 1. Packing Characteristics of Three Anhydrous Polymorphs

	α	β	δ
Dimerization effect	high	medium	low
Terminal packing	medium	compact	loose

In light of above comparison, the nature of interactions in different D-man solutions was investigated in terms of the radial distribution function (RDF) calculation. The solvation effect on D-man¡¯s body parts (-O3H3) and terminal parts (-O1H1) were both investigated. The RDF results together with snapshots for the molecular distribution within different solutions are shown below:

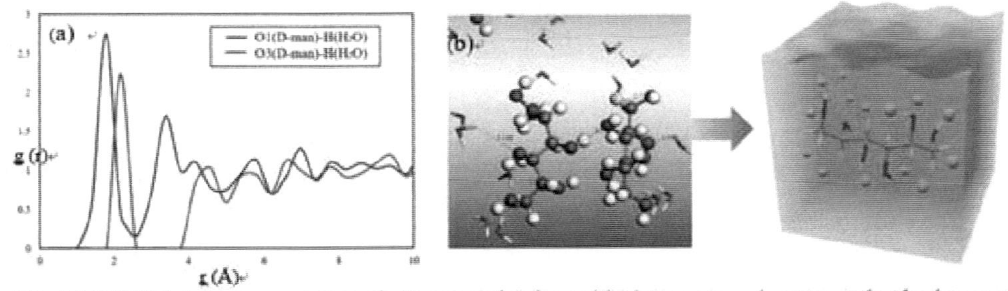

Fig. 4 (a) RDFs between oxygen atoms in D-man molecules and hydrogen atoms in water molecules in aqueous solution; (b) Snapshot of D-man molecules solvated in water.

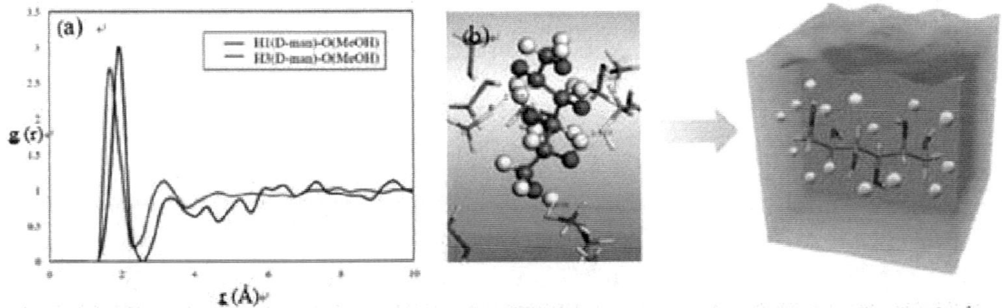

Fig. 3 (a) RDFs between hydrogen atoms in D-man molecules and oxygen atoms in methanol molecules in methanol solution; (b) Snapshot of D-man molecules solvated in methanol.

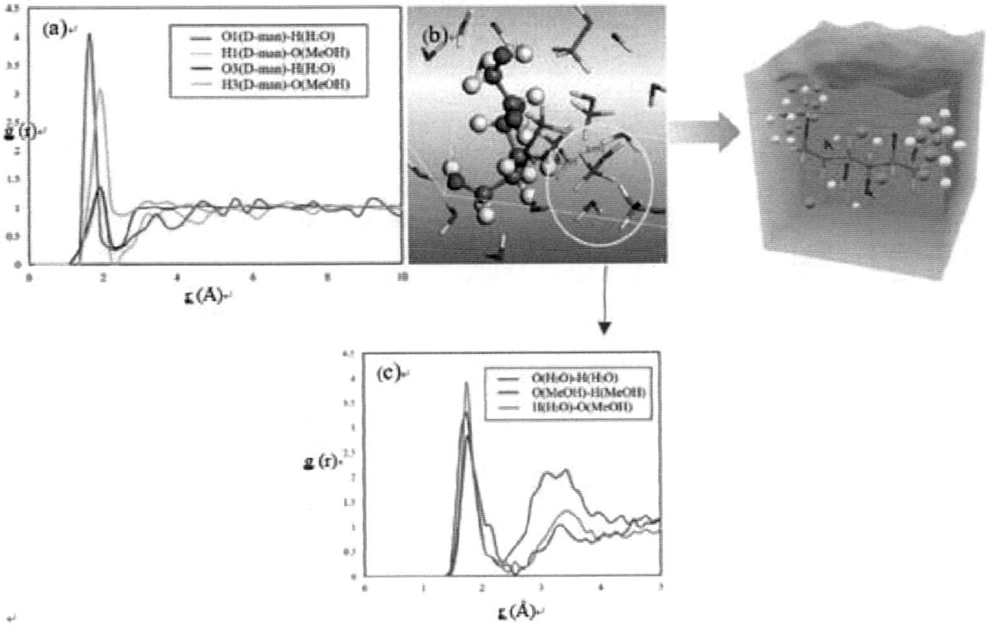

Fig. 5 (a) RDFs between atoms of D-man molecules and atoms in solvent molecules in binary solvent solution; (b) Snapshot of D-man molecules solvated in binary solvent; (c) RDFs between solvent molecules in binary solvent solution

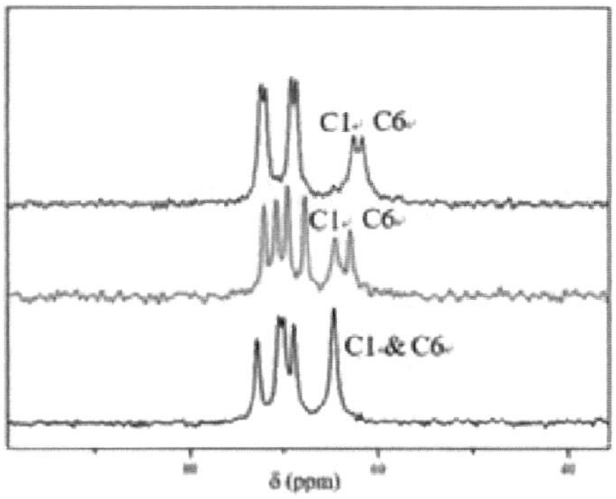

Fig. 6 ^{13}C solid-state NMR spectrum of three D-man anhydrous polymorphs.

From above results, no link was found between the solvation effect on D-man¡¯s body part and the dimerization effect in three forms, indicating that the dimerization strength cannot be a probe for the nucleation behavior explanation. While a good correlation was observed for terminal packing characteristics: in pure methanol and water, the solvation intensities at the end part are 3 and 2.7, respectively, while in the mix solvent, much higher solvating power on terminals was observed, with the terminal-water in 4 for g(r) and terminal-methanol in 3. A double-layer solvent shell with water molecules in the inside track while methanol in outsides was formed, considering the higher probability of methanol-water assemblies than the homologous solvent aggregates (as illustrated in Fig. 5(c)), we have reason to believe that those outside methanol may be bonded to the inside water shell to some degree. In the following process of solute molecule assembly, this distribution will serve as a shield for solute aggregation, create limited space for solute-solute interactions, thus facilitate a head-to-tail connection due to the relatively small volume of both ends, and meanwhile bring about a lower local packing compactness in final structure, consequently leads to the appearance of delta form. On the other hand, the light solvation degree in other two solutions can provide favorable condition for more solute molecules to occupy the area adjacent to terminals. As lots of solutes accumulate around the ends, one exposed hydroxyl group at the molecular terminal may be likely to bond with others in more orientations, contribute a lower selectivity for terminal groups, which could be supported by the separate peaks of C1 and C6 (two terminal carbon atoms) obtained from solid state NMR characterization, as shown in Fig. 6. The broader space in solution together with the optional bonding orientation finally stabilized an approximate herringbone linkage network and a dense packing motif exist in alpha and beta.

Thus we have found that the packing characteristics in the lattice of D-man should be a direct reflection of the solvation state of D-man in solution, the solvating power of different solvents on the both ends of D-man can provide distinct selectivity for terminal ¨COH groups to interact with each other. To verify the assumption, same experiments were carried out in the acetone-water system. The results showed that in water-acetone solvent, the delta form was not crystallized exclusively, instead a mixture of alpha and delta was usually yield. This phenomenon to a certain extent supports our conclusion, as acetone possesses lower HB donor and acceptor abilities than methanol, cannot construct a firm double-solvent-shell as observed in methanol-water, as a result a local environment being suitable for both alpha and delta was created.

This study, for the first time, investigated the solvent-dependent nucleation behavior of D-man which is one of the most important excipients in pharmaceutical and food industry [3, 4], and revealed the significant roles played by solvation manners in determining polymorphic outcomes of D-man. In this regard, the use of local packing compactness as a probe might constitute a potential way could be extended to other systems for polymorphic nucleation research.

Reference:

[1] Carespacheco, M. G., Vacamedina, G., Calvet, R., Espitalier, F., Letourneau, J. J., & Rouilly, International Journal of Pharmaceutics, 475 (2014) 69.

[2] M.A. Spackman, J.J. McKinnon, CrystEngComm. 4 (2002) 378.

[3] Wagner, C.M., Pein, M., Breitkreutz, Pow. Tech. 270 (2015) 470.

[4] Rowe, R.C., Sheskey, P.J.S., Owen, S.C., 2006. Handbook of pharmaceutical excipients, 449.

524665 Bio-Inspired Immobilization of Casein-Coated Silver Nanoparticles on Cellulose Acetate Ultrafiltration Membranes for Biofouling Control

Abstract:

The focus of this project was on immobilizing Ag-NPs on ultrafiltration membranes to provide the membranes with antimicrobial properties while preventing nanoparticle leaching and membrane degradation. To this end, low-biofouling nanocomposite membranes were loaded with casein-coated silver nanoparticles (casein-Ag-NPs) using two approaches, physical blending of Ag-NPs in the dope solution (^{P}Ag-NP/CA membranes) and chemical attachment of Ag-NPs to cast membranes (^{C}Ag-NP/CA membranes). The functionalization of Ag-NPs onto the CA membranes was achieved via attachment with functionalized thiol groups with the use of glycidyl methacrylate (GMA) and cysteamine chemistries, which was inspired by the affinity of silver to the thiol groups of cysteine proteins in bacteria. The performance of two different membranes were characterized via structural, morphological, permeation, and biofouling studies. The ^{P}Ag-NP/CA membranes displayed more pore channels in their cross-sections, exhibited higher average flux values, and had the highest average flux decline during short-term filtration studies. There were minor differences in pore channel morphology between CA and ^{C}Ag-NP/CA membranes, but the latter displayed a tighter pore network compared to the other membranes due to GMA crosslinking, which contributed to its lower average flux. While the ^{C}Ag-NP/CA membranes had a similar average flux decline compared to the CA membranes, the chemically attached Ag-NPs expressed the highest salt rejections. Lastly, ^{C}Ag-NP/CA membranes appeared to possess better biofouling control as compared to CA membranes, based on the reduction of observable cells. There was minimum leaching of silver during cross-flow filtration of ^{C}Ag-NP/CA membranes, and the leached water could still meet drinking water regulations.

530720 Concurrent Desalination and Boron Removal Via Reverse Osmosis

Abstract:

Desalination of saline water is a critical technology for supplying the potable and irrigation water needed for the rapidly increasing world population. The boron content of seawater averages 4.6 ppm and is as high as 9.6 ppm in the Eastern Mediterranean. A small amount of boron is required for all plants; however, it affects the yield and can become toxic to plants at levels ranging between 0.5 and 20 ppm. The World Health Organization (WHO) recommends that the boron concentration in drinking water and irrigation water be 0.5 ppm or less. Since boron is present in water as boric acid, its concentration cannot be reduced adequately by single-stage reverse osmosis (SSRO) using commercially available reverse osmosis (RO) membranes. This can be done using commercially available RO membranes by using a two-pass SSRO at the cost of reduced water recovery. Increased water recovery is possible by raising the pH that causes boric acid to dissociate into hydrated borate ions for which commercially available RO membranes have a good rejection. However, this requires adding chemicals and an additional process step to reduce the pH again. Desalination followed by ion exchange or dialysis can also reduce the boron concentration at the cost of increased process complexity. None of these processes can achieve desalination and boron removal with a high water recovery using just commercially available membranes. This paper describes the Concurrent Desalination and Boron Removal (CDBR) process for water desalination and reduction of the boron concentration to 0.5 ppm or less using commercially available membranes. (Krantz, 2016; Kürklü, 2017)

In the CDBR process, the saline water is fed to one or more SSRO stages in series whose retentate is fed to a countercurrent membrane cascade with recycle (CMCR) and whose permeate is fed to a low pressure membrane stage (LPMS) for further boron removal. The permeate from the LPMS is blended with that from the CMCR to obtain the desired product water. This process configuration employs countercurrent flow of the retentate and permeate, permeate recycle, and retentate reflux in the CMCR to reduce the required osmotic pressure differential (OPD) significantly. This reduces the pumping, maintenance, pretreatment and brine disposal costs. The performance metrics of the process have been predicted for operation at the thermodynamic limit.

The performance metrics for this novel CDBR process are compared with those for using two SSRO stages in series for desalination and boron removal. The assumptions are a seawater feed containing 35,000 ppm salt and 10 ppm boron producing water containing 0.5 ppm boron and less than 100 ppm salt and, pump and energy recovery device (ERD) efficiencies of 85% and 90%, respectively. For these conditions the CDBR process employing commercially available RO membranes with a 90% boron rejection can reduce the OPD relative to two SSRO stages in series by 35% at a water recovery of 55% and 37% at a recovery of 75%, and can reduce the net specific energy consumption (SEC_{net}) by 0.87% at 55% recovery and 39% at 75% recovery. The SEC_{net} and OPD can be further reduced by recycling and blending the retentate of the LPMS stage with the feed to the SSRO.

A total cost analysis of water production for the CDBR configurations is also carried out in terms of the fixed and operating costs, relative to a two-pass SSRO process. In the comparison, both processes are designed using commercial membranes to yield a water product with less than 0.5 ppm boron concentration from a saltwater feed containing 10 ppm boron and 35000 ppm TDS. While the two-pass RO process yields an overall water recovery of 46%, the yield of the CDBR process is 61%. The gross specific energy consumption of the CDBR process 4.44 kWh/m^3 in comparison to 4.76 kWh/m^3 for the two-pass RO process. If energy recovery devices are used, then the net specific energy consumption of the CDBR process is reduced to 3.51 kWh/m^3 in comparison to 3.19 kWh/m^3 for the two-pass RO process. The cost comparison is based on a water production capacity of 10000 m^3/day.

A novel CDBR process for achieving concurrent desalination and boron removal has been developed. The CDBR process uses only membrane technology without the need for increasing the pH or using other technologies such as ion exchange that are required for current commercial processes for desalination and boron removal. As such, the CDBR process permits concurrent desalination and boron removal rather than requiring sequential processes.

This work is supported by the Scientific and Technological Research Council of Turkey (TUBITAK) through Project No. 114Y165.

References

Krantz, W.B., S. Velioğlu, S. Kűrklű, M.G. Ahunbay, S.B. Ersolmaz, "Concurrent Desalination and Boron Removal (CDBR) Process," Patent application no. PCT/TR2016/050387, October 19, 2016.

Kürklü, S.; Velioglu, S.; Ahunbay, M. G.; Tantekin-Ersolmaz, S. B.; Krantz, W. B., Concurrent Desalination and Boron Removal (CDBR) Process, ***Desalination***, 423,79-94, 2017.

532764 Surface Nano-Structuring with Hydrophilic Polymer Brush Layers for Tailored Performance of Fouling Resistant RO and UF Membranes

Abstract:

Surface nano-structuring with hydrophilic polymer brush layers is an effective approach for imparting fouling resistance to membrane surfaces due to improved surface wettability as well as partial chain mobility. In the present approach, a base membrane of a suitable level of rejection and permeability is selected, and its surface is activated via treatment with atmospheric pressure plasma (APP) followed by free radical graft polymerization in an aqueous vinyl monomer solution. The resulting polymer brush layer grafted onto the membrane surface alters the base membrane separation performance (i.e., permeability and solute rejection), which can be tailored by the plasma and graft polymerization conditions. Therefore, selection of the base membrane and the conditions at which surface nano-structuring is performed is critical to achieve a fouling resistant membrane with the target membrane performance. Accordingly, the present study focused on evaluating the impact of plasma surface activation (i.e., APP type and treatment time) and graft polymerization (i.e., monomer type and concentration, grafting temperature and reaction time) conditions on the performance of surface nano-structured (SNS) polyamide (PA) RO and polysulfone (PSf) UF membranes with respect to water permeability and solute rejection. Polymer surface grafting was confirmed via X-ray photoelectron spectroscopy (XPS), and surface wettability and topography of the SNS membranes were evaluated via contact angle measurement and AFM, respectively. Permeability and solute rejection of the SNS-PA and SNS-PSf membranes were evaluated using laboratory scale plate-and-frame and stirred cells, respectively. Fouling propensity of the SNS-PA and SNS-PSf membranes was evaluated using model foulants (i.e., BSA and alginic acid), and membrane cleaning effectiveness was evaluated with DI water flushing at the end of foulant filtration. Results of the present study indicate that the present membrane surface nano-structuring approach is suitable for tailoring the performance of both RO and UF membranes in addition to mitigating membrane fouling and improving membrane cleaning effectiveness.

528629 Dual-Functionalized Nanofiltration Membranes Exhibit Multifaceted Anti-Fouling and Ion Rejection Performance

- Download/View Extended Abstract

Abstract:

As interest in decentralized water treatment units that enable localized water reuse grows, the development of nanofiltration (NF) membranes as a complementary technology to reverse osmosis membranes has gained traction. Furthermore, the use of advanced materials (e.g., graft copolymers), which form uniform nanoscale pores that are decorated with tunable chemical moieties can provide a ready means for post-assembly functionalization. For example, the incorporation of charged functionalities increases ion rejection by NF membranes due to repulsive electrostatic interactions. However, the incorporation of charge functionality tends to exacerbate the process of fouling, which is a continued hurdle in the development and application of NF membranes. Fouling is the process whereby attractive dispersive, electrostatic, and hydrodynamic interactions between dissolved solutes and the membrane leads to the attachment and accumulation of material on the surface of the membrane. This process reduces the throughput and performance of the membrane. This study evaluates novel NF membranes that exhibit both high ion rejection and high fouling resistance through the use of a post-fabrication multi-functionalization process performed on the membrane template.

Membranes cast from a poly(acrylonitrile-*co*-oligo-(ethylene glycol) methyl ether methacrylate-*co*-(3-azido-2-hydroxypropyl methacrylate)) P(AN-OEGMA-AHPMA) copolymer are the parent substrates utilized in this study. The azido groups lining the pore wall of the parent substrate lead to multiple functionalization avenues. Here, a dual functionalized membrane containing both ion-selective and anti-fouling functionality is attained through a controlled reaction between propiolic acid and the membrane. This copper-catalyzed azide-alkyne cycloaddition (CuAAC) reaction occurs rapidly and by varying the concentration and the exposure time the penetration depth of the initial reaction can be controlled. This control is confirmed through a decrease in the signal intensity associated with the azido group in the Fourier transform infrared spectroscopy (FTIR) spectrum. This layer achieves an anti-fouling character through a 1-ethyl-3-(3-dimethylaminopropyl)-carbodiimide (EDC) coupling reaction of 1H,1H-perfluorooctylamine with the propiolic acid. The azido moieties not exposed to propiolic acid in the initial stage are then charge-functionalized through a CuAAC reaction of propargyl amine. When challenged with a 0.5 g/L Bovine Serum Albumin (BSA) solution, these membranes express increased anti-fouling properties compared to a fully charged membrane, while maintaining similar ion rejection performance.

This study begins the analysis of incorporating multiple, useful functionalities into a single membrane template. The rapid exposure of a reactive solution works to prove that through readily controllable reactions, the required depth for inhibiting foulant interactions could be reduced down to the minimal distance in which the attractive forces between the foulants and the surface is shielded. Minimizing this layer incorporates more charged-groups on the pore wall, increasing the overall resistance to ion-transport. This membrane fabrication process can be further extended to ion-exchange membranes, in which the inclusion of a top surface anti-fouling layer could increase the lifespan of the membrane while maintaining a high ionic throughput.

530611 Inorganic Microfiltration Membranes Modified with Hydrophilic Silica Nanoparticles for Oil-in-Water Emulsion Separation

Abstract:

On a global scale, approximately three barrels of produced water are produced for every barrel of oil extracted. The existence of oil droplets can result in the formation of stable oil-in-water emulsions. In order to meet the discharge requirements, there is a significant need for efficient, cost-effective strategies to remove hydrocarbons from produced water. Our work aims at developing new membrane strategies that result in high water flux and low membrane fouling for produced water applications. To accomplish this, alumina microfiltration membranes were modified by incorporating silica nanoparticles into the membrane structure. The modified alumina tubular membranes were tested using cyclohexane-in-water emulsions. The experimental results showed that the alumina membrane treated with a 0.5 wt.% silica nanoparticle solution yielded a water flux of 350 L/m^2-h and oil rejection efficiency of 93%. This is a 21% enhancement in the flux and a 6% enhancement in the oil rejection efficiency compared with the unmodified membrane. The cyclohexane concentration of the permeate was less than 40 ppm when the cyclohexane concentration of the feed was 500 ppm at 40 â„ƒ with a feed pressure of 43.7 psia.

532326 Effect of the Surface Charge of Monodisperse Particulate Foulants on Cake Formation

Abstract:

In microfiltration and ultrafiltration, particulate foulants are inevitably deposited on the membrane surface, forming a cake whose structure and behavior play crucial roles in the subsequent filterability of the suspensions. This study investigated the impact of fouling by three types of latex particulate foulants, which were of the same size (3 Î¼m) but with different surface charges. Surprisingly, although the positively charged aminated latex was expected to perform the worst in the flux-decline experiments due to attractive electrostatic interaction with the negatively charged membrane, this latex displayed the best performance relative to the two negatively charged latex. To understand these counter-intuitive results, a novel network model and three-dimensional (3D) optical coherence tomography (OCT) image analysis were employed to reveal the underlying reasons for the different fouling behaviors. Two mechanisms were found to contribute to the worse performance of the negatively charged latex. Firstly, these particles tended to deposit on the pore rather than non-pore region of the membrane due to the repulsive particle-membrane electrostatic interactions, which led to a more complete pore blockage and thereby greater initial cake resistance. Secondly, these particles had a greater tendency to cluster and deposit on other deposited latex due to similarly repulsive particle-membrane and particle-particle interactions, which led to a more inhomogeneous cake and thereby greater specific cake resistance.

532835 Zwitterion-Substituted Polysulfones As Fouling-Resistant Desalination Membranes

Abstract:

We have demonstrate a novel synthesis method to prepare a zwitterion-modified copolymer: poly(arylene ether sulfone-*co*-sulfobetaine arylene ether sulfone) (PAES-*co*-SBAES), which was blended with unmodified polysulfone (PSf) to prepare asymmetric, free standing membranes. The polymer's chemical structure was analyzed by ^{1}H-NMR spectroscopy and the molecular weight of polymer was determined using size exclusion chromatography. Then, asymmetric PSf/PAES-co-SBAES blend membranes with various zwitterionic SBAES segment contents were fabricated via the non-solvent induced phase separation process. The cross-section morphology, surface hydrophilicity, and surface roughness of the membranes were analyzed by scanning electron microscopy, water contact angle measurements, and atomic force microscopy. The results indicated that both the porosity of the support layer and surface hydrophilicity increased drastically due to the incorporation of hydrophilic zwitterionic (or SBAES) segments. Surface roughness remained low for all membranes. The water permeance and antifouling ability of the PSf/PAES-co-SBAES blend membranes were both remarkably improved, with permeance values significantly higher than the native PSf and at or above levels of commercial polyamide membranes. Also, salt rejection was maintained at a high level, despite increases in hydrophilicity. The flux recovery when challenged with protein-rich solutions was ~93%, and complementary fluorescence microscopy experiments revealing very small levels of protein adhesion. Finally, no decrease in performance was observed when the membranes were challenged with chlorinated solutions. In summary, the strategy outlined to prepare zwitterion-modified polysulfones yielded asymmetric membranes with a high water permeance and salt rejection that also exhibit fouling resistance and tolerance toward chlorinated solutions. This will talk will discuss these developments and ongoing experiments to further improve the performance and scalability.

531120 Probing Metal-Organic Framework (MOF) Design for Adsorptive Sour Natural Gas Purification

Abstract:

Parent and amine-functionalized analogues of the metal-organic frameworks (MOFs) UiO-66(Zr), MIL-125(Ti), and MIL-101(Cr) were evaluated for their hydrogen sulfide (H_2S) adsorption efficacy and post-exposure acid gas stability. Adsorption experiments were conducted through fixed-bed adsorption breakthrough studies, utilizing multicomponent 1% H_2S/99% CH_4 and 1% H_2S/10% CO_2/89% CH_4 test gases as simplistic representative sour gas mixtures. Pelletized MIL-101(Cr) adsorbents were found to be unstable through diffraction and porosity measurements following H_2S exposure, while other materials retained their characteristic properties. Linker-based amine functionalities increased H_2S breakthrough times and saturation capacities from their parent MOF analogues. Competitive CO_2 adsorption effects were mitigated in mesoporous MIL-101(Cr) and MIL-101-NH_2(Cr), in comparison to microporous UiO-66(Zr) and MIL-125(Ti) frameworks. This result suggests the installation of H_2S binding sites in large pore MOFs could potentially enhance H_2S selectivity. *In-situ* FTIR measurements in 10% CO_2 and 5000 ppm H_2S environments illustrate framework hydroxyl and amine moieties serve as selective H_2S physisorption sites. Results from this study elucidate design strategies and stability considerations for engineering MOFs in sour gas purification applications.

533929 The Role of Solvent in the Room Temperature Synthesis of Two Isomeric Metal-Organic Frameworks

Abstract:

Two isomeric Metal-Organic Frameworks (MOFs) with the chemical formula $Zn_2(X)_2(TED)$ (X = terephthalic acid, 2,3,5,6-tetramethyl terepthalic acid, dimethyl terephthalic acid, and anthracene dicarboxylic acid) have been synthesized via a fast, room temperature synthesis procedure to produce DMOF-1 or ZnBD which differ by topology. It was found that functionalization of the BDC ligand results in the formation of the DMOF-1 structure. This is a result of the imposed steric effects that prevent the formation of the Kagome lattice. Solvent choice was found to play a vital role in the synthesis of ZnBD when using BDC as a ligand and is the determining factor in the MOF topology that is formed. When dimethylformamide (DMF) was used as the solvent, the synthesis resulted in the formation of ZnBD, whereas methanol produced DMOF-1. The synthesized MOFs were found to have similar water stability, both degraded upon exposure to 50% relative humidity at 25 °C, and both are hydrophobic. Additionally CO_2 adsorption isotherms of ZnBD were collected and compared to those previously reported for DMOF-1. ZnBD adsorbs less CO_2 at low pressures, but reaches similar capacity at 20 bar. This can be explained by the nature of the large hexagonal pore in ZnBD (15 Å) compared to the smaller pore opening (7.5 Å) in DMOF-1. The heat of adsorption for CO_2 on ZnBD was calculated to be ~20 kJ/mol at zero coverage, which is slightly lower than what has been reported for DMOF-1.

525738 Insight on Fluorinated MOFs Structural Properties-Gas/Vapor Adsorption/Sensing Relationships

Abstract:

Insight on Fluorinated MOFs structural properties-gas/vapor adsorption/sensing relationships

<u>Mohamed Rachid Tchalala</u>, Youssef Belmabkhout, Karim Adil, Prashant. M. Bhatt, Karumbaiah Nanaiah, Khaled Nabil Salama, Mohamed Eddaoudi

Functional Materials Design, Discovery & Development (FMD3), Advanced Membranes and Porous Materials Center, Division of Physical Sciences and Engineering, King Abdullah University of Science and Technology, 4700 King Abdullah University of Science and Technology (KAUST), Thuwal 23955-6900, Kingdom of Saudi Arabia.

The understanding of gas/vapor sorption properties of materials in relation to their applications in separation, catalysis and sensing in extremely important to achieve intensified processes. In this paper, we report very recent advances on the use of MOF thin films as a medium for detecting gases/vapors, with respect to their concentration toxicity range or uncomforting zones. We show for the first time a set of rarely unveiled MOF structural properties-gas/vapor adsorption /sensing relationships study, that allowed unveiling a very sensitive MOF based sensors for CO_2, H_2S, H_2O and SO_2. In particular, the recent discovery of ultra-microporous fluorinated MOF with adjustable inorganic pillars, allowed unveiling powerful MOF sensors capable of detecting gases in the presence of atmospheric moisture. For, illustration purpose, this talk will focus on the exceptional concurrent selective sensing of CO_2 and H_2O (Figure 1) from atmospheric air. The effect of transduction mechanism (QCM or IDE) will be also discussed.

a)

b)

c)

d)

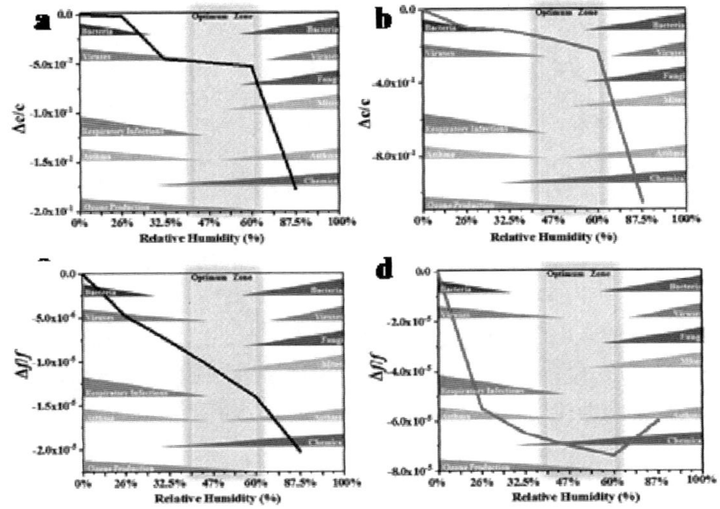

Figure 1: Variation in the capacitance with RH change in a) NbOFFIVE-1-Ni and b) AlFFIVE-1-Ni. Frequency shift as function of RH for QCM coated with c) NbOFFIVE-1-Ni and d) AlFFIVE-1-Ni

532966 Methane Adsorption on Zeolitic Imidazolate Framework-8 (ZIF-8)

Abstract:

ZIFs constitute a class of MOFs that have a high degree of thermal and chemical stability.[1] This makes them appealing candidates for practical application. In this study we report on methane adsorption on Zeolitic Imidazolate Framework-8 (ZIF-8) at low temperatures.

ZIF-8 is a porous MOF that has the sodalite structure. Structural studies at room temperature and high pressures, using a diamond anvil cell, found that at 1.47 GPa ZIF-8 undergoes a structural phase transition.[2] What occurs is that the pressure-transmitting fluid is forced into the ZIF-8 resulting in a reorientation (gate-opening) of the imidazolate windows in the ZIF-8 that increases their size. The larger windows allow more fluid to get into the ZIF-8; this additional fluid produces an expansion of the ZIF-8. ZIF-8 at 1.47 GPa has the same space-group symmetry that it had before any fluid was loaded in it.[2] Fairen-Jimenez and his collaborators conducted a combined structural, adsorption isotherm, and computer simulation study of N_2 sorption in ZIF-8 at 77 K.[3] Using ***in situ*** XRD measurements they found that the same gate-opening transition that was observed in ZIF-8 at room temperature and high pressures was also present upon N_2 sorption in the ZIF-8 at low temperatures (77 K) and much lower pressures[3]. The 77 K N_2 in ZIF-8 isotherm exhibits a substep (i.e. a region of sharp increase in the loading with little pressure change in the isotherm). The GCMC simulations agree with the adsorption isotherm data only if the structure for the empty ZIF-8 (i.e., the as-produced or AP structure) is used for the simulation of the lower loadings, and, the structure of ZIF-8 after the transition at room temperature (i.e., the high-pressure or HP structure) is used for the simulation of the higher loadings. The isotherm substep falls between these two regimes; hence, the substep was associated with the occurrence of the structural transition.[3]

Several other gases (including Ar,[4] Xe,[5] CO[4,6] and O_2[4,7]) sorbed in ZIF-8 have substeps in their adsorption isotherms measured at low temperatures; while other sorbents, e.g., CH_4[8] and CO_2[9], do not exhibit substeps in the isotherms that have been reported. The nature of the substeps present in the adsorption isotherms for some sorbates in ZIF-8 has been the subject of much study and some controversy. [1,3,10]

Methane was one of the first sorbates studied in ZIF-8. Yildirim's group reported adsorption isotherms for temperatures between 125 and 300 K and pressures of up to 65 bar.[8] Fairen-Jimenez' group compared Yildirim's experimental data at 125 K to their GCMC computer simulations.[11] Fairen-Jimenez used both the AP and the HP structures of ZIF-8 in their simulations. They found that the best agreement between the experimental data and simulations was obtained when the AP-structure was used for lower loadings and the HP structure was used for higher loadings. This led Fairen-Jimenez to conclude that gate-opening was occurring for CH_4 in ZIF-8, even if there were no subteps in the isotherm data.[11]

For Xe adsorbed in ZIF-8 we found that there are substeps associated with gate-opening in isotherms measured at low temperatures, and that as the isotherm temperature increases eventually the substeps are no longer present.[5] The relative pressure of the substeps (i.e. $P_{substep}/P_o$ where P_o is the saturated vapor pressure and $P_{substep}$ is the pressure at the midpoint of the substep) increases as the isotherm temperature increases, eventually reaching $P_{substep}/P_o = 1$ (i.e., no substep in the isotherm).[5] We note that for Xe adsorbed in ZIF-8 other experimental techniques show that gate-opening is occurring at the higher temperatures.[12,13] We wanted to explore whether a similar behavior was present in CH_4 adsorbed in ZIF-8; so we explored CH_4 adsorption in ZIF-8 at temperatures lower than those which had been previously studied.

We will present the results of low temperature adsorption isotherms of methane in ZIF-8 over a range of temperatures between 85 K and 107 K. We have found that all five isotherms measured (85.00 K, 92.12 K, 97.49 K, 106.31 K and 107.37 K) exhibit substeps. The loading interval for the substep for all five isotherms was approximately 2000 μmoles/g. We have observed that the relative pressure of the substep approaches 1 as the isotherm temperature increases.

In addition to presenting results for the isotherm characteristics as a function of temperature we will present the loading dependence of the isosteric heat of adsorption for CH_4 in ZIF-8. At saturation the isosteric heat of adsorption for a system should be equal to the bulk latent heat for the corresponding transition. We have verified that this is the case for our CH_4 adsorption data on ZIF-8; the value of the isosteric heat of adsorption at saturation agrees with methane's bulk latent heat of vaporization.[14]

We have also explored the kinetics of adsorption for methane in ZIF-8 as a function of loading. Results on the loading dependence of the equilibration time for a fixed temperature will be presented.

We will compare our results for methane with those that others and us have obtained for the sorption of other gases in ZIF-8.

References

1. S. Park, Z. Ni, A.P. Cote, J.Y. Choi, R. Huang, F.J. Uribe-Romo, H.K. Chae, M. OKeeffe, O.M. Yaghi, Exceptional chemical and thermal stability of zeolitic imidazolate frameworks, Proc. Natl. Acad. Sci. 103 (27) (2006) 10186–10191.

2. A. Moggach, T.D. Bennett, A.K. Cheetham, The effect of pressure on zif-8: increasing pore size with pressure and the formation of a high- pressure phase at 1.47 GPa, Angew. Chem. 121 (38) (2009) 7221–7223.

3. Fairen-Jimenez, S. Moggach, M. Wharmby, P. Wright, S. Parsons, T. Duren, Opening the gate: framework flexibility in ZIF-8 explored by experiments and simulations, J. Am. Chem. Soc. 133 (23) (2011) 8900–8902.

4. O. Ania, E. Garcıa-Perez, M. Haro, J. Gutierrez-Sevillano, T. Valdes- Solıs, J. Parra, S. Calero, Understanding gas-induced structural deformation of ZIF-8, J. Phys. Chem. Lett. 3 (9) (2012) 1159–1164.

5. H. Gallaba, A.G. Albesa, A.D. Migone, Evidence of gate-opening on Xenon adsorption on ZIF-8: an adsorption and computer simulation study, J. Phys. Chem. C 120 (30) (2016) 16649–16657.

6. Gallaba, Dinuka H., Jhonny Villarroel-Rocha, Karim Sapag, and Aldo D. Migone. "Carbon monoxide adsorption in ZIF-8: Kinetics and equilibrium." Microporous and Mesoporous Materials 265 (2018): 227-233.

7. Russell, J. Villaroel, K. Sapag, A.D. Migone, O2 adsorption on ZIF- 8: temperature dependence of the gate-opening transition, J. Phys. Chem. C 118 (49) (2014) 28603–28608.

8. Zhou, H. Wui, M.R. Hartman and T. Yildirim "Hydrogen and Methane Adsorption in Metal-Organic Frameworks: A High-Pressure Volumetric Study", J. Phys. Chem. C **111**, 16131-16137 (2007).

9. A. Russell, A.D. Migone, Low temperature adsorption study of CO2 in ZIF-8, Microporous Mesoporous Mater. 246 (2017) 178–185.

10. E. Casco, J. Fernandez-Catala, Y. Cheng, L. Daemen, A.J. Ramirez-Cuesta, Cuadrado-Collados, J. Silvestre-Albero, E.V. Ramos- Fernandez, Understanding ZIF-8 performance upon gas adsorption by means of inelastic neutron scattering, ChemistrySelect 2 (9) (2017) 2750–2753.

11. Fairen-Jimenez, R. Gavelis, A. Torrisi, A.D. Gellan, M.T. Wharmby, P.A. Wright, C. Mellot-Draznieks, T. Duren, "Flexibility and swing effect on the adsorption of energy-related gases on ZIF-8: combined experimental and simulation study" Dalton Trans. 41, 10752-10762 (2012).

12. Magdysyuk, F. Adams, H.-P. Liermann, I. Spanopoulos, P. Trikalitis, M. Hirscher, Morris, M. Duncan, L. McCormick, R. Dinnebier, Understanding the adsorption mechanism of noble gases Kr and Xe in CPO-27-Ni, CPO-27-Mg, and ZIF-8, Phys. Chem. Chem. Phys. 16 (43) (2014) 23908–23914.

13. -A. Springuel-Huet, A. Nossov, F. Guenneau, A. Gedeon, Flexibility of ZIF-8 materials studied using 129Xe NMR, Chem. Commun. 49 (67) (2013) 7403–7405.

14. https://webbook.nist.gov/cgi/cbook.cgi?ID=C74828&Mask=4#Thermo-Phase

531421 Incorporation of a Dioxo-Molybdenum(VI) Complex into a Titanium-Functionalized Zr(IV)-Based Metal-Organic Framework

Abstract:

Metal-Organic Frameworks (MOFs) are very porous, crystalline and well-ordered materials that can be used in different applications such as catalysis, gas adsorption-desorption, and photocatalysis among others due to their properties [1]. Therefore, it is not surprising that many researchers are studying the design, synthesis and surface modification of new types of MOF [2]. In fact, recent studies have focused on the synthesis of new MOFs based on titanium as a substitute for titanium dioxide for photocatalytic applications [3-5]. However, these studies have reported different problems mainly due to the use of the metal precursor, in that sense, in many cases, it is not possible to obtain a porous material, or the obtained product is TiO_2 [1]. Nowadays, the possibility of a metal exchange using an ion exchange within a Zr-MOF to build a material that combines the properties of both metals seems highly promising [6]. These new Zr/Ti-MOFs should be used as supports for the immobilization of a Mo(VI) catalyst [7, 8]. In consequence, such materials have the potential to catalyze the epoxidation of alkanes using UV-Vis irradiation, and molecular oxygen as an oxidizing agent, which is quite challenging [9, 10]. In this work, the synthesis of new Ti-MOFs that incorporated a dioxo-molybdenum (VI) complex was studied. First, a molybdenum complex was anchored to a bipyridine-dicarboxylate ligand, which was then immobilized in a UiO-67-type MOF. Finally, Ti^{+4} ion exchange was performed at Zr^{+4} of the Zirconium MOF nodes. This material was characterized by sorption of N_2, XRPD, FTIR, RAMAN, TGA, XPS, atomic absorption and SEM analysis.

REFERENCES

1. Bosch, M., S. Yuan, and H.-C. Zhou, ***Group 4 Metals as Secondary Building Units: Ti, Zr, and Hf-based MOFs***, in ***The Chemistry of Metal–Organic Frameworks: Synthesis, Characterization, and Applications, First Edition.***, S. Kaskel, Editor 2016, Wiley-VCH Verlag GmbH & Co: KGaA, Germany. p. 130-170.
2. Liu, J., et al., ***Applications of metal-organic frameworks in heterogeneous supramolecular catalysis.*** Chem Soc Rev, 2014. **43**(16): p. 6011-61.
3. Fu, Y., et al., ***An amine-functionalized titanium metal-organic framework photocatalyst with visible-light-induced activity for CO2 reduction.*** Angew Chem Int Ed Engl, 2012. **51**(14): p. 3364-7.
4. Hendon, C.H., et al., ***Engineering the optical response of the titanium-MIL-125 metal-organic framework through ligand functionalization.*** J Am Chem Soc, 2013. **135**(30): p. 10942-5.
5. Hu, S., et al., ***Solvothermal synthesis of NH2-MIL-125(Ti) from circular plate to octahedron.*** CrystEngComm, 2014. **16**(41): p. 9645-9650.
6. Nguyen, H.G.T., et al., ***Comparative study of titanium-functionalized UiO-66: support effect on the oxidation of cyclohexene using hydrogen peroxide.*** Catal. Sci. Technol., 2015. **5**(9): p. 4444-4451.
7. Neves, P., et al., ***Incorporation of a dioxomolybdenum(VI) complex in a ZrIV-based Metal–Organic Framework and its application in catalytic olefin epoxidation.*** Microporous and Mesoporous Materials, 2015. **202**: p. 106-114.
8. Leus, K., et al., ***A MoVI grafted Metal Organic Framework: Synthesis, characterization and catalytic investigations.*** Journal of Catalysis, 2014. **316**: p. 201-209.
9. Páez, C.A., et al., ***Oxygen atom transfer photocatalyzed by molybdenum(VI) dioxodibromo-(4,4′-dicarboxylate-2,2′-bipyridine) anchored on TiO2.*** Catalysis Today, 2008. **133-135**: p. 619-624.
10. Martínez, H., et al., ***Photo-epoxidation of cyclohexene, cyclooctene and 1-octene with molecular oxygen catalyzed by dichloro dioxo-(4,4′-dicarboxylato-2,2′-bipyridine) molybdenum(VI) grafted on mesoporous TiO2.*** Journal of Molecular Catalysis A: Chemical, 2016. **423**: p. 248-255.

533256 Moisture-Enhanced Feature for Potential CO_2 Capture Under Humid Conditions within Microporous PCN-250 Frameworks

Abstract:

Developing metal-organic frameworks (MOFs) with moisture-resistant feature or moisture-enhanced adsorption for the practical applications of CO_2 capture remains a challenge. In this work, we investigate the CO_2 adsorption behavior of two iron-based MOFs materials under humid conditions, PCN-250(Fe_3) and PCN-250(Fe_2Co). An interesting and unusual adsorption behavior is observed that the two materials demonstrate an unusual moisture-enhanced adsorption of CO_2. H_2O molecule eccentrically induces a remarkable increase in CO_2 uptake of the MOFs for the dynamic CO_2 capture from CO_2/N_2 (15: 85) mixture. For PCN-250(Fe_3), its CO_2 adsorption capacity increases by 54% under wet conditions (50% RH) in comparison with that under dry conditions (from 1.18 mmol/g to 1.82 mmol/g). Similarly, PCN-250(Fe_2Co) also exhibits an increase of 69% in CO_2 adsorption capacity, from 1.32 mmol/g to 2.23 mmol/g. Even under the conditions of 90% RH, the CO_2 adsorption capacities of PCN-250(Fe_3) and PCN-250(Fe_2Co) still have increase of 43.7% and 70.2% in comparison of dry conditions, respectively. Molecular simulations indicate that the hydroxyl functional groups (*Î¼*-OH) within the framework play a determined role in improving CO_2 uptake in the presence of water vapor. Furthermore, partial substitution of Fe^{3+} by Co^{2+} ions in the PCN-250 framework leads to a significant improvement in CO_2 adsorption capacity and selectivity. These combinational properties including excellent moisture stability (stable in exposing 90% RH humid air for 30 days), superior recyclability and moisture-enhanced feature make PCN-250 materials one of the best adsorbent materials for CO_2 capture under humid conditions. This study provides a new paradigm that the PCN-250 frameworks can not only be moisture-resistant but also subtly convert the common detrimental impact of moisture into a positive impact on improving CO_2 capture performance.

533803 Advanced Froth Flotation Using Oil-Coated Bubbles and Its Application in De-Inking

Abstract:

Froth flotation is a highly efficient solid-liquid separation technique with a history of more than one century. It has been applied in many industrial processes, such as wastewater treatment, mining operations and paper recycling. The separation mechanism is that hydrophobic particles can attach to rising gas bubbles and are lifted to the top, creating a layer of froth which can be skimmed off, whereas more hydrophilic particles cannot attach to gas bubbles and will remain in the mixture. In many cases, however, the particles that we want to separate are less hydrophobic and cannot attach to gas bubbles. To overcome this limitation, our strategy is to coat the gas bubbles with a thin layer of oil, making the bubble surface more polar compared to uncoated bubbles; as a result, less hydrophobic or even mildly hydrophilic particles can attach as well. In the present work, we show that oil-coated bubble flotation has thermodynamic, kinetic and hydrodynamic advantages over the traditional method. With regard to flotation de-inking as a practical application, we find that oil-coated bubbles are far more efficient than uncoated bubbles at removing hydrophilic ink pigment particles from aqueous solutions.

Tribo-charging of Binary Mixtures Composed of Coarse and Fine Particles in Gas-Solid Pipe Flow

Farzam Fotovat[1,2*], Haifeng Wang[1,3], Xiaotao T. Bi[1] and John R. Grace[1]

[1]Fluidization Research Centre, Department of Chemical and Biological Engineering, University of British Columbia, Vancouver, Canada V6T 1Z3

[2]Department of Chemical and Petroleum Engineering, Sharif University of Technology, Azadi Avenue, P.O. Box 11365-9465, Tehran, Iran

[3]Department of Chemical Engineering, China University of Mining and Technology, Xuzhou, China, 221166

1. Introduction

Despite the studies carried out on tribo-charging of particle mixtures composed of fine and coarse particles, it is not clear yet how fines alter the tribo-charging behavior of coarse particles in gas-solid flows. Moreover, the effect of the physical, chemical and electrical properties of particles on the electrification of powdery mixtures has not been well understood, requiring further investigation. In this study, tribo-charging of mixtures of coarse glass beads and conductive or non-conductive fines of different compositions is investigated in a pneumatic conveying system. Understanding the relationship between charge density and composition of particle mixtures, as well as determining the impact of fine particle properties on electrification of binary mixtures of fine and coarse materials are the key objectives of this investigation.

2.

Modeling electrification of binary mixtures in dilute gas-solids pipe flow

According to the model developed by Matsusaka & Masuda [1] based on a condenser model [2,3], the mass charge density of an initially neutral single spherical particle charged in a dilute gas-solids pipe flow is

$$q_m = -k_c V_c \varepsilon_0 \frac{6nS}{\pi d_p^3 \rho_p z_0} \qquad (1)$$

where S and z_0 are the contact area and the critical gap between the contact bodies (particle and pipe wall in our case). ρ_p and d_p are the particle density and diameter, respectively. n denotes the number of particle-wall collisions, and ε_0 is vacuum permittivity (8.854×10^{-12} F/m). k_c is the charging efficiency and V_c is the potential difference based on the surface work functions of contacting bodies:

$$V_c = \frac{(\phi_p - \phi_w)}{e} \qquad (2)$$

where φ_p and φ_w are the work functions of the particle and pipe wall, respectively, and *e* is the elementary charge. Note that the potential differences arising from space and image charges are assumed to be negligible in the systems studied here. It is also assumed that V_c of a binary mixture is obtained from the superposition of the potential differences between fine/coarse particles and the pipe wall, i.e. V_{cf} and V_{cc}, respectively:

$$V_c = xV_{cf} + (1-x)V_{cc} \tag{3}$$

To adapt Eq. (1) to a binary mixture of dissimilar particles, we assume that *n* is proportional to the total number of particles in the mixture:

$$n = k(N_f + N_c) \tag{4}$$

where *k* is the proportionality factor. N_f, and N_c are the number of fine and coarse particles in the unit mass of the mixture, respectively, calculated as follows:

where

x is the weight fraction of fine particles in the mixture. The c and f subscripts denote the coarse and fine particles, respectively. From Eqs. (1) and (3-6), the net charge of all fine and coarse particles in a unit mass of a binary mixture is given by:

$$q_{m,mix} = -\frac{k k_t \varepsilon_0 S}{\pi z_0}\left(x V_{cf} + (1-x) V_{cc}\right)\left\{\frac{x}{\rho_f d_f^{\,3}} + \frac{1-x}{\rho_c d_c^{\,3}}\right\} \qquad (7)$$

The coefficient of Eq. (7) can be obtained by fitting the equation with the measured mass charge densities of the mixtures.

3. Experimental

3.1. Materials

Glass beads were chosen in this study as the coarse material for all mixtures. Conductive silver-coated fine glass beads (SFGB) and uncoated dielectric fine glass beads (FGB) were used as fine materials added to coarse glass beads (CGB). These fine materials were similar in terms of mean size, size distribution, and density. Table 1 provides the key properties of CGB and fines used in this study.

3.2. Charge measurements

The electrostatic test system used in the present work consisted of a cylindrical feeder, a grounded stainless steel spiral tube (3.0 m in length, 0.64 cm in diameter, 2 coils (d_s=0.15 m) per meter) for tribo-charging, a Faraday cup for charge measurement, and control valves. Dry building air was used as the particle-carrying gas. During the tests, temperature and relative humidity were varied in the ranges of 16.9 to 25.7°C and 0.2% to 7.2%, respectively. Based on the preliminary experiments carried out to obtain reproducible experimental conditions, 1.0 g particle was conveyed in the pipe by air flowing with the velocity of 16.5 m/s in all tests. The air inlet pressure was about 600 kPa.

In order to prepare 1.0 g mixtures consisting of 0.2, 1.0, 2.0, 5.0, 10.0, 20.0, 30.0, 70.0, and 90.0 wt.% fines, predetermined amounts of fines and CGB were mixed together. In addition, 1.0 g of CGB and 1.0 g fines were tested in separate experiments. The charge measurements of each mixture and pure materials were repeated at least three times.

4.

Preliminary Results

Charge

densities of CGB-SFGB mixtures of various compositions are shown in Fig.
1. Unlike the negative polarity of pure
CGB, pure silver-coated fine glass beads were charged positively, consistent with the negative V_{cf} value
obtained from Eq. (2), knowing the work functions of the stainless steel pipe wall and SFGB provided in Table 1. Compared to the pure coarse particles, the magnitude of charge densities of mixtures to which less than 10 wt. % of fines had been added increased with increasing fines content, and the charge polarities remained negative. With increasing weight
proportion of SFGB in the mixture beyond 10%, the charge polarity changed from negative to positive and the charge density then slightly increased.

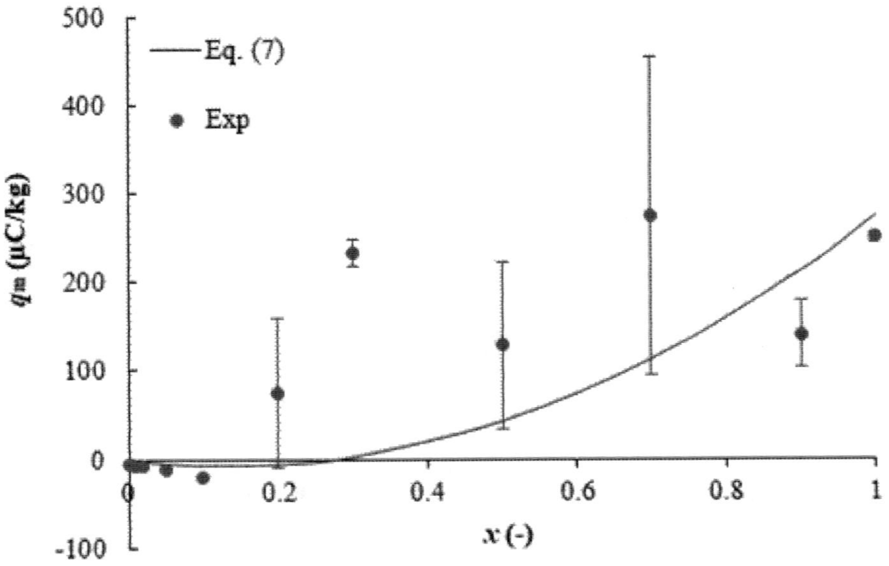

Fig.1. Mass charge density of CGB- SFGB mixtures vs. weight fraction, *x*, of fine particles. Error bars show ± one standard deviation from the average experimental measurements.

Fig.
2 depicts the effect of added fine glass beads (FGB) on tribo-charging of CGB. As per SFGB alone, uncoated fine glass beads by themselves were charged positively,

despite the negative polarity of pure CGB. This shows that the fine and coarse glass beads used in our study likely differed in chemical composition or surface properties, leading to dissimilar contact potential differences with the pipe wall.

Adding FGB to coarse glass beads resulted in an increase in the absolute value of mass charge density, relative to pure CGB. The charge polarity remained negative when the weight percentage of fines was less than 10 wt.%. Further addition of fines reversed the mixture polarity, and the charge density then leveled off. Despite the lower magnitude of charge densities for CGB-FGB mixtures than for corresponding CGB-SFGB mixtures, charge densities of both systems followed similar trends when changing the weight fraction of fines from 0 to 1.

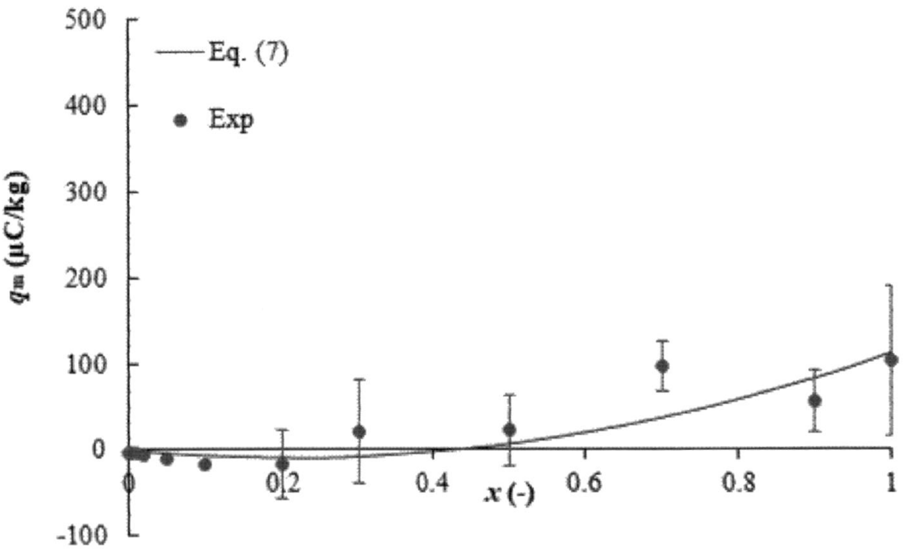

Fig. 6. Mass charge density of CGB-FGB mixtures vs. weight fraction, *x*, of fine particles. Error bars show ± one standard deviation from the average experimental measurements.

As seen in Figs. 1 and 2, $|q_m|$ experienced a local maximum at $x \approx 0.1$, beyond which contacts between fine particles and the pipe wall became the dominant charging mechanism, as inferred from the change in the charge polarity. Thus, observation of charge polarity reversal in gas-solid pipe flows involving mixtures of fine and coarse particles can signify a change in the type of particles dominantly in contact with the pipe wall. Figs. 1 and 2 depict that the measured charge densities of mixtures with $x > 0.1$ were widely scattered, likely due to the opposite charge polarities of fine and coarse particles and the major influence of turbulence on the dynamics of fine particles, as discussed below.

Figs. 1 and 2 also suggest that
Eq. (7) could successfully predict the trend in the mass charge densities of all tested mixtures as a function of the fines content. The discrepancy observed between the experimental results and the proposed model in Figs. 1 and 2, especially in the range with low fraction of fine particles, is likely attributed to negligence of particle-particle interactions, which should be further investigated in the future.

References

1. Matsusaka S, Masuda H. Simultaneous measurement of mass flow rate and charge-to-mass ratio of particles in gas-solids pipe flow. *Chem Eng Sci.* 2006;61(7):2254-2261.

2. Matsusaka S, Maruyama H, Matsuyama T, Ghadiri M. Triboelectric charging of powders: A review. *Chem Eng Sci.* 2010;65(22):5781-5807.

3. Matsusaka S, Ghadiri M, Masuda H. Electrification of an elastic sphere by repeated impacts on a metal plate. *J Phys D Appl Phys.* 2000;33(18):2311-2319.

4. Trigwell S, Grable N, Yurteri CU, Sharma R, Mazumder MK. Effects of surface properties on the tribocharging characteristics of polymer powder as applied to industrial processes. *IEEE Trans Ind Appl.* 2003;39(1):79-86.

5. Dweydari AW, Mee CHB. Work function measurements on (100) and (110) surfaces of silver. *Phys status solidi.* 1975;27(1):223-230.

Topics:

533997 Simultaneous Separation of Protein and Starch Particles in Oat Flour *Via* Dry Fractionation Approaches

Abstract:

Concurrent classification of cereal grains into their major constituents of protein and starch through dry fractionation technologies is a promising approach towards sustainable food industry. Conventional wet technologies that are commonly used for plant protein separations are water-, chemical- and energy-intensive processes thus reducing the nutritional quality and bio-functionality of the extracted proteins. We have developed a novel dry fractionation approach by combining dry sieving and electrostatic separation techniques to effectively enrich oat (*Avena sativa*) flour with proteins. Milled oat flour with ~14% protein was initially passed through a sieving system where the starch-rich fraction with ~10% protein and the bran-rich fraction with ~17% protein were obtained. The starch-rich fraction was then enriched through novel electrostatic separation approach utilizing a lab-scale pneumatic tribo-electrostatic separator consisted of a fluidized bed, tribo-charging tube, and a fractionation chamber with two oppositely charged electrodes and a high voltage supply. In this approach, oat flour was suspended in the air stream inside the fluidized bed and then pneumatically passed through a tribo-charger tube where protein particles and starch granules were selectively charged to different levels and separated based on their acquired charges in the fractionation chamber under the influence of an electric field. We have systematically studied the chargeability of protein and starch particles in the air stream and their separation behavior as a function of air flow rate and electric field strength. Optimum sieving and electrostatic separation conditions resulted in the production of protein-rich fraction with ~20% protein accounting for ~35% of the total protein of the starch-rich fraction. This novel sustainable technique not only resulted in rapid simultaneous separation of protein and starch particles, but also produced protein-enriched fractions with preserved native bio-functionality.

537242 Investigation of *Shewanella Oneidensis* MR-1 and Community 31 for Microbial Reduction of Iodate at the Hanford Site

Abstract:

Investigation of *Shewanella Oneidensis* MR-1 and Community 31 for Microbial Reduction of Iodate at the Hanford Site

Radioiodine-129 (I^{129}) from leaking nuclear waste storage tanks is one of the major contaminants of groundwater at the Hanford Site in Washington State. Speciation of I^{129} in the contaminated groundwater is predominantly as iodate, organo-iodide and iodide. A forthcoming bioremediation solution is to use microbial dissimilatory reduction of iodate to iodide. *Shewanella oneidensis* MR-1, a strain of bacteria known to reduce silver, lead and uranium, was investigated in this study for its potential to serve as a bioremediation agent against iodate contaminants. For this purpose, MR1 was grown in minimal media in the presence of iodate under both aerobic and anaerobic conditions for 24 hours. During each experiment, the amount of iodate present in the growth medium was monitored by colorimetric assays with UV Spectrophotometry. Data suggest that, within 24 hours, MR-1 could reduce iodate content in the medium by 33.8% in aerobic conditions, whereas no reduction was observed in anaerobic conditions. Additionally, iodate reduction experiments using a community of bacteria collected from groundwater at the Hanford Site, named Community 31, suggest that there may be local bacteria capable of reducing iodate. Similar experiments as those carried out with MR-1 were conducted using Community 31. The results showed that the community may be capable of reducing iodate by 45.8% in aerobic conditions. No reduction was observed in anaerobic conditions.

Future work would explore MR-1â€™s reduction capabilities in concert with Community 31 in suspension as well as in biofilm cultures. Findings from these experiments could lead to the development of novel methods for in situ bioremediation of iodate and to the design of novel bioreactors for the bioremediation of soils and sediments.

525048 Spatial Confined Capillary Flow and Precisely Controlled Crystallization Via 3D Printed Platform: A Comprehensive View

Abstract:

Spatial Confined Capillary Flow and Precisely Controlled Crystallization via 3D Printed Platform: A Comprehensive View

Mingguang Han, Gaohong He, Xiaobin Jiang*

State Key Laboratory of Fine Chemicals, Research and Development Center of Membrane Science and Technology, School of Chemical Engineering, Dalian University of Technology, Dalian, P. R. China

*Corresponding author: Email: xbjiang@dlut.edu.cn

Abstract

Precision crystallization control is crucial problem for crystallization progress design and industrial crystallization optimization in crystallizer designing, which was widely applied in pharmacy. Pacing the development of high precision manufacture method, micro-level or even nano level microfluidic chip and stable microwell array chip was developed and applied in recent state of the art crystallization processing research[1-4]. This chips-based crystallizer provides manageable environment control and repeatable crystal formation or nucleation results rapidly and accurately.

Here, we developed material-free micro pillar array chip as crystallizer utilizing high precise 3D printer (Digital Light Process (DLP)) with systematic dripping module (shown in Fig. 1) and dripped crystal solution drops onto the chip with different drop volume and different position in mild evaporation surrounding condition. We also found diversified initial contact angles when dripping different volume of drops, which possess different liquid-air interface initially and influence capillary flow inside system impactful for crystal formation (shown in Fig. 2).

For sessile drop, to understand evaporation condition and crystal formed position, concentration gradient, evaporation time and distance between formation and pillar center was encountered as variables to measure concentration change, vapor diffusion flux and nucleation energy barrier, and indicated the relationship of crystal form and capillary micro flow (shown in Fig. 3); For cover drops, we raised inside capillary micro flow by coffee ring inspiration to explain coffee-ring disappearance and guide the stable diffusion crystal formation in concave structure with observation of crystal form (shown in Fig. 3); For sitting drops and missing drip, some irregular and unexpected crystal form appeared without explanation, which need to be avoided. This method was well popularized in commercial crystal compounds with facile management and modification for high throughout crystal production. Meanwhile, lower concentration or even, polymorph system was probably suitable in crystallization process.

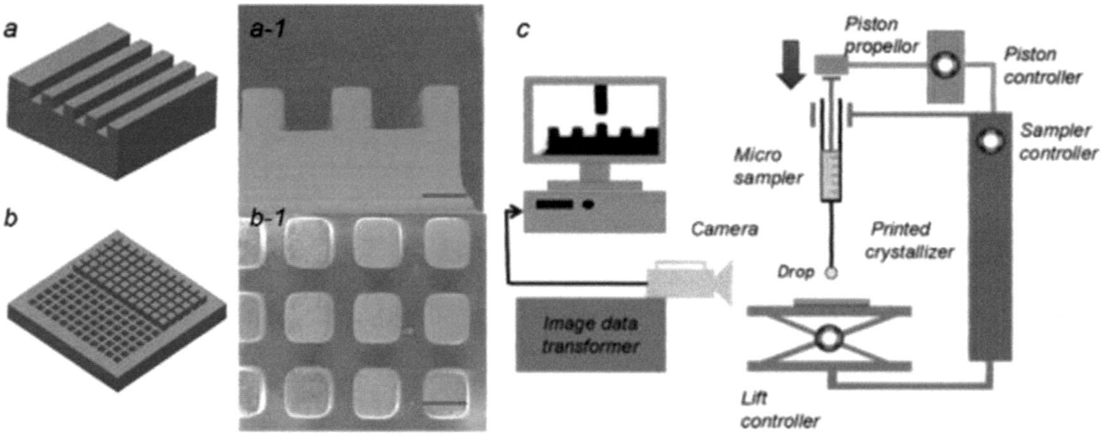

Fig 1: (a) The closed-up review of tunnel concave structure and (b) lattice concave structure of designed crystallizer. (a-1) SEM image of the cross section of (a). (b-1) SEM image of lattice concave from (b) (scale of the bar, 500 ¦Ìm). (c) The process of dripping drop system.

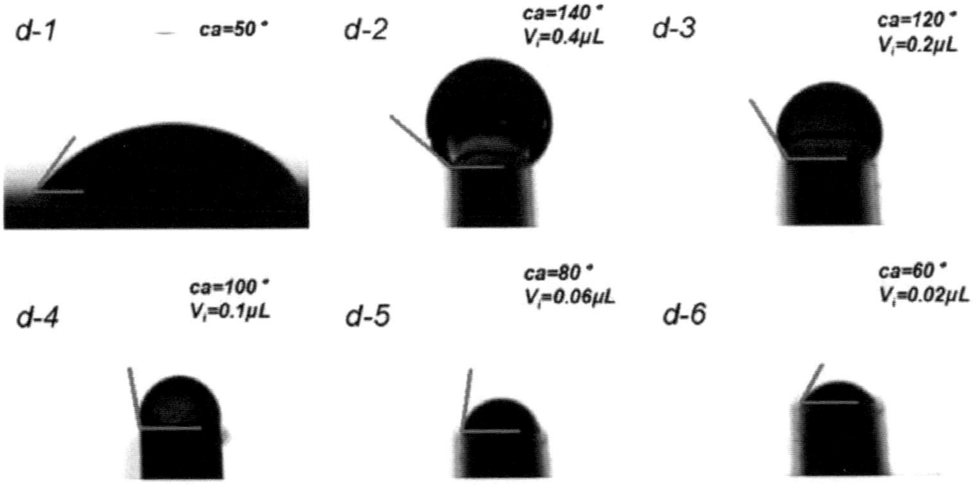

Fig 2: (d-1): The hydrophilic property of commercial 3D material shown in 50¡ãintrinsic angle. (d-2)~(d-6): gradient contact angle results when dripping numerous volume liquid drops on concave structure

Fig 3: The relationship of crystal form and capillary micro flow in sessile drop condition and cover drop condition

Acknowledgment

We acknowledge financial contribution from National Natural Science Foundation of China (Grant No. 21527812, 21676043, U1663223, 21606035), Changjiang Scholars Program (T2012049), the Fundamental Research Funds for the Central Universities (DUT16TD19, DUT17ZD203) and Education Department of the Liaoning Province of China (No. LT2015007).

References

1. Duffy DC, C MJ, A SOJ. Rapid Prototyping of Microfluidic Systems in Poly(dimethylsiloxane). *Anal. Chem.* . 1998;70:4974-4984.

2. Leng J, Salmon JB. Microfluidic crystallization. *Lab Chip.* Jan 7 2009;9(1):24-34.

3. Sultana M, Jensen KF. Microfluidic Continuous Seeded Crystallization: Extraction of Growth Kinetics and Impact of Impurity on Morphology. *Crystal Growth & Design.* 2012;12(12):6260-6266.

4. B Z, S RL, F IR. Screening of protein crystallization conditions on a microfluidic chip using nanoliter-size droplets. *Journal of the American chemical society.* 2003;125(37):11170-11171.

5. Bourgks-Monnier C, M. E. R. S. Influence of Evaporation on Contact Angle. *Langmuir.* 1995;11:2820-2829.

6. Shanahan MER, Bourg C. Effects of evaporation on contact angles on polymer surfaces. *INT. J. ADHESION AND ADHESIVES.* 1994;14(3):201-205.

7. Deegan RD, Bakajin O, Dupont TF, Huber G, Nagel SR, Witten TA. Capillary flow as the cause of ring stains from dried liquid drops. *Nature.* 1997;389(6653): 827-829

8. Cira NJ, Benusiglio A, Prakash M. Vapour-mediated sensing and motility in two-component droplets. *Nature.* Mar 26 2015;519(7544):446-450.

9. Mersmann A, ed *Crystallization technology handbook.* New York and Basel: Marcel Dekker, Inc.; 2001.

527642 Polymorph Screening of L-Glutamic Acid By Anti-Solvent Crystallization in Easy-to-Use Microfluidic Device

Abstract:

Most pharmaceuticals have polymorphs. Since different polymorphs exhibit distinct physical properties, such as solubility, bioavailability, density, mechanical properties, color and morphology, which result in difficulties in the downstream processing, including separation, handing and so on. Hence, it is crucial to control the crystallization of targeted polymorph. As an amino acid commonly used in pharmaceuticals and food, L-glutamic acid has two polymorphs: prismatic or granular shaped metastable Î± form and four different shaped stable Î² form, including need-, rod-, lozenge- and plate-shaped. In practical applications, Î± form has higher utilization efficiency. In previous work, to obtain a specific polymorph form of L-glutamic acid, the commonly adopted methods includes control of larger crystallization temperature difference, in quiescent and flow state through membrane crystallizers, adding amino acid additives, sonocrystallization and so on.

In this work, based on anti-solvent crystallization, solutions and antisolvent were loaded into separate syringes which were connected to channels. By varying the initial concentrations of solutions, temperature, and velocities of solution and antisolvent, the resulted crystals were flitered directly, and dried in an air oven. Then crystals were characterized by SEM and PXRD for morphology evaluation and form confirmation. The polymorph selectivity of L-glutamic acid was investigated. This method was simple in operation and low consumption was required. And the crystallization temperature was kept constant throughout the process. The results showed that this method was more flexible in the selectivity of L-glutamic acid polymorphs without other assistive technologies or additives which might introduce impurities.

533156 Thermodynamics of Co-Crystal Systems

Abstract:

Thermodynamics of Co-crystal Systems

Dipali. Ahuja,[1] and Åke. C. Rasmuson[1]

[1]Synthesis and Solid State Pharmaceutical Centre, Bernal Institute, Department of Chemical Sciences, University of Limerick, Limerick, Ireland

Introduction.

The pharmaceutical industry has a significant interest in co-crystals for their ability to alter the physical properties of the active pharmaceutical ingredient (API) without affecting its chemical structure ***[1]***. Co-crystallization may enhance a wide range of pharmaceutical properties like hygroscopicity, thermal stability, solubility and bioavailability. Most research in this field has been focused towards discovering new co-crystals. Much less work has been done on studying the co-crystal thermodynamic properties and manufacturing techniques. Ternary phase diagrams are crucial for the design of the manufacturing process ***[2]*** as they highlight the stability regions for various solid crystalline phases. Factors like solvent and temperature significantly affect the shape of the phase diagram and the solubility differences between the co-crystal components can cause a system to be congruent or incongruent ***[3]***. It is most likely that a large solubility difference causes a system to show incongruent dissolution. The main interest in the ternary phase diagram is the size of the region where co-crystal is the most stable solid phase.

Methods and Experiments.

For this work, thermodynamic studies for 1:1 sulfamethazine (SMT)-salicylic acid (SA) co-crystal system were carried out. SMT is a sulfonamide drug and is used an antibacterial and anti-infective agent; SA is used to relieve pain and fever and is used for anti-acne treatments. Ternary phase diagrams were constructed in three solvent systems (methanol, acetonitrile and DMSO-methanol) at three temperatures. The methodology behind the phase diagram construction involved determination of the pure component solubilities by gravimetric technique. The invariant points were determined by mixing the solid and solvent and allowing sufficient time for equilibration. Solids were then analysed by PXRD (Powder X-ray Diffraction) or DSC (Differential Scanning Calorimetry) and the solution composition was determined by HPLC (High Performance Liquid Chromatography).

Results and Discussion.

Solvent dependent congruent and incongruent dissolution was observed, highlighting how solvent choice can affect the phase diagram. The co-crystal system showed congruent dissolution in acetonitrile but showed an incongruent behavior in methanol and DMSO-methanol. The effect of temperature over the range studied was found to be weak. Based on the solubility data for the pure components and the co-crystal, the Gibb's free energy of co-crystal formation was determined, and ranged from -5.5 to -7.7 kJ/mol, changing with temperature. The negative values reveal the thermodynamic stability of the co-crystal and spontaneity of the co-crystal formation from SMT and SA. The enthalpy and entropy associated with co-crystallization were determined to be +23 kJ/mol and +0.1 kJ/mol, , respectively. From this data, co-crystal formation seems to be an entropy driven process. A relationship between the solubility ratio between API and coformer and the size of the region where the co-crystal is the most stable phase has been observed. In addition, it was found that the solubility ratio between the pure components does not allow for confident prediction of whether a system will show congruent or incongruent behaviour. The co-crystal yields and volumetric productivities in the three solvent systems were determined using the slurry crystallization technique.

References:

[1] J. Urbanus, C. P. M. Roelands, D. Verdoes, P.J. Jansens, J.H. ter Horst, Co-Crystallization as a Separation Technology: Controlling Product Concentrations by Co-crystals, Cryst. Growth Des., 10, 1171 (2010).

[2] D.M. Croker, M. E. Foreman, B. N. Hogan, N. M. Maguire, C.J. Elcoate, B. K. Hodnett, A.R. Maguire, Å. Rasmuson, Understanding the p-Toluenesulfonamide/Triphenylphosphine Oxide Crystal Chemistry: A New 1:1 Cocrystal and Ternary Phase Diagram, Cryst. Growth Des., 12, 869 (2012).

[3] S. Zhang, Å. Rasmuson, Cryst. Thermodynamics and Crystallization of the Theophylline–Glutaric Acid Cocrystal, Cryst. Growth Des., 13, 1153 (2013).

532935 Membrane Mineral Scaling in Semi-Batch and Steady State Reverse Osmosis Desalination - a Comparative Study

Abstract:

RO membrane desalination has proven to be a robust technology for desalting both brackish water and seawater. However, despite the popularity of RO technology, membrane mineral scaling remains a major impediment to achieving high recovery, particularly in brackish water desalination, and also leads to increasing operational cost (due to required membrane cleaning cleaning) and shortening membrane lifetime. Recent studies have proposed the use of semi-batch RO as an alternative operational approach that could enable high recovery operation at reduced level of antiscalant dosing which would provide more effective means of mitigating mineral scaling. In semi-batch RO operation, continuous concentrate recycling is adopted in a filtration period to enhance the system water recovery. During the filtration period salt concentration (including the concentration of sparingly soluble salts) rapidly increases in the feed flow channel as salt accumulates in the feed flow channel. Once the osmotic pressure in the feed channel reaches the upper allowed limit, the concentrate holdup is discharged and can be flushed with the raw-feed water. Such intermittent flushing after each filtration period has been said to reduce membrane mineral scaling by resetting the nucleation induction period (or â€œinduction clockâ€) for the onset of mineral scaling on the membrane surface. Current studies have argued that the above operational mode would lead to more effective RO operation with respect to scale mitigation relative to conventional steady state RO. In order to assess the above hypothesis, an investigation was undertaken to evaluate the impact of repeated filtration-flushing cycles in semi-batch RO operation on mineral scaling. In order to unambiguously assess mineral scaling under semi-batch operation, a direct real-time membrane surface imaging system was developed to detect the onset of membrane scaling, with gypsum (calcium sulfate dihydride) selected as the model mineral scalant. The surface monitoring system was interfaced with a spiral-wound pilot scale RO system, capable of achieving up to 90% water recovery. RO desalination tests were then carried out in both semi-batch and steady-state operational modes to compare the above two operational modes with respect to resetting of the scaling induction period and quantify mineral scale kinetics at various water recoveries. Real-time membrane surface image analysis enabled quantification of gypsum scaling kinetics in terms of both the number density of gypsum crystals and the extent of membrane surface scale coverage. The above information enabled evaluation of: (a) impact of intermittent raw-feed water flushing for semi-batch RO operation on the formation of nuclei and rate of crystal growth, (b) effect of convective residence time in the RO system on the scale induction period, (c) mineral scaling at different levels of raw feed supersaturation with respect to gypsum, and (d) mineral scaling at semi-batch RO ad steady state RO at comparable levels of mineral saturation in the RO feed channel. Given the ability for early detection of mineral scaling, via the current direct and real-time membrane surface imaging, the present work provides an unambiguous demonstration that, contrary to hopes raised regarding the superiority of semi-batch RO, mineral scaling can be significantly exacerbated in this operational mode relative to steady state RO operation. However, there are certain limited operational domains in which semi-batch RO may be advantageous. The fundamental results of the current study provide a basis for assessing the recovery limits that are imposed by mineral scaling and possible operational approach to scale mitigation for semi-batch and steady-state RO.

521862 Water Desalination and Purification through Mixed-Matrix Membranes: An Atomistic Simulation Study

Abstract:

Development of technologies for water desalination is critical to cope with the global inadequate supply of clean water. Membrane-based separation is an efficient method to produce fresh water suitable for human consumption and irrigation. In particular, reverse osmosis (RO) has become the most widely used for water desalination. Currently, RO comprises 60% of the capacity of desalination and represents 75-85% new desalination projects worldwide. RO is also significant in purifying polluted water due to its low energy cost and simple operational process, especially in mining areas. In RO process, a selective membrane with high water ï¬,ux is of vital importance to its performance. Compared with conventional polymeric membranes and zeolites, mixed-matrix membranes (MMMs) formed by incorporating organic/inorganic fillers into a polymer matrix have attracted considerable interest since they combine the merits of each type of material.

In this work, an atomistic simulation study is conducted for water desalination and purification through MMMs constructed by PB-1A organic cage and a polymer of intrinsic microporosity (PIM-1). PB-1A is a newly reported analogue of porphyrin box PB-1 with intrinsic porosity. The swelling of the MMMs in water and its effect on water permeation are examined. For desalination of seawater, the MMMs exhibit high water permeability and 100% salt rejection. For purification of wastewater (containing Pb^{2+}, Cu^{2+} and Cd^{2+} ions), high water permeation and ion rejection are also achieved at various conditions. For both applications, the dynamic and structural properties of water molecules in the membranes are investigated. This simulation study provides microscopic insights into water transport in PB-1A/PIM-1 MMMs and suggests their potential use for water desalination and purification.

538824 Metal Oxide Functionalized Graphene Oxide Membranes for Advanced Oxidation of Pollutants

Abstract:

Fast water transport, ease of membrane fabrication, separation on nanoscale, mechanical and thermal stability and stability in solvent media has made Graphene Oxide (GO)-based membranes a promising candidate for nanofiltration. Additionally, their ability to catalyze redox reactions have attracted much attention in environmental remediation of toxic organics. Aim of this study is to combine the ability of GO membranes for nanoscale separations with pursulfate mediated removal of smaller chlorinated organics from water. A novel membrane platform consisting of partially reduced GO and polyacrylic acid was fabricated using layer by layer assembly approach. Iron nanoparticles were embedded in the membrane domain by in-situ synthesis from ferrous ion precursors immobilized on the introduced carboxylic groups of polyacrylic acid. Extensive exclusion studies with various salts and dyes showed that the fabricated membranes retained separation abilities of the conventional GO membranes. Catalytic property of the GO membranes were investigated in a cross flow mode of operations. Extended study over 20 hours showed effective decomposition of persulfate on passing its aqueous solution (500 ppm) through the fabricated membranes for an operating flux of 2 LMH, pH 6.5 and 23°C. Evaluation of effectiveness of the fabricated membranes for removal of trichloroethylene (TCE) under same operating conditions showed near 100% removal of TCE (from 30 ppm feed). The GO-based platform with its capabilities of separation on nanoscale and catalyzing oxidative removal of toxic organics can further help in achieving more efficient water purification. This research is supported by the NIEHS-SRP (Grant no: P42ES007380) and NSF KY EPSCoR (Grant no: 1355438).

535985 Enhanced Biocidal and Antifouling Properties of Thin Film Composite Membranes Via Active Layer Modification with Polyrhodanine Nanoparticles

Abstract:

The demand for fresh water has been increasing and is expected to rise continuously for the foreseeable future. A freshwater scarcity can negatively affect the dynamics of ecosystems[1], food security[2], and so on. Forward osmosis (FO) is a high-water-recovery and low-cost membrane-based technology that has potential applications in desalination and wastewater treatment[3-5]. However, the lack of FO membranes with effective anti-fouling and antibacterial properties is still a challenge[3]. The incorporation of hydrophilic and antimicrobial nanomaterials into the active layers of membranes has been found to be a promising strategy to improve anti-fouling and anti-biofouling properties of nanocomposite membranes[6].

In this work, polyrhodanine (PRh) nanoparticles are synthesized via oxidization polymerization and incorporated into the active layers of thin film composite (TFC) membranes. High compatibility of the PRh with polyamide (PA) chains is corroborated using several analysis methods and molecular dynamics (MD) simulations. TFC-PRh membrane containing 0.01 wt% PRh displays enhanced water flux and sieving ability when tested in a FO process. Compared to the pristine TFC, smoother and more hydrophilic surfaces of TFC-PRh membranes provide more fouling resistance against a sodium alginate solution. Furthermore, outstanding biocidal activity of the TFC-PRh-0.01 membrane decreases bacteria adhesion hindering biofilm formation and leading to an enhancement in membrane anti-biofouling properties.

Keywords: Polyrhodanine, Thin-Film Composite Membrane, Forward Osmosis, Antimicrobial, Antifouling

References:

1. Barkaoui, K.; Navas, M. L.; Roumet, C.; Cruz, P.; Volaire, F., Does water shortage generate water stress? An ecohydrological approach across Mediterranean plant communities. *Functional Ecology* **2017,** *31* (6), 1325-1335.
2. Kummu, M.; Guillaume, J.; De Moel, H.; Eisner, S.; FlÃ¶rke, M.; Porkka, M.; Siebert, S.; Veldkamp, T.; Ward, P., The worldâ€™s road to water scarcity: shortage and stress in the 20th century and pathways towards sustainability. *Scientific reports* **2016,** *6*, 38495.
3. Lutchmiah, K.; Verliefde, A.; Roest, K.; Rietveld, L. C.; Cornelissen, E. R., Forward osmosis for application in wastewater treatment: a review. *Water research* **2014,** *58*, 179-197.
4. Shaffer, D. L.; Yip, N. Y.; Gilron, J.; Elimelech, M., Seawater desalination for agriculture by integrated forward and reverse osmosis: Improved product water quality for potentially less energy. *Journal of membrane science* **2012,** *415*, 1-8.
5. Chung, T.-S.; Zhang, S.; Wang, K. Y.; Su, J.; Ling, M. M., Forward osmosis processes: yesterday, today and tomorrow. *Desalination* **2012,** *287*, 78-81.
6. Zirehpour, A.; Rahimpour, A.; Arabi Shamsabadi, A.; Sharifian Gh, M.; Soroush, M., Mitigation of thin-film composite membrane biofouling via immobilizing nano-sized biocidal reservoirs in the membrane active layer. *Environmental Science & Technology* **2017,** *51* (10), 5511-5522.

538249 Modifying TiO_2 MagnÃ©li Phase Reactive Electrochemical Membranes with Pyrogenic Carbonaceous Materials for Adsorption and Electrochemical Removal of Water Contaminants

Abstract:

This study was focused on the synthesis, characterization, and performance evaluation of Ti_4O_7 reactive electrochemical membranes (REMs) amended with pyrogenic carbonaceous materials (PCMs). Among the Magnéli phases of TiO_2 (Ti_nO_{2n-1} n= 4 to 10), Ti_4O_7 (n=4) is the most conductive phase and produces the highest yield of hydroxyl radicals. Ti_4O_7 possesses an electrical conductivity in the range of 1000-1500 S/m, but has limited adsorption capacity for organic contaminants. In this work, a method was investigated to improve the conductivity and adsorption capacity of Ti_4O_7 REMs and their performance for removal of different classes of water contaminants.

Two types of PCMs were added to REMs, 10 wt% multi-walled carbon nanotube (MWCNT) and 10 wt% powdered activated carbon (PAC). The composite PCM/Ti_4O_7 REMs were synthesized and characterized by a series of techniques (e.g., X-ray diffraction, thermal gravimetry analysis, Raman spectroscopy). Conductivity analysis showed that by adding MWCNTs and PAC, REM conductivity increased 3.3-fold and 2-fold respectively. Preliminary results indicated that a 10 wt% PAC and 10 wt% MWCNT loading to a Ti_4O_7 REM increased the adsorption capacity towards N-nitrosodimethylamine (NDMA)) by > 4.7-fold and > 7.1-fold, respectively. These results were determined by analyzing the time for permeate breakthrough curves to reach 50% of feed concentration.

By employing two different types of REMs, simultaneous chemical reduction/oxidation reactions and adsorption were used to separate and destroy different groups of dissolved water contaminants. For example, results showed for the Ti_4O_7 REM that steady state NDMA reduction was 60% during the three day experiment. By contrast, using 10 wt% PAC/Ti_4O_7 and 10 wt% MWCNT/Ti_4O_7 REMs, NDMA was no longer detected in the permeate during the three day experiment. These results indicated that the PCM/Ti_4O_7 composite REMs under cathodic potentials have potential for long-term removal of contaminants of concern.

Keywords: Separation, Water treatment, Reactive electrochemical membranes, pyrogenic carbonaceous materials.

534972 Quantification of Thermal Energy Delivery to Water-Membrane Interface in Membrane Distillation

Abstract:

Membrane Distillation (MD) is an attractive technology for treatment of high salinity waters (e.g. waste brines, produced water, saline ground waters) that can take advantage of low grade heat or solar thermal energy, reducing its operational costs. However, the MD process performance (i.e. flux, water recovery, and thermal efficiency) which is governed by the heat transfer to the membrane-water interface cannot be directly quantified in experimental systems. Past efforts to quantify energy transport to the water-membrane interface have used thermal transport models that rely on bulk temperature and flow rate measurements, and the inaccuracies from this approach have led to high uncertainty in translating process performance from the bench to commercial scale. Herein, we demonstrate through analysis of literature and experimental data the effect of system size on MD performance, and current limitations in our ability to predict its performance. We present a novel approach that leverages our fundamental understanding of vapor transport to correct existing heat transport correlations and significantly reduce the uncertainty in estimated MD process performance.

532965 Ceramic Membranes for Desalination By a Vacuum Flow-through Evaporation

Abstract:

Purification of salty water is a huge challenge in the present times. Several membranes like GO, zeolite, and GOF membranes have been used for the purification of salt water. But limitation in terms of water flux and ion rejection demands for better methods for water purification. The atomic layer deposition (ALD) process is one type of chemical vapor deposition process based upon surface limiting chemical reaction between precursors and substrate. ALD uses alternating, self-limiting chemical reactions between gaseous precursor and a solid surface to deposit thin films in an atomic layer by layer fashion. In this work, we deposited ZnO, Al_2O_3 and TiO_2 using ALD on macroporous Î±-alumina disk membranes having 1 in. diameter, 1 mm thickness, and 25% porosity for the purification of salt water. We used these membranes for salty water desalination by flow-through membrane evaporation. The interaction between ALD layer and water molecules enhanced the water flux and ion rejection values. A high water flux (~100 L. m^{-2}. h^{-1}) and ion rejection rate (~99%) were obtained. Moreover, performance of ZnO, Al_2O_3 and TiO_2 ALD disk were also compared and it was found ZnO showed better performance than Al_2O_3 and TiO_2

525830 Thermodynamic Properties of Paracetamol Impurities and Their Impact on the Crystallisation of Paracetamol from Solution

Abstract:

The impact of additives and impurities on active pharmaceutical ingredients is an essential yet poorly understood area in crystallization research. Paracetamol, which is one of the most widely used model systems in crystallization, is typically produced with trace amounts of unwanted impurities. However, it remains unclear how the main impurities 4-nitrophenol (NP) and 4â€™chloroacetanilide (CA) affect the solubility and polymorphism of paracetamol.

This work describes the solid- and solution characterisation of the main impurities of paracetamol. Temperature-dependent solid-liquid properties of 4-nitrophenol (NP) and 4â€™chloroacetanilide (CA) were obtained in four different alcohols and their effect as impurities on the crystallisation and solubility of paracetamol (PA) will be presented. HPLC measurements were used to determine the solution stability of these compounds. The solubility of NP appeared to be significantly higher than PA whereas the solubility of CA was lower than PA. The solubility difference between the impurities could be rationalised based on their molecular structure and hydrogen bonding interactions. Despite these significant differences between the impurities, their effect on the solubility of PA is the same, as each impurity slightly increases the solubility of PA. The solubility data was modelled using the empirical modified apelblat equation and thermodynamic models, which include the Margules, van-Laar, Wilson and NRTL models. The quality of fit of the tested thermodynamic models to the experimental data increases in the order Margules < van-Laar < Wilson < NRTL < Apelblat.

XRPD measurements were conducted to establish the polymorhpism of the compounds used in this study. Recrystallisation of PA from solution containing the highly soluble NP impurity resulted in small uniformly sized high purity PA form I crystals whereas the presence of the poorly soluble CA impurity induced the formation of large needle shaped crystals of PA form I. These differences in crystallisation are a consequence of the solubility difference and the different functional groups of PA and its impurities. The effect of the impurities on the nucleation kinetics of paracetamol was investigated using induction time measurements and showed that the nucleation process is inhibited for both impurities.

The presentation will cover experimental and modelling data and describes the effect of impurities on the solubility, crystallization and nucleation kinetics of paracetamol. Overall this work presents fundamental information for the development of crystallisation approaches for the purification of paracetamol from its main impurities.

525983 A Novel Approach into Secondary Nucleation and Crystal Growth, Controlled Particle Size Distribution, Using a Large Single Seed Crystal in Solution Crystallization

Abstract:

Secondary nucleation by means of seed crystals is extensively employed in crystallization process and remained a noteworthy consideration of research in recent years, but its in-depth understanding particularly at molecular level remained elusive until recently, molecular dynamic simulations provided an insight to it, i.e. crystal nuclei breeding. However, seed crystals are typically dispersed in solution, it becomes difficult to decouple nuclei breeding from attrition-enhanced secondary nucleation. In this work, a novel approach was used to experimentally investigate the mechanism of crystal nuclei breeding as an origin of secondary nucleation and its link with fluid shear, using a large single crystal of paracetamol (size = 15mm x 12mm) in propan-2-ol. The crystal was held stationary in the stirring vessel using a specially design crystal holder, to eliminate the possibility of enhanced secondary nucleation due to crystal attrition or breakage during agitation. The experiments were done in a low supersaturation (S_o = 1.0) at stirring speed of 300 and 370 rpm. Particle imaging velocimetry (PIV) was used to quantify the fluid shear rate at each stirring speed using a high-speed camera and laser illuminated sheet. Laser diffraction was used to perceive the onset of secondary nucleation using a focused beam of red laser passing through the vessel which was calibrated to detect the particle size as small as 20 Æžm. Focused beam reflectance measurement (FBRM) was used for real time tracking of particles which displayed a significant increase in particle counts with an increase in shear rate at higher rpm. At a high fluid shear, the MSZW exhibited a decrease in secondary nucleation threshold with a relative supersaturation ratio of 1.17 at 300 rpm and 1.08 at 370 rpm. Below this threshold, a growth only zone was apparent which was validated using the scanning electron microscopy (SEM). PSD data was measured on a volume basis with increasing number of fines at a higher share rate as expected. Since, no crystal attrition and breakage were involved in the onset of secondary nucleation, these results provided experimental evidence that support the hypothesized nuclei breeding mechanism. Further evidence from the SEM images of the seed crystal, showed the formation of alternate layered pattern of the surface growth and the existence of very fine crystallite layers attached to it. It is proposed that the surface of seed crystal served as a catalyst for the pre-existing clusters in the solution to attach and nucleate to form crystallites, which then sheared off by the fluid flow. The experiments presented an excellent repeatability over MSZW and growth only zone at specific shear rate, particularly the revelation of controlled particle size distribution elucidates the significance of fluid shear on the secondary nucleation using a single seed crystal. Controlled particle size distribution is an imperative quality factor in crystallization and the quality of crystalline product. This approach can offer a more precise modelling in process scale-up and manufacturing in industry.

533602 Crystal Growth Kinetics of Salicylamide Investigated Under Different Crystallisation Processes and Also Environmental Conditions

Abstract:

Crystal Growth Kinetics of Salicylamide Investigated under Different Crystallisation Processes and also Environmental Conditions

Crystallisation is a very important technique, especially in the Pharmaceutical Industry, as it is used in purification and recovery of the crystalline active pharmaceutical ingredients (API) [1]. Despite this, crystallisation fundamentals are still insufficiently understood. Efforts have been made by researchers to understand the kinetics' of salicylamide nucleation [2]. This research focuses on improving knowledge of crystal growth process and kinetics of salicylamide by investigating under rotating disk and also seeded desupersaturation agitated tank experiments. Observed the growth of individual crystals as well as populations of crystals. Crystal growth is also studied under different environmental conditions to detect any influence of the solvent, temperature and supersaturation.

The rotating disc experiments uses inverted optical microscope integrated with a camera, combined with a PC with image/ video capture and analysis software to capture micrographs of each crystal before and after growth. Multiple individual seed salicylamide crystals are fixed onto a rotating glass disc, and placed into a supersaturated solution for a growth period of 1 hour and rotated at 200rpm. Crystals are observed in-situ, so allows for the growth of specific faces to be measured and also view overall habit. The agitated tank experiments use a reactor vessel to perform seeded isothermal desupersaturation experiments. The concentration decay is recorded with a FTIR probe and first principles data treatment is applied to extract the concentration information from IR spectra; besides, FBRM is used for detection of nucleation during crystal growth. In the agitated tank experiments the average growth rate of a certain mass of seeds (better statistics) can be determined at hydrodynamic conditions closer to an industrial process.

The rotating disk crystal growth of salicylamide crystals at 15°C was measured as a function of relative supersaturation (S-1) in the range from 0.01 – 0.04, and in four different solvents (acetone AC, acetonitrile MeCN, ethyl acetate EA and methanol MeOH), rotating the disc at 200rpm. Different rotation speeds where tested in the range of 50 - 350rpm, and 200 rpm was found to be sufficient to essentially eliminate the film mass transfer resistance. The growth of a minimum of 25 crystals was determined for each system. The growth rates of the length and width of each individual crystal was measured using the maximum and minimum feret diameter respectively, also growth of the (200) face was measured. The studies revealed that salicylamide crystals grown from seed in different solvents, at different supersaturations and also within each solvent receive different shapes. There is a significant spread in the crystal growth rates obtained within each system, e.g. 0.05 μm/s up to 0.55 μm/s in one system; this is due to the crystal quality, where lower quality results in increased growth. The average growth rates increase with increasing supersaturation in each solvent. The average individual crystal growth rate behaviour, as a function of solvent and supersaturation, is similar for length and width. The average crystal growth rate of the (200) face of individual seed crystals decreases in the order MeOH>MeCN>AC>EA at supersaturation of 0.03. Salicylamide crystal surface was examined before and after the growth process using scanning electron microscopy; noticed growth steps and layers clearly present after growth indicating rough interfacial growth mechanism. The agitated tank experiments were also performed in all four solvents (AC, MeCN, EA and MeOH) and at 4 temperatures (10, 15, 20, 25°C). The empirical power law equation, spiral growth mechanism (BCF) equation and the birth & spread mechanism (B+S) equation were fitted to the data. In all four solvents, the growth rate increases with temperature as expected. The average crystal growth rate decreases in the order AC>MeOH>MeCN>EA at supersaturation of 0.03. The interfacial energy values for crystal growth depend on solvent, and where obtained by fitting the B+S model; the order is EA>AC>MeCN>MeOH. Due to the fact that low interfacial energy tends to lead to high crystal growth rate, which is in agreement with the result from the rotating disc experiments. The growth rate is faster in the agitated tank experiments as compared to rotating disc which possibly reflects an influence of mass transfer resistance in the latter.

In all growth processes, salicylamide was found to have the slowest growth kinetics in ethyl acetate. Molecular modelling was performed which high lightened the strong hydrogen bond propensity of ethyl acetate and due to its high molecular weight results in slow growth. These crystallisation studies with salicylamide in different solvents, supersaturations and temperatures have shown that crystal growth rates are affected significantly by the use of different environmental conditions.

Reference:

1. Lacmann, R., ***Crystallization, Third Edition. J. W. MULLIN, Butterworth-Heinemann, Oxford 1997, 527 Seiten, zahlr. Abb. und ISBN 0-7506-3759-5.*** Chemie Ingenieur Technik, 1998. **70**(11): p. 1468-1468.
2. Nordstrom, F.L., M. Svard, and A.C. Rasmuson, ***Primary nucleation of salicylamide: the influence of process conditions and solvent on the metastable zone width.*** CrystEngComm, 2013. **15**(36): p. 7285-7297.

539930 Fast Temperature Cycling via Microwave Heating Enables Enhanced Deracemization

Abstract:

Fast Temperature Cycling via Microwave Heating Enables Enhanced Deracemization

Christos Xiouras[1], Fabio Cameli[1], and Georgios D. Stefanidis[1]

[1] Process Engineering for Sustainable Systems, Department of Chemical Engineering, KU Leuven, Belgium;

Keywords: Deracemization, Crystallization, Temperature Cycling, Viedma ripening; Microwave Heating;

In 2005, Viedma [1] demonstrated a remarkable non-classical crystallization phenomenon according to which an initially racemic crystal population converts to a single chirality upon grinding the crystals in their saturated solution. The process was initially demonstrated for the achiral $NaClO_3$ which crystallizes as chiral crystals, but was later extended to intrinsically chiral compounds that crystallize as conglomerates (*i.e.* each individual crystal contains a single enantiomer) [2]. The underlying mechanisms of the phenomenon are still under discussion, but the process is considered to have high potential for application in the pharmaceutical industry, where separation of enantiomers is often required. However, the application of grinding at a larger industrial scale is rather complicated and could lead to problems in downstream processing. Recently, it was discovered that grinding could be replaced by periodic temperature fluctuations (temperature cycling), which is more convenient at a larger scale [3].

In contrast to conventional heating, which is slow and requires a heat transfer surface, microwave heating is rapid and volumetric in nature. Therefore, if combined with fast cooling, microwave activation can enable novel process operating windows for temperature cycling as compared to conventional heating. Figure 1 (left) shows that the application of such operation for the model system glutamic acid leads to an order of magnitude faster heating and cooling rates compared to conventional operation, which in turn leads to significant reduction in the total deracemization time required (Figure 2, right). Besides intensification of the deracemization process, the extreme operating windows applied here also lead to new insights into the obscure driving mechanisms of solid-state deracemization phenomena.

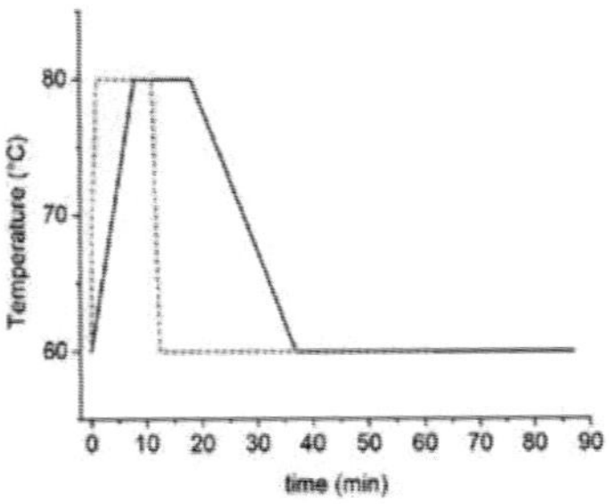

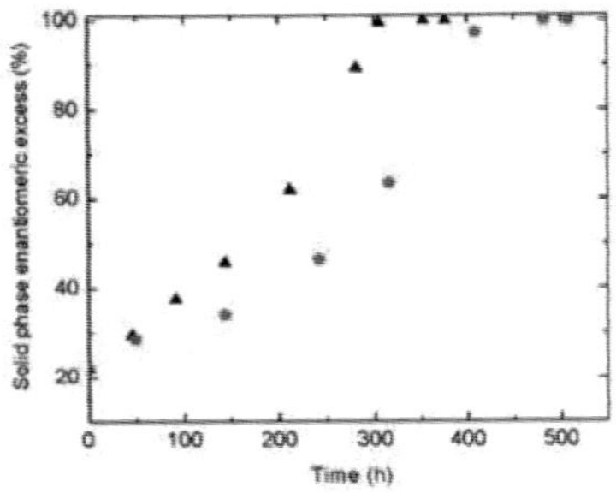

Figure 1. Left: Temperature profile using microwave heating (red dashed line) and conventional heating (black line). Right: Evolution of solid phase enantiomeric excess for in time in: microwave reactor (black points) and conventional heating configuration (red points). Model system: glutamic acid conglomerate crystals in acetic acid and catalytic salicylaldehyde.

References

[1] C. Viedma; Chiral Symmetry Breaking During Crystallization: Complete Chiral Purity Induced by Nonlinear Autocatalysis and Recycling *Phys. Rev. Lett.* 94 (2005), 3–6

[2] W.L. Noorduin, T. Izumi, A. Millemaggi, M. Leeman, H. Meekes, W.J.P. Van Enckevort, R.M. Kellogg, B. Kaptein, E. Vlieg, D.G. Blackmond, Emergence of a single solid chiral state from a nearly racemic amino acid derivative, *J. Am. Chem. Soc.*, 130 (2008), 1158-1159.

[3] K. Suwannasang, A. E. Flood., C. Rougeot, & G. Coquerel. Using programmed heating–cooling cycles with racemization in solution for complete symmetry breaking of a conglomerate forming system. *Cryst. Growth Des.*, 13 (2013), 3498-3504.

530887 Resolution of Conglomerates Using Preferential Crystallization and Enzymatic Racemization

Abstract:

Producing pure enantiomers is of great interest to pharmaceutical and fine chemicals industries, since these molecules interact differently with living organisms. Preferential crystallization (PC) is an important resolution technique to purify enantiomers. Nonetheless the separation can only reach maximum 50% yield. The addition of racemization of the non-desired enantiomer increases the theoretical process yield to 100%. Free and immobilized enzymes are attractive catalysts for this recycling step coupled with chiral separation processes. In this contribution, the separation of a model conglomerate forming system was investigated experimentally and the results were evaluated using a shortcut mathematical model for the coupling of enzymatic racemization with PC.

The amino acid racemase used was overproduced in bacteria and purified using affinity chromatography. The kinetic properties of the enzyme were investigated in preliminary studies. Low enantiomeric excesses produced by PC generate low driving forces for the racemization. This causes severe requirements related to the enzyme dosage. The optimal immobilization conditions were established to enable reuse of the catalyst and its easy separation from the products. Besides performing systematic experiments, a simple consistent mathematical model was developed assuming a lumped crystallization mechanism. The effective growth rate and the racemase kinetics were used to simulate the process combination. The performance of the biocatalyst was evaluated for the proposed application based on the experimental and simulated results.

Paper Number: 528336

ALKALINE CRYSTALLIZATION OF $CaCO_3$ IN A DIRECT AIR CAPTURE PROCESS

Caroline E. Giacomin[1], Thomas Holm[2], Luisa Burhenne[2], and Walter Mérida[2]

1) Chemical and Biological Engineering, University of British Columbia, Vancouver, BC, Canada
2) Clean Energy Research Centre, University of British Columbia, Vancouver, BC, Canada

Abstract

Crystallization of $CaCO_3$(s) from solution in an alkaline environment was studied as a function of temperature, $CaCO_3$ pellet loading, and $CaCO_3$ pellet size. Occurrence of crystallization as spontaneous nucleation or growth on pellets are both possible where growth is the desired outcome. Growth is quantified as retention. The crystallization on pellets was found to be mass transport controlled, while spontaneous nucleation was found to be kinetically controlled. The highest reported retention, 73%, was found at the conditions where mass transport was favored by high pellet loading and small pellet diameter, and kinetics were hindered by low temperature.

Introduction

Crystallization of $CaCO_3$ from alkaline solution is relevant for a process capturing CO_2 directly from air. Recently, the economic feasibility of this process was assessed, and it was found to be possible at a cost of less than $100/ton$CO_2$ [1]. A sketch of the overall industrial process is shown in Figure 1, and the pelletizer (central in Figure 1) is a fluidized bed reactor that precipitates dissolved carbonate as $CaCO_3$(s). Precipitation can occur either on the solid $CaCO_3$ pellets or spontaneously in solution, where spontaneously nucleated particles are significantly smaller than the pellets and are referred to as fines. Precipitation on pellets is the desired outcome. Both forms of precipitation have the same driving force, the supersaturation, F_d, and this is defined in Equation (1) [2]:

$$F_d = [\mathrm{Ca}^{2+}][\mathrm{CO}_3^{2-}] - \frac{K_{sp}}{f_D^2} \qquad (1)$$

where, $[\mathrm{Ca}^{2+}]$ and $[\mathrm{CO}_3^{2-}]$ are the concentrations of calcium and carbonate in solution, K_{sp} is the thermodynamic solubility product, and f_D is the activity factor for divalent ions [3].

Crystallization of $CaCO_3$ has been studied in the water softening industry, where addition of dissolved carbonate is used to reduce the calcium content of water by precipitation on $CaCO_3$ pellets [3][4][5][6][7]. Tai et al. found that increased temperature, supersaturation, and pH increase the spontaneous nucleation rate and reduce pellet growth [4]. In works where the effect of surface area was studied, seed pellets with diameters between 0.6 and 4 μm were used, and it was found that the total available surface area significantly impacts calcium removal performance whereas pellet size showed no impact [6]. Also, Ševčík et al. used beaker experiments to show that increasing the probability of contacting a solid particle, via higher stirring velocities, increases pellet growth [5]. Therefore, conditions expected to be favorable for pellet growth are large total surface area, low temperature, low supersaturation, high stirring rate, and low pH.

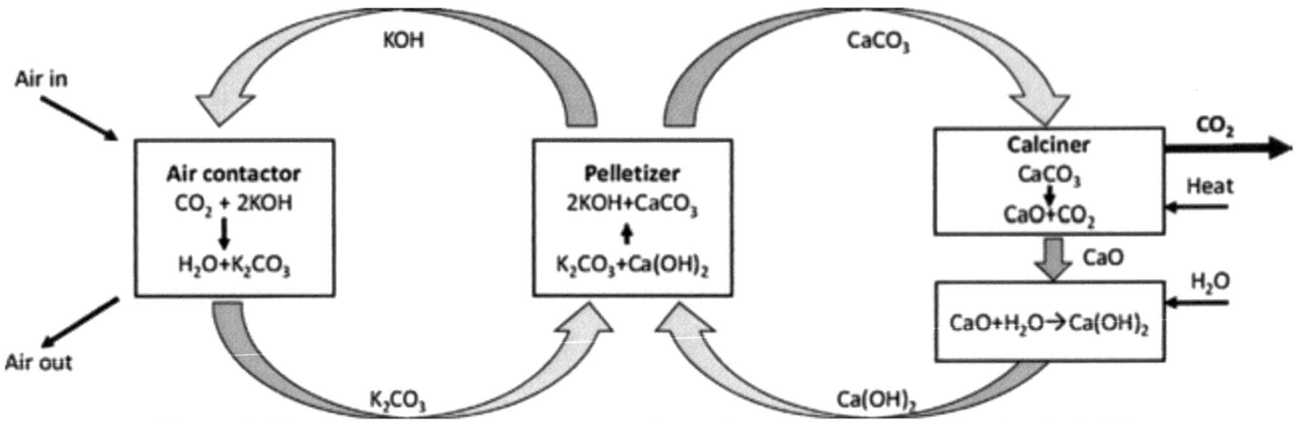

Figure 1: The direct air capture process and reactions of Carbon Engineering Ltd. [8].

The crystal growth on solid pellets was successfully modelled by van Schagen et al. [9] and the equation for the crystallization rate, C, is given in Equation (2):

$$C = \kappa A_s\left([\mathrm{Ca}^{2+}][\mathrm{CO}_3^{2-}] - \frac{K_{sp}}{f_D^2}\right) \quad (2)$$

where, κ is the rate constant, and A_s is the surface area. The reaction was described as a two-stage process where the first stage is the transportation of ions to the solid surface and the second stage is the crystallization on the pellets. Depending on the local conditions in the reaction vessel, the pellet growth reaction is limited by either mass transport, kinetics, or a combination of the two.

In contrast to the water softening process, the reaction studied in this work is operating at high pH (between 14-14.6), high carbonate concentration, and high supersaturation due to calcium being fed as solid $Ca(OH)_2$ particles. In addition, the pellets are large, more than 150 μm in diameter [8]. All these factors are expected to increase production of fines. However, high retention, of approximately 85%, has been demonstrated with high loading of $CaCO_3$ pellets even at high supersaturations [1][8]. Therefore, it is interesting to study pellet growth in a controlled environment to understand how the experimental parameters influence the retention.

In this work, $CaCO_3$ pellet growth was performed in beaker experiments controlled for pellet size, mass loading, total pellet surface area, and temperature in order to understand their effect on the overall retention. The most favorable conditions for high retention were for experiments with the smallest pellet size, highest mass loading and lowest temperature. Total pellet surface area had a limited effect on the retention. The experimental results suggest that crystallization on pellets is mass transport controlled, while spontaneous nucleation is kinetically controlled, and guides directions for further optimization of the direct air capture process.

Experimental Procedure

The reaction consists of three components: process solution, seed pellets, and lime slurry. The process solution used in these beaker experiments is a 1.0 M KOH (Sigma Aldrich 99% 45wt% solution) and 0.5 M K_2CO_3 (Sigma Aldrich 99% reagent grade) aqueous solution. The seed pellets consist of $CaCO_3$ and are sieved to yield a uniform size sample (see Table 1) and washed with deionized water to ensure all fine powder is removed (see Figure 2). Sieving is done using a Fritsch Analysette 3 PRO to shake approximately 150-200 g of pellets at an amplitude of 1.0 mm for 30 minutes. The lime slurry consists of a 2.0 M KOH solution that has $Ca(OH)_2$ powder (Sigma Aldrich, 95+%, A.C.S.) added to comprise 20wt% of the final slurry.

Table 1: These size groups represent the most common pellet sizes found within a bench scale reactor post-operation [8]. Seed is the average size fed; mid-size is about halfway grown and makes up the predominate size range in the reactor; and mature pellets are fully grown and ready for removal.

	Mesh	Diameter range [μm]	BET surface area [m^2/g]
Seed	60 - 50	250 – 300	39.7
Mid-size	45 - 40	355 - 425	27.6
Mature	25 - 14	700 - 1410	24.5

Figure 2: SEM images of mid-size pellets (a) and their surface (b) after being sieved to uniform diameters and washed of residual powder.

First, 250 mL of process solution is added to a 500 mL Pyrex beaker. The beaker was stirred using an overhead mixer (JJ-1 Precise Strength Power Mixer) with a two-bladed stainless steel impeller. A measured mass of calcium carbonate pellets is added from one size group listed in Table 1. Stirring is constant and sufficiently fluidizes the sample for all pellet sizes and loadings.

An aliquot of approximately 6.0 g of lime slurry was added. The precise mass and volume of lime slurry is recorded. After 30 minutes the beaker contents were poured through a wet sieving device consisting of a size 70 mesh (210μm) encased in PVC pipe. Effluent from the mesh was vacuum filtrated in a 9.0 cm Buchner funnel with a type 1 filter paper from Whatman. Solids found in the filter paper are the fines. By looking at the mid-sized pellets (see Figure 2) using a Nova NanoSEM their irregular shape and porous surface was observed. Surface area measurements and data for each size group were reported in Table 1.

Pellets and fines recovered are placed in separate glass petri dishes. These samples are dried overnight (at least 12 hours) at 105-110°C. Samples are allowed to cool to room temperature before mass gain is recorded. To account for relative humidity fluctuation, a control 50 g mid-size trial is run on all experiment days.

For temperature dependent trials, a type-K thermocouple attached to National Instruments LabView software was added to the reaction beaker and a Corning PC 420D heat plate was used as a heat source and temperature controller. The process solution was brought to the reaction temperature prior to pellet and slurry addition. Three trials were run for each set of conditions.

Results are reported as retention, R, calculated as in Equation (3):

$$R = \frac{m_{\text{pellets,after}} - m_{\text{pellets,before}}}{m_{\text{pellets,after}} + m_{\text{fines}}} \tag{3}$$

where $m_{\text{pellets,after}}$ is mass of pellets after reaction, $m_{\text{pellets,before}}$ is the original mass of $CaCO_3$ pellets added, and m_{fines} is mass of fines collected after reaction.

After reactions process solution concentrations are analyzed for their ion ratio, $[OH^-]:[CO_3^{2-}]$. SEM images were taken of both the pellets and fines produced. Powder X-ray diffraction (XRD) was also performed on both using a Bruker D8 Advance with a 2θ range of 10 to 90, step size 0.1, and scan time of 0.3s per step.

Results and Discussion

Ambient temperature

For all ambient temperature conditions reported, a positive retention rate was observed and $CaCO_3$ was formed both on pellets and as fines. In these experiments process solution concentration, amount of fed $Ca(OH)_2$ particles, and stirring rate were constant meaning that the supersaturation, F_d, was constant. Therefore, this set of experiments evaluates the influence of mass transport.

For varied mass loading with a constant pellet size (mid-size), Figure 3a, the retention varies from 52% at 25 g loading to 64% at 100 g loading. Higher loadings are expected to increase the surface area of $CaCO_3$ in the beaker and thus lead to higher retention. This trend was confirmed in our experimental data, but the effect is limited as lower loading did not seem to affect the retention. This suggests that other factors, such as pellet size, are more important.

When the pellet size was varied at constant mass loading, Figure 3b, the retention changed from 67% at seed size to 22% at mature size. The change of pellet from seed to mature increases the particle size and decreases the total surface area of $CaCO_3$ resulting in fewer particles present in the system. Both of these effects are expected to reduce the retention rate and therefore explains the trend observed in Figure 3b. In an attempt to deconvolute the two effects, a third set of experiments was done where the total surface area confirmed through BET analysis was kept constant (at 1,215 m^2). The results of these experiments are shown in Figure 3c. Here, it was evident that the pellet size has a large effect, and the retention changes from 58% at seed size to 22% at mature size. Notably, the seed size pellets and mid-size pellets give retentions within one standard deviation of each other.

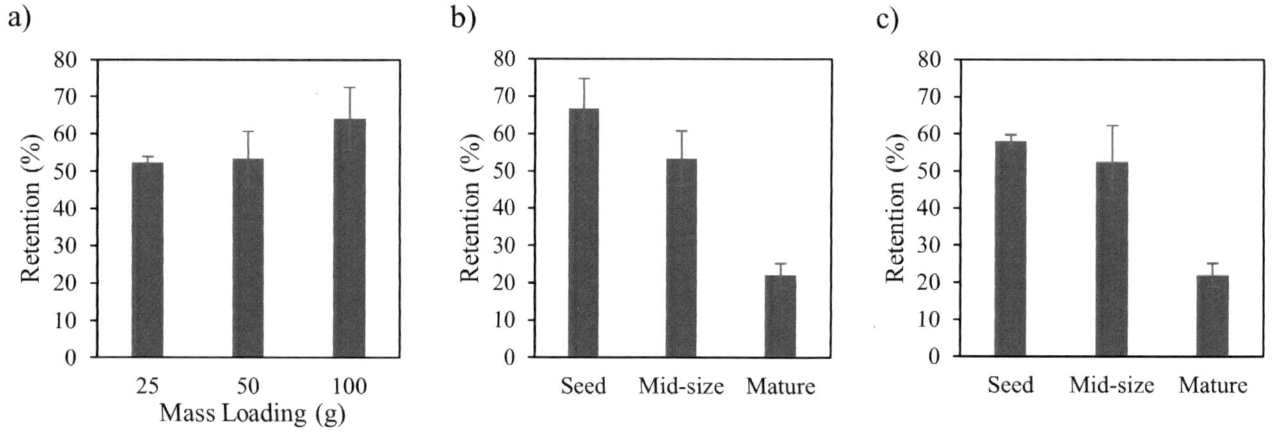

Figure 3: Experiments run at room temperature 20°C with the following variables being varied: a) mid-size pellets at different mass loadings of 25, 50, and 100 grams, b) constant mass loading with varied pellet size, and c) constant surface area with varied pellet size.

In order to understand this result and to separate the effect of pellet size and surface area, the retention rate was plotted as a function of surface area for all results reported in Figure 3. This is shown in Figure 4. Here, the trend in Figure 3c is confirmed where mature pellets give a low retention rate, while the seed and mid-size pellets get comparable retention rates. This can be expected if the actual pellet size is considered (Table 1), where the seed and mid-size pellets are close in diameter while the mature pellets have a diameter approximately 2.5-6 times the seed size and 1.75-4 times the mid-size pellets. From these results, it is evident that while Equation (2) can explain the overall reaction in some cases, care should be taken when interpreting the results solely with this equation. In particular, when $Ca(OH)_2$ is fed as solid particles, as is the case here, the actual supersaturation, F_d, is largely confined to the volume around the solid $Ca(OH)_2$ particles. Thus, to achieve high retention, one must ensure that solid $CaCO_3$ is present near the $Ca(OH)_2$ particles. From the results reported in Figure 3 and Figure 4, high mass loading and small pellet sizes is the best approach. To confirm this, an additional experiment was performed using 100 g of seed pellets resulting in 73% retention. This data is shown in Figure 4 as the rightmost triangle. Overall, this suggests that collisions, or near collisions, between the $Ca(OH)_2$ particles and $CaCO_3$ pellets are key to high retention and this gives a guideline for how to achieve an even higher retention rate at the pH and concentration used in this work.

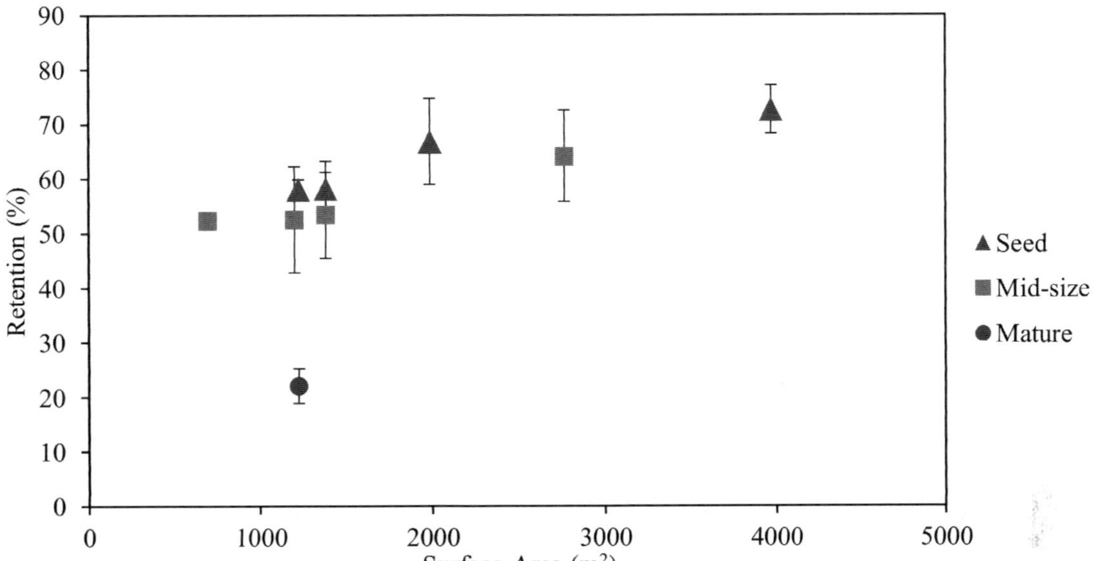

Figure 4: Retention for all trials shown in Figure 3 as well as best conditions experimental set. All shown with respect to surface area as calculated by BET isotherm measurements from the pellet size groups.

Elevated temperatures

Elevated temperature is expected to modify the kinetics of crystallization [3]. With mass transport variables held constant, increased crystallization rate is expected to increase spontaneous nucleation. In earlier work [5][10], at high temperatures, different $CaCO_3$ morphologies were reported and thus temperature can change the precipitation mechanism both for spontaneous nucleation and pellet growth. For example, the aragonite morphology has a needle-like structure that can help reduce the effect of increased temperature on the retention by assuring that a large surface area of $CaCO_3$ is present in the beaker. This morphology was reported at temperatures higher than 50°C [10].

In Figure 5, the retention rate at four temperatures is reported for mid-size pellets. Here, it is evident that the higher temperature gives significantly lower retention going from 53% at 20°C to -1.8% at 80°C. The negative retention at 80°C was within one standard deviation of no retention, and was assumed to be due to experimental uncertainty. If the limiting mechanism for pellet growth is physical proximity of $CaCO_3$ pellets to $Ca(OH)_2$ particles, an increase in temperature is expected to mainly influence the spontaneous nucleation rate. An increased spontaneous nucleation rate is expected to produce fines that are both more numerous and smaller in diameter than a slower spontaneous nucleation rate. In order to investigate this, SEM imaging was done on the fines at the various temperatures and the results are shown in Figure 6. Here, it is evident that the size of the fines decreases as the temperature is increased.

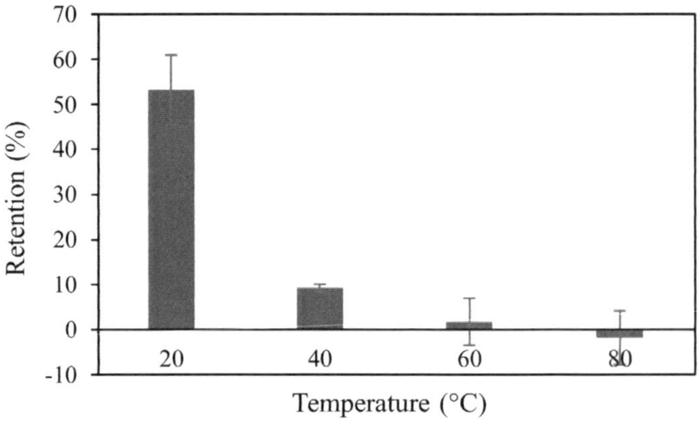

Figure 5: Percent of mass found on pellets over total mass collected after reaction for trials at different temperatures for mid-sized pellets.

Figure 6: $CaCO_3$ fines grown at a) 20°C, b) 40°C, c) 60°C, and d) 80°C.

SEM imaging was also done on the pellets after growth and is shown in Figure 7. No clear trend is visible in these images, and this can be expected as the amount of $CaCO_3$ deposited on a single pellet is relatively small compared to the overall pellet size (mass gain at 50% retention is 0.80 g or only 1.6% of total mass for a loading of 50 g). A long-term experiment might give different results and possibly different $CaCO_3$ morphologies, however that is outside the scope of this work but will be investigated in the future.

Figure 7: Pellet surfaces after reaction at a) 20°C, b) 40°C, c) 60°C, and d) 80°C.

The different $CaCO_3$ morphologies result in different shape of the particles so it may be visible in SEM images. However, if only small amounts of a morphology are present, it is hard to distinguish. Therefore, XRD was run to confirm the results of the SEM images. The XRD results are shown in Figure 8 for the a) pellets and b) fines and compared with reference spectra for three different $CaCO_3$ morphologies (calcite, vaterite, and aragonite) and $Ca(OH)_2$. For the pellets, this could be compared to the unreacted pellets. Here, it is evident that at all temperatures, the pellets are present mainly as calcite. Only a small peak appears in the 80°C results at a 2θ of 61. The peak position is not one of the main peaks for the vaterite or aragonite phases. Therefore, it is assumed that this is an impurity peak and that $CaCO_3$ is only present as calcite on the pellets.

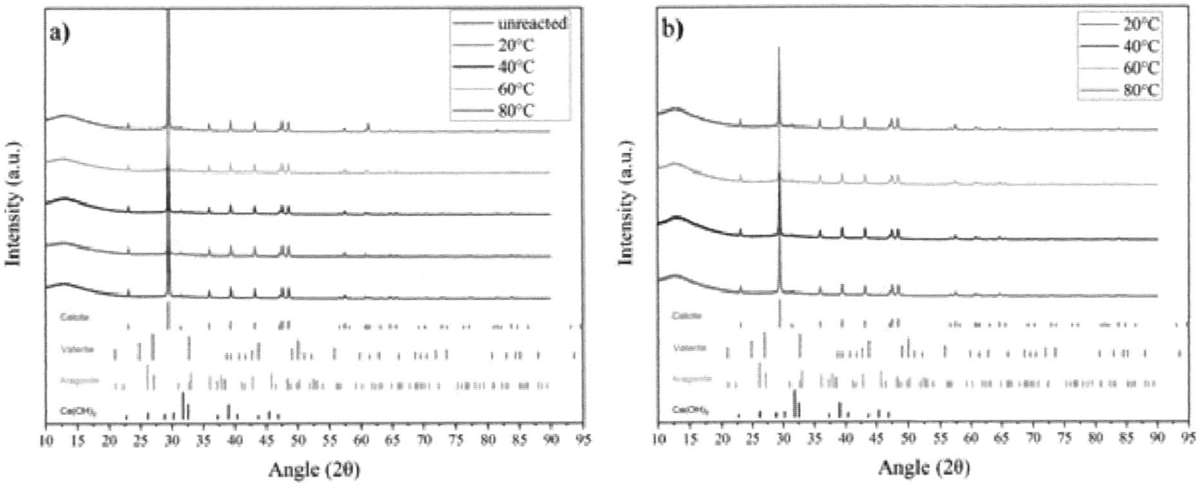

Figure 8: XRD results for (a) pellets, and (b) fines at different temperatures. Reference spectra for calcite (pink), vaterite (green), aragonite (orange), and $Ca(OH)_2$ (blue) are shown at the bottom.

For the fines, in Figure 8b, the XRD results were consistent with calcite at all temperatures. The fines are not seeded and consequently will form the morphology that is most kinetically and thermodynamically favorable. In our results, only calcite is observed, and this suggests that at the high pH used in this work, only calcite will be formed both through spontaneous nucleation and as precipitation on $CaCO_3$ pellets. Therefore, in contrast to work at lower pH values the temperature does not change the morphology of $CaCO_3$ produce, and only modifies the nucleation rate resulting in lower retention.

When results at ambient and elevated temperatures are compared, a framework for achieving high retention can be found. This builds on the understanding that high retention is achieved by improving the conditions for precipitation on $CaCO_3$ particles while hindering the spontaneous nucleation process. These two processes have equal driving force, the supersaturation, F_d. In the conditions used in this work, the supersaturation is high but locally confined around the $Ca(OH)_2$ particles. Thus, increased precipitation on the $CaCO_3$ pellets and resulting higher retention is improved by increasing the chance of a $CaCO_3$ pellet being present close to the $Ca(OH)_2$ particle. This is a pure mass transport constraint. On the other hand, the rate of spontaneous nucleation is increased by increasing the kinetic driving force, demonstrated here by the temperature. High retention is achieved when the mass transport conditions are favorable and the kinetic driving force is low. The highest retention reported in this work is 73% and is achieved for seed size pellets at 100 g loading and at 20°C. A further improvement of the retention can be expected for smaller particle sizes, higher mass loadings of $CaCO_3$ pellets, and lower temperatures.

Conclusion

Crystallization of $CaCO_3(s)$ from solution in an alkaline environment relevant for a direct CO_2 capture from air reactor was studied as a function of temperature, $CaCO_3$ pellet loading, and $CaCO_3$ pellet size. The results were evaluated in terms of retention: the fraction of $CaCO_3$ formed that grows on pellets. Higher mass loading and smaller pellet size resulted in higher retention. Higher temperatures resulted in lower retention. Higher temperatures also were shown to reduce the diameter of fines produced. The crystallization on pellets was found to be mass transport controlled, while spontaneous nucleation was found to be kinetically controlled. The highest retention, 73%, was found at high pellet loading, small pellet diameter, and low temperature.

The fluidized crystallization will next be looked at in a fluidized bed constructed at the lab scale to model the version used by Carbon Engineering Ltd. The work within fluidized bed reactor will provide longer reaction times, allowing larger net mass growth and conditions closer to the industrial process.

Acknowledgement

This work was supported by Carbon Engineering Ltd., the National Sciences and Engineering Research Council of Canada (NSERC), and Mitacs through the Mitacs Accelerate Program. The support of all colleagues and students at the University of British Columbia and Carbon Engineering Ltd. is greatly acknowledged, in particular we thank Jenny McCahill for many discussions pertaining to the direct air capture process of Carbon Engineering Ltd.

References

[1] D. W. Keith, G. Holmes, D. St. Angelo, and K. Heidel, "A Process for Capturing CO2 from the Atmosphere," *Joule*, 2018.

[2] N. Spanos and P. G. Koutsoukos, "Kinetics of Precipitation of Calcium Carbonate in Alkaline pH at Constant Supersaturation. Spontaneous and Seeded Growth," *J. Phys. Chem. B*, vol. 102, no. 34, pp. 6679–6684, 1998.

[3] H. N. S. Wiechers, P. Sturrock, and G. v R. Marais, "Calcium carbonate crystallization kinetics," *Water Res.*, vol. 9, no. 9, pp. 835–845, 1975.

[4] C. Y. Tai and H. P. Hsu, "Crystal growth kinetics of calcite and its comparison with readily soluble salts," *Powder Technol.*, vol. 121, no. 1, pp. 60–67, 2001.

[5] R. Ševčík, M. Pérez-Estébanez, A. Viani, P. Šašek, and P. Mácová, "Characterization of vaterite synthesized at various temperatures and stirring velocities without use of additives," *Powder Technol.*, vol. 284, pp. 265–271, 2015.

[6] K. L. Mercer, Y. P. Lin, and P. C. Singer, "Enhancing calcium carbonate precipitation by heterogeneous nucleation during chemical softening," *Journal / American Water Works Association*, vol. 97, no. 12. 2005.

[7] G. H. Nancollas and M. M. Reddy, "The crystallization of calcium carbonate. II. Calcite growth mechanism," *J. Colloid Interface Sci.*, vol. 37, no. 4, pp. 824–830, 1971.

[8] L. Burhenne, C. Giacomin, T. Follett, J. Ritchie, J. S. J. McCahill, and W. Mérida, "Characterization of reactive CaCO3 crystallization in a fluidized bed reactor as a central process of direct air capture," *J. Environ. Chem. Eng.*, vol. 5, no. 6, pp. 5968–5977, 2017.

[9] K. M. Van Schagen, L. C. Rietveld, and R. Babuška, "Dynamic modelling for optimisation of pellet softening," *J. Water Supply Res. Technol. - AQUA*, vol. 57, no. 1, pp. 45–56, 2008.

[10] Y. F. Ma, Y. H. Gao, and Q. L. Feng, "Effects of pH and temperature on CaCO3 crystallization in aqueous solution with water soluble matrix of pearls," *J. Cryst. Growth*, vol. 312, no. 21, pp. 3165–3170, 2010.

536581 Integrated Forward Osmosis-Membrane Distillation Process for Sustainable Treatment of High Tds Produced Waters

Abstract:

Horizontal drilling and hydraulic fracturing make the extraction of tightly bound hydrocarbons from shale formations economically feasible. Brine management has become a key issue for the future of unconventional onshore oil and gas industry. In recent years, forward osmosis (FO) and membrane distillation (MD) have been widely investigated for treating wastewaters containing high levels of total suspended solids (TDS). Here we aim to demonstrate the feasibility of a hybrid FO-MD process for water recovery from high TDS produced waters, obtained from Southwestern Energy operations in the Marcellus shale. In FO-MD hybrid system, the FO process draws water from the feed solution while MD process reconcentrates the diluted draw solution and produces pure water. The integration of FO-MD processes combines the advantages of both processes and can provide high permeate quality and low membrane fouling. Cellulose triacetate (CTA) obtained from HTI (Albany, OR) and ethylene chlorotrifluoroethylene (ECTFE) obtained from 3M (Maplewood, MN) were employed as FO and MD membranes, respectively.

Experiments with raw and pretreated produced water, containing 57,000 mg L^{-1} TDS, were carried out in the integrated FO-MD system. Sodium chloride in de-ionized water was used as FO draw solution and was continuously regenerated using the MD system. Although the individual FO process had relatively high rejection to TDS and total organic carbon (TOC), these contaminants could also accumulate in draw solution after a long-term experiment. The MD process provided an effective rejection to contaminants in the draw solution after FO process and the integrated system revealed nearly complete rejection of TDS and TOC. Feed water was concentrated up to 50% in the hybrid system. Next, the concentrated feed was treated using an individual MD and was further concentrated up to 300,000 mg L^{-1} TDS under variety of operating conditions. The results of this study highlight the potential for employing FO-MD to treat high salinity wastewaters from unconventional gas extraction.

533547 Optimizing Sand Control in Wellhead with Advanced De-Sander Technology

Abstract:

Optimizing Sand Control in Wellhead with Advanced De-sander Technology

Banchao Shu, Oluwatosin Oyelakin, Priyanka Shahi, Isaac Snyder, Mike Fredrick, Jeevan Dahal, Certified Holdings LLC- Certified Technologies

Abstract

The well testing equipment usually suffers from severe erosion caused by sand production in conventional and unconventional oil and gas wells. The choke manifold, gate valves, and two-phase/three-phase separators are required to repair and clean up frequently, otherwise, it will cause equipment failure and operation problems. Tremendous time have spent on shut-in wells and repair equipment also.

It is critical to remove the sand from oil and gas production stream and ensure continues operation. The wellhead de-sander can remove sand to protect all the downstream equipment, which enables the easier design and operation of solid-handling systems. This novel multiphase cyclone geometry for wellhead was optimized with Computational fluid dynamics (CFD), and Finite Element Analysis (FEA) methods are presented in this paper. An integrated control system was designed with Ladder Logic for automated and continuous operation.

Methodology

The multiphase cyclonic de-sander geometry was designed and optimized by CFD software ANSYS 18.2. The Volume of Fluid (VOF) model, with the kε-RNG method, was used for multiphase flow simulation.

The nonlinear FEA was used to determine the potential failure locations in the pressure vessels. The failure point was identified in the vessel where the total stress, strain, and elongation exceeds the actual limit of the material. The control system design was designed in Ladder Logic.

Results

A 5000-psi rating sand separation system is designed using Boiler and Pressure Vessel Code (BPVC) Section VIII-Rules for Construction of Pressure Vessels Division 2-Alternative Rules. A cyclonic housing and two sand accumulator are be designed based on this code (Figure 1).

Figure 1 Full diagram of the de-sander system

The separator geometry dimensions and separation efficient can be calculated based on stocks' law, from the following equations. In this model, the multiphase flow spins through a number ***N*** of revolutions in the outer vortex. The value of ***N*** can be approximated as the sum of revolutions inside the body and inside the cone (***Equation 1***):

$N = (1/H)\,(Lb + Lc/2)$ ***Equation 1***

where

N = number of turns inside the device (no units)

H = height of inlet duct (m or ft)

Lb = length (vertical) of cyclone cone (m or ft).
Lc= length of cyclone body (m or ft)

Lapple developed a semi-empirical relationship to calculate a "50% cut diameter" d_{pc}, which is the diameter of particles collected with 50% efficiency. The expression is (***Equation 2***)

$d_{pc} = 9\mu W/(2*N*V_i*(\rho_p - \rho_a)$ Equation 2

where

d_{pc}= diameter of particle collected with 50% efficiency.

W = width of inlet (m or ft)

V_i = multiphase flow inlet velocity (m/s or ft/s) = Q/WH

ρ_p = density of the particle (kg/m^3)

ρ_a = air/fluid density (kg/m^3)

μ = air viscosity (kg/m.s).

Theodore and DePaola then fitted an algebraic equation to the curve, which makes Lapple's approach more precise and more convenient for application to computers. The efficiency of collection of any size of particle is given by (***Equation 3***)

$= 1/(1+(d_{pc}/d_{pj})^2$ ***Equation 3***

where

$_j$ = collection efficiency of particles in the ***j***th size range ($0 < j < 1$)

d_{pc} = diameter of particle collected with 50% efficiency

d_{pj} = characteristic diameter of the ***j***th particle size range (in μm).

The overall efficiency, called performance, of the cyclone is a weighted average of the collection efficiencies for the various size ranges, namely (***Equation 4***)

$= \Sigma_j\, m_j/M$ Equation 4

where

= overall collection efficiency (0 << 1)

$\boldsymbol{m_j}$ = mass of particles in the ***j***th size range

M = total mass of particles.

The cyclonic insert geometry was then optimized by ANSYS 18.2 with multiphase flow simulation based on critics such as higher sand separation efficiency and lower pressure drop. The cyclone separation system can treat up to 100 MMSCFD of natural gas and up to 13,000 bbl/d of liquid with 1-2% of sand. The multiphase simulation result suggests more than 99% sand remove efficiency for a particle size larger than 5 μm.

The FEA analysis was further adapted to determine the failure locations in the pressure vessels. Using the FEA analyses to optimize the shape and the wall thickness of the pressure vessel thereby have prevented failure and reduced the cost of over-engineering in the designed.

Figure 2 FEA analysis for the sand Accumulator

The FEA analysis result suggest the potential failure point in the two-inch water inlet hole (Figure ***2***). We then have added fillet to reduce the stress result from the sharpen are.

The control system would allow for a maximum 75 psi pressure drop in the de-sander system to ensure a longer product lifetime. The advanced automation control system was implemented in the designed thereby elimination possible human interference of the whole assembly operation.

Conclusion

The optimized wellhead cyclonic de-sander with CFD, FEA technology will be able to separate the gas, liquid, and sand with the highest efficiency of 99 percent which has never been approached before. It will provide great protection for all the downstream equipment and reduce the maintenance cost. The automated and continuous process will reduce the shutdown time, reduce human error, and improve the overall production efficiency.

525099 Exploration of Polysaccharide Derivatives-Based Anionic Polyelectrolytes As Novel Draw Solutes in Forward Osmosis

Abstract:

In this study, a series of anionic polyelectrolytes, sodium carboxymethyl dextrans with different molecular weights (CM-dextran-1000/5000/20000), are facilely synthesized via a one-step reaction and employed as novel draw solutes in FO process. Various characterizations (FTIR, NMR, conductivity titration and cytotoxicity test) are conducted to confirm the chemical structure, substitution degree and toxicity of the synthesized CM-dextrans. Effects of the concentration, molecular weight and substitution degree on the physicochemical properties (solubility, relative viscosity, conductivity and osmotic pressure) of CM-dextrans as well as their corresponding FO performance are investigated systematically. Compared with 1 M NaCl draw solution, 40 wt% CM-dextran-1000 draw solution shows a higher FO water flux (24.9 LMH) and a much lower reverse solute flux (0.97 gMH) in PRO mode. The recyclability of the diluted CM-dextran draw solutions is evaluated by an ultrafiltration process. Furthermore, the developed CM-dextrans with low reverse solute diffusion and nontoxicity can be potentially employed as promising draw solutes for the protein concentration via FO.

525227 Assessing the Selective Adsorption of Contaminants Using Spectroscopy and Chemometrics

Abstract:

The National Academy of Engineering has identified a global grand challenge for mankind is the accessibility of clean water for all human beings. Approaches for achieving this goal vary; however, one approach entails the use of carbon nanomaterials to remove contaminants from water systems. Carbon nanomaterials (e.g. graphene and carbon nanotubes) have been shown to provide a versatile means for achieving high uptake of organic, biological, and metallic materials from aqueous environments. A challenge with using carbon nanomaterials is tailoring them to selective removal of specific compounds. Given their strong non-covalent interactions, especially towards organic species, it can be difficult to assess the degree to which a specific compound has been removed from water systems containing multiple organic species. This is especially true when using spectroscopy techniques where contaminants have similar spectral signals. One alternative is to combine spectroscopy with chemometric analysis tools to discern between different species in the system. Parallel Factor Analysis (PARAFAC) provides the means to which multiple species can be identified when dealing with realistic water systems. In this study, the use of hybrid graphene/single wall carbon nanotube adsorbents towards the removal of organic compounds from natural water is assessed. These carbon nanomaterial adsorbent systems have already been shown to effectively remove 80-90% of organic contaminants from artificially contaminated water. Using natural water containing dissolved organic matter along with prevalent organic compounds (e.g. 2,4-dichlorophenoxyacetic acid), adsorption studies were completed to understand the removal uptake of these species. UV-Vis and multidimensional fluorescence spectroscopy were used to analyze the samples. PARAFAC was then utilized to assess which compounds were removed by the adsorbent. A discussion of the removal effectiveness of the prevalent compounds from natural water will be discussed.

533554 Functionalized Poly(methacrylic acid) Membranes with Bimetallic Nanoparticles: From Bench Scale to Water Remediation Applications

Abstract:

The detoxification of toxic chlorinated organics (chlorophenol, trichloroethylene) in groundwater has been widely investigated using bimetallic (Fe/Pd) nanoparticle methods [1,2]. With the integration of catalytic particles and functionalized commercial membranes, this membrane platform shows improved scalability and ease of recycle. This integration was achieved in both commercial flat-sheet membranes and spiral membrane modules. The Fe/Pd nanoparticle was formed inside the pores of membrane through in situ reduction after chelating by introduced carboxylate groups (methacrylic acid (MAA) was polymerized inside the membrane). The functionalized MAA could prevent nanoparticle aggregation and metal ion loss (recapture after dechlorination treatment). The metal nanoparticle-based membrane system showed great performance on the treatment of lab synergic polychlorinated biphenyls and field water samples (from a Superfund site, which includes trichloroethylene, tetrachloroethylene and carbon tetrachloride). At 2 seconds residence time, 95% of chlorinated organic species (ppm level) in the field samples were dechlorinated in the convective flow mode. The pilot studies with functionalized commercial spiral membrane modules are in progress.

Advanced characterization methods, such as FIB, TEM, XPS were applied to understand the correlation between nanoparticle properties and depth inside membrane pores, leading to the optimization of membrane design and treatment performance. The depth profile of functionalized membrane indicated that particle size was uniform inside membrane pores (particle size: 24±6 nm) but slightly smaller than those nanoparticles located on the surface (39±9 nm). Besides, the distribution of particles inside the membrane as well as the composition of single palladized-iron nanoparticles were analyzed by using EDS and EELS. Furthermore, the effects of pH, temperature as well as water hardness on the dechlorination treatment were studied. These investigations illustrated the advantages of the Fe/Pd particle based membrane system method, compared to the conventional applications of the metal nanoparticles.

In XRD analysis, the Fe/Pd particle samples (which were deliberately oxidized and then reduced) exhibited the same crystalline patterns as the original samples. The membrane maintained reactivity after four degradation cycles with regeneration between each cycle. The increase of surface particle size of 22% resulted in a decrease of 9.7% PCB conversion for the 4 hr reaction time [3].

This research is supported by the NIEHS-SRP grant P42ES007380. Partial support is also provided by NSF KY EPSCoR grant (Grant no: 1355438).

536397 Effect of Membrane Surface Chemistry on Grafting Density and Molecular Weight Cutoff

Abstract:

Water membranes usually suffer from fouling issue which can lead to a dramatic decrease in membrane productivity. This phenomenon can be mitigated through hydrophilizing the membrane surface. Zwitterions (ZWs) and poly (ethylene glycols) (PEGs) have been widely used to make antifouling membranes by enhancing the surface hydrophilicty. A facile method to graft the antifouling ZWs and PEGs onto the membrane surface is co-depositing with dopamine, which can easily be polymerized to an adhesive agent, polydopamine (PDA). PDA can enhance the membrane surface hydrophilicity, and, simultaneously, bind ZWs or PEGs onto the surface. However, increasing the content of grafted ZWs or PEGs into the coating layer still remains a challenging issue which needs to be studied. PDA can react with either acrylate or amine functional groups through Michael Addition. In this work, two series of ZWs and PEGs (acrylate-terminated and amine-terminated compounds) are co-deposited with dopamine onto the silicon wafers and ultrafiltration (UF) membranes. Acrylate-terminated compounds include sulfobetaine methacrylate or SBMA and PEG-diacrylate or PEGDA and amine-terminated compounds include a sulfobetaine ZW, functionalized with the primary amine, or SB-amine and PEG-diamine. To achieve the highest grafting density of ZWs or PEGs, co-deposition times and concentrations of ZWs or PEGs were varied, under a constant dopamine concentration (1 g/L). Results obtained from thickness measurements, conducted on the silicon wafers, exhibited significant contents of grafted ZWs or PEGs into the coating layers. Tuning the grafting conditions led to enhance the SBMA or PEGDA contents up to 38% and 64%, respectively. In the case of PEG-diamine, even when co-deposited under a lower concentration, compared with that of PEGDA, its content reached about 72% in the coating layer. Such this behavior demonstrates the PDA-amine interactions can graft a higher content of polymers onto the surface, than that of the PDA-acrylate interactions, under the same co-deposition time and dopamine concentration. The proposed surface modification was also applied to a polysulfone UF membrane (PSf) with the molecular weight cutoff (MWCO) of 25 kDa. The modified membranes with dopamine and ZWs or PEGs (for both cases of acrylate- and amine-terminated compounds) indicated lower water contact angles, compared with those of uncoated ones, suggesting increasing the membrane surface hydrophilicity. We will also systematically compare the effect of the surface chemistry on the MWCO and water permeance of the PSf membarne, aiming to derive the structure/property relationship.

537409 Cost Optimization of Osmotically Assisted Reverse Osmosis

Abstract:

Safe and cost-effective management of high salinity brines is a critical enabler of inland brackish water desalination, produced water treatment, geothermal energy production, and CO_2 sequestration. While reverse osmosis is considered the most energy efficient and cheapest desalination technology, the process has limited effectiveness for brines with salinities greater than seawater. In this work, we assess the techno-economic viability of a membrane-based process capable of desalinating high salinity brines, osmotically assisted reverse osmosis (OARO). We develop a non-linear optimization model to minimize the product water costs and extract generalizable guidelines for low cost design and operation of the OARO process. We estimate that the OARO process has lower or similar costs to mechanical vapor recompression, the current benchmark for high salinity brine desalination technology. We find that membrane costs account for the majority of OARO process costs, which suggests that advances in membrane properties and manufacturing may reduce OARO process costs significantly.

537985 Advanced Electrokinetic Desalination for Brackish Water: Water and Energy Nexus

Abstract:

This work explores the state-of-the-art electrokinetic processes for brackish water desalination, as well as identifying emerging technologies and potential advances via materials science, process engineering, and system integration. We focus on two major types of electrokinetic processes: electrodialysis/electrodeionization and capacitive deionization. We discuss the balance between energy efficiency and throughput in the context of large-scale deployment of a desalination plant to address the associated challenges in water and energy nexus. Thermoelectric power plants traditionally have required huge volumes of water to condense steam from the turbine exhaust. The complex interdependency between water and energy poses new challenges for policy makers to achieve a secure and sustainable supply of water and energy in the future. In order to facilitate the use of impaired water in cooling systems, we suggest the key technical issues and available water technologies for brackish water desalination.

528254 Filtration of Multi-Component Aerosols Using Polymeric Nanofiber Membranes

Abstract:

Aerosols are known to pose a threat to both human health and the environment. Submicron aerosols are abundant, hazardous, and difficult to remove. Among various particle control technologies, filtration is the most economical method for removing submicron aerosols. Most real-life aerosols, including tobacco smoke, coal gas, and combustion products, are complex in composition and exhibit compositional heterogeneity. For example, the distribution of toxic components in cigarette smoke has been shown to vary with particle size. Despite the ubiquity of multi-component aerosols, model filtration studies tend to be performed on single-component aerosols. These studies give meaningful insights into the effects of filtration on aerosol concentration and size distribution, but do not address whether or how the filtration process affects the composition of the aerosol stream. In our work, we aim to understand the effects of filtration on the chemical composition of submicron liquid aerosols and to explore the possibility of achieving chemical selectivity through filtration. We have performed filtration tests on binary liquid aerosols using polymeric nanofiber filters of different fiber sizes, and our results demonstrate that preferential filtration of specific components can be achieved through adjusting the structure and properties of the nanofiber filters.

The filters used in this work were fabricated using the electrospinning technique. Nanofiber membranes have low basis weights, large surface to volume ratios, and are thus great candidates for submicron aerosol filtration. Filters comprising fibers with diameters ranging from 300 nm to 1 Î¼m were produced and tested in filtration experiments.

Binary liquid aerosols consisting of bis(2-ethylhexyl) sebacate (DEHS) and bis(2-ethylhexyl) phthalate (DEHP) were generated and used as model bi-component aerosols. The size and compositions of these aerosols were controlled by adjusting conditions in the condensation monodisperse aerosol generator (CMAG). The size distribution of aerosols was measured with a differential mobility analyzer (DMA) paired with a condensation particle counter (CPC). The chemical composition was obtained from GC-MS measurements.

Results of our filtration experiments show that the both the particle size distribution and the composition of the aerosol stream change from upstream to downstream of the filtration process, and that the magnitude of these changes depends on the average fiber diameter in the filter. The magnitude of the compositional change also correlates with that of the change in size distribution of the aerosols. Our results demonstrate that, for compositionally heterogeneous aerosols, filtration serves not only as means for removing particles from a gas stream, but can also be used as a chemical separation process, resulting in the selective removal of certain chemical components. This study provides valuable insights into both the filtration mechanism of submicron particles and the possibility of designing chemical separation processes for aerosols through filtration with nanofiber membranes.

534788 Dynamic Split Flow Separation of Micron-Sized Slurry Fischer-Tropsch Catalyst Particles

Abstract:

The rising demand for transportation fuels, and the depletion of fossil fuel resources has attracted the production of transportation fuels from renewable energy sources such as biomass. The biomass is first converted to syngas, which is further converted to liquid fuels via Fischer-Tropsch (FT) Synthesis. Iron and cobalt based catalysts are commonly used for Fisher-Tropsch Synthesis. Different technologies were developed for Fischer-Tropsch Synthesis, namely fixed bed, fluidized bed, and slurry reactors. Of the technologies mentioned, the slurry reactors are preferred because of high heat and mass transfer. Generally, the FT reactors are operated at a pressure > 20 bar and temperature > 220°C.

In a typical slurry FT reactor, the feed consists of 16 to 20 wt% catalyst. A fresh catalyst particle size ranges from 20 to 400 microns (James, N. et al., 2005; Sergio, M.,Harold, W., 2014; Basha, O., 2015). In a slurry FT reactor, the syngas will be bubbled through the catalyst slurry. The high rates of heat and mass transfer are due high degree of mixing. Due to high degree of mixing, the catalyst particles will undergo attrition and the exit catalyst particles size were reported in range of 5 to 120 micron depending on the operating conditions of the reactor (James, N. et al., 2005; Oluwaseyi, O., et al., 2006). The exit stream of a slurry reactor consists of liquid fuel and the catalyst. The catalyst particles need to be remove before they are charged into internal combustion engines, they create problems otherwise.

Different methodologies were proposed for the separation of FT catalyst from catalyst-wax slurry. The proposed methods include, batch settling, lamella settling, filtration (using cake filtration and filter press), magnetic field assisted settling (for iron catalyst).

The proposed work addresses a dynamic separation of catalyst particles from a slurry in a continuous split flow process, where the catalyst slurry fed is split in to upward and downward flows. The ratio of upward and downward velocity is a critical parameter. This dynamic split flow. For a proof of concept, the split flow settler is modelled using fundamental mass and momentum conservation equations considering the axial convection and dispersion under steady state conditions. The individual particle velocities were calculated using Richardson-Zaki model (Richardson, J.F., and Zaki, W.N., 1954). The solution of the modelled equations was obtained by using first order difference for the convection term and second order difference for the dispersion. The equations were coded in MATLAB®.

The developed model was validated against a static settler and compared results with the existing experimental results on particle segregation (Asif and Petersen, 1994). Different parameters for the split flow settler was studied namely, the slurry feed rate, particle size variation (5 to 93 microns) in the slurry, catalyst wt% in the slurry, average particle size in the feed, and the ratio of velocities between up and downward flow. It was found that, the ratio of velocities between upward and downward flow is an important parameter. It was found that all the most of the fines were removed from the fuel. The catalyst wt% in the upward flow can be brought down to less than 0.15%.

533931 High Throughput in Situ Cultivation and Isolation of Unculturable Bacteria Using Microfluidic Devices

Abstract:

Microorganisms can be found everywhere, from the air we breathe, to the food we eat. It is widely known that the majority of microorganisms cannot grow in artificial media that is prepared in laboratories. This problem was first realized over a century ago with the phenomenon known as â€œgreat plate count anomalyâ€ when the direct bacterial counts from environmental samples did not correlate with the number of resulting colony forming units (CFUs). Approaches, such as rRNA sequencing, have revealed that there is an immense number of uncultivated species in nature.

Over the past several years, microfluidic devices have been used for self-sorting and isolation of bacterial species. Microfluidic devices have had success, but they use complex instrumentation and need to remove the sample from the original environment for processing, which potentially introduces sample bias and loss of diversity. Other approaches use a polymer microfluidic device with sub-micrometer constrictions to isolate single bacterial species from heterogeneous mixtures, however there is a limitation on the number of isolation chambers that fit on a given device, which leads to the use of a significant number of devices to isolate a sufficient number of bacteria to be able to compare results with conventional culture techniques.

In this project, we use new materials and devices geometries to improve on the existing state-of-the- art. Using a SOI (silicon on insulator) wafer, it is possible to increase the density of holes on a device and also further remove sample collection bias.

The new design will have 24 sub-micrometer constrictions in the SOI wafer, and it will be connected to a polycarbonate plastic part (isolation chamber) where the bacteria will be cultured after isolation.

Once the bacteria are trapped in the constriction, it will reach the isolation chamber by cell division. This way, only one species can enter one isolation chamber. This method is very important and practical because it is an autonomous process and can be easily deployed for a variety of applications. It can be used in isolating and obtaining pure cultures from environmental samples and in water purification processes.

522241 Interfacial Transport in Nanocellulose-Based Nanocomposite Membranes for Improved Reverse Osmosis Performance

Abstract:

Improvement of current reverse osmosis membranes poses an intricate challenge that, if solved, could affect global access to clean water. Thin film composite (TFC) membranes are used commonly in industrial desalination processes, relying on a thin polyamide skin layer for selectivity. Recent studies in our research group involving the addition of nanoparticles to the selective polymer layer of existing TFC membranes have shown promising results, yielding higher water permeance (flux), improved chemical resistance, anti-biofouling properties, and mechanical property enhancement[1]. It is desired to understand how the addition of nanoparticles to the polyamide skin layer influences polymer formation, thereby altering the transport properties of the overall membrane. In addition, these thin film nanocomposite (TFN) membranes are often cost-ineffective, prompting the search for an economically viable alternative. Nanocellulose has become a popular field of study given that it is produced on large scales globally in many different varieties. In this study, cellulose nanocrystals (CNCs) of varying lengths are explored as additives to TFN membranes. The high aspect ratios of the CNCs may assist the formation of transport pathways for water molecules, as well as improve the mechanical properties of the membranes. CNCs were deposited in situ during polyamide skin layer formation in various amounts and the resulting transport properties of the membranes were observed. Inclusion of CNCs has been shown to increase flux by 20% while maintaining acceptable levels of salt rejection, however, developing a consistent deposition method remains a challenge.

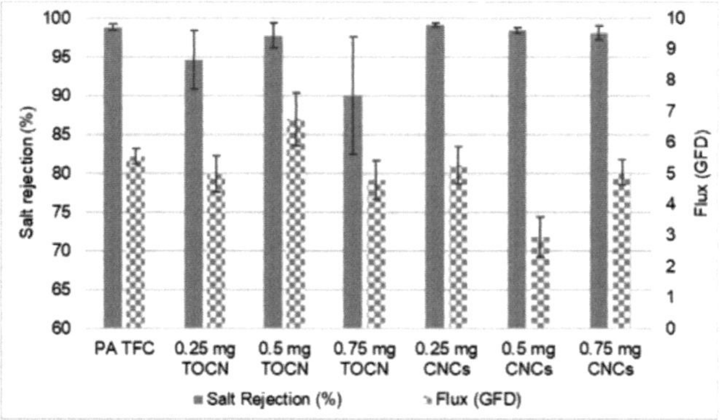

Figure 1. Comparison of TFN membrane performance between control (PA TFC) and various loading amounts of TEMPO-oxidized CNCs (TOCNs) and CNCs as obtained. The best performance was observed at a loading of 0.5 mg TOCNs.

[1] W. Chan, E. Marand, S.M. Martin, Novel zwitterion functionalized carbon nanotube nanocomposite membranes for improved RO performance and surface anti-biofouling resistance, J. Memb. Sci. 509 (2016) 125–137. doi:10.1016/j.memsci.2016.02.014.

527547 Predictive Models for Ion Removal in Brackish Water Desalination Using Electrodialysis Process

Abstract:

Electrodialysis (ED) is a membrane-based desalination technology which is widely used to desalinate brackish water. The performance of this technology is affected by operating parameters, such as temperature, velocity, applied voltage, and feedwater composition. In this research a form of empirical model was developed using dominant dimensionless numbers in the ED/EDR process and showing the Stanton number as a function of Peclet number and the dimensionless applied voltage. This dimensionless form of the model extends the modelâ€™s applicability to large scales. Additionally, a new model was developed for the selective removal of ions in ED/EDR based on the presented dimensionless ion removal correlations, showing how operating factors affect the removal of divalent counter ions against monovalent counter ions.

528656 Separation of Oil-in-Water Emulsions Stabilized By Different Types of Surfactants Using Electrospun Fiber Membranes with Surface Modification

Abstract:

The compositions and properties of discharged oil-in-water emulsions in industry vary widely according to their sources. Such variety poses a challenge for membrane separations. Electrospun fiber membranes have gained attention in emulsion separation due to their high permeability and improved robustness against fouling, but the interaction between the emulsion and the membrane, and the fouling mechanism that results, remains unclear. Therefore, this work seeks to understand membrane-foulant interactions and fouling mechanisms, and to develop novel membranes through surface modification to reduce the adverse effects of membrane fouling.

Electrospun polyamide membranes were challenged by model emulsions of dodecane stabilized by anionic, cationic, non-ionic and zwitterionic surfactants, in both dead-end and cross-flow filtration configurations. Analysis of permeate flux and oil rejection revealed that the types of surfactants influenced the membrane fouling in both dead-end and cross-flow systems but in different ways. Fouling in dead-end filtration was found to be a function of the electrostatic interactions between the oil droplets and the membrane, while fouling in cross-flow filtration was mainly determined by the hydrophilic/hydrophobic interactions due to the adsorption of surfactants at the interfaces. Blocking filtration models were used to corroborate these findings and to illustrate the transition between modes of fouling in dead-end filtration.

In order to modify the membrane-foulant interactions and thereby enhance separation performance for specific types of oil-in-water emulsions, surface modification on the electrospun polyamide membranes was performed by plasma treatment followed by deposition of a polycationic branched polyethylenimine from solution. The permeate fluxes of the modified membranes were shown to improve when the membrane was oppositely charged to the emulsified oil droplets, while retaining the oil rejection. Membrane fouling was successfully reduced .

534160 Methods for Direct Surface Temperature Measurement for Quantification of Membrane Distillation Process Performance

Abstract:

Membrane Distillation (MD) is a thermally driven process that can treat high salinity waters (e.g. waste brines, produced water, saline ground waters) utilizing low grade heat, providing a low-cost treatment option. The driving force in MD process is defined by the membrane surface temperature. However, traditional methods (e.g. thermocouples) cannot be used to measure the surface temperature as they can affect the bulk fluid flow, inducing turbulence and mixing. Thus, to-date most efforts to estimate the surface temperature have used thermal transport models that rely on bulk temperature and flow rate measurements. The inaccuracies from these approaches have led to high uncertainty in estimating membrane performance and translating process performance from the bench to commercial scale Herein, we present two separate approaches to measure the membrane surface temperature directly: a thermoreflectance method and an electrical impedance method. The thermoreflectance techniques leverages the change in material reflectivity with temperature, while impedance approach correlates changes in properties of water and permittivity of space to changes in temperature. These approaches make it is possible to directly measure the driving force in MD and validate standard heat transport correlations.

538332 A Simple Approach for Functionalization of Poly(vinylidene fluoride) (PVdF) Membranes for Desalination of Oil-Contaminated Saline Water Using Membrane Distillation

Abstract:

Desalination of oil-contaminated water by using membrane distillation is a promising approach to recover valuable products. The membranes needed for this process must possess special wettability. More specifically, the whole membrane should be hydrophobi, yet the surface in contact with the oil-water emulsion must be oil-repellent. To achieve this, introducing chemical heterogeniety and the desired surface topography to the hydrophobic membranes is required. Creating a hydrophilic surface with the desired surface topography will allow the water to wet the surface. Protecting the surface with a film of water prevents oil adhesion to the surface of the membrane and creates underwater oleophobicity. Here, we developed an approach to control the structure of Poly(vinylidene fluoride) (PVDF) membranes and selectively functionalize the surface intended to be exposed to the oil-water emulsion. We demonstrated that the morphology of the PVDF membrane can be simply adjusted by changing the composition of the polymer solution and the coagulation bath used during the phase inversion process. Additionally, we introduced a one-step process to chemically modify the membrane surface, rendering it more hydrophilic. We confirmed that the membrane surfaces are functionalized using a variety of characterization methods such as x-ray photoelectron spectroscopy (XPS), flourier transformed infrared spectroscopy (FTIR), zeta potential, and under-water oil contact angle. The scanning electron microscopy images showed that alkaline treatment did not damage the membrane structure. Also, we observed an underwater contact angle (c.a. 150°) for the Hexane droplets in contact with the modified surface. The membranes were tested in a direct contact membrane distillation process to desalinate a saline solution including 500 ppm oil. The results showed that oil did not wet the membrane during 12 hours of continuous MD operation; the average water flux and salt rejections were measured at 1.6 $Lm^{-2}h^{-1}$ and 99.99%, respectively.

538779 In-Situ pH Control for Selective Removal of Toxic Elements to Sustain Water Supply for Cooling and Process Water

Abstract:

Water quality determines the feasibility of using non-traditional water for current cooling equipment in thermoelectric power plants. The "tough toxic impurities" in low quality water are the most frequent existing species in cooling water that limits the number of water recycle. Selective separation remove the key saline constitutes selectively for the recycled water therefore reduces the energy consumption. One of the toxic elements in impaired water heat-exchanger is the silica. Conventionally, it was removed by adding chemicals to adjust aqueous pH. The toleration of silica in cooling is around 20 - 100 ppm depending on the heat-exchange efficiency. However, acidification of the feed cooling water is applied to filtrate silica in coagulation was not an efficient process. Silica will form multiple compounds with other constitutes in aqueous at different pHs. At high pH, ~10, silica will stay as ionic form that opens an opportunity to use electrochemical driven force to selectively separate silica from aqueous. However, raise the feed water pH using chemicals is not a viable solution because the cost as well as the increase of TDS that needs more energy to treat. Electrodeionization (EDI) offers the prospect of desalination with a low energy footprint. It is a modular technology based on electrically driven membrane separations platform and is capable of producing clean water fit for consumption and numerous industrial processes such as energy production, semiconductors, and pharmaceuticals. ANL's resin wafer electrodeionization (RW-EDI) is developed from commercial EDI technology with higher energy efficient and better economic viability in producing portable water. A bench-scale RW-EDI has been demonstrated capable to desalinate brackish water (i.e., one of the sources of impaired water) with 90% water recovery at 0.70 kwh/m^3 or ~ 30% energy efficiency. In-situ pH control of the process stream during desalination is a unique capability using the RW-EDI that enables low cost removal of "tough toxic impurities" in cooling system. We use RW-EDI technology to extract silica from aqueous.

AUTHOR INDEX

AUTHOR INDEX

AUTHOR INDEX

AUTHOR INDEX

AUTHOR INDEX

AUTHOR INDEX

AUTHOR INDEX

AUTHOR INDEX

AUTHOR INDEX

AUTHOR INDEX

AUTHOR INDEX

AUTHOR INDEX